Innovation without R&D

Oliver Som

Innovation without R&D

Heterogeneous Innovation Patterns of Non-R&D-Performing Firms in the German Manufacturing Industry

With a foreword by Prof. Dr. Carsten Dreher

Springer Gabler

RESEARCH

Oliver Som
Karlsruhe, Germany

This publication has been accepted as a dissertation at the University of Flensburg, Internationales Institut für Management (IIM), 2011.

ISBN 978-3-8349-3491-8 ISBN 978-3-8349-3492-5 (eBook)
DOI 10.1007/978-3-8349-3492-5

The Deutsche Nationalbibliothek lists this publication in the Deutsche Nationalbibliografie; detailed bibliographic data are available in the Internet at http://dnb.d-nb.de.

Springer Gabler

Cover design: KünkelLopka GmbH, Heidelberg

Printed on acid-free paper

Springer Gabler is a brand of Springer DE.
Springer DE is part of Springer Science+Business Media
www.springer-gabler.de

Foreword

"Research and development" were regarded in the past decades as the driver of productivity growth in economies. Inevitably, therefore, investments in research and development were described as economically meaningful and a sign of normative political will. This was reflected not least in the 3% target of the European Union. With this target, an indicator gained importance which was regarded as a success guarantee for sustainable growth: the so-called R&D ratio, which is described in economic terms as R&D expenditure relative to gross domestic product and from a business administration perspective as R&D expenses related to sales (R&D intensity). This seemingly paradigmatic correlation, however, has been increasingly questioned in the past years. For instance, Arundel et al. established in 2008[1] that half of the firms surveyed in the Community Innovation Survey do not conduct any research or development at all. Srholec and Verspagen (2008)[2] point to the different innovation strategies within the so-called high-tech sectors which have been successful despite their varying expenditures on R&D.

This debate is potentially explosive as it casts doubts on accepted causal links in innovation and research policy and gives additional weight to alternative innovation strategies. In his thesis the author deals with exactly these relationships by specifically targeting this group of companies in his investigation. Oliver Som addresses a truly relevant issue in the area of innovation research, innovation economics and innovation policy which calls for a holistic view of both the operative level and also the economic interconnections.

The main research question in this thesis is whether the non-R&D-performing firms in German manufacturing industry display heterogeneous innovation patterns. For this purpose, it is important to observe both the structural dimension - namely, whether and why heterogeneity in the non-R&D-performing firm populations can be expected – as well as the substantive dimension - to what

---

1 Arundel, A., Bordoy, C., Karneva, M. (2008): Neglected Innovators: How Do Innovative Firms That Do Not Perform R&D Innovate? Results of an analysis of the Innobarometer 2007 survey No. 215. INNO-Metrics Thematic Paper.

2 Srholec, M., Verspagen, B. (2008): The Voyage of the Beagle in Innovation Systems Land. Exploration on Sectors, Innovation, Heterogeneity and Selection. UNU-Merit Working Paper Series #2008-008, United Nations University, Maastricht.

extent and in which dimensions heterogeneity of innovation patterns on the part of non-R&D-performing firms can really be distinguished.

For this he developed a conceptual framework which draws on evolutionary innovation economics and the resource-based view of strategic business management. This emphasises the heterogeneity and the use of different resources and thus the possible elaboration of competitive advantages for companies. The empirical analysis is unique, because Mr. Som succeeds in coherently organising the presently available, partly contradictory analytical findings from previous studies, and for the first time paints a differentiated picture of various innovation strategies of non-R&D-performing firms based on quantitative data. The heterogeneous innovation patterns identified in his cluster analysis provide a consistent explanation of why firms, despite the fact that they do not conduct research and development, are still competitive and contribute to economic growth. I therefore wish this thesis, which was both presented and successfully defended at the University of Flensburg, many interested readers.

Carsten Dreher

Preface

As though it were a corporate innovation project, writing a doctorate thesis is a rocky road paved with all kinds of challenges. To start with, there are large uncertainties with regard to the expected final result. And if a result is actually reached, will anybody be interested in it, in other words: is there any demand? Furthermore, the necessary financial and human resources have to be provided, where especially the latter are particularly susceptible to all kinds of distractions during the writing phase of such a paper and are therefore in need of special care and attention. Bureaucratic hurdles have to be managed, external impulses and ideas have to be acknowledged and evaluated, knowledge stocks and data have to be gathered and tied together. In addition, the marketing concept has to be constantly modified and revised for the doctoral advisor. Not least, the project has to be protected from the persistent emergencies and peculiar logics of the day-to-day normal work schedule. To ensure the smoothest and perfectly timed interplay of all these factors, many people contributed decisively to the success of this doctorate thesis. I would like to express my thanks to them at this point.

First of all, I must say thank you to my doctoral advisor, Professor Dr. Carsten Dreher. He not only created the necessary conditions for the project by adopting me as an "extern" in his department, but also contributed to the success of the undertaking with his scientific expertise and personal advice. Thanks also go to Professor Dr. Hartmut Hirsch-Kreinsen for taking on the role of co-supervisor, as well as to Professor Dr. Dr. Willi Petersen for acting as the chairman of the examination committee. As well as all the co-workers of Professor Dreher at the Universities of Flensburg and Berlin (FU), I am especially grateful to Dr. Alexander Martin and Dr. Anette Biedermann for their valuable advice and support.

This paper was written in the framework of my professional activity as a scientific researcher at the Fraunhofer Institute for Systems and Innovation Research ISI in Karlsruhe. I am deeply indebted to numerous colleagues at Fraunhofer ISI. I would especially like to mention Professor Dr. Eva Kirner, who during her time at ISI accompanied the project as my mentor from its first steps up to its oral disputation, who honed the study at decisive points using her scientific expertise and who managed to motivate me, the author, again and again at critical phases with her confidence. I also thank Dr. Steffen Kinkel, Head of the Competence Center Industrial and Service Innovations, for putting in place the very good framework and working conditions which enabled me to write this thesis. Special thanks go to Spomenka Maloca for her valuable and patient advice and support on statistics as well as Dr. Christoph Zanker and Katharina Mattes for many fruitful discussions and shouldering some of my other project responsibilities during the final crucial phase of writing this thesis. For taking on the job of editing and proofreading the final manuscript, I would like to thank Christine Mahler-Johnstone and Gillian Bowman-Köhler, as well as Sabine Schöller, who was in charge of its publication at the Gabler Verlag.

The biggest thanks by far must go to my family. Above all, I would like to express my heartfelt appreciation to my partner, Antje Bierwisch. She was at my side through all the ups and downs of this project, accompanying me with understanding and love and also contributed significantly to its success with her many suggestions and discussions. Finally, I would like to thank my parents Rosemarie and Otto Som from the bottom of my heart. Their unconditional and unswerving support made it possible for me to pursue this course of study and they were always there for me during its writing. Sadly, my father did not live long enough to see this work completed. It is dedicated to him.

Oliver Som

Table of contents

List of figures

List of tables

List of abbreviations

AMT	Advanced manufacturing techniques/technologies
BERD	Business expenditures for research and development
CIS	Community Innovation Survey
destatis	German Federal Statistical Office
ed.	Editor
eds.	Editors
e.g.	Exempli gratia (for example)
EMS	European Manufacturing Survey
et al.	Et alia (and others)
etc.	Et cetera
EU	European Union
f.	Following page
ff.	Following pages
Fig.	Figure
GDI	Gross domestic income
GDP	Gross domestic product
GMS	German Manufacturing Survey
GNI	Gross national income
GNP	Gross national product
Ibid.	Ibidem (in the same place)
ICT	Information and communication technology
i.e.	Id est (that means)
NACE	Nomenclature générale des activités économiques dans les Communautés Européennes

OECD	Organisation for Economic Co-operation and Development
OMC	Organisational and management concepts
PRS	Product-related services
R&D	Research and Development
RFID	Radio frequency identification
ROS	Return on sales
SME(s)	Small and medium-sized firm(s)
Tab.	Table
U.S.	United States of America
Vol.	Volume

1 Introduction

During the last decades, research and development (R&D) activities have become the most substantial pillar of innovation research in the attempt to explain the innovation ability and competiveness of enterprises, sectors and economies (TEECE 1986; BROWN/EISENHARDT 1995; FREEMAN 1994a, 1994b; FREEMAN/SOETE 1997; ROSENTHAL 1992; SAVIOTTI/NOOTEBOOM 2000; STOCK ET AL. 2002; OECD 2005; RAMMER ET AL. 2009; SANTAMARÍA ET AL. 2009; RAYMOND/ST. PIERRE 2010). There is a vast amount of empirical literature that supports the importance of firm-internal, institutionalised R&D activities as the main source of many, productivity-enhancing, technological innovations to improve competitiveness, especially of fast growing industries such as pharmaceuticals, automobiles, computers, communications, instruments, and machinery (BECHEIKH ET AL. 2006; FREEMAN 1994a; FREEMAN/SOETE 1997). "There is ample empirical evidence supporting the hypothesis that R&D expenditures are a sine-qua-non for the firm's level of innovation activities" (SHEFER/FRENKEL 2005: 25).

Furthermore, the application of R&D to produce technologically advanced products for export can also improve the terms of trade at the national level. In addition, R&D activities create a demand for highly skilled and qualified human resources, which provides an impetus to develop and improve educational systems and leads to potential benefits throughout an economy (HUANG ET AL. 2010: 5). Empirical studies also show that a high technological level of an economy results from high innovation activities conducted by R&D, which in turn result in a higher share of new goods and an extended use of new techniques in production (FAGERBERG 1987: 88).

Subsequently, the terms of "innovation" and "R&D", respectively, the level of R&D intensity and the innovativeness of firms and economies have been increasingly equated by mainstream innovation literature in the past. At the economic level, this prevailing R&D focus is prominently rooted in the works of SOLOW (1956, 1957) and the subsequent scholars of endogenous growth theory (ROMER 1986, 1990; LUCAS 1988). They have substantially explained economic growth and competitive advantage in international markets through endogenous, intentional R&D activities of firms as the most important source of technological progress. The relationship between technological progress and economic growth is thereby characterised as a linear, steady-state growth pattern which can be adjusted relatively easily by "turning the knobs of the R&D process" (VERSPAGEN 2005: 504).

However, empirical studies showed early on that a significant number of firms does not invest in own R&D. For example, COHEN ET AL.'S (1987) study of large U.S. firms reported that 24% of them did not invest in formal R&D. Similarly, BOUND ET AL.'S (1984) study indicated that 40% of U.S. firms did not report R&D expenditures, and GALENDE and SUAREZ'S (1999) analysis of Spanish firms revealed that 71% of their surveyed companies did not undertake formal R&D.

Not long ago, ARUNDEL ET AL. (2008) showed, based on data from the Community Innovation Survey (CIS), that slightly over half of all innovative firms in Europe do not perform R&D. Even more surprising, they could not find any differences in the economic performance between innovative firms with and without in-house R&D, measured by profit development. Most recently, RAMMER ET AL. (2011) supported this finding for the German manufacturing industry. Their analyses reveal that about 44% of all innovative manufacturing firms (i.e. firms that introduced new products or processes during the past years) did not perform internal R&D. Moreover, they also could not identify any significant differences in the economic outcome between non-R&D-performing innovators and their R&D-performing counterparts. Regarding efficiency performance, KIRNER ET AL. (2009a, 2009b) additionally showed that firms in the German manufacturing industry with low R&D intensity are able to obtain at least the same, or even a slightly higher level of productivity compared to R&D-intensive firms.

Therefore, firms that do not invest in institutionalised, in-house R&D activities and which nevertheless manage to survive in market competition over long periods of time pose a major challenge to mainstream innovation theory, because the lack of any R&D or low R&D intensity is usually associated with the stagnation or decline of firms.[3] This applies all the more as firms with low or zero R&D intensity can be found in relevant shares in all sectors of the manufacturing industry, even in sectors that are usually associated with very high levels of R&D, such as medical engineering or process measuring and control technologies (KIRNER ET AL. 2009a). Hence, there is a relevant group of manu-

3 "(A) sustained high rate of growth depends upon a continuous emergence of new inventions and innovations, providing the basis for new industries whose high rates of growth compensate for the inevitable slowing down in the rate of invention and innovation, and upon the economic effects of both, which retard the rates of growth of older industries. A high rate of overall growth in an economy is thus necessarily accompanied by considerable shifting in relative importance among industries, as the old decline and the new increase in relative weight in the nation's output" (Kuznets 1959: 33).

facturing firms which do not perform any in-house R&D whose innovation strategies, competitive advantage and economic success cannot be explained sufficiently by the predominant R&D-focused approaches of mainstream innovation theory.

Following NELSON and WINTER (1982), researchers of the evolutionary tradition in economics have criticised this R&D paradigm. They suggest that innovation ability is more likely to be based on firm-specific routines and firm-individual heuristics instead of merely single, homogeneous R&D-based innovation strategies. One of the key arguments of evolutionary economics points to the fact, that enterprises show considerable heterogeneity in their innovation behaviour and strategies, even within similar framework conditions of sectors or innovation systems (SRHOLEC/VERSPAGEN 2008: 5; NELSON 1991). Following the large strand of theory labelled the "resource-based view of the firm" (BARNEY 1991; WERNERFELT 1984; PETERAF 1993)[4] which originates in an evolutionary perspective, this heterogeneity is related to the differences routines, capabilities, skills and experiences of firms (NELSON/WINTER 1982; NELSON 1991; CHRISTENSEN 2002; TEECE ET AL. 1997; MASSINI ET AL. 2005). The innovativeness and economic success of firms are thus not necessarily simply about high technology or excessive R&D expenditures. R&D-focussed approaches often overlook that a major part of firms' innovation does not necessarily originate from institutionalised, internal R&D activity (KLINE/ROSENBERG 1986; FAGERBERG 2005; NELSON/ROSENBERG 1993; HANSEN/SERIN 1997; BENDER ET AL. 2005; MARSILI/SALTER 2006).

In addition to technological, R&D-based innovation, non-technological dimensions or "modes of innovation" (JENSEN ET AL. 2007), like services, marketing, organisational innovation, and experience-based sources of know-how are also increasingly recognised as distinct kinds of innovation which contribute to firms' economic success (DAMANPOUR/EVAN 1984; WOMACK ET AL. 1990; PIVA/VIVARELLI 2002; TOTTERDELL ET AL. 2002; SMITH 2005). This corresponds to SCHUMPETER'S broad definition of innovation which has already included the development of new product-related services, the implementation of innovative organisational or marketing concepts in addition to new products, or the use of innovative manufacturing technologies at an early stage (SCHUMPETER 2006: 158f.). TIDD and BESSANT (2009: 5) exemplify this in the German firm of Würth which is the largest manufacturer of screws and other fastenings like nuts and

---

4 See also chapter 3.3.3.

bolts in the world. Despite low-cost competition from China, the company has managed to stay ahead of its competitors through product and process innovation across a supplier network similar to the model used by Dell computers (FINANCIAL TIMES 2008). In the first half of 2010, Würth achieved a total turnover of 4.2 billion euros in the first half of 2010.[5] Another example is the German fashion company of Gerry Weber[6] which belongs to the textile sector and is thus a typical "low-tech" industry. But currently, it is investing strongly in process technologies like computer aided design (CAD), new information and communication technology in sales and distribution or the implementation of RFID-chips to enhance logistics (VDI NACHRICHTEN 2010).

However, these insights have hardly been translated into new standard indicators, although the latest edition of the OSLO Manual (OECD 2005) shows some promising improvements regarding innovation sources other than R&D. Today, the majority of scholarly research and policy documents on innovation still focus almost entirely on R&D, ignoring other methods that firms use to innovate (ARUNDEL 2007: 50; BECHEIKH ET AL. 2006: 655f.). Thus it has to stated, that – in particular – empirical, academic innovation research has not managed to overhaul the linear science-push model of innovation (ARUNDEL 2007: 51). This is partly due to the belief that non-R&D innovators are either infrequent or have little to offer, with most productivity improvements and performance outcomes due to more advanced innovative activities based on R&D (ARUNDEL ET AL. 2008: 32). As stated by ARUNDEL ET AL. (2008: 5) and CUERVO-CAZURRA/UN (2010: 759), if only R&D indicators are used to measure innovation capability, such firms will not be captured adequately, and in consequence will be classified as non-innovative or even be excluded from innovation studies. In consequence, firms that do not perform formal R&D have been largely neglected by academic research and the policy community.

For these reasons, non-R&D-performing firms are still regarded as some kind of a "black box" in the scientific, managerial, and especially in the current public and political debate. As a result, the majority of today's innovation and technology policy instruments based on neoclassical models of mainstream innovation research still aim to increase industrial R&D activities in some selected fields of high technology. But as only little is known about the complex interplay between R&D- and non-R&D-intensive manufacturing sectors and the

---

5 See: http://www.wuerth.de/web/de/awkg/unternehmen/portraet/portraet.php

6 See: http://www.gerryweber-ag.de/

underlying, individual innovation strategies of non-R&D-intensive firms, innovation and technology policy may risk disregarding important competitive potentials of other innovation paths besides R&D. Hence, contemporary innovation and technology policy needs more insight into the innovation strategies of non-R&D-performing firms to be able to increase the innovation capability of firms and the economy in general, instead of only focussing on the specific niche of R&D as the alleged "one best way" to generate innovation.

This thesis aims to open this "black box" of non-R&D-performing firms to enhance scientific knowledge about their different kinds of innovation behaviour and thereby to increase the scientific, managerial and political awareness of competitive and innovation strategies beyond mere R&D. By exploring heterogeneous innovation patterns of non-R&D-performing firms, this thesis aligns itself with a growing volume of innovation literature that focused on the sources of innovativeness of non-R&D-intensive industries and non-R&D-intensive firms (low-tech firms) during the last years. The increasing interest in this issue is also reflected by the fact that the journal "Research Policy" published a special issue on "Innovation in Low-and Medium-technology Industries" last year.[7] What all these studies have in common is the question: how do non-R&D-intensive firms innovate?

However, the existing research into non-R&D-intensive or non-R&D-performing firms is characterised by some limitations and shortcomings which form the starting point of this thesis. Hence, in order to describe the specific research gap which this thesis aims to address, as well as depict the underlying motivation and research objectives in a more detailed manner[8], it is necessary to provide a short overview of current empirical research on non-R&D-intensive industries and firms that has been conducted during the past years.

1.1 Economic relevance of non-R&D-intensive industries

At the industry or sector level, the differentiation between R&D-intensive and non-R&D-intensive industries follows the classification of manufacturing industries into high-tech, medium-tech and low-tech sectors introduced by the OECD (1994) or its recent modification (LEGLER/FRIETSCH 2007). The division

7 Research Policy (2009), special issue on "Innovation in Low-and Medium-technology Industries", Vol. 38 (3), April, pp. 441-570.

8 This will be done in chapter 1.4.

into low-, medium-, and high-tech sectors is based on the respective sectors' average share of business expenditures for research and development (BERD). Sectors with a less than 3%, respectively less than 2.5% (LEGLER/FRIETSCH 2007) share of BERD are classified as non-R&D-intensive or "low-tech" sectors (OECD 1994). Based on this, industries such as aerospace, computers, semiconductors, telecommunications, pharmaceuticals and instruments are commonly classified as "high-tech", while "medium- tech" typically include electrical and non-electrical machinery, transport equipment and parts of the chemical industries. The remaining, "low-tech" category with low R&D intensity then comprises for example industries such as textiles, clothing, leather products, furniture, metal products and so on.[9]

Comparisons of OECD and European economies' manufacturing industries showed that growth performance across countries is not correlated with shares of high-tech in manufacturing. Low-tech and medium-low-tech sectors still account for the largest part of manufacturing output and employment in OECD and European economies, and their preponderance is decreasing only at a very slow rate (SANDVEN ET AL. 2005; BENDER 2006; HIRSCH-KREINSEN ET AL. 2003; for Germany: GEHRKE/LEGLER 2010; SOM ET AL. 2010; HAHN 2009; HIRSCH-KREINSEN 2004). For instance, non-R&D-intensive industries among eleven OECD countries on average still employ about half of the total workforce in these manufacturing industries and account on average for almost half of industrial value-added (SANDVEN ET AL. 2005: 36ff.). Thus, there has hitherto been no trade-driven "hollowing out" of low-tech sectors by moving their production into low-wage countries as predicted by some authors (e.g. LU 2007; HEIDENREICH 2009). As BENDER (2006) summarises, these findings suggest the existence, and obvious still importance of non-R&D based types of innovation in OECD economies, and they bulk far too large in total output to be relegated to the scrap-heap of stagnation (VON TUNZELMANN/ACHA 2005: 430; likewise: IIZUKA 2009; FREDDI 2009; MASKELL 1998). Moreover, in the case of Germany, some very recent findings partially indicate even an increasing economic importance of non-R&D-intensive sectors (GEHRKE/LEGLER 2010: 1; SOM ET AL. 2010).

As ROBERTSON and PATEL (2007) argue, the role of low-, medium-, and high-tech sectors is also undeniable as they are not only generators of innovation,

9 Sectoral classifications of R&D intensity based on both, OECD (1994) and LEGLER/FRIETSCH (2007) are provided in tables Tab. A-31 and Tab. A-32 in the annex (A.2).

but also key customers of innovation generated in high-tech industries. An input-output analysis conducted by HAUKNES and KNELL (2009) revealed the direct and indirect flows of knowledge between different R&D-intensive industries in France, Germany, Norway, Sweden, and the United States. Following the prominent sectoral classification developed by PAVITT (1984), they show that technology flows between high-tech and low-tech industries are dominated by the flows into low-tech sectors. However, this view may be too one-sided, as it can reasonably be assumed that low-tech industries stimulate or even influence new innovation activities in high-tech sectors, either by giving customer feedback on their practical experiences with high-tech products or by their role of suppliers to high-tech industries. For instance, MENDONÇA (2009) found out that, based on patent analysis, firms in mature industries contributed significantly to the development of cutting-edge technologies such as ICT, new materials, and even biotechnology. Therefore, the levels of economic performance of low-, medium-low, and high-tech industries are highly independent (HIRSCH-KREINSEN 2008a; HIRSCH-KREINSEN ET AL. 2006).

Thus, it cannot be suggested that low-tech sectors lack technological opportunities or appropriability or other factors associated with benefiting from technological innovation (VON TUNZELMANN/ACHA 2005: 429f.). Instead, a varying degree of permeation of high technologies into low-tech and medium-tech as well as into high-tech sectors can be observed. This suggestion corresponds closely to the findings of KIRNER ET AL. (2009a) who presented ample empirical evidence that LMH-tech industries themselves comprise a heterogeneous mixture of low-, medium-, and high-tech firms.[10]

1.2 Innovation and competitive strategies of non-R&D-intensive firms

Unsurprisingly, firm level studies reveal that non-R&D-intensive firms generally perform more weakly with regard to product innovation compared to R&D-intensive firms (HUANG ET AL. 2010; KIRNER ET AL. 2009a, 2009b; ARUNDEL ET AL. 2008; HEIDENREICH 2009; COX ET AL. 2002). They also show weaker performance in the field of product-related services, probably because innovative product-related services are often developed in close association with a new

---

10 An analysis of the macro-economic relevance of non-R&D-intensive industries in Germany based on more recent data is provided by this thesis in chapter 2.7, respectively A.3 in the annex.

product (KIRNER ET AL. 2008, 2009b). Thus, innovation in low-tech firms is assumed to be mostly in the form of improvements to existing products (RAYMOND/ST-PIERRE 2010). This picture fits to some extent into the picture of low-tech firms drawn by the famous taxonomy of PAVITT (1984): firms of traditional (= non-intensive) sectors of manufacturing consist largely of small, "supplier dominated firms" with weak in-house R&D and weak engineering capabilities.

"They appropriate less on the basis of a technological advantage, than of professional skills, aesthetic design, trademarks and advertising. Technological trajectories are therefore defined in terms of cutting costs. Supplier dominated firms make only a minor contribution to their process or product technology. Most innovations come from suppliers of equipment and materials, although in some cases large customers and government-financed research and extension services also make a contribution (PAVITT 1984: p.356)".

This first, rather negative appraisal of non-R&D-intensive firms has been qualified to some extent today. On the one hand, KIRNER ET AL. (2009b) found that a relevant share of non-R&D-intensive firms is definitely able to successfully develop new products or even market innovations. On the other hand, many studies emphasise the role of engineering knowledge and technical process innovation for the innovation success of non-R&D-intensive firms (EVANGELISTA ET AL. 2002; NASCIA/PERANI 2002; HIRSCH-KREINSEN 2004; BENDER 2004, 2006; KIRNER ET AL. 2007, 2009a), as innovation activities of non-R&D-intensive firms are often more focused on production efficiency, product differentiation and marketing (VON TUNZELMANN/ACHA 2005; HEIDENREICH 2009). Regarding technical and organisational process innovations, KIRNER ET AL. (2007, 2009a, 2009b) has shown that non-R&D-intensive firms are able to organise their production processes at least as efficiently as R&D-intensive firms, resulting in comparable productivity and process speeds. Low-tech firms seem to place a higher emphasis on the quality of their production processes, enabling them to differentiate themselves from their global competitors via excellent product quality and reasonable costs. The higher prevalence of process innovation among non-R&D-performers suggests that there are more options for developing process innovations without performing R&D (ARUNDEL ET AL. 2008). Thus, also organisational innovation (e.g. in the fields of human resource management and team work) can help non-R&D-performing firms to gain similar innovation success as R&D-performers, especially when combined with each other (RAMMER ET AL. 2009; HEIDENREICH 2009; BARGE-GIL ET AL. 2008).

The acquisition and purchase of innovative machinery and equipment or intermediate goods from external sources with little or no further work required represents another source of innovation for non-R&D-intensive firms (BENDER 2004, 2006; ARUNDEL ET AL. 2008; SANTAMARÍA ET AL. 2009; BARGE-GIL ET AL. 2008; TSAI/WANG 2009; HEIDENREICH 2009). They are thereby relatively more dependent on the diffusion of external knowledge than R&D-performing firms, particularly through knowledge embodied in acquired products and processes (ARUNDEL ET AL. 2008; HEIDENREICH 2009). However, such adoptions of external technologies may also involve the incorporation of high-technology products and technologies (HIRSCH-KREINSEN 2008b; ARUNDEL ET AL. 2008). The development and availability of absorptive capacity plays a major role for non-R&D-intensive firms, as it enables them to manage and implement such technology adoption successfully (BENDER/LAESTADIUS 2005; BENDER 2008; HIRSCH-KREINSEN 2008b).

HAUKNES and KNELL (2009) take the arguments of inter-industry linkages further. Their results based on CIS data indicate that firms within low- and medium-low-tech industries collaborate with external actors in their innovation projects to a greater extent than firms in high-tech industries. BENDER ET AL. (2005) and HIRSCH-KREINSEN (2004, 2008a) also underline the importance of network relations between companies and supportive social networks at a regional level as important knowledge resources, particularly for increasing non-R&D-intensive firms' innovation capabilities (BENDER 2006). Thus, non-R&D-performing or less R&D-intensive firms may gain innovation success from external knowledge sourcing or formal cooperation with external partners, particularly with customers, suppliers and competitors (HUANG ET AL. 2010; RAMMER ET AL. 2009; SANTAMARÍA ET AL. 2009; BARGE-GIL 2010; COX ET AL. 2002; BAYONA ET AL. 2001; PITTAWAY ET AL. 2004). However, ARUNDEL ET AL. (2008) found no statistically significant difference between low- and high-tech firms regarding the percentage of firms that report technology adoption. Twice as many firms that perform R&D in-house collaborate in product or process innovations, compared to non-R&D innovators. This is also supported by TSAI and WANG (2009) who found that the level of internal R&D investment contingently impacts the innovation performance that is achieved by collaborating with different types of external partners. Firms with higher internal R&D investment thereby gain higher innovation returns from collaboration with external partners than firms with fewer internal R&D investments. More thoroughly, KIRNER ET AL. (2009b) show that non-R&D-intensive firms are indeed less likely to cooperate with external partners than R&D-intensive firms, particularly in the fields of R&D and service. But there are no differences to be stated regarding collaborations in production or supply. Based on matched-pair analyses, they provide novel evidence that non-R&D-intensive firms participating in R&D cooperation are more likely to develop product innovations than their structurally similar counterparts without R&D cooperation.

A third major point refers to the aspect of knowledge that is primarily used by non-R&D-intensive firms. In contrast to "science-based" modes of innovation (MOWERY/ROSENBERG 1998), innovation in low-tech enterprises is suggested to be rather driven by customer-related and practical knowledge (HEIDENREICH 2009; VON TUNZELMANN/ACHA 2005; HIRSCH-KREINSEN 2004, 2008a, 2008b; SUNDBO 1996). Non-R&D-intensive innovation is usually not the outcome of the latest scientific or technological knowledge. The centre of the debate about low-tech innovations involves attempts to include innovation processes that are not primarily based on systematic research and technological development, but on practical, experience-based, often implicit knowledge. Especially creativity and innovation-enabling capabilities, which are related to the abilities required to identify and assimilate the potentially relevant knowledge and translate it to fit the specific conditions of the firm, are seen as major assets of these firms (BENDER/LAESTADIUS 2005). KIRNER ET AL. (2009b) have found empirical evidence that non-R&D-intensive firms are characterised by an above-average share of low- and unskilled employees; a result which may suggest that these firms put forth alternative paths of knowledge generation and accumulation than science-based modes of expert knowledge. In the terminology proposed by LUNDVALL and JOHNSON (1994), "learning-by-doing, learning-by-using, learning-by-interacting, learning-by-producing and learning-by-searching" should be recognised as essential sources of non-R&D-based innovation. As ARUNDEL ET AL. (2008) show, non-R&D innovators are likely to source ideas from alternative sources within the firm, like production engineers or design staff. Internal organisational practices, knowledge management and personnel policy, in particular, play a vital role for innovation and the innovation performance of low-tech firms (BENDER 2006; HIRSCH-KREINSEN 2008b; RAMMER ET AL. 2009). Based on a sample of 4,500 firms from 13 European countries, GRIMPKE and SOFKA (2009) found that search patterns in low-technology industries focus on market knowledge and that they differ from technology-sourcing activities in high-technology industries. Thus, imitation or reverse engineering by non-R&D-intensive firms (KIM/NELSON 2000) may probably also be interpreted as a kind of market-oriented search pattern along these lines.

1.3 Identifying the research gap

Throughout the studies, low-tech or high-tech firms are most frequently identified according to their sector affiliation to low- or high-tech industries[11], while just a few studies have identified non-R&D-intensive firms at the firm level, by considering their individual intensity of R&D either by transforming the sectoral OECD classification to the firm level (e.g. KIRNER ET AL. 2007, 2008, 2009a, 2009b) or by selecting firms that do not perform (intramural) R&D at all (HUANG ET AL. 2010; RAMMER ET AL. 2009; ARUNDEL ET AL. 2008; BARGE-GIL ET AL. 2008). Moreover, the existing empirical field work on low-tech firms can be distinguished into qualitative, case-study-based approaches[12] which dominated the early phase of research on non-R&D-intensive firms. Although they provided estimable and precious insights which serve as fruitful starting points for further analysis, their results are frequently based on anecdotal evidence and can therefore hardly be generalised. Moreover, many studies are limited to specific segments of low-tech industry[13] which again makes them difficult to interpret from a more general point of view. Meanwhile, the number of quantitative studies based on broad, quantitative firm level surveys has increased

---

11 See for example: Corbett (2008); Cuervo-Cazurra/Un (2010); Spithoven et al.(2009); Cox et al.(2002); Grimpke/Sofka (2009); Heidenreich (2009); Santamaría et al.(2009); Bender (2004, 2006), Tsai/Wang (2009); Hirsch-Kreinsen (2008a); Hirsch-Kreinsen et al.(2003); Sunbo (1996); Alfranca et al.(2004); Bender/Laestadius (2005, 2007); Robertson/Smith (2008); Schmierl/Köhler (2007); Laestadius (2007); von Tunzelmann/Acha (2005); Hall et al. (2009); Chamberlin/Doutriaux (2010)

12 See for example: Bender (2004); Bender/Laestadius (2005, 2007); Schmierl/Köhler (2007); Laestadius (2007); Robertson/Smith (2007); Hirsch-Kreinsen (2007); Corbett (2008); Spithoven et al. (2009); Lichtenthaler (2009); von Tunzelmann/Acha (2005); Pedersen (2005)

13 See for example: MASKELL (1998 - Danish wooden furniture industry), CHRISTENSEN, J.L. (2010) – Danish fur production, FILIPPAIOS ET AL. (2009 - food and beverage multinationals), FREDDI (2009 - mechatronics industry in northern Italy), IIZUKA (2009 - Chilean salmon farming industry), KIM/PENNINGS (2009 - tennis racket industry), ROBERTSON/SMITH (2008 - food processing), VON TUNZELMANN/ACHA (2005 - British textile/clothing and food processing industry), NIELSEN (2008 - Danish agricultural, forestry and fishery industry), ALFRANCA ET AL. (2004 - multinational agriculture and food industry), JACOBSON/MUSYCK (2008 - airline industry), CAPUTO/PALUMBO (2005 – textile industry), PEDERSEN (2005 - Norwegian mechanical engineering), or CHAMBERLIN/DOUTRIAUX (2010 - Canadian forest sector)

during the last two years.[14] Nevertheless, empirical quantitative findings about the innovation behaviour of non-R&D-intensive firms are still scarce (ARUNDEL ET AL. 2008: 5; SANTAMARÍA ET AL. 2009: 507f).

Considering the existing empirical work on the innovation behaviour of non-R&D-intensive firms, the following limitations and shortcomings can be summarised:

- ***No common definition of non-R&D-intensive firms:*** Most studies define non-R&D-intensive firms according to their sector affiliation; firms that are located in the OECD classification of low-, or/and medium-low-tech industries are summarily defined as non-R&D-intensive. Other studies instead apply sectoral classifications of low-, medium-, and high-tech sectors (OECD 1994; LEGLER/FRIETSCH 2007) at the firm level by building firm level taxonomy out of the firms' individual share of R&D expenditure on total sales. In both alternatives, non-R&D-intensive firms may still incur R&D expenditures, and it can be questioned whether it is really reasonable to assume that a firm that performs less than 2.5% R&D expenditure is less R&D-intensive than a firm with 3% R&D expenditure. The solution of RAMMER ET AL. (2009) and BARGE-GIL ET AL (2008) OF focussing on firms that do not perform intramural R&D at all seems more practicable. An even more detailed approach is taken by HUANG ET AL. (2010) who distinguish between two types of innovation without R&D: firms with in-house creativity work without performing formal R&D and firms which only innovate by buying advanced machinery, hardware and software, or licences from other firms or organisations. However, this inconsistency in the definition of non-R&D-intensive firms implies that their findings can hardly be compared with each other.

- ***Linear relationship between non-R&D intensity and innovation:*** This aspect is closely connected to the prevailing sector-based approaches to low-tech firms, but applies also to firm level studies. Despite the fact that many of the studies criticise and want to overthrow the linear R&D paradigm at the beginning, they are based on the implicit assumption that firms which systematically lack a common resource (R&D) must show a systematically identical innovation behaviour which replaces the missing R&D resource. This appears as an inversion of the originally criticised linear R&D paradigm of innovation. Thus, research on "typical"

14 See for example: Peneder (2010); Arundel et al. (2008); Rammer et al. (2009); Kirner et al. (2007, 2008, 2009a, 2009b); Cox et al. (2002); Grimke/Sofka (2009); Hall et al. (2009); Heidenreich (2009); Mendonça (2009); Santamaría et al. (2009); Thornhill (2006); Tsai/Wang (2009); Barge-Gil et al. (2008)

low-tech innovation strategies often yields no results because not enough room is left for the possible heterogeneity of low-tech firms. This problem may be also reflected in the circumstance that some findings concerning non-R&D-intensive patterns of innovation contradict each other. At a first glance, such conflicting findings may indicate that there is probably more than one single strategy of non-R&D innovation.

- ***Neglect of possible intra-sectoral heterogeneity of firms:*** As shown above, most studies define non-R&D-intensive firms by their sector affiliation to so-called low-tech industries or, alternatively, by the fact that they belong to one of PAVITT'S (1984) "supplier dominated" industries. Due to the similar market and technological frame conditions of sectors, a certain type of common innovation strategy seems to promise economically successful development. However, this approach contradicts other studies that provide empirical evidence that low-, medium-, and high-tech industries themselves comprise a heterogeneous mixture of low-, medium-, and high-tech firms (VON TUNZELMANN/ACHA 2005; KIRNER ET AL. 2007, 2009a). Therefore it has to be assumed that recurring principles and similarities with respect to innovation patterns are of cross-sectoral character (HIRSCH-KREINSEN 2008a). Moreover, the very recent innovation taxonomy by PENEDER (2010) revealed four clusters of innovation intensity.[15] Most interestingly, three clusters also encompass typical low-tech industries; a fact which underlines the potential firm heterogeneity that can be assumed for the residual "low-tech industry".[16] This supports the suggestion of VON TUNZELMANN and ACHA (2005) that low-tech industries resist easy classification, precisely because many of them are not very distinctive or singular in technological terms. Likewise, HUANG ET AL. (2010) found that innovating firms without R&D do not form a homogeneous group with several important differences between contract R&D-performers, non-R&D innovators, and technology adopters. Thus, by pre-selecting firms or grouping them a priori by their sector affiliation, a lot of firm level heterogeneity may be lost (SRHOLEC/VERSPAGEN 2008; NELSON 1991).

---

15 His paper presents an integrated set of innovation taxonomies of firms and sectors. The clusters are: a) intermediate-to-high innovation intensity (e.g. chemicals, motor vehicles, other transport equipment, or telecommunication), b) intermediate innovation intensity (e.g. wood and wood products, pulp and paper, metal products), c) Intermediate-to-low innovation intensity (e.g. food sector), and d) low innovation intensity (e.g. wearing apparel, leather products, recycling).

16 According to both the original OECD (1994) classification and its modification by LEGLER/FRIETSCH (2007), the category of low-tech industries represents the residual of those industries that do not fit into the categories of medium or high R&D intensity.

First steps towards a more differentiated view of different modes of low-tech innovation strategies have been taken by KÖHLER (2008), HIRSCH-KREINSEN (2004, 2007), PENEDER (2010), and HUANG ET AL. (2010). KÖHLER (2008) identified two innovation strategies in the food industry, one of automation of production, quality and safety control and the invention of new products for growing market segments, and one flexible customer-specific product development. These types of strategies are very similar to those of HIRSCH-KREINSEN (2004) who distinguished non-R&D-intensive firms which produce simple and traditional products in large batch sizes by highly automated manufacturing processes, on the on hand, and low-tech firms which produce their simple products by labour-intensive manufacturing in rather small batch sizes, on the other hand. More recently, HIRSCH-KREINSEN (2007) identified three characteristic types of non-R&D-intensive firms' innovation strategies on the basis of case studies. They range from firms that predominantly aim at incremental improvements of existing products to radical and extensive process, product and technological innovation. He named them "producer of standardised goods", "problem solver" and "process specialist". Last but not least, PENEDER'S (2010) convincing approach of an innovation taxonomy based on both firm and sector level, provides three out of four clusters which consist partially of LT manufacturing industries ("intermediate innovation intensity", "Intermediate-to-low innovation intensity", and "low innovation intensity"). Finally, HUANG ET AL. (2010) show that non-R&D-performers do not form a consistent block, with several notable differences between firms without internal R&D, technology adopters that purchase technological machinery and equipment from outside, and firms that contract out R&D activities without performing in-house R&D.

1.4 Motivation, research objectives and approach of the thesis

As this short overview of previous empirical research show, non-R&D-intensive sectors still represent a substantial pillar of developed economies in terms of their value-added or employment rates. Many enterprises with no or low R&D intensity obviously manage to survive in their markets and even achieve economic success, which raises the question of how they are able to do this. In this context, the first major motivation for this research work is ***to overcome the R&D focus of empirical innovation research*** by explicitly addressing the innovation activities of the evolutionary subset of manufacturing firms that do not perform any formal R&D at all. As a large part of the existing findings on the innovation activity of these firms is based on anecdotal evidence and case study research (either on single firms or on single, non-R&D-

intensive industries), it is also the aim to generate additional **quantitative data** and thus gain insights into these firms that can be generalised.

However, the research on non-R&D-intensive firms has hitherto suffered from a major shortcoming: it is frequently based on the implicit assumption that non-R&D-intensive firms represent a more or less homogeneous group of firms, either because they are affiliated to a certain category of "low-tech" industries, or they all lack the same innovation resource of formal R&D. Ironically, despite the fact that most of these authors criticise it in their writings, this appears just as an inversion of the originally criticised linear R&D paradigm of innovation. As a result, the existing findings about non-R&D-performers are scattered, fragmented, and in some cases even contradictory. For example, some studies point to their technical expertise, others to their non-technical expertise. Some say that they have a higher propensity to collaborate with external partners, while again others state that they have a lesser propensity to participate in innovation cooperation than R&D-performing firms. Last but not least, the findings are even more blurred and hard to compare as the studies are based on numerous different definitions of either non-R&D-intensive or non-R&D-performing firms, with some of them also including firms with low R&D intensity, others that only consider firms without any internal R&D or studies that select firms with no formal R&D at all.

The second major motivation of this thesis is to **address the identified shortcomings of previous research and to take them as a starting point for a more detailed theoretical and empirical investigation of heterogeneous firm level innovation patterns of non-R&D-performers**. Based on the first empirical hints mentioned above (KÖHLER 2008; HIRSCH-KREINSEN 2004, 2007; PENEDER (2010), and HUANG ET AL. 2010), the basic assumption of this thesis is that the scattered, fragmented and partially inconsistent findings so far are an indicator of a high level of inter-firm heterogeneity of non-R&D-performing firms' innovation strategies. It is suggested that these existing findings are only part of a broader set of heterogeneous innovation patterns of non-R&D-performing firms and thus probably could be integrated by a more detailed empirical analysis that explicitly takes firm heterogeneity into consideration.

Besides this rather academic motivation, there are ***two more reasons*** why the author believes this study to be import. From a ***managerial perspective***, the first reason is that, due to the overwhelming presence of R&D in past innovation research, many firms may have adopted this narrow view. As a consequence, just as many firms intuitively first think about reducing the number of employees to lower their costs, they may only think about R&D as a means to achieve innovativeness. In turn, the overemphasis placed on R&D as a source of innovation may also have discouraged many firms from attempting to be innovative, because they perceive R&D as a very costly and uncertain process, which demands very specific investments (Barge-Gil 2008: 3; Rammer et al. 2009: 36f.). However, if the role of other activities that are close to daily routines is highlighted as a source of innovation, managers could decide to enter into the innovation process more frequently. Without anticipating the results, a deeper knowledge of possible innovation sources beyond formal R&D can also contribute towards enhancing the innovativeness of R&D-performing firms in the sense of a more holistic, "generic" understanding of their innovation process, so that innovation is not relegated only to a specialised department within the firm.

Last but not least, my motivation lies also in the assumption that a modern, "21st century" technology and ***innovation policy*** cannot afford to solely rely on R&D activities as the "Holy Grail" of innovation, competitiveness and economic growth. As industrial structures are becoming more and more interwoven in regional, national, or even international innovation systems of multiple economic and non-economic actors, a sole focus on only one distinct group of firms may turn out as unsatisfactory in the long run. It neglects large parts of industrial value chains in manufacturing industry. Thus, substantial potentials for the economy's competitiveness and growth may be overlooked. But if innovation policy is to be increasingly targeted towards activities other than R&D, more specific knowledge about these firms' innovation behaviour is needed. The results of this thesis will hopefully contribute to this aim.

These motivating factors mentioned above can be translated into the main question of the thesis:

> ***Do heterogeneous innovation patterns of non-R&D-performing firms exist in German manufacturing industry?***

In more detail, this research question is composed of two main sub-dimensions:

> ***A structural dimension:*** *Can inter-firm heterogeneity generally be expected in non-R&D-performing firms' innovation patterns? If so, why?*

> ***A substantive dimension:*** *To what extent and in which dimensions can heterogeneous innovation patterns of non-R&D-performing firms in manufacturing industry be distinguished?*

Basically, these questions will be addressed by both theoretical and empirical analysis. The thesis can be further divided into the following subgoals:

- **Consideration of the theoretical and historical origins of R&D as a driver of economic growth in macro-economic growth theory, description of the emergence of the linear, R&D-based model of innovation, discussion of the definitions and measurement of R&D, as well as a short overview of the empirical state of research considering the role of R&D as a driver of firms' competitive advantage and economic success.** On the one hand, these points should enhance the reader's basic understanding of the R&D paradigm's roots in economic theory and its manifestations in theoretical and empirical innovation research. But on the other hand, discussion about R&D definition and measurement also provides basis for the further theoretical and in particular empirical analyses by pointing out what is to be understood by the term "R&D" and – even more important – what is not.

- **Development of a theoretical explanation a) if and why inter-firm heterogeneity of non-R&D-performing firms in manufacturing industry may be expected, and – if so – b) in which dimensions these heterogeneous innovation patterns may be distinguished?** Most theorists (e.g. NELSON/WINTER 1982; NELSON 1991; PRAHALAD/HAMEL 1990) argue that inter-firm heterogeneity is caused by the firms' heterogeneous behaviour (e.g. in terms of individual routines, competences, abilities) and justify the heterogeneity in terms of substantive dimensions which can be observed as ex post manifestations of heterogeneity. In contrast, this thesis will primarily rely on original evolutionary arguments as to explain why inter-firm heterogeneity can be principally expected on the basis of the three main evolutionary core mechanisms of variation, selection and replication. Then, it discusses in which dimensions of innovation behaviour this heterogeneity may become observable. These two steps build the core of the theoretical work of this thesis, with the first one elaborating on the structural dimension, and the second one dealing with the substantive dimension of inter-firm heterogeneity.

- **Development of a conceptual framework for subsequent empirical analysis of non-R&D-performing firms' heterogeneous innovation patterns.** Due to the explanatory nature of this thesis this step is necessary to ensure that the empirical analysis is strongly oriented to the previous insights of theoretical discussions and thereby to avoid analytical arbitrariness. This part will integrate the previous findings from theoretical discussions and develop a conceptual framework to define the central terms, specify the methodological design, and form a few guiding theses for the subsequent empirical analysis.

- ***Empirical analyses of innovation patterns of non-R&D-performing firms in German manufacturing industry.*** This step aims at the descriptive and multivariate exploration of heterogeneous innovation patterns of non-R&D-performing firms. No systematic and comprehensive overview of the innovation behaviour of non-R&D-performing firms in German manufacturing industry based on quantitative data is available until today. Therefore, a first descriptive analysis will present novel information about their innovation input and output. It also serves as a descriptive test if and which of the identified innovation sources are indeed used by non-R&D-performing firms and whether they already reveal a "typical" non-R&D innovation pattern or not. In the second step, a cluster analysis will be conducted to clarify whether and which heterogeneous innovation patterns of non-R&D-performing firms' innovation behaviour can be identified across these data.

Altogether, by the interplay of these subgoals the present thesis aims to provide comprehensive theoretical and empirical insights into the innovativeness of non-R&D-performing firms, and thereby to close the research gaps identified above. Due to the availability of empirical data, these questions and goals will be exemplarily addressed for non-R&D-performing firms in German manufacturing industry.

In order to address the research question and to achieve the subgoals described above, the thesis is basically divided into two major parts. Part A develops the theoretical argumentation of the study, while part B deals with empirical analyses. In this context, chapter four (see below) which is located at the end of the first part serves as a "bridgehead" between the theoretical and empirical part by integrating the theoretical findings derived in part A and developing a guiding conceptual framework for the empirical analyses conducted in part B. In detail, the thesis is divided into seven main chapters (Figure 1):

Chapter 1 contains a brief introduction which outlines the research topic of this thesis and summarises the current state of research on non-R&D-intensive sectors and firms. Based on this, the underlying motivation of this thesis, its research objectives and its detailed work methods are pointed out.

Firstly, it is necessary to understand why mainstream innovation research in the past has been shaped by such a strong R&D focus and how the linear, R&D-based model of innovation has emerged. ***Chapter 2*** opens the theoretical discussion by considering the theoretical and historical origins of R&D as a driver of economic growth in macro-economic growth theory. It will elaborate on the emergence of the linear, R&D-based model of innovation and discuss the definitions and measurement of firm level R&D. Furthermore, it will provide a short overview of the empirical state of research considering the role of R&D as a driver of firms' competitive advantage and economic success. By discussing the explanatory limitations of firm level R&D, this chapter also sets the cornerstone for further theoretical discussion of heterogeneity of non-R&D-performing firms' innovation behaviour in Chapter three.

Chapter 3 takes up the two dimensions of the research questions and develops a theoretical framework setting out the major concepts and aspects of explanation. To address the structural dimension of the research question, it refers to an evolutionary approach of innovation theory and provides evolutionary arguments why inter-firm heterogeneity of non-R&D-intensive firms can be expected, even - or particularly because - they are located within homogeneous market conditions. Secondly, by developing a resource-based perspective which allows for multiple modes of non-R&D based innovation, the second half of the third chapter deals with the substantive dimension and seeks to put some flesh on the bones of inter-firm heterogeneity.

By integrating the major findings of theoretical discussion of Chapter three, ***Chapter 4*** sets out important conceptual and methodological guidelines for empirical analysis. This entails defining key concepts, developing a corresponding analytical and methodological design, discussing some measurement issues as well as establishing some guiding theses for empirical analysis.

Having discussed the requirements of the empirical analysis in the previous chapter, **Chapter 5** contains the complete empirical analyses. The chapter introduces and discusses the database underlying the German Manufacturing Survey 2009 (GMS) and elaborates on the selection of indicators and measurements for empirical analysis as well. Afterwards, the descriptive analysis is linked to the existing analytical approaches of empirical literature and will provide a first systematic overview of the innovation behaviour of non-R&D-performing firms. By additionally comparing these findings to the group of R&D-performing firms, it is possible to identify whether they already constitute some kind of a non-R&D-specific pattern of innovation or not. Moreover, the descriptive analysis ensures that the identified indicators for innovation input are adequate and feasible to descripte the innovation behaviour of these firms. The subsequent cluster analysis then goes one step further than existing studies and will elaborate on how these innovation characteristics constitute heterogeneous innovation patterns among non-R&D-performing firms.

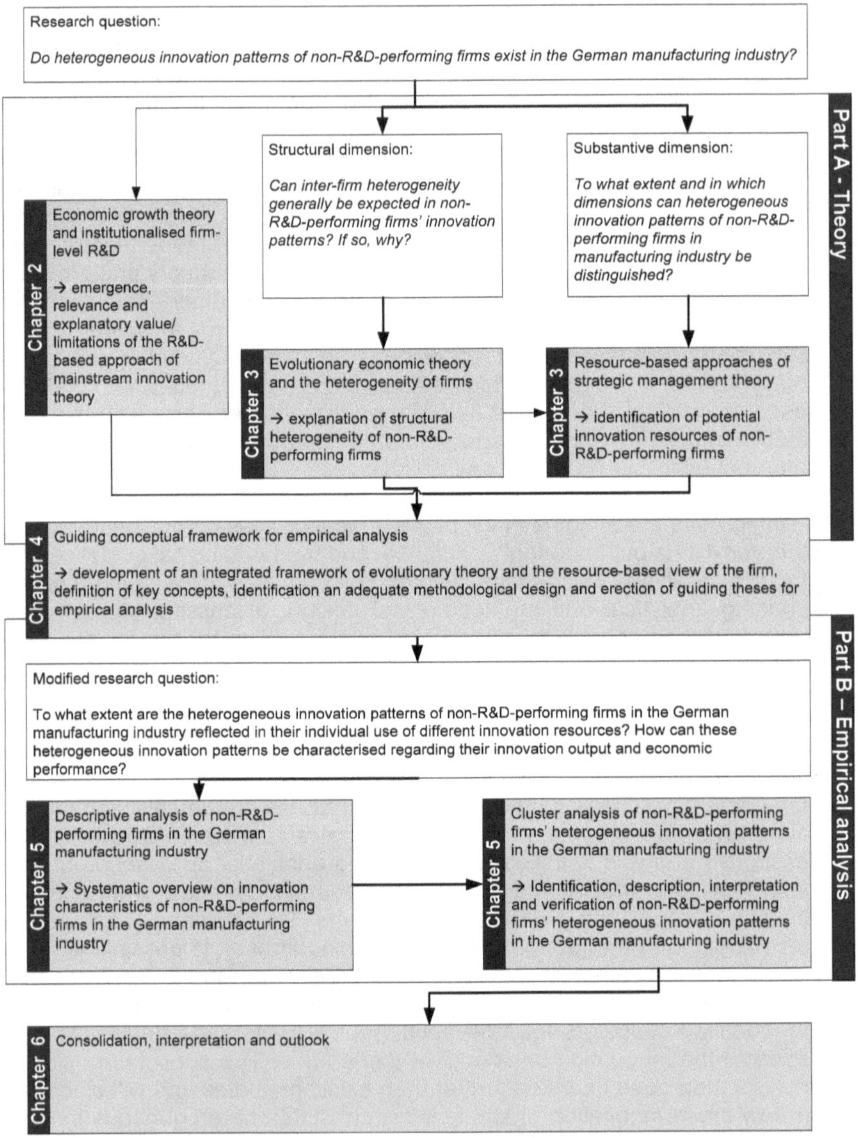

Figure 1: Structure and methods of the thesis
(Source: own illustration)

Subsequently, ***Chapter 6*** will summarise the main theoretical and empirical findings of the thesis before they will be conclusively interpreted and integrated in to the state of research. This interpretation of results will be devided into the three perspectives of academic innovation research, innovation management, and innovation policy.

Last but not least, a list of references and an annex which provides additional information which otherwise would have led beyond the scope of the discussion can be found at the end of this thesis.

1.5 Positioning in theory of science

After having presented the research objectives and the work schedule, it is necessary to make some basic remarks on the general understanding of the thesis and its location within the spectrum of different approaches of scientific theory.

The present thesis recognises business economics as applied research.[17] This means that the thesis aims at problems within the practical sphere of actioning individuals or aggregates of individuals (e.g. firms, organisations) for which no satisfactory knowledge exists so far (ULRICH 1981: 5). For further substantiation, this term can be deconstructed into three, logically not interconnected dimensions (BEHRENS (1993: 4769f.).

- The ***context of discovery*** refers to the emergence of hypotheses or theories. It aims to explore relevant questions, problems or methods of problem solving that have so far not being recognised (CHMIELEWICZ 1993: 87).
- The ***context of justification*** falls into the field of science logic which aims to critically test hypotheses or theories by analysing their consistency and probability of falsification (BEHRENS 1993: 4770).
- The ***context of application*** finally aims to apply and implement approved hypotheses or theories on problems of entrepreneurial practice (Behrens 1993: 4770).

17 For an overview of different definitions of the term "science" see ULRICH/HILL (1979: 161ff.), for the demarcation of "applied science" see ULRICH (1981: 3ff.).

Due to the fact that the intra-heterogeneity of non-R&D-performing firms has hardly been addressed by previous research, this research work is primarily devoted to the context of discovery. However, this orientation towards discovery and exploration implies some consequences for the function of empirical data. Unlike in the context of justification in which empirical data are used for to test already existing theories, empirical data in the context of discovery are supposed to help with the formulation and development of hypotheses and theories ("explorative research strategy") (WOLF 2008: 36; KUBICEK 1977: 12f.; WOLLNIK 1977: 42f.).

The use of empirical data within the context of discovery requires developing a conceptual frame of reference to systematically assess reality (WOLF 2008: 37; WOLLNIK 1977: 44f.). Otherwise, explorative research remains arbitrary, led by mere subjective experience or interest. Considering this thesis, it is thus necessary to develop a frame of reference in the first step which guides and structures the subsequent empirical analysis.[18] Such a framework should provide maximum heuristic potential which means that it leaves enough room for possible alternatives in reality (WOLLNIK 1977: 45).

Moreover, it has to be explicitly mentioned that findings that have been explored in the context of discovery cannot be readily transferred into the context of justification (MARTIN 1989: 224; BEHRENS 1993: 4770). This means that all findings, conclusions or practical guidance drawn from empirical analysis are of a hypothetical character as their testing in the context of justification remains to be done by future research.

---

18 For the issue of explorative research within a frame of reference see also WOLF (2008: 37ff.), GROCHLA (1978: 53ff.) and KRETSCHMANN (1990: 121ff.); for critical comments MARTIN (1989: 221ff.)

2 R&D as a determinant of technological progress, economic growth and firms' competitiveness

The purpose of the following chapters is to introduce the theoretical discussion by considering the theoretical and historical origins of R&D as a driver of economic growth, both from the macro-economic and micro-economic perspective. It will describe the emergence of the linear, R&D-based model of innovation, discuss the definitions and measurement of R&D as well as provide a short overview of the empirical state of research taking into account the role of R&D as a driver of nations' and firms' competitive advantage and economic success. Most importantly, this chapter will clarify why mainstream innovation research has been shaped by such a strong R&D focus in the past. By discussing the main explanatory limitations of the R&D concept, it paves the way for further theoretical discussion on the heterogeneity of non-R&D-performing firms' innovation strategies in Chapter three.

2.1 The concept of "economic growth"

Since the 18th and 19th century, when industrialisation led to noticeable economic growth in northern Europe, the explanation of the long-term development of economies has been one of the core issues of economics (SMITH 1937; RICARDO 1951; MALTHUS 1986; MARX 1987). The starting points were especially the considerable differences in the economic status and living standards between nations which represent the results of specific growth processes in the past. Since then, it has become standard in growth research to explain important empirically observed regularities which are often named "stylized facts" (REHME 2007: 1)[19] by identifying the determinants and influencing factors of processes of economic development.[20]

---

19 KALDOR (1957: 593ff.) was the first author to provide a record of "stylized facts", a set of six fundamental empirical constancies in the history of economic growth processes across different countries: a) the shares of national income received by labour and capital are almost constant over time, b) the growth rate of the capital stock is almost constant over time, c) the growth rate of output per worker is almost constant over time, d) the ratio of capital and output is almost constant over time, e) the return on capital is almost constant over time, and f) the real wages grow at an almost constant rate over time.

20 The technical term of modern economic growth goes back to KUZNETS (1973).

But against the background of World Wars I and II, the paramount task of macro-economic research in the first half of the twentieth century shifted to providing a theoretical background for economic policy to compensate the social impacts caused by economic cycles, mainly the (unequal) distribution of income and employment.[21] Hence, it was not until the 1950s that economic growth had its revival and a broad range of economic schools of thought dealing with technological change and innovation was established. Although they differ in their basic assumptions and methodologies to a great extent, their relationships are not inevitably characterised by rivalry or exclusiveness. Instead, all strands of these theories are supplementary to each other by contributing a special insight into the understanding and explanation of economic processes (GRUPP 1997: 49; MILLER/MAIR 1991: 18).[22]

The term "economic growth" stands for the quantitative increase of the amount of available goods and services within a national economy which is described by the gross domestic/national income (GDI/GNI) or the gross domestic/national product (GDP/GNP) per head of population or per head of the (potential) workforce population (labour productivity) (FRENKEL/HEMMER 1999: 1; BRETSCHGER 2004: 1).[23] More specifically, the literature distinguishes between "extensive" (the growth of the general amount of goods) and "intensive" growth (the growth of the available amount of goods per head) (REHME 2007: 1). However, "modern economic growth" (KUZNETS 1973) is usually understood as the increase of individual welfare which is approximately measured "per head". As intensive growth can only take place on the basis of extensive growth within a given constant population, both terms are often used synonymously and will also be used synonymously in this thesis.

If a certain good is produced by two nation states with identical production functions and constant economies of scale, the output per head can only differ

---

21 E.g. by the prominent work of KEYNES (1923, 1936, 1937, 1939)

22 "(...) as in all exciting subject areas there are a multiplicity of theoretical approaches with little universal agreement as to the superiority of one over another. It is often the case that the choice of framework is a matter of "horses for courses" and the most appropriate framework depends crucially on the questions being asked." (STONEMAN 1995: 9)

23 The indicators of Gross National Income (GNI) and Gross National Product (GNP) are very similar. The major difference is in the concept of residents. While the Gross National Income describes the value of all final goods and services produced by the residents of a nation, the Gross National Product reports the value of all final goods and services produced within the nation, whether by residents or foreigners (SIEG 2007: 172).

if the capital expenditure per head is different from each other country. Thus, the early, "classical" growth theorists explained economic growth by the continuous accumulation of real capital (e.g. machinery). But as economic research could not find any empirical evidence for this convergence, the current understanding of the processes of economic development at that time was questioned and revisited (FLEMMING/GÖTZ 1993: 203). It turned out that the fundamental failure of the existing theories was the non-consideration of the problem of decreasing returns to scale (GRIEBEN 2001: 1). The value of each additionally accumulated unit of capital decreases according to its declining shortage. For example, the addition of one euro to an already existing capital stock of 1 billion euro does not create the same value as the initial investment of the first euro. National economies thus do not grow by just accumulating more and more capital in kind.

2.2 From SCHUMPETER to economic growth theory

One of the first authors to introduce the role of technological progress and innovation for economic growth was SCHUMPETER (2006, 2008a, 1942), who argued that capitalist economies rely on an ongoing change of production based on new machinery and new methods of production. Following SCHUMPETER, technological innovations are the key to economic growth (ARNOLD 1997: 109; GRUPP 1997: 55). Thus, SCHUMPETER anticipated the importance of technological progress through innovation for economic theory long before neoclassical theories of growth and is hence rightly considered as the most significant author of innovation research.[24] This appears even more remarkable, as SCHUMPETER'S original interest was not to explain economic growth in the long run by innovation[25], but instead to shed light on the phenomenon of economic cycles (GRUPP 1997: 55). His main effort was to develop a theory of economic change that could complement WALRAS' theory of general equilibrium. As he argued, the main reason for the existence of economic cycles is rooted in the discontinuous appearance of new combinations or re-combinations of produc-

---

24 SCHUMPETER is often denoted as the "classical author" (GRUPP 1997: 49, 55; DREHER 1997: 43), the "founder" (HÜBNER 2002: 3) or even the "godfather" (TIDD/BESSANT 2009: 15) of innovation research.

25 In his famous book "Theorie der wirtschaftlichen Entwicklung" of 1912, he ironically did not even use the term of "innovation". Instead, he paraphrased innovation or technological progress with the expression "the carrying out of new combinations" of production factors (SCHUMPETER 2006: 164ff.).

tion factors invented by the entrepreneur. Thus, the dynamics of the economy are fundamentally based on the dynamics of technological change. Technological progress or innovation hence is the independent, explanatory variable.

SCHUMPETER established the role of the "entrepreneur" as the central source of technological progress (SCHUMPETER 2006: 171). But he did not define the entrepreneur by a certain formal hierarchical status within the enterprise. Instead, he used a behavioural definition by assigning him the characteristics of risk-taking, courageous but also careful and prudent behaviour (SCHUMPETER 2006: 176ff.). But the literal generation of technological progress does not take place by the entrepreneur himself. Instead, SCHUMPETER followed the historical and romantic imagination of that time, namely that technological inventions are randomly developed by some kind of an ivory-towered, genius inventor outside the economic system, who does not intend economic goals of profit maximising and efficiency (SCHWITALLA 1993: 4). Thus, in the early work of SCHUMPETER, technological progress is conceptualised as an exogenous variable. The entrepreneur is the one who adopts and implements these inventions and enables their economic applicability, market diffusion and economic success. With the achievement of economic applicability and economic success, the invention becomes what SCHUMPETER understands as innovation.[26] Later, SCHUMPETER refreshed his interpretation of technological progress. There, the process of invention is no longer dedicated to the random genius of a single inventor, but to the intentional and purposeful strategy of (large) enterprises and companies and becomes institutionalised in R&D departments within the firms (SCHUMPETER 1942: 215).

The essential work of SCHUMPETER significantly contributed to the development of several strands of economic innovation research and economic growth theory which will be discussed in Chapter three (GRUPP 1997: 56). Of course, the certainly most prominent influence of SCHUMPETER can be found in the strand of evolutionary economic theory (NELSON/WINTER 1982; DOSI 1982; 1988a, 1988b; WINTER 1990; NELSON 1991; DOSI/MARENGO 1994; METCALFE 1998; DOSI ET AL. 2000). Especially the absence of a steady state of economic equilibrium in favour of a dynamic modelling of the economy is one of the keystones of evolutionary theory and can directly be traced back to the ideas of

26 SCHUMPETER'S differentiation between invention and innovation as well as the characteristics of his definition of innovation are explicitly discussed in chapter 4.3.2.

SCHUMPETER (SCHWITALLA 1993: 39).[27] By characterising the capitalistic economic process as "evolutionary" (SCHUMPETER 2008b: 82; ANDERSON 2009: 3; WITT 1987: 88; VON HAYEK 1969: 142)[28], he probably donated the evolutionary theory its name.

But as GRUPP (1997: 57f.) states, SCHUMPETER may even have influenced the development of neoclassical theory at the micro and the macro level. The idea of technological progress and innovation as an endogenous source of growth in the later work of SCHUMPETER (2008a: 93; 2008b: 31; also: 1942) has been recovered by the authors of the Endogenous Growth Theory (e.g. ARROW 1962; LUCAS 1988; ROMER 1986, 1990). GRUPP further justifies this rather unusual point of view by referring to the focus on competition and competitive structures in SCHUMPETER'S later work. The second of the so-called "Schumpeter-hypotheses" reflects on the relationship between market structure and innovation behaviour, by arguing that monopolistic market structures reduce the risk and uncertainty of the firm's innovative behaviour and thus increase the chance of higher innovation rates and higher profits from innovation (SCHUMPETER 2008a: 62ff.; WITT 1987: 48).[29] The prospect of higher returns from innovation represents the bait which encourages the enterprises to invest their capital in innovation projects (SCHUMPETER 1942: 148). Following this argument, several neoclassical approaches raised doubt about the assumption that perfect competition is the optimal frame condition for innovation (e.g. KANTZENBACH 1967; GROSSEKETTLER 1985) or tried to falsify this hypothesis by stressing the role of competition for innovation (e.g. ARROW 1962a; VON HAYEK

27 "Capitalist economy is not and cannot be stationary" (SCHUMPETER 2008b: 31).

28 "The worries that Schumpeter, in 1926, formulated about the German word 'Entwicklung' re-emerged when he helped (...) in translating his magnum opus into English. Even the chosen title of the book— "The Theory of Economic Development"—was problematic. Schumpeter, quickly after the publication of Development, began to switch from the term 'economic development' to 'economic evolution'; and he used the latter term consistently in Business Cycles from 1939" (ANDERSON 2009: 3).

29 These "Schumpeter-hypotheses" have been the origin of many theoretical and empirical studies (see e.g. SCHWITALLA 1993; also the various contributions in: DODGSON/ROTHWELL 1994; MALERBA/ORSENIGO 1996; AGHION ET AL. 2005; RAA/MOHNEN 2008).

1969).[30] Furthermore, the micro-economic perspective, established by the crucial role of the innovative entrepreneur, has found its way into the neoclassical micro-economic models of decision theory (e.g. KAMIEN/SCHWARZ 1982) as well as game theory (e.g. SCHERER 1967; REINGANUM 1981, 1982; DASGUPTA/STIGLITZ 1980).[31] Therefore, it is not surprising that innovation in terms of technological change plays the central role in many economic growth theories.

However, the relevance of SCHUMPETER'S work was neglected for a long time by economists. Like NELSON and WINTER (1974: 890), GRUPP (1997: 56) and GERYBADZE (1982: 115) suppose this was mainly due to its lack of mathematical formalisation. SCHUMPETER'S theory of economic development was never exposed in mathematical formulas, perhaps because mathematical instruments for modelling complex, dynamic developments were not available at that time. On the other hand, the absence of mathematical formalisation may have possibly permitted the recognition and diffusion of SCHUMPETER'S work in other scientific disciplines (e.g. social and political science).

Hence, SCHUMPETER laid out the foundations for economic growth theory by introducing the determining role of technological progress and innovation as the driving force of economic growth. To provide a better understanding of the central importance of R&D as a driver of economic growth and firms' competitive advantage, the following paragraphs will firstly discuss the theoretical roots of the predominant R&D focus in mainstream economic research during the last decades and how they shaped the understanding of the innovation process. Secondly, as the role of R&D as a driver for firms' competitive advantage has been supported by a vast amount of empirical evidence, an overview of the most important findings on the relationship between R&D and the innovation success of firms will be given. Based on this impressive history of innovation research dealing with the positive effects of R&D on innovation and competitive advantage, this chapter will conclude by asking whether non-R&D-

---

30 However, what SCHUMPETER had in mind when he spoke about the inherent monopoly power of successful innovators was concerned explicitly with *transient* monopoly as a consequence of, and trigger for, innovation and *not* structural monopoly as a base for innovation. Despite the fact that there is a more structural argument in his late work (SCHUMPETER 1950), the desirable market structure that he proposed is not that of sheltered monopoly (NELSON/WINTER 1977: 70).

31 For an elaboration of the main features of these approaches see GRUPP (1997), SCHWITALLA (1993), and DREHER (1997).

performing firms play a relevant role in German manufacturing industry today at all.

Before that, it should be mentioned that the following explanations represent a very brief overview of the central argument of neoclassical growth theory. If the reader is interested in their further elaboration and detailed mathematical derivation, the books of ARNOLD (1997), BLANCHARD (1999), SAMUELSON and NORDHAUS (1987); SALA-I-MARTIN (1990a, 1990b); BRETSCHGER (2004), BOLLMANN (1990), SCHWITALLA (1993), STADLER (1993) or MÜNT (1996) will provide a very good overview.

2.3 The classical start - SOLOW's theory of exogenous economic growth

Spontaneously, most people would agree with the idea that the systematic generation of new knowledge and economic development are closely interrelated and that thus the access to knowledge can be regarded as a determining factor of the economic growth of national economies. However, this is not how the development and growth of economies used to be explained by economists. More than two hundred years ago, economists instead focused on accumulated capital per worker when trying to explain differences in income or productivity. Similarly, differences in economic growth were seen as reflecting different rates of capital accumulation. This perspective arguably reflects the important role played by mechanisation as a mean for advancing productivity during the (first) Industrial Revolution, the period during which the frame of reference for much economic reasoning was formed (FAGERBERG ET AL. 2009: 9).

But with advancing growth and proceeding capital accumulation (i.e. the capital stock increases), each extra unit of investment generates less and less growth. Thus, growth theory had to find another explanation for the empirical phenomenon of steady economic growth. While the early growth theorists located the driving force of technological progress outside (exogenous) the economic system, more recent advocates of growth theory view technological progress as a factor which is intentionally created by economic actors and thus within (endogenous) the economic system. In the following, both major strands of growth theory are discussed.[32]

SOLOW'S articles "A Contribution to the Theory of Economic Growth" (1956) and "Technical Change and the Aggregate Production Function" (1957) paved the way for the incorporation of technological change in the model-based theories of growth (BOLLMANN 1990: 18). "In the late 1950s Solow's work (...) made this notion an intellectually respectable part of neoclassical thinking about economic growth", (NELSON/WINTER 1982: 197). SOLOW'S model assumes that only one single product is produced in the economy. The production of this good can be described by a production function which accounts for variable amounts of invested capital and labour. This production function was developed by COBB and DOUGLAS (1928). Contrary to post-Keynesian models (HARROD 1954; DOMAR 1957), neoclassical growth theory postulates the possibility of stable economic growth, which adapts itself to an exogenously given population and technological change by the variability of these factor inputs (BOLLMANN 1990: 18; RÖTHELI 1993: 2). Apart from this, the assumptions of a constant aggregate saving rate and the synonymy of savings and investment was taken over from the Harrod-Domar approach (BRETSCHGER 2004: 25). The produced good (output) can be consumed or reinvested, which means that capital is increased by non-consumed output. The inhabitants of this theoretical model consume a constant rate of their income and their number also increases at constant rate. The model takes full employment as well as perfect competition as granted (RÖTHELI 1993: 2).

As one of the main founders of the neoclassical growth theory, SOLOW added the variable of technological change into the hitherto established models based on the general production function of COBB and DOUGLAS:[33]

(2.1.) $$Y_{(t)} = F\left(A_{(t)}, K_{(t)}, L_{(t)}\right)$$

32 Beside the concepts discussed here, there are also models of growth based on the economic theory of MARX. But as they are focussed on the distribution of income as a main driver of economic growth, instead of technological change and innovation (MARGLIN 1984) they will not be discussed in this context.

33 Just a few months later than SOLOW, SWAN (1956) also published his contribution "Economic Growth and Capital Accumulation" which contains practically the same model of economic growth. Therefore, this approach is also frequently named as "Solow-Swan model" in literature. Further early studies were also conducted by MEADE (1961) and PHELPS (1961).

with *Y* representing the whole production output of an economy, *K* describing an index of the whole capital and *L* standing for the engaged amount of labour at the time of t. Unlike capital, labour as a factor of production cannot be expanded as wanted and thus cannot be accumulated or saved. But the major improvement consists in the production factor *A* which stands for an assumed state of technology at a point in time. The state of technology depends in turn on the economy's stock of technological knowledge. By considering technological change in the shape of the difference between two levels of technology at different points in time ($A_{(t+1)}>A_{(t)}$), the production output can be increased at constant rates of factor input of capital and labour. In other words, a country with a more advanced technology will produce more output from the same quantities of capital and labour than an economy with only a primitive technological level. The production function defines the maximum production output which can be achieved by any and each possible combination of given input factors (SAMUELSON/NORDHAUS 1987: 247).

As emerges from the formula, each of the input factors is essential. If one of them is not at all available in an economy, the output turns to zero. Thus, output depends on both capital and labour and on the state of technology. But in contrast to the model of HARROD and DOMAR, the input factors of capital and labour are assumed to be substitutable. With given capital and labour, an improvement in the state of technology leads to an increase in output. In turn, the necessary (demanded) amount of each input factor depends on the availability respectively the prices of all input factors (SAMUELSON/NORDHAUS 1987: 246f.). For example, if labour becomes more expensive (e.g. through wage increases), the demand for machine tools and automated manufacturing machinery increases. Thus, there is a high degree of interdependency between the production output and the input factors.

To expose the effect of technological progress on the relation among output, capital and labour, the production function can be rewritten in a slightly more restrictive form (BLANCHARD 1999: 228), namely:

$$(2.2) \qquad Y_{(t)} = F(K_{(t)}AL_{(t)})$$

This equation emphasises the dependency of production on capital and labour multiplied by the state of technology. On the one hand, given an existing stock of capital, technological progress reduces the number of workers needed to achieve a given amount of output. On the other hand, technological progress increases the factor of labour in the sense of the amount of effective labour in the economy. For instance, if the technology progress doubles, it is as if the economy had twice as many (effective) workers. Subsequently, output is being produced by the two factors of capital and effective labour in the form of output per worker (BLANCHARD 1999: 199, 229).

Further, as already mentioned above, it is assumed that the factors labour and capital feature constant returns to scale. In the long run, this implies that no capacitive limitations in the invested capital and labour exist. If the scale of operation is doubled for a given state of technology – that is, if the quantities of capital and labour are doubled – then output will also double. The production function is thus of homogeneous linearity (ARNOLD 1997: 49).[34]

However, constant returns to scale refer to what happens to production output when both capital and labour are increased. But what happens if only one input factor is increased? Again, it is reasonable to assume decreasing returns to each of the input factors capital and effective labour (REHME 2007: 6; SAMUELSON/NORDHAUS 1987: 248ff.). Given effective labour, an increase in capital is likely to increase output, but at a decreasing rate. For example, if a given population of workforce (labour) is provided with a machine tool (capital), the introduction of the new machine tool will substantially increase the workforce's production. But if the workforce is equipped with more machine tools than it is able to handle at once, the production output is unlikely to increase very much. In the worst case, additional machine tools remain unused and lead to no increase in output whatsoever. Symmetrically, with given capital, an increase in effective labour is likely to increase output, but at a decreasing rate. In the context of the example above, this would mean increasing the workforce for a given number of machine tools. The central assumptions of the production function – a) only one input factor is endogenously accumulated (capital), with decreasing marginal revenues, and b) technological change is taking place merely autonomously – were consolidated by the Japanese economist INADA (1963) and since then have been named "Inada conditions".[35] Technological progress can equally or differently impact on both input factors of capital and labour. Based on the production function of neoclassical

34 $2Y_{(t)} = F(2K_{(t)}, 2AL_{(t)})$ or more generally for any multiplier : $xY_{(t)} = F(xK_{(t)}, xAL_{(t)})$

35 From this property of decreasing marginal revenues also follows the so-called "catch-up" effect (MATHEWS 2006; FAGERBERG/GODINHO 2005; KUMAR/RUSSEL 2002). It describes the circumstance, that emerging markets, despite their lower capital resources, often feature higher growth rates than industrialised countries. Assuming the international mobility of capital, each invested capital unit leads to a relatively high growth effect in production output under the condition of meagre capital resources.

growth theory, technological progress is said to be neutral in terms of its effect on the functional distribution of income. This applies to all three cases of technological impact: constant coefficient of capital = labour increasing ("Harrod-neutrality"), constant coefficient of labour = capital increasing ("Solow-neutrality") and constant intensity of capital = product increasing ("Hicks-neutrality") (FRENKEL/HEMMER 1999: 112ff.; BRETSCHGER 2004: 32ff.).

Economic growth in neoclassic theory is thus determined by three major sources (REHME 2007: 6):

- accumulation of capital by saving and investing
- changes in the deployed amount of labour
- changes in technology, for example through technological innovation

Assuming a roughly constant ratio of workers (employment) to the population over time as an exogenous given variable, which means that the numbers of workers also grows at an annual rate (gL) (BLANCHARD 1999: 200, 230ff.; REHME 2007: 8), economic growth comes from capital accumulation and from technological progress (innovation).

Firstly, capital accumulation alone (saving rate) cannot sustain growth in the long run because of its decreasing return to capital. But a higher saving rate can sustain a higher level of output as the output per worker can be written as an increasing function of capital per worker. The state in which output per effective worker and capital per effective worker are no longer changing is called the "steady state of the economy" (FRENKEL/HEMMER 1999: 38). Secondly, as capital cannot sustain growth forever, sustained growth requires sustained technological progress. When the production function allows for technological progress, the number of effective workers increases over time. Thus, maintaining the same ratio of capital to effective workers requires an increase in the capital stock proportional to the increase in the number of the effective workers. Technological change is modelled in the fashion of growing by a constant growth rate (*gA*) over time. Brought together with the number of workers (*gL),* this implies that the growth rate of effective labour (*AL*) equals *gA+gL*. Under the condition of a constant depreciation rate of capital, a certain level of investment is needed to maintain a given level of capital per effective worker. This level of investment must at least compensate the rate of depreciation. An additional amount of investment is further needed to ensure that the capital stock increases at the same rate as the effective labour, respectively the technological progress (BRETSCHGER 2004: 28ff.; BLANCHARD 1999: 230f.; FRENKEL/HEMMER 1999: 40ff.) If an economy tries to achieve output growth in excess of *gA+gL*, capital would have to grow faster than output because of decreasing returns to capital.

Subsequently, the economy would have to devote more and more output to capital accumulation until there would be no more output left and growth would come to an end. Thus, the economy cannot permanently grow faster than *gA+gL*. Because output, capital and effective labour can all just grow at the same rates in a steady state economy, this state is also called a state of "balanced growth" which describes the state of economic equilibrium. If the relevant growth rates of the factors diverge from each other, the pricing mechanism adapts the growth rates of capital and national income to the growth rate of effective labour using adjustments in the ratio of factor inputs. The system is thus stable as the modelled economy pursues this equilibrium regardless of its initial situation (FRENKEL/HEMMER 1999: 57; RÖTHELI 1993: 2).

To summarise, the model of SOLOW supplies insight that is fundamental for comprehending economical growth. The model succeeds in explaining its main objective by emphasising technological change (in absentia of adaptation processes) as the necessary and sufficient condition for the continuous growth of production output (ARNOLD 1997: 78, 117). "The neoclassical model that was developed and applied by Robert Solow and others constituted a giant step forward in the process of constructing a formal model of growth" (ROMER 1994: 13).

But simultaneously it limits itself in its explanatory value by neglecting to consider the nature of technological change:

- ***Technological progress as a residual:*** Modelled as such, the growth rate of technological change is just the residual ("Solow-residual" - FRENKEL/HEMMER 1999: 136) of the difference between the growth rates of the national product and the production factors of capital and labour (NELSON/WINTER 1982: 197; GRUPP 1997: 65; BOLLMANN 1990: 19f.) By defining technological change as "any kind of shift in the production function" (SOLOW 1957: 312), SOLOW put the emphasis of the explanation rather on the effects of technological progress than on its processes of development (GRUPP 1997: 64f.; BRETSCHGER 2004: 30). In his book "New Review of Economic Growth", SCOTT (1989) developed an alternative growth theory which is not based on a production function. Thereby he responded to KALDOR'S criticism of the way technological progress is conceptualised in the production function. "Any sharp or clear-cut distinction between the movement along a 'production function' with a given state of knowledge, and a shift in the 'production function' caused by the change in the state of knowledge is arbitrary and artificial" (KALDOR 1957: 595f.).

- ***Technological progress as an exogenous variable:*** As the probably most serious shortcoming, technological change is considered as an exogenous factor, which means it does not require any factor input for its production (ARNOLD 1997: 49; REHME 2007: 8). It is thus treated as a given external factor of the economic system (GRUPP 1997: 64f) which just shifts the production function to a higher level. How technological progress and change develop in reality and which frame conditions like entrepreneurial motivations affect its emergence remain unanswered. The results derived are determined without making explicit assumptions about the institutions concerned with property rights, the market behaviour of technology-producing (or purchasing) firms, and the problems of uncertainty related to the R&D-process. What should be imagined under the term of "technological progress" or "technological change" also remains open. Is it human capital, organisational issues, knowledge, new or improved machinery? The central force of economic development is understood as an unspecified function of time (SCHNEIDER 1995: 433f.). As NELSON and WINTER put it, "in the case of technical change, neoclassical theory did not specify very well how "large" or important technological change must be – only that there was "something" there" (1982: 199). Technological progress is assumed to be provided by external actors and institutions outside the economy, namely institutions of scientific research (e.g. universities, research laboratories). Thus, technological knowledge is seen as a public good which can be adopted by the economic system (FRENKEL/HEMMER 1999: 142, 166; SCHNEIDER 1995: 433f.).

- ***Technological knowledge as a public good:*** The exogenous character of technological change causes another serious problem: treated as an external factor input, technological knowledge is characterised as a public good. Under the condition of the global availability of knowledge (e.g. through patent data bases, the World Wide Web, electronic libraries) and following the argumentation of the model, it can be reasoned that the growth rates of technological change will equalise internationally in the long run. In consequence, all nations ought to show consistent rates of production growth over time. Obviously, this conflicts strongly with the empirical fact of persistently differentiating growth rates between nations (ARNOLD 1997: 78), as divergent levels of development imply unrealised profit opportunities which conflict with the assumption of the model. The theory thus ultimately leaves no room for an economic development other than world-wide convergence unless differences in the parameters and exogenous variables between different countries are assumed (SCHNEIDER 1995: 434).

- ***Neglect of the aging rate of technologies:*** Further objections can be raised against the argument that technological progress increases the production output of the workforce, no matter how old the manufacturing equipment is (ARNOLD 1997: 49). This would mean that technological progress affects both newly installed manufacturing equipment as well as already existing capitalisation. In his article "Investment and Technical Progress" (1959), SOLOW admittedly assumed that the manufacturing equipment in each case represents somehow the technological state-of-the-art only at moment of its installation. Existing stocks of capital are affected by technological change only when they are modified or replaced by new investments.

Hence, the economic growth theory in the tradition of SOLOW already pointed to the central role and importance of technological progress for the growth of economies. However, because it treats technological progress as an exogenous variable, it provides no reasons why firms should undertake private R&D activities and thus does not explain the role of R&D for the competitiveness and innovativeness of firms.

2.4 The advancement of SOLOW's inheritance - The endogenous growth theory

The starting point of endogenous growth theory[36] was the theoretical shortcoming of the neoclassical models regarding the explanation of globally different income per head and the development of particular economies (e.g. described by LUCAS (1988) and ROMER (1986)). While the "old" growth theory remained rather vague about the contents of technology or accumulated knowledge, the new growth theory has associated this aspect with spillovers of technology and an increasing division of labour (ROMER 1990) both provided by the private sector (SCHNEIDER 1995: 440).

---

36 The Endogenous Growth Theory is usually named the "New Growth Theory". But since - in distinction from the neoclassical growth theory - technological change is modelled as an endogenous variable, the author prefers the notion of endogenous growth theory.

2.4.1 Different strands of endogenous growth theory

Apart from KALDOR and MIRRLEES (1962)[37], the first important step towards models of endogenous growth theory was made by ARROW (1962a, 1962b) (SCHWITALLA 1993: 20ff; BRETSCHGER 2004: 9.).[38] The origins of endogenous growth are based on the consideration of technological spillovers from the production of goods respectively research activities. Building upon the work of WRIGHT (1936)[39], ARROW referred to the necessity to include learning effects in neoclassic models (ARROW 1962: 156). He analysed to what extent learning effects in manufacturing processes could cause the development of technical knowledge. Technological progress appears therein, that – due to these learning effects – machines require less staff or – with a constant engagement of the workforce – are able to achieve higher output. ARROW (1962a: 158f.) chose the cumulated capital stock *K* as index of the cumulated stock of experiences (learning effects). Hence, technological change is no longer reduced to an input factor, but is also a kind of output factor of learning processes. This formulation stresses the idea that the installation of new capital opens up for new learning effects (ARNOLD 1997: 82). Despite some serious shortcomings of ARROW'S approach of sector-specific learning effects[40], it nevertheless raised the awareness of subsequent authors that the element of technological change cannot simply be reduced to an external, exogenous factor in growth theory (BRETSCHGER 2004: 9).

---

37 They explained growth through learning externalities by using a technical progress function. The rate of growth of output per head depends on the rate of growth investment per head. They showed that technical progress phases out if the technical progress function has decreasing returns to the rate of growth of investment. Only an added element of exogenous technical progress can guarantee long-run growth (SCHNEIDER 1995: 434). Because they modelled endogenous growth in a very abstract way without using the production function, their early contribution to an endogenous growth theory attracted just little interest in contrast to Arrow's approach (ARNOLD 1997: 82).

38 As GRUPP (1997: 59) states, the contribution of ARROW tends to be overseen in many economic textbooks. For a detailed consideration on ARROW'S early work on the economics of innovation see HOUNSHELL (2000).

39 Wright introduced a quantitative model of "learning by doing" to describe the time savings and associated cost reductions achieved in manufacturing aircraft. He found that the time required for assembling an aircraft decreased with increasing production levels.

40 For an overview of the various points of criticism see for instance ARNOLD (1997: 92f.), BOLLMANN (1990: 33f.), SCHWITALLA (1993: 23f.) or HELPMAN (1994: 369f.), who entirely rejected the approach of "learning by doing".

Neither in the SOLOW model nor in ARROW'S model of learning-by-doing does technological progress result from purposeful activities of economic actors. While SOLOW considered it as a completely exogenous variable, ARROW'S model regards it as an unintended by-product of capital accumulation (ARNOLD 1997: 95; BOLLMANN 1990: 33; DREHER 1997: 38). This means that no one earns his living by developing inventions or technological improvements (RÖTHELI 1993: 5). First approaches to modelling economic growth as an intended (endogenous) result of rational actors were made by UZAWA (1965), LUCAS (1972), and SARGENT and WALLACE (1976).[41] UZAWA for example argued that constant growth is possible if workers can spend part of their working time for qualification and training, which means they are able to accumulate technical knowledge, respectively human capital (UZAWA 1965: 18f.; ARNOLD 1997: 95). Contrary to the mere input factor of labour, the concept of human capital allows for the purposeful accumulation through qualification and training. As a result of modelling rational expectations of economic actors on technological progress, SOLOW'S assumptions of an exogenous given technological stock of knowledge were questioned and the way was open for the development of a new strand of theory in the 1980s, called "endogenous growth theory".

The most innovative and central assumption of the new growth theory and its founders like ROMER (1986, 1990), LUCAS (1988), REBELO (1991), AGHION and HOWITT (1992), BARRO (1990), and GROSSMAN and HELPMAN (1990, 1991a, 1991b) is that new technological knowledge is partially generated by adequately qualified employees within the enterprise. The generation of new technological knowledge and technological progress is thus conceptualised as an economic calculation of rational agents and not as an external function of time or as a publically provided good. Profit-oriented enterprises balance the costs of R&D against its expected profits on the market. The production of personally independent technological knowledge (like new or significantly improved products or processes) through purposeful R&D activities is the crucial and determining source of growth in production output and thus economic growth in general.

Because of the variety of endogenous growth models today, it is hardly possible to speak about one single theory of endogenous growth (FRENKEL/HEMMER 1999: 176). Nevertheless, they are connected by their goal of explaining con-

41 At that time, the hypotheses of LUCAS and SARGENT/WALLACE still referred to the financial market but can also be directed to the goods market (GRUPP 1997: 65f.).

tinuous economic growth by non-exogenous variables (like technological progress in neoclassic models). They search for internal, endogenous variables and drivers and determinants of economic growth. Thereby, the neoclassical postulate of decreasing marginal productivity of capital is rejected.

According to FRENKEL and HEMMER (1999: 176f.), two main strands of endogenous growth theory can be distinguished (Figure 2):

- The first one argues that even without the consideration of technological progress no decreasing returns of capital exist
- The second one focuses on endogenous technological change to explain economic growth

These approaches extend SOLOW'S original neoclassic model by various independent variables and differ, for example, with respect to whether a competitive equilibrium or monopolistic competition equilibrium is used as the basic set-up, although they are of course all built on some notion of economies of scale as the ultimate driver of persistent growth.

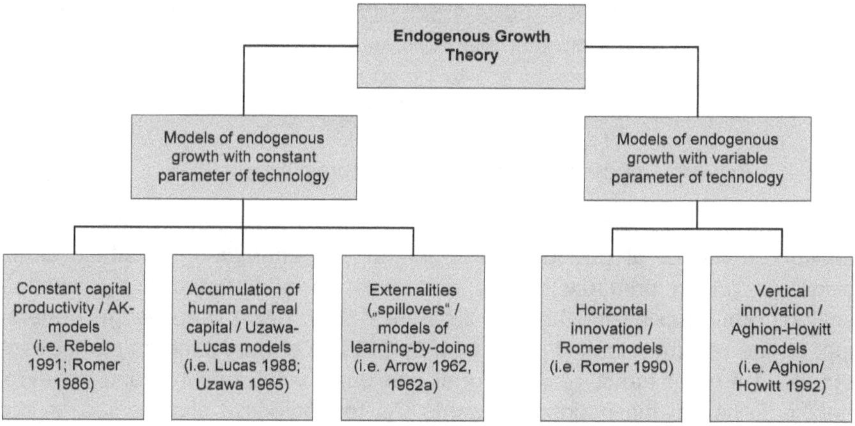

Figure 2: Strands of endogenous growth theory
(Source: adapted from FRENKEL/HEMMER 1999: 177)

Within the first strand of models, the so-called "AK-models", a linear relationship between capital and national income is underlying. This linearity compulsorily excludes decreasing returns of capital accumulation. The production factor of labour remains unconsidered for the reason that the definition of capital is very widespread and comprises human capital as well as physical capital. Thus, this family of models can be classified as investment-oriented.

Regarding the second major group of models, in the ones of AGHION and HOWITT (1992) as well as GROSSMAN and HELPMAN (1991), innovation leads to the improvement of consumer goods. For example, video recorders were improved by DVD players which in turn are currently being more and more substituted by "Blue Ray" players. Such a chain of improvement-innovations within one product family can be labelled as "vertical innovation". In contrast, innovation in ROMER'S model leads to an increased product variety. This can be illustrated by the example of yoghurt. At the beginning there was only natural yoghurt without any flavour. Over time yoghurt with more and different flavours has become available. This widening of product variety can be denoted as "horizontal innovation".[42] This family of models is unified by its assumption that ongoing (endogenous) technological progress is the main driver of economic growth. These models can be hence summarised as being R&D-based.

As probably the most seminal representative of the innovation-related endogenous growth theory (FRENKEL/HEMMER 1999: 263), and because he was one of the first authors who explained economic growth directly by endogenous technological progress (FOSS 1998: 228f.), I would like to focus the following paragraph on the contribution of ROMER. Elaborative descriptions of the other models and approaches are to be found for example in FRENKEL and HEMMER (1999: 181ff.) or ARNOLD (1997: 119ff.).

2.4.2 ROMER'S model of endogenous economic growth

In his first paper, ROMER underlines that his model "(...) can be viewed as an equilibrium model of endogenous technological change, in which long-run growth is driven primarily by the accumulation of knowledge by forward-looking, profit-maximizing agents" (ROMER 1986: 1003). Knowledge generation, on the one hand, takes place as a by-product of manufacturing processes and, on the other hand, by intentional and purposeful research and development activities of the economic agents. As technological change lies "at the heart of economic growth" (ROMER 1990: 72), his model is based on Solow's model with a neoclassical production function as described in the previous chapter. But now it is assumed that the level of technological knowledge A and

42 For a closer look on the relationship between horizontal and vertical innovation see COZZI/SPINESI (2006).

its growth rate (γ_A) is an endogenous variable caused by the rational decision-making of economic agents. Technological change arises in large part because of intentional actions taken by people who respond to market incentives and no longer as a given public externality like in Solow's model. ROMER acknowledges that not every R&D activity is motivated by market incentives. An academic scientist who is supported by government grants is most likely not affected by market incentives. But market forces are the driver which promotes the translation of scientific knowledge into goods with practical knowledge (IBID.).[43]

2.4.2.1 The nature of knowledge

The modelling of economic growth by the accumulation of technological knowledge has deep implications for the understanding of the nature of the good "knowledge". Economists usually distinguish between two fundamental attributes of any economic good: the degree to which it is rivalrous/non-rivalrous and the degree to which is excludable/non-excludable (CORNES/SANDLER 1986). Non-rival goods can be simultaneously used by different persons any number of times without losing their productivity (FLEMMING/GÖTZ 1993: 204; FRENKEL/HEMMER 1999: 240). Excludability instead is a relationship between technology and the legal system (ROMER 1990: 74). A good is called non-excludable if the owner cannot exclude others from using it (FRENKEL/HEMMER 1999: 240; FLEMMING/GÖTZ 1993: 204; ROMER 1990: 74). For example, a certain product can be made excludable in its use by means of a legal system that prohibits copying the product, or by means of specific encryption or copy protection schemes.

Unlike conventional economic goods which are both rivalrous and excludable (they are provided by private actors and can be traded in competitive markets), public goods are defined as both non-rival and non-excludable as they are not being provided by the private sector or can be traded in markets. ROMER (1990: 72) argues that technological knowledge is, on the one hand, a non-

43 With this argument, ROMER shows noticeable parallels to the reasoning of SCHUMPETER and the central function of his entrepreneur, and later the role of the R&D departments within big companies. The strand of endogenous growth theory which deals with imperfect, monopolistic competition is often named as "Neo-Schumpeterian" (e.g. ROMER 1994: 17ff.). Although ROMER followed some of SCHUMPETER'S central reflections, his models were not significantly influenced by SCHUMPETER or any other famous economist. As he stated in an interview (SNOWDON/VANE 1999: 298), he started developing his model from a blank sheet.

rival input of the production function. If the cost of developing new knowledge has once been expended, the results (the knowledge) can be used over and over again at no additional cost. For example, the knowledge of how to solve a mathematical equation can be applied by many different users at the same time without hindering each other from using the knowledge. In contrast, consumer goods like cars or industrial goods like machines, for example, can only be used by a single individual or a certain number of persons at each point in time. Continuous economic growth through continuous expansion of production and consumption can only be explained by endogenous technological change, if the activities of knowledge accumulation do not underlie the principle of decreasing returns to profit. Without increasing returns to scale, enterprises would have no incentive to steadily invest in technological progress. ROMER titled the increasing marginal productivity of the intangible capital good knowledge the "key feature in the reversal of the standard results about growth" (ROMER 1986: 1004).

But this argument implies that the firms must be able to cover their expenditures for research and development by correspondingly pricing their products. Moreover, abandoning constant returns to scale in the production function entails that each input factor of production could no longer be remunerated according to its productivity, which means that each of the factor markets can no longer be characterised by perfect competition (GRUPP 1997: 66; ARNOLD 1997: 122; FLEMMING/GÖTZ 1993: 203f.) As firstly coined in SCHUMPETER'S second hypothesis, imperfect competition works as a driver of economic growth as it motivates enterprises to continuously invest in R&D to develop new products and manufacturing processes. If increasing returns to scale remained completely firm-internal, the market share would gain central importance for innovation activities and general tendencies of monopolisation would become observable. Imperfect competition however holds more appeal to steadily develop new products and manufacturing processes. According to ARNOLD (1997: 122), this is the main reason why it took growth theorists so long to handle the element of R&D in their models. It was only after SPENCE (1976), DIXIT and STIGLITZ (1977), and ETHIER (1979) developed their model of monopolistic competition that R&D-models became manageable in new growth theory (ARNOLD 1997: 122; FOSS 1998: 233; ROMER 1994: 17).

Hence, rational firms which intentionally invest in R&D activities to develop new technologies and products expect that these developments will provide them with competitive advantage and additional profit. Firms would not be willing to invest in any R&D and to bear the enormous cost of these activities, if they do not feel certain to actually benefit from their R&D. If knowledge were a completely public and non-excludable good, other enterprises ("free riders" -

FRENKEL/HEMMER 1999: 240) could freely adapt the developments of their competitors and would be able to profit from them without any own R&D cost. If innovative firms should have the chance to realise a reasonable rate of return on their innovations, a mechanism is needed which allows the firms to secure their competitive advantage and additional profit, at least for a certain period of time (ARNOLD 1997: 33). This mechanism is provided by the principle of intellectual property rights, especially the patent law. Protection by patent leads to a monopolistic competition of innovating enterprises. The innovator holds the monopoly for its specific product (and its profit) and competes with other monopolists and their products. Competition is thereby not characterised by the pricing pressure on one and the same product (perfect competition), but is based upon the market implementation of new products that appear substitutable, from the customers' perspective.

Besides their protective function, patents also have an information function (ARNOLD 1997: 34; BURR ET AL. 2007: 39f.). The patent certificate compulsorily contains a detailed description of the invention. This accelerates the knowledge diffusion and counteracts the principle of exclusivity. Thus, patent protection is imperfect as it cannot prevent the incorporated knowledge from its diffusion into the economy's stock of knowledge (DREHER 1997: 27), and only preserves the commercial benefit for the innovator to cover its cost of R&D and to skim off profit.

The knowledge which is inherently immanent in each new product or new manufacturing process is contained in the patent and published through the patent specification which can be freely assessed by others. ROMER further distinguishes between two types of knowledge: product-specific knowledge and general knowledge (ROMER 1990: 75ff.). The product-specific knowledge is mainly developed through R&D activities and closely tied to a single product and is thereby to somewhat extent protected by the patent application. On the other hand, the general knowledge included in each product can be seen as an unintended, non-excludable by-product of manufacturing processes ("learning-by-doing") and spills over into the economy's common stock of knowledge. Knowledge is thus defined as a non-rival but not completely excludable good (ROMER 1990: 75; FRENKEL/HEMMER 1999: 240).[44] The diffusion of knowledge thereby could employ various mechanisms. Patents, (scientific) publications,

44 ROMER explicitly points out that "the distinctions between rival and non-rival inputs, and the distinction between excludable and non-excludable goods, are of absolutely fundamental importance in modelling and in policy formulation." (ROMER 1994: 16)

employees who change their occupation, imitation and even reverse-engineering are just some examples of such diffusion mechanisms (FRENKEL/HEMMER 1999: 204; DREHER 1997: 27).

2.4.2.2 The relationship between knowledge and human capital

Technological knowledge thus is characterised differently from the production factor of human capital which describes the available entity of all personal competences and capabilities. Thus, humans that incorporate knowledge are defined as human capital. Human capital as such is both a rival and excludable good. A person cannot work in more than one place at a given point in time, nor can this person solve many (unlimited) problems at once. Hence, human capital can be privately provided and traded in competitive markets (labour market). Unlike human capital, technological knowledge like "designs" or "blueprints" is – being a non-rival good - not tied to a physical object. It can be copied at comparably low-cost[45] and used in as many different activities as desired (ROMER 1990: 75f.). Considering human capital, training a second person for a certain competence or capability is almost as cost-intensive as training the first one. When the person dies, all his/her skills and competences are lost. But any knowledge that this person has produced (e.g. principles of mechanical, electrical or chemical engineering, a mathematical proof, a piece of software, or a patent) during his/her professional career lives on and remains accessible after the person is gone. The innovative knowledge thereby loses its physical tie to single persons.

ROMER'S concept of human capital understood as years of education or training that are person-specific is close to the one that has been used in labour market contexts[46] and corresponds to approaches of economic growth which account for changes in the level of education and experience (ROMER 1990: 79). His concept of human capital is characterised by decreasing returns to scale (it can be assumed that the benefit of adding another R&D engineer is more and more decreasing over time) and is thus more limited than the notion used in the models of growth of KING and REBELO (1987), REBELO (1991), or

45 "...the cost of replicating it [*the design - author's note*] with a drafter, a photocopier, or a disk drive is trivial compared to the cost of creating the design in the first place (ROMER 1990: 75).

46 For instance by HECKMAN (1976) or ROSEN (1976).

LUCAS (1988) that explained economic growth by the possibility of unlimited human capital accumulation.

2.4.2.3 The model

In his first model, ROMER (1986) argued that the generation of knowledge through the R&D activities of firms contributes to improvements in the macro-economic stock of knowledge due to its non-rival characteristic. This complementary diffusion of knowledge is described by technological spillovers. "The creation of new knowledge by one firm is assumed to have a positive external effect on the production possibilities of other firms, because knowledge cannot be perfectly patented or kept secret" (ROMER 1986: 1003).[47] Subsequently, parts of the new knowledge are available for other competitors at the same time. For instance, the discovery of a certain technological solution tells the other competitors which other solutions they have probably been considering will not work. This type of positive external effects (externalities) gives rise to positive continuous economic growth in ROMER'S first model.[48]

But ROMER himself by and by recognized that his modelling of technological progress was unsatisfactory (ROMER 1994: 15f). "When I look back on my work on growth, my greatest satisfaction comes from having rejected the first round of external effects models that I tried" (ROMER 1994: 20). The reason for this self-critical judgment was the neglect of the second characteristic of knowledge: its partial non-exclusivity. "I reduced the dynamic model to one with a single-state-variable model by assuming that the excludable good *A* [*technological knowledge - author's note*] that the firm produces intentionally is used in fixed proportions with physical capital" (ROMER 1990: 77). As a result, ROMER'S first model ended up having similar dynamics to those of ARROW'S learning-by-doing model. He realised that the learning-by-doing formulation indeed has the advantage that it makes the rate of accumulation of non-rival knowledge endogenous, but on the other side is unsatisfactory as it takes the strict propor-

---

47 "Since newly produced private knowledge can be only partially kept secret and cannot be patented, we can represent the technology of firm (...) in terms of a twice continuously differentiable production function (...) that depends on the firm-specific inputs (...) and on the aggregate level of knowledge in the economy" (ROMER 1986: 1015).

48 "This seemed appealing because it recognized that firms did research and development on purpose and that the relevant spillovers or incomplete property rights were associated with the results from research and development." (ROMER 1994: 15)

tionality between knowledge and physical capital or knowledge and education as an unexplained and exogenously given feature of technology.[49]

When refining his model, ROMER separated the rival component of knowledge (incorporated in human capital - *H*) from the non-rival, technological component (*A*) (ROMER 1990: 79). The aggregate production function used captures the main ideas of the different modelling strategies in the new growth theory and may be written as follows (ROMER 1990: 78f.,80; FOSS 1998: 227):

$$Y = F(K, L, H, A) \quad (2.3)$$

with *K* standing for capital measured in units of consumption goods, *L* for unqualified labour skills[50] measured by counts of people, *H* as a distinct measure of the cumulative effect of activities such as formal education and training named as human capital, and *A* representing the level of technology and knowledge. Labour is thus divided into two aspects: the production of output (*L or* L_Y *= unskilled labour)* and the search for technological knowledge and innovation *(H or* L_A *– skilled labour)*. Thus, $L_Y + L_A = L$ represents total labour in the economy. According to the model of Romer, *L* is assumed to be constant.

The improved model of the economy thus has three sectors: an R&D sector, an intermediate goods sector, and an end products sector which produces the consumer goods (ROMER 1990: 79ff.). This three-sector model, as outlined in Figure 3, allows for internal economies, which means that price-taking competitive equilibrium cannot be supported. Monopolistic competition is introduced to reflect the presence of market power.

The R&D sector uses human capital and the existing stock of technological knowledge to produce new knowledge like designs and blueprints for producers' goods under increasing returns to scale. Technological knowledge is nonrival and can be simultaneously used in multiple contexts without losing its productivity. Hence, the current level of technology is available for all researchers without causing any or almost any further cost for the R&D sector. The economic incentive for the R&D sector for the continuous development of

49 "It preserves the public-good character of knowledge assumed by Solow (...) but makes it a public good that is privately provided as a side effect" (ROMER 1990: 77).

50 "(...) such as eye-hand coordination that are available from a healthy physical body" (ROMER 1990: 79).

new knowledge lies in selling and licensing patents that contain the product-specific knowledge. Thus, this form of knowledge can only be used by producers if they buy the permission for production from the R&D sector. The other type of knowledge produced through learning-by-doing processes is publicly available for further R&D activities as a positive externality. Subsequently, new knowledge increases the level of technology and thus in turn increases the productivity of the R&D sector. According to ROMER (1990: 79), R&D can take place either internally or in a separate firm that sells its patent to the firm that will produce the actual good.

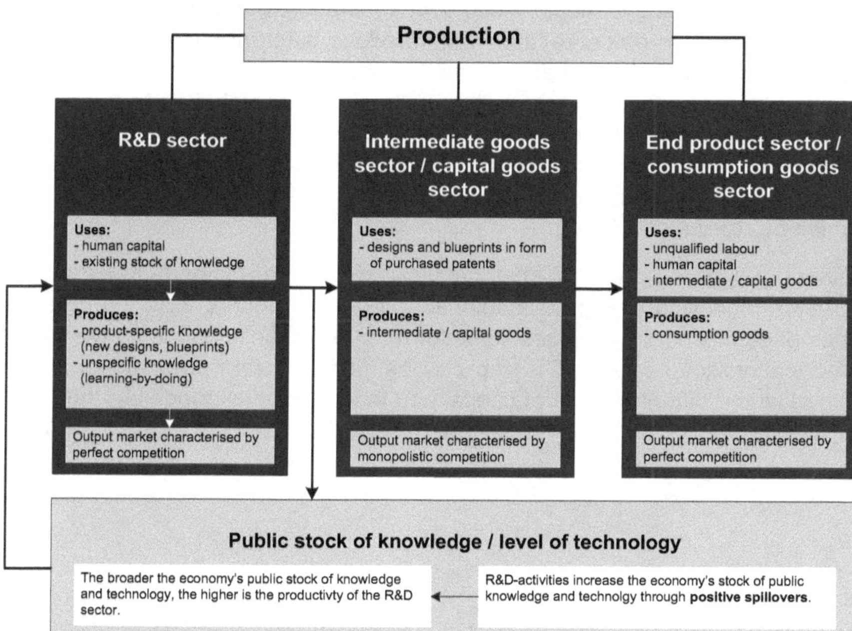

Figure 3: Overview of the sectoral structure in ROMER'S model
(Source: own illustration referring to FRENKEL/HEMMER 1999: 244)

The intermediate goods sector (or capital-goods sector) uses the designs and blueprints from the research sector together with the foregone output to produce the large number of producer durables (product innovations) that are available for use in final goods production at any time. On the supply side, the price of a patent results from the marginal cost of its production. On the demand side, the patent prices are determined by the payment reserves of the purchasers. The purchasers from the intermediate-goods sector compete for the purchase of a patent as long as its price is below the profit that can be expected from production and purchase of their intermediate good in the sector of final goods. The patent's price, and thus the incentive to undertake R&D efforts, follows from the production elasticity of the industrial goods sector and the demand for capital goods.[51]

Because a patent secures the utilisation for its purchaser, the firms which own a patent gain monopolistic market power for the production of a specific intermediate good. Monopolistic competition allows the firms to price their intermediate goods in such a way that the purchase cost of the patent can be absorbed. Subsequently, the producers of intermediate goods only indirectly compete with each other, because they produce different variants of intermediate goods which can neither be directly substituted nor are directly complementary to each other, and thus all provide similar contribution to the overall economic production (FRENKEL/HEMMER 1999: 243). This disaggregation of capital into an infinite number of distinct types of intermediate goods respectively producer durables, is an unusual feature of the production function assumed by ROMER (1990: 80).

Without going too far into mathematical details, an important aspect of ROMER'S modelling lies in its assumption about the degree to which different types of capital goods are substitutes for each other while, in the conventional specification, total capital K is implicitly defined as being proportional to the sum of all the different types of capital and that, all capital goods are thus per-

---

51 However, LEVIN ET AL. (1987) have shown that, in the case of process innovation, firms prefer to protect their knowledge by secrecy instead of patents. Thus, patent protection seems mainly to work for product innovation in certain industries (LEVIN ET AL. 1987: 796ff.). Moreover, high dynamics in technological progress may also be one reason why patents are not sufficient to ensure the appropriation of innovation rents (HANUSCH/CANTNER 1993: 33). The necessity to improve the patent law is thus probably not as relevant as certain authors try to make us believe.

fect substitutes. By rewriting the production function[52], ROMER points out that output is an additively separable function of all of the different types of capital goods, so that one additional dollar capital good “A” has no effect on the marginal productivity of capital good “B” (ROMER 1990: 81).

Finally, the sector of end-products uses unqualified labour, highly skilled human capital, and purchases the set of intermediate goods that is available as input factors to produce the final output, the consumer good. The assembly of intermediate goods into consumer goods is thus attached to cost, which means that input factors have to be invested to produce the consumer goods. Contrary to the intermediate goods sector, the market for end products is again characterised by perfect competition.

2.4.2.4 Steady state of economic growth in ROMER'S model

Endogenous growth means that no economic growth would be possible without the development of new and the improvement of existing products and processes. By considering the accumulation of technological knowledge, the production function will show increasing returns to scale.[53] Doubled investments in capital and labour lead to a doubled production output, even if the level of technological knowledge is given as constant. If technological knowledge is also doubled, the production output would increase more than double (FLEMMING/GÖTZ 1993: 207; JONES 1995a: 514f.). This represents the basic mechanism of how increases in R&D are transformed into increases in the consumption output.

The determining aspect of the R&D sector's description is that R&D causes positive external effects. An innovation does not only lead to the production of a new product and to monopoly profits on the part of the firm, but it also contributes to the increase of the economy's stock of knowledge and technological

52 ROMER expresses the final output Y in his model as a function of physical labour *L*, human capital devoted to final output H_y, and physical capital: $Y(H_\gamma, L, x) = H_\gamma^\alpha L^ß \sum_{i=1}^{\infty} x_i^{1-\alpha-\beta}$,with the producer durables indexed by i. Only a finite number of these potential inputs, the ones that have already been invented and designed, are available for use at any time. Thus if $x = \{x_i\}_{i=1}^{\infty}$ represents the list of inputs used by a firm that produces final output, there is some value *A* so that $x_i = 0$ for all $i \geq A$. Because *A* changes as new producer durables are invented, it is important to be able to describe final outputs as a stationary function of all conceivable input lists (ROMER 1990: 80).

53 $Y = K^{1-\alpha}(AL_Y)^\alpha$

level. This public stock of knowledge of an economy thereby represents the stock of capital which is placed at the disposal of future generations (FLEMMING/GÖTZ 1993: 208).

Steady state or equilibrated economic growth is present if all independent variables simultaneously grow at the same rate. Which equilibrated growth rate is achieved in ROMER'S model depends on the amount of human capital which is bound to the R&D sector for the production of new knowledge. This underlines an important feature of the ROMER model: the growth rate of innovations linearly increases with the human capital engaged in the R&D sector, whose productivity in turn also increases with its amount of human capital.[54]

The R&D equation is the "heart" of ROMER'S model (JONES 1995a: 515) and relates the labour engaged in R&D to the growth rate of knowledge. The available amount of human capital is determined by the competition between the R&D sector and the end-product sector as both are competing for the human capital of an economy. Dependent on current wage differentials, human capital is said to be mobile between both of these sectors. In an equilibrated state of growth, the rewards for human capital have to be identical in both sectors (FRENKEL/HEMMER 1999: 245ff.). Unqualified or low-skilled labour is just deployed in the end-product sector, whereas knowledge is only used in the R&D sector.[55] The allocation of the existing human capital between the R&D and the end-product sector results from the interplay of production characteristics of the economy and the preferences of economic actors.

Like all approaches of the new growth theory, the ROMER model emphasises the central driver of growth in the economy, namely, that a certain share of human capital is no longer engaged in the production of consumer goods and is rather deployed in the R&D sector to produce new knowledge. This share of human capital indeed not longer contributes to the ongoing economic production and therefore the short-term output of production is lower than without R&D activities. But in the long-run, R&D activities give rise to new products and to an improvement of knowledge and technology, which in turn lead to

54 $\frac{\dot{A}}{A} = \delta L_A$

55 Of course, the end-product sector also indirectly benefits from the newly created knowledge through the number of available intermediate goods which in turn are based on patents.

economic growth. Thus, future possibilities of production and consumption will increase (ROMER 1990: 254).[56]

The driver of economic growth is thus closely interlinked to the positive externalities of R&D activities. This increase of productivity by the R&D sector, for which the particular research organisations do not have to pay, leads to more and more innovations and guarantees the continuous process of growth. Without these externalities of R&D, the percental increase of new designs and blueprints would more and more decrease over time to the point of deadlock. In the model of ROMER, R&D externalities build the prerequisite of the continuous process of economic growth. The more R&D, the more innovations, the higher the number of variants of intermediate, capital goods, and, in consequence the higher the increase of production of consumer goods in each period (FLEMMING/GÖTZ 1993: 208; FRENKEL/HEMMER 1999: 254f.).

2.4.3 Summary and criticism

ROMER succeeded in developing a growth model which proves that economies cannot grow in the long run just by saving and accumulating physical and human capital. Instead, the necessary condition for continuous growth is technological progress whose properties require abandoning the notion of perfect market competition. The endogenous character of technological progress lies in the circumstance that new technological knowledge is intentionally and purposefully produced within the R&D departments of rationally behaving firms by specifically qualified personnel (human capital).

Thus, new growth theory accomplishes the purpose of explaining long-term economic growth by endogenous R&D activities and their positive externalities in form of spillovers from learning-by-doing processes. The latter is of particular importance when it comes to explaining how less industrialised countries with lower R&D intensity can achieve economic growth by imitation or copying (ARNOLD 1997: 40, 217; DREHER 1997: 27). Moreover, the ROMER-model provides an answer to the question, why growth rates of different countries can persistently diverge from each other, which means that they do not converge in the course of time (one of the most serious shortcomings of neoclassical

56 ROMER'S model also contains the theoretical chance of stagnation or zero growth. If the stock of human capital of an economy is so low that the deployment of human capital in the " not productive" R&D sector will cause heavy losses in the level of production, even low R&D activities would not be profitable any more (FRENKEL/HEMMER 1999: 255).

growth models). Following ROMER, a large stock of human capital and a high productivity of the R&D sector have a favourable effect on the economy's development. Economic growth can additionally be accelerated if the economic actors are more secretive regarding their consumption, and have a higher propensity to save (FRENKEL/HEMMER 1999: 260). The reflection of aspects like technological progress, innovation, patent development, the different natures of knowledge, as well as monopolistic market structure in the intermediate goods sector shifts the neoclassical analysis of mere impact to an analysis of causes which considers the expectations and intentions of rational actors and is thus oriented more micro-economically. In so far, the endogenous growth theory constitutes a substantial improvement compared to neoclassical models (CHRISTIAANS 2004a: 202f.).[57]

However, the followed points of criticism are frequently raised:

- ***Is the New Growth Theory really "new"?*** It is occasionally asked what is actually "new" about the (New) Endogenous Growth Theory and whether this theory really does contain any novel facts (FOSS 1998: 224; FREEMAN/SOETE 1997: 327; FRENKEL/HEMMER 1999: 175; also: SCHNEIDER 1995). The critics argue that the basic theoretical mechanisms that are highlighted in the Endogenous Growth Theory have already been known for a long time. Even very early on, theorists like YOUNG (1928) who is rooted to some extent in the work of ALFRED MARSHALL, pointed to the centrality of externalities and increasing returns in the growth process. Likewise, SCHUMPETER shaped the endogenous character of technological change (CANTNER/HANUSCH 1998:1) and the role of innovations implemented by the dynamic entrepreneur (SCHUMPETER 2006) and later the R&D departments of large enterprises (SCHUMPETER 1942) long before ROMER or even SOLOW. SCHUMPETER even established the thought that a monopolistic market structure may favour the innovation activity of economic actors.[58] As NELSON underlines, another thought leader about the rudiments of Endogenous Growth Theory (e.g. the role of

57 „(...) these formal models are different than most of the earlier ones, and in ways that make them more "realistic" in the sense of capturing, at least in stylized form, features of growth that many economists studying the topic empirically long have known to be important. Incorporation of these features certainly makes it easier for formal growth theorizing to engage effectively with the empirical work of economists trying to come to grips with the puzzling features of experienced economic growth" (NELSON 1994: 2).

58 However, SCHUMPETER may have rather assumed a transient monopoly through innovation advance instead of a structural market monopoly.

technological change, externalities and economies of scale) was MOSES ABRAMOVITZ with his article of 1952 (NELSON 1994: 2). Aside from that, it was described in the previous paragraph that a great number of the ideas contained in the endogenous growth theory were already proposed by ARROW (1962) who emphasised the relevance of learning-by-doing processes for the process of economic growth. Thus, the question seems justified whether the central aspects of Endogenous Growth Theory may be more than just "old wine in new bottles" (FOSS 1998: 224), whereby "new bottles" can be interpreted as the addition of mathematical formalisation to these already known ideas (NELSON 1994: 3).

- ***Neglect of institutions:*** Related to the central importance of knowledge spillovers to the growth process, it seems surprising that the role of institutions and firm dynamics underlying the process of knowledge accumulation is not treated theoretically by the Endogenous Growth Theory (LANGLOIS/ROBERTSON 1996: 22). "They inherently suffer from lack of attention to institutions; and [...] they are vulnerable to problems of institutional change" (PASINETTI 1994: 356). This neglect of institutions makes the theory vulnerable from both an empirical and a theoretical point of view. Following COASE'S contributions (1960, 1974), it can be asked why institutions, like property rights and transaction costs, do not arise in Endogenous Growth Theory to internalise the knowledge externalities that are so important (WEDER/GRUBEL 1993; LANGLOIS/ROBERTSON 1996). Other authors go even further by arguing that social institutions matter to a theory of economic growth for reasons that go well beyond the mere internalisation of knowledge spillovers.[59] As SOLOW (1994b: 378) put it, social institutions affect motivations, attitudes, judgements and thus respond to many more influences and affect behaviour in many more ways than can usefully be classified as transaction costs. Institutions in the end determine the effects of knowledge spillovers as crucially as does the nature of knowledge itself (LANGLOIS/ROBERTSON 1996: 2). Instead, theories of economic growth either assume institutions to be unimportant or reduce institutional questions to questions of relative price. Institutions are thus "just a matter of 'getting the prices right'" (LANGLOIS/ROBERTSON 1996: 23f.).

59 For example: NORTH (1981, 1990), LANGLOIS/ROBERTSON (1995), or ROSENBERG/BIRDZELL (1986).

- ***Knowledge spillovers only within national boundaries:*** A question which is closely connected to the issue of spillovers appears to be whether the benefits of knowledge investments can be appropriated domestically or whether they will diffuse globally (FREEMAN/SOETE 2009: 588). For instance, FAGERBERG (1991) and VERSPAGEN (1991) emphasised that the catching up by lagging countries depends very largely on the import and transfer of technology and knowledge. Against the background of the current global world economy, it seems obvious that increasing R&D expenditures are unlikely to benefit only the domestic economy.[60] "In short, the link between the location of 'national' firms' private R&D activities and national productivity gains appears today to be increasingly tenuous." (FREEMAN/SOETE 2009: 588)

- ***Too favourable assumptions:*** Another point of criticism refers to the risk of tautology which the endogenous growth theory runs into by making too "favourable" assumptions (FOSS 1998: 224; SOLOW 1994: 53). This means that the basic explanatory structure of the Endogenous Growth Theory is partially characterised by a very short distance from assumptions to conclusions. For example, it is assumed that knowledge is the driving force behind economic growth because of its externalities. Behind the background of this assumption, the result that different rates of growth in knowledge accumulation lead to different rates of long-run growth seems to be rather trivial and tautological (SOLOW 1994: 53). Further, SOLOW also entertains the suspicion that some of the assumptions made by endogenous growth theory are not sufficiently grounded in the empirical reality. Spending more resources on R&D will cause more innovations per year, and the growth rate of A will be higher. But if one supposes that an innovation generates only an absolute increase in technological knowledge, then a greater allocation of resources to R&D buys a onetime jump in productivity, but not a faster rate of productivity growth. The problem is that it is not immediately apparent which assumption is the reasonable one, and that this needs to be checked against empirical reality, but that proponents of the NGT are just making the "un-

60 Thus, as MEISTER/VERSPAGEN (2004) pointed out, achieving the "3% Barcelona target" in the EU by 2010 would probably not reduce the income gap between the EU and the USA, the benefits of the increased R&D efforts accruing not only to Europe, but also to the USA and the rest of the world (FREEMAN/SOETE 2009: 588).

earned", but very specific, assumption of proportionate increases in the level of technology.[61,62]

- ***Models of "knife-edge character":*** As SOLOW further argues, such questionable assumptions in the models of Endogenous Growth Theory lead to a somehow "knife-edge character" (SOLOW 1994: 50; also: GROTH/SCHOU 2002). This means that they are designed in such a way that the equilibrium state of growth with constant shares of input factors just results in exponential growth. According to ROMER, exponential growth is only possible, as long as constant returns to scale of technological knowledge's accumulation exist. If just small deviations in the assumption of constant returns to scale occur (like minimally decreasing/increasing returns), the models would collapse and continuous growth would be impossible (ARNOLD 1997: 259f.).

- ***No vertical innovation:*** Another shortcoming in the axiomatic structure of ROMER'S model is rooted in the circumstance that the variety of intermediate (capital) products does not underlie any function of obsolesce (FRENKEL/HEMMER 1999: 262). ROMER'S hypothesis of horizontal innovation simply describes the coexistence of old and new capital goods. The empirical evidence that old technologies and products based on them are also gradually replaced by new products and technologies is not considered by ROMER. GROSSMAN and HELPMAN (1991b) as well as AGHION and HOWITT (1992) have provided refined models which account for such quality improvements and product replacements (CHRISTIAANS 2004a: 203). This idea of vertical innovation approaches SCHUMPETER'S notion of the "creative destruction" of old factor combinations by new ones (FRENKEL/HEMMER 1999: 262).

- ***Overemphasis of growth policy:*** Pursuing the argument of positive knowledge spillovers, the social benefit of R&D activity must necessarily exceed the private benefit (LANGLOIS/ROBERTSON 1996: 1). Subsequently, R&D-activity is undersupplied relative to the social optimum (in terms of the optimal maximum growth rate) because firms

61 „The idea of endogenous growth so captures the imagination that growth theorists often just insert favorable assumptions in an unearned way; and then when they put in their thumb and pull out the very plum they have inserted, there is a tendency to think that something has been proved" (SOLOW 1994: 53).

62 "The number of scientists engaged in R&D in advanced countries has grown dramatically over the last 40 years (because of population growth and an increase in the intensity of R&D), and growth rates either have exhibited a constant mean or have even declined on average" (JONES 1995b: 760).

are not able to realise the full rate of return on their innovation activities (GEROSKI 1995: 89). Governments thus ought to address this problem of appropriability by subsiding or encouraging private R&D to promote social welfare. The rate of steady state growth in models of endogenous growth depends on all sorts of subsidy policies to adequately control the innovation activities of the private sector and to stimulate optimal long-term growth (ARNOLD 1997: 217). According to the models, the economy's equilibrated growth rate could be theoretically increased for instance through policy measures like the introduction of R&D-subsidies, a lowered taxation of capital income or the funding of basic research. Thus, if ROMER'S argumentation is consistent, the problem of low or stagnating economic growth rates could easily be solved. But here, endogenous growth theory seems to be overemphasise the scope of growth policy.

Another group of economists (e.g. KITCH 1977; DASGUPTA/STIGLITZ 1980; DASGUPTA/MASKIN 1987), however, argue that all enterprises take part in some kind of "patent race" (GRUPP 1997: 63; HANUSCH/CANTNER 1993: 16) in technological innovation, which can only be won by a single firm. Thus, the total R&D-input should be estimated higher than the one corresponding to the social optimum; that is to say in ROMER'S words that the R&D sector ties up so much human capital that the economy's production of end-products decreases. In consequence, any regulatory action by policy makers in the form of R&D subsidies would exacerbate this situation.

As ARNOLD (1997: 41) further points out, these views contradict all past experiences with growth policy. Policy action to stimulate growth rates seems to be a very new and challenging approach for developed industrial countries. Instead, it has to be assumed that policy instruments play only a minor role for economic welfare (KRUGMAN 1990: 41). The studies of LEVINE and RENELT (1992) and SALA-I-MARTIN (1997) admittedly present some results which suggest a significant correlation between specific fiscal indicators and economic growth, but the coefficients become insignificant as soon as the independent variables are even marginally changed (LEVINE/RENELT 1992: 952). Similarly, EASTERLY and REBELO (1991: 442) suggests that the evidence that tax rates matter for growth is disturbingly fragile and in sharp contrast to the robustness of the theoretical predictions. Moreover, JONES (1995a: 521) concludes that nothing seems to have caused a long-lasting impact on the economic growth process in the USA.

– ***Scale effects of human capital in the R&D sector:*** This leads us to certainly one of the most serious criticisms, particularly of ROMER'S endogenous growth model: the implicit scale effects of human capital in the R&D sector. Additional human capital in the R&D-sector leads to a proportional increase of the innovation rate respectively the output rate of new technological knowledge (FRENKEL/HEMMER 1999: 261). This has been criticised especially by JONES (1995, 1995a), who has provided evidence that the presence of scale effects in the R&D-based growth models of ROMER (1990), as well as those of GROSSMAN and HELPMAN (1991a, 1991b), and AGHION and HOWITT (1992), is inconsistent with time series data from OECD countries (JONES 1995a: 497, 1995b: 762; ARNOLD 1997: 36). According to the models of ROMER, GROSSMAN/HELPMAN, and AGHION/HOWITT, the exponential trend in the labour force (namely human capital), which can be observed over the period 1950-1988, should lead to an exponential trend in per capita growth rates.[63] But "(...) the endogenous growth literature appears inconsistent with time series evidence in documenting the lack of an increase in per capita growth rates. (...) The R&D-based models are rejected by this evidence (...) very strongly" (JONES 1995a: 519).

But JONES (1995a: 519) also emphasises that, apart from the prediction of inter-temporal scale effects, the R&D-based growth models are intuitively very appealing, because growth occurs as a result of innovation activities by profit-maximising actors. Therefore, it appears desirable to maintain their basic structure while eliminating the prediction of scale effects. To achieve this goal, JONES (1995b: 765) suggests considering that the most obvious ideas are probably discovered first so that the probability that a person engaged in R&D discovers a new idea is decreasing with an increasing level of knowledge. Furthermore, it should be taken into account that the duplication and overlap of research reduce the total number of innovations produced by L_A units of human capital.

To account for these considerations and to eliminate the scale effects in ROMER'S linear production function of the R&D sector, Jones modified ROMER'S equation for the R&D-sector.[64] Individuals are the critical input in the discovery of new designs, and the economy's growth rate crucially depends on the growth rate of the exogenous variable of the labour force (JONES 1995b: 779). While endogenous growth theory treats human capital as a given quan-

63 For a detailed description of the unsatisfactory time series evidence of the Endogenous Growth Theory see JONES (1995a: 516ff.)

64 For the detailed mathematical derivation see JONES (1995b: 764ff).

tity, JONES' model closes this gap and includes the accumulation of human capital in the economic explanation; it endogenise the variable of human capital (ARNOLD 1997: 222). In doing so, it abandons the strict proportionality between the stock of knowledge and the growth rate of knowledge.[65]

JONES' model is perfectly consistent with the empirical time series data of OECD-countries he presented (1995a). His extended model predicts that the growth rate is determined by parameters that are usually taken to be invariant to policy manipulation such as R&D subsidies, subsidies for basic research or taxation of capital income. Growth in the economy is tied directly to growth in productivity, which in turn depends on the discovery of new designs through R&D activities.

Similar to models of ROMER, GROSSMAN/HELPMAN and AGHION/HOWITT, and in contrast to SOLOW'S model, growth in JONES' extended model arises endogenously through R&D. Growth in the model is endogenous in the sense that technological progress, which generates long-run growth, results from R&D undertaken by profit-maximising agents. However, long-run growth is not endogenous in the sense that traditional policy interventions have no long-run growth effects. The source of economic growth (the individuals that are working on the discovery of new designs) is endogenous, but the resulting observable growth rate is independent of growth policy measures. "A term such as "semi-endogenous" growth may usefully be applied to the model developed here" (JONES 1995b: 761). Strictly speaking, growth in JONES' model is determined by exogenous parameters and its implication of invariance to policy instruments is similar to that of SOLOW (1956) (CHRISTIAANS 2004a: 207; JONES 1995b: 760).[66] Nevertheless, technological progress remains an endogenous variable as in the models of ROMER. This approach of "semi-endogenous growth" has been further developed by others.[67] For example, ARNOLD (1997, 1998) has further developed Jones' suggestions. He links the approach of

65 Respectively, it *must* be abandoned, because otherwise this would result in counterfactual exploding growth rates.

66 This circumstance has lead to the question whether the inter-temporal scale effect of R&D can be eliminated without causing policy invariance. Models with this characteristic have been developed, for example, by YOUNG (1998), PERETTO (1998), and AGHION/HOWITT (1998). The author would like to forgo a detailed elaboration of these models at this point.

67 See for example LI (2000), GROTH/SCHOU (2002), CHRISTIAANS (2004b). For critical contributions on semi-endogenous theory see e.g. HA/HOWITT (2007), or MADSEN (2008).

GROSSMAN and HELPMAN (1991a) with the models of UZAWA (1965) and LUCAS (1988) in order to endogenously explain the accumulation of human capital.

2.5 Economic growth theory in the context of this thesis

Today, there is broad acceptance among economists about the fact that innovations in terms of technological progress are the crucial driver of economic growth in developed countries (ARNOLD 1997: 37f.; SCHMOOKLER 1965: 333, 1966; KUZNETS 1973: 249, 251; GROSSMAN/HELPMAN 1994: 32; also: FAGERBERG 1987, 1988, 1994). Having briefly discussed some of the major criticisms of the endogenous growth theory, the success of this theory may appear surprising to the reader. Why is a theory that incorporates a significant amount of tautological reasoning, neglects almost completely the aspect of social institutions which a large number of economists have seen as crucial to economic growth, and basically restates older ideas in formal terms, so successful (FOSS 1998: 234)?

To really appreciate the achievements of neoclassical theory in general, and endogenous growth theory in particular, it is helpful to remember that this kind of theory is based on a longitudinal and full employment perspective. The central goal of growth theory is to identify the main factors or drivers that determine the level of economic growth in the long run. For this reason, factors of influence which only imply temporary growth effects are intentionally excluded. The steady state of an economy's equilibrium growth is determined only on the basis of long-term sustainably effective factors assuming full employment and complete clearing of factor markets (KOOTHS 2004: 33). Temporary ups and downs of the growth rate caused by economic cycles are thus not a subject of interest. Even a permanently low unemployment rate and a resulting increase in production output would – from the viewpoint of economic growth theory - only imply temporary or short-term growth effects (IBID.).

In my opinion, some of the criticisms of classical growth theory are thus probably misdirected because the explicitly long-term models as well as their included variables are directly measured against the everyday reality of comparatively short-term innovation processes, which in fact describe a quite different level of abstraction. The dominant methodology of neoclassical macroeconomic theory can be described as an "exact, model-based theory" (BRETSCHGER 2004: 6f.), where "exact" does not imply the accurate coverage of empirical reality, but rather that the underlying models should be based on consistent (which means "not self-contradictory") assumptions. By abstracting from reality and following the principle of a powerful, universal theory, these models are characterised by their strong reduction of the empirical complexity. In combination with this high degree of reduction and universality, these macro-economic theories are also marked by a high level of mathematical formalisation and "heroic assumptions" (GRUPP 1997: 9, 67). These "heroic assumptions" have been constantly subjected to numerous criticisms, which will not be repeated here.[68]

Neoclassical and endogenous growth theory show that economic growth, measured in terms of welfare by income per head, results from applying of new technical methods to increase the efficiency of scarce economic input factors such as labour and capital. Both, exogenous (first generation) and endogenous (second generation) studies (FREEMAN 1994b: 84f.) explain the increase in efficiency by technology progress based on research and development activities. No more, no less, this the admirable achievement of SOLOW and ROMER, who were both pioneers of their time. By explicating the source of technological progress as intentional R&D performed by firms, the models of endogenous growth theory and their recent improvements[69] - illustrated above using example of JONES - account for some serious shortcomings and may help to bring this strand of theory more in line with reality.

When looking at these strands of economic growth theory, it becomes obvious why past innovation research has focused so strongly on R&D as a determinant of firms innovativeness and competitiveness. And again, it becomes obvious why these strands of economic theory are hardly an adequate tool for analysing the innovation patterns of non-R&D-performing firms. There are three main points in economic growth theory which make it difficult to apply this strand of theory in the context examined in this thesis:

- ***No room for heterogeneity in performance among firms:*** Studies by economic historians like LANDES (1970), HABAKKUK (1962) or ROSENBERG (1972) as well as authors studying industrial organisation and technical change like SCHMOOKLER (1966), GRILICHES (1957),

---

68 These criticism can be referred for instance in the contributions of NELSON (1973: 462ff.; 1991: 64), NELSON/WINTER (1974: 886ff.; 1982: 195ff.), BOLLMANN (1990: 33ff.), SCHWITALLA (1993: 37ff.), GRUPP (1997: 67f.), DREHER (1997: 42f.), RÖPKE (1977:152ff.), and IWAI (2000: 167f.).

69 Other very recent contributions of endogenous growth theory emphasise the role of expert knowledge gained from higher education as a driver of the creative entrepreneur (e.g. VANDENBUSSCHE ET AL. 2006; AGHION/HOWITT 2006). They assume that the closer an economy moves itself to the global technological frontier, the higher the relevance of higher education and expert knowledge.

MANSFIELD (1968a), and FREEMAN (1974) have revealed interesting facts about the technological change process, some of which are in harmony with neoclassical concepts, while others are quite conflicting (NELSON/WINTER 1974: 888; 1982: 202f.). The latter refer to considerable differences among firms in terms of the technology they use or their productivity and profitability. It is hard to believe that a sharply defined body of common technological knowledge exists that is on the one hand accessible to all subjects, and, on the other hand, quasi automatically governs their production possibilities (NELSON/WINTER 1974: 888). In the neoclassical framework with property rights protection, all firms in an economy would be intrinsically motivated to invest in R&D, to be able to access the common stock of technological knowledge, to use this knowledge to develop new technological solutions, and thus strengthen their competitiveness and maximise their profit and return of investment. As a consequence, all firms would perform similarly using similar technologies based on the same technological knowledge or have the same capability to exploit this knowledge. In other words, there would not be much room for differences or heterogeneity in the innovation activities of firms as each would be pursuing the same strategy to become more competitive. Thus, it is very difficult for neoclassical concepts, which assume that actors behave in a profit maximising and perfectly rational way, to reconcile the differences revealed by the empirical reality that there is no best way for firms to innovate, or, at least this is not recognisable without considering differences in aspects like experience, knowledge, maximising capabilities or even luck. Instead, firms are assumed to choose the profit maximising element, where the "one-best way" to proceed is seldom apparent and the firm has to be satisfied with finding the most promising way instead of the best way (NELSON/WINTER 1977: 50ff.). That means that the concept of innovation is not adequately characterised in terms of an induced change in choice within a given and constant set of choices (NELSON 1991: 64f.).

- ***Innovation activities are equated with routine operation:*** By simply integrating R&D as another input factor in the rectified production function, it is assumed that innovation activities can be equated to activities increasing other kinds of capital (plant or equipment) through physical investment to enhance production output (NELSON/WINTER 1982: 201f.). The underlying assumption of perfectly rational, profit-maximising actors implies that R&D is an activity whose outcome and effects can be predicted in advance. However, as innovation always implies the investigation of something new and the generation of new knowledge, it is inseparably linked with future risk, uncertainty, and ambiguity. This uncertainty about the results and future pay-offs of innovation activities disables firms from predicting or even estimating their expected cost and profit returns (GERYBADZE 1982: 44; SCHWITALLA 1993: 38; FORAY 2006: 52f.). Thus, it can be reasonably assumed that the underlying behavioural model (e.g. heuristics of cognition, decision and action) of innovation activities differs significantly from the routine operations of the day-to-day business (NELSON/WINTER 1982: 201). It is the combination of uncertainty and the individual behavioural models of firms which accounts for different innovation strategies and, as a result, for the heterogeneity among firms with regard to their innovation behaviour.

- ***Economy as a stationary concept:*** As noted earlier, the issue of innovation has particular relevance in the context of capitalistic economic system with free and fair competition among the economic actors. But as SCHUMPETER (1950: 31) stated 60 years ago: "Capitalist economy is not and cannot be stationary". Nevertheless, neoclassical growth theory is primarily a stationary theory, because it seeks to describe an equilibrium state in time when reproduction and growth has a constant rate ("steady state growth"). Economic growth is thus a steady movement of in an equilibrium state as a response to an increase in population and the accumulation of savings (IWAI 2000: 168). The Schumpeterian understanding of innovation, however, is completely different. For SCHUMPETER, innovation itself is a factor of change without which the modus operandi of the capitalist world cannot be understood, and which includes "the introduction of new commodities, etc., the technological change in the production of commodities already in use, the opening-up of new markets or of new sources of supply, Taylorization of work, improved handling of material, the setting-up of new business organisations, etc. – in short, any 'doing things differently' in the realm of the economic life" (SCHUMPETER 1950: 59). When following this line of argumentation, innovation becomes a generic action within capitalist economies and a generic behavioural pattern of capitalist firms. Innovation is not assumed to maintain or restore economic equilibrium; on the contrary: innovation creates economic "disequilibria" by a "process of creative destruction" that "incessantly revolutionalises the economic structure from within, incessantly destroying an old one, incessantly creating a new one" (SCHUMPETER 1950: 72, 83). Thus, technological progress, R&D expenditure and, more generally innovation, should not be treated as given but as variables which have to be explained (NELSON 1991: 64; NELSON/WINTER 1977: 50ff.). From this follows that, in an evolutionary understanding, innovation activity is not somehing firms can choose to do or refrain from doing. Instead, a firm would not be able to survive over time in market competition without being innovative in terms of being able to continuously adapt to changing environments or to influence its external environment by "active" adaptation. This means, that even non-R&D-performing firms are – by definition – innovative if they manage to survive in market competition.

Having described the conceptual origins of the R&D focus of previous innovation theory research, I would like – according to the tradition of economic growth theory – to supplement this theoretical perspective with some "stylized facts" of empirical innovation research about the role and impacts of R&D on the micro-economic level of firms. Thus, the following chapters elaborate the extent to which the described macro-economic models are reflected on the level of the organisational actors who are ultimately responsible for generating innovations – the firms themselves.

2.6 The firm level perspective - institutionalised in-house R&D

Until now, there has been a broad agreement among economists that innovations developed through R&D are the determining source, the "engine" (PESSOA 2010: 152) of a developed economy's growth (ARNOLD 1997: 37f.; KUZNETS 1973: 249ff.; GROSSMAN/HELPMAN 1994: 32; SANDVEN ET AL. 2005: 2; see also: SCHMOOKLER 1965, 1966; FAGERBERG 1987, 1988, 1994; ASHEIM/ISAKSEN 1997; MICHIE 1998).

The following subchapters will shed some light on industrial R&D and its positive effects and impacts on firms' innovativeness and economic success, a topic which has been extensively investigated by empirical innovation research. To start with, I will discuss how R&D is conventionally defined in economic and innovation research, which different types of R&D can be distinguished, how they are typically interrelated, and how the R&D intensity of firms is usually measured.

2.6.1 Definition of Research and Development (R&D)

In parallel with the rise of neoclassical growth theory, the measurement of science and technology also emerged in the early fifties with the US government taking the lead (GODIN 2001b: 2). Following the concept and definitions of the American National Science Foundation (NSF), the OECD decided to produce a first methodological manual on research and development statistics in 1963, called the "Frascati Manual" (OECD 1963).[70] Since then, this has been revised several times[71] and its so far latest 6th edition was published in 2002. To establish an international standard of coherent science and technology measurement, it proposes precise definitions and classifications of the concepts and activities to be measured and makes recommendations for the figures and indicators produced (GODIN 2006b: 63).

The main purpose of these R&D indicators is to measure and illustrate the science and technology endeavour of a country, to demonstrate its strengths and weaknesses and follow its changing character with the aim of providing early warning of events and trends which may impair its capability to meet the country's needs (OECD 1976: 6).

The OECD defines R&D as "(...) creative work undertaken on a systematic basis to increase the stock of knowledge, including knowledge of man, culture and society and the use of this stock of knowledge to devise new applications" (OECD 2002: 30). Thereby, R&D covers both formal R&D in designated R&D units or laboratories and informal or occasional R&D performed in other units of an organisation (OECD 2002: 17).

The term "systematic" refers to the requirement of purposeful effort, rather than accidental discovery (ARUNDEL ET AL. 2008: 6). As GODIN (2001a: 3ff.) discusses, the essential idea of systematicity pervades most of the 20th century definitions of R&D in dictionaries or international conventions. Originally, the systematicity requirement referred to processual aspects of scientific regularities, the use of scientific methods, the intersubjectivity of the applied methodology, and the logical consistency (in terms of objectivity, reliability, and validity). But over time, the meaning of systematic in definitions of research has shifted from an emphasis on the scientific method to an emphasis on institutionalised research (GODIN 2001a: 2). R&D must be properly structured, that means it must meet the minimum requirements of a systematic activity: the persons performing R&D must work a significant number of hours per year; there must be a programme of work for which certain financial resources (R&D expenditure) are earmarked. In turn, diffuse or scattered R&D activities (e.g. activities carried out sporadically, or from time to time, within the various ser-

70 For an excellent overview of the history of science and technology indicators see GODIN (2001b).

71 Revisions of the OECD Frascati Manual: 1963 (1st edition), 1970 (2nd edition), 1976 (3rd edition), 1981 (4th edition), 1993 (5th edition).

vices or functional areas of an organisation) are not taken into account (MESSMAN 1977: 10). In fact, the concept of systematic research had considerably more influence than another innovation that appeared around the same time, namely the concept of "scientific and technical activities" (STA) invented in Canada and used for some time by the NSF (GODIN 2001b: 17).

The Frascati Manual (OECD 2002: 30) divides R&D into three types of activities:

- ***Basic research:*** Experimental or theoretical work undertaken primarily to acquire new knowledge of the underlying foundation of phenomena and observable facts, without any particular application or use in view
- ***Applied research:*** Original investigation undertaken in order to acquire new knowledge. It is, however, directed primarily towards a specific practical aim or objective
- ***Experimental development:*** Systematic work, drawing on existing knowledge gained from research and/or practical experience, which is directed to producing new materials, products or devices, to installing new processes, systems and services, or to substantially improveing those already produced or installed

As described in more detail in the Frascati Manual (OECD 2002: 77ff.), basic research aims at analysing properties, structures and relationships with a view to formulating and testing hypotheses, theories or laws. Thereby, the reference to "no particular application in view" is crucial, because the purpose of this type of research activity is just to generate new scientific knowledge which is not sparked off a concrete practical problem or goal that has to be solved. Basic research can be oriented or directed towards some broad fields of general interest, with the explicit goal of a broad range of applications in the future (ROSENBERG 1990: 168f.).[72] This implies that the results of basic research are not generally sold but are usually published in scientific journals or circulated to interested colleagues. Basic research is expected to take place mainly in the institutions of the higher education sector (e.g. universities), the government sector, or private, non-profit research organisations such as Max Planck

72 "(...) the route from discovery to application is often long and tortuous, involving the movement of knowledge, techniques and instruments from one discipline to another. All this makes the forecasting and planning of applications of basic research a difficult, if not impossible task." (PAVITT 2001a: 769).

or Helmholtz Society (SPECHT ET AL. 1996: 16f.). In contrast to the definition in the Frascati Manual, the NSF has blurred the boundaries between basic and applied research in its modified definition of basic research to account for the empirical observation that basic research is also performed in some industrial firms (ROSENBERG 1990: 171).[73] However, commercialy oriented private firms usually do not undertake basic research (SPECHT ET AL. 1996: 16) because its results are generally characterised as public goods, which means other individuals or competitors can almost not be excluded from using them.[74] Thus, basic research corresponds to the element of publically available technological knowledge in Endogenous Growth Theory that diffuses through spillover processes among the economic actors and stimulates them to transform this public knowledge into commercial products, or commercially advantageous production methods, by applied research.

Thus, applied research is undertaken either to explore possible uses for the findings provided by basic research or to determine new methods or ways of achieving specific and predetermined objectives (OECD 2002: 78; ROSENBERG 1990: 170). It is suggested to draw upon already available stocks of knowledge and extend them to solve particular problems. In consequence, the results of applied research are primarily valid only for a single or limited number of products, operations, methods or systems. This type of research obviously corresponds most closely to the type of research that is connoted by the term "industrial" or "private sector" research, as it intentionally aims to generate a new commercial product, process, or method which helps the enterprise to overcome a certain economic problem or unsatisfactory situation which could not be solved by utilising existing knowledge (GRUPP 2008: 87; DREHER 1997: 22). In this context, R&D conducted in the private sector is commonly oriented towards concrete problems as well as the direct applicability of the solutions and the quick market exploitation of the newly generated knowledge (BURR 2004: 34).

---

73 "To take into account industrial goals, NSF modifies this definition for the industry sector to indicate that basic research advances scientific knowledge not having specific commercial objectives, although such investigations may be in fields of present or potential interest to the reporting company" (ROSENBERG 1990: 171).

74 BURR (2004: 35ff.) discusses some arguments why firms of the private sector may be motivated to engage themselves in basic research and how they may be able to appropriate at least some of its results (e.g. first mover advantages, strong market power, by-products or accidental luck of basic research may generate specific problem solutions, as basic research is a necessary precondition to successfully perform applied research in certain fields).

Additionally, the category of experimental development[75] became more and more relevant when industrialists, consultants and academics in business schools began to study industrial, applied research (GODIN 2006b: 68).[76] By this category, they described the enterity of innovation process' elements that deal with the improvement, trouble shooting, or technical control of process and quality. Moreover, this category corresponded to research divisions in firms and organisations that defined themselves as research and development (GODIN 2006b: 70). Thus, development first described a series or a list of activities before it came to be identified as a subcategory alongside basic and applied research. Finally, it became a separate category and constituted the well-known acronym of "R&D".

Thus, industrial research has also led to some kind of "semantic innovation" (GODIN 2001b: 12). Previously, people simply spoke of science or inquiry or investigation. As Godin states, research as a term came into general use because "science" was often considered inappropriate when applied to industry, because industrial research has blurred the boundaries between basic and applied science. Thus, R&D is an activity which may act not only as the original source of inventive ideas but also as a means of problem solving which can be called upon at any point up to implementation (OECD 2002: 18).

Originally, these three dimensions of R&D represented sequential steps along a linear model of the innovation process. The model postulated that innovation starts with basic research, is followed by applied research and development, and ends with production and diffusion (Figure 4).

75 "Since the 1970 edition of the Frascati Manual, the OECD suggests adding the adjective 'experimental' to 'development' to avoid, so it is argued, a confusion between development, a phase of R&D, and the same term in economics, and to use the same term as eastern European countries and UNESCO." (GODIN 2006b: 73; note #34).

76 "In the 1940s and 1950s, these individuals began developing conceptual frameworks, or what would later be called models of innovation. The models, usually illustrated with diagrams, portrayed research as a linear process starting with basic research, then moving on to applied research, and then development." (GODIN 2006b: 67).

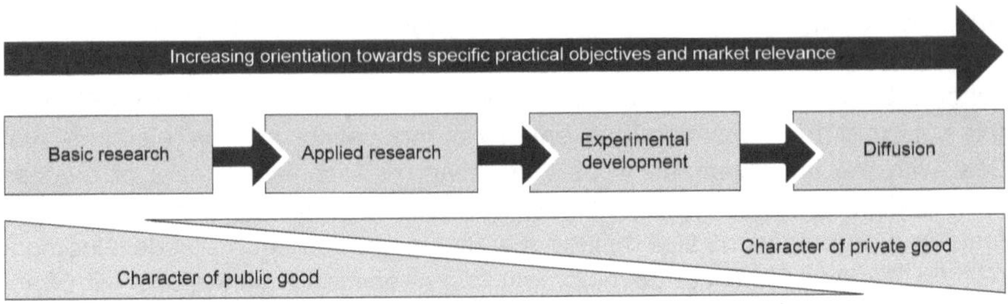

Figure 4: Linear model of innovation process
(Source: own illustration, adapted and modified from GODIN 2006: 639; BURR 2004: 32; SPECHT ET AL. 2002: 16) [77]

It appears from the figure that with each further step, the character of R&D as a public good decreases and the orientation towards a specific problem solution and its commercial realisation in the market increases. Moreover, this model emphasises the importance of initial basic research because without such basic research all the subsequent steps become meaningless as they have nothing to draw on.

STONEMAN (1995: 2f.) aligns this linear model with SCHUMPETER'S trilogy of invention, innovation, and diffusion. The first stage (basic research) corresponds to the invention process, encompassing the generation of new ideas without thinking about their commercial or practical applicability. The second stage (applied research) represents the innovation process comprising the development of new ideas into marketable products and processes. The third stage finally is the diffusion stage, in which the new products and processes spread across the potential market.

This model has been very influential as academic organisations as a lobby for research funds and economists as expert advisors to policy makers have widely disseminated this model, and have justified government support to science by using such a model (GODIN 2006: 640). As a consequence, science policies permeated a linear conception of innovation for many decades (Mowery 1983a: 27f.). Despite its widespread use, the linear model of innova-

77 The precise source of the model remains nebulous and has never been documented. Several authors who have used, improved, or criticised the model in the past fifty years rarely acknowledged or cited any original source (GODIN 2006: 639). In his recommendable article "The Linear Model of Innovation", BENOÎT GODIN (2006) traces the history of the linear model.

tion was not without its opponents. As early as the 1960s, numerous criticisms were levelled concerning, among other things, the linearity of the model (e.g. SCHMOOKLER 1966; PRICE/BASS 1969; MYERS/MARQUIS 1969).

2.6.2 How can firm level R&D be measured?

As SMITH (2005: 151) suggests, R&D is a knowledge creating process for which both activities and outcomes are radically incommensurable because there is no meaningful way either to assess the dissimilar actions and events that are part of the R&D process, or to compare the increments to knowledge that follow from R&D. As, according to SMITH, this problem ca not be overcome, it can only be handled by carefully specifying elements of the R&D process that can be measured. The solution adopted by the authors of the Frascati Manual (the operating statistical manual for R&D data collection provided by the OECD) was define research-comprising activities, and then seek data on either expenditure or personnel resources devoted to such activities. "The measurement concept for R&D is therefore economic in character, and the datasets that result are collections of economic indicators compatible with industrial datasets, and indeed with the national accounts" (SMITH 2005: 151).

Basically, the Frascati Manual (OECD 2002: 20) describes two indicators by which R&D activities can be measured as innovation input of firms: monetary expenditures devoted to R&D during a specific period, and the personnel working in R&D. The latter encompasses "all persons employed directly on R&D should be counted, as well as those providing direct services such as R&D managers, administrators, and clerical staff" (OECD 2002: 92). These employees are measured either in terms of full-time equivalents (FTE) or person-years spent on R&D (OECD 2002: 20).

Regarding the monetary indicator of R&D expenditure, the basic measure is "intramural expenditures", that means "all expenditures for R&D performed within a statistical unit (e.g. the firm) or sector of the company" (OECD 2002: 21). Another measure called "extramural expenditures" on R&D covers payments for R&D performed outside the statistical unit (e.g. external partners or contractors outside the firm) or sector of the economy during a specific period (OECD 2002: 21). "R&D expenditure data should be compiled on the basis of performers' reports of intramural expenditures" (OECD 2002: 108). Moreover, to assess the extent to which R&D plays a central role in a firm in the sense of an institutionalised activity characterised by constantly used routines and capabilities, the indicator can be deployed if the firm conducts R&D on a continuous basis or not (OECD 2009: 28).

R&D indicators are frequently classified or categorised according to multiple criteria. Beyond the mentioned distinction between basic research, applied research, and experimental development, R&D data are often structurally classified by its sector of performance (business enterprise, government, higher education, private non-profit, its sources of finance (domestic, international), as well as more content-oriented, for instance according to its socio-economic objectives or fields of research (BRONWYN ET AL. 2010: 5; SMITH 2005: 154). While such detailed classifications are usually not considered by researchers, the R&D intensity of firms, industries or countries is one of the indicators most widely used by researchers, policy analysts and policy makers (SMITH 2005: 155). The R&D intensity is defined as the ratio of R&D expenditure to some measure of output. For a firm, it is usually the ratio of R&D expenditure divided by sales. For an industry or a country, it is the ratio of business expenditure on R&D (often known as BERD) to total production or value added. For a country, it is the gross expenditure on R&D divided by the gross domestic product (GDP) (SMITH 2005: 155; LEGLER/FRIETSCH 2007: 7ff.).[78]

2.6.3 The historical emergence of R&D activities at firm level

SCHMOOKLER (1966) showed that professionalised corporate R&D accounted for about half of industrial inventions in the United States and probably for a higher proportion of those which were economically exploited by transfer into innovations. Many of the other inventions also originated from professional R&D in government and universities. Moreover, the development and exploitation of inventions arising from university, government or private inventors probably also involved some professional R&D work in industry in the great majority of cases (FREEMAN/SOETE 1997: 220).

For a long period, economic research on international competitiveness was mainly centred on the issue of price- and cost-related factors of competitiveness. Productivity turned out to be important factor in improving the trade performance of an industry, and the industry's trade performance in turn is a relevant driver of growth of value-added and productivity (CASTELLACCI 2008a: 985). Ever since the 1980s, the focus of economic research has shifted from price- and cost-related factors towards the role of technological change and innovation as non-price factors of competitiveness. Inspired by the seminal work of SCHUMPETER on the role of innovation and technological change, dif-

78 For a detailed discussion of R&D indicators see GRUPP (1997: 146ff.).

ferent strands of empirical research provided new insights into the relationships between innovation and international competitiveness. As outlined in the second chapter, a large body of economic growth theory analysed some of the Schumpeterian insights within a neoclassical equilibrium framework pointing to the existence of increasing returns and spillovers effects related to the R&D activities of private firms.

At the firm level, early scholars of innovation research understood innovation ability as a characteristic attribute of single individuals like inventors or the creative entrepreneur (SCHUMPETER 2006: 171). In this context, innovation always remained an external dimension of the economic system which was assumed to be naturally or randomly given. The recognition of R&D as an important internal driver of firms' innovativeness did not arise until the emergence of large companies and their increased internal division of labour at the beginning of the industrial revolution (DODGSON ET AL. 2008: 77f.).[79] The former external innovation activity was transferred, institutionalised and systematised within the enterprises and was thus no longer left to randomly distributed entrepreneurial genius. Since then, professional education, the establishment of laboratories and improvements in techniques of measurement and experimentation have progressively increased the efficiency and controllability of discovery, invention, and innovation (MOWERY/ROSENBERG 1989: 98ff.).[80] By the first decade of the twentieth century, the focus of inventive effort was shifting from the contract laboratory to the in-house industrial R&D departments established by firms like Kodak, General Electric or Du Pont (FREEMAN/SOETE 1997: 198).[81] Most of the major inventions of that time were the result of professional R&D activities, often over long periods. Even if inventor-entrepreneurs were involved or played a key-role in the innovative process, such individuals were

---

79 An interesting historical overview of the emergence of institutionalised firm level R&D can be found in DODGSON ET AL. (2008: 77ff.) and particularly in FREEMAN/SOETE (1997: 191ff.).

80 This transition from the great "individualists" to the large-scale R&D laboratories is probably best embodied in the person of Thomas Edison. Although he made a host of inventions and was granted more patents than any other single individual, this success was partly possible because he set up large contract research laboratories. Among Edison's staff were some of the most outstanding engineers and scientists who later helped to build up corporate in-house R&D in Germany and Britain, as well in the United States (FREEMAN/SOETE 1997: 198).

81 Concerning this issue, MOWERY conducted an interesting study on "The Relationship between Intra-firm and Contractual Forms of Industrial Research in American Manufacturing, 1900-1940 (MOWERY 1983b).

usually scientists or engineers who had the facilities and resources to conduct sustained R&D work. Moreover, university scientists or inventors worked closely as consultants with the corporate R&D departments of the innovating firms. As FREEMAN/SOETE (1997: 198) state, intimate links with basic research through one means or another were normal for R&D in these industries. Thus, the technologies at that time were strongly science-based, because they could not have been developed at all without a foundation in theoretical principles.[82]

SCHMOOKLER (1966: 12ff.) compared the innovation process to the blades of a pair of scissors: on the one hand, it involves the recognition of a need or a potential market for a new product or process (market pull). On the other hand, it necessarily involves technical knowledge which may also often include new scientific and technological knowledge as the result of original R&D activities (technology push). Thus, establishing systematic R&D activities within industrial firms represents an institutional response to this two-sided character of innovation, by organising the matching of market needs to technological possibilities, which was earlier performed by the single individual of the entrepreneur.[83]

Subsequent to the establishment of industrial R&D departments, the nature of R&D activities developed differently, based on functional specialisation and division of labour. PAVITT (2001b: 13220ff.) identifies three complementary forms of specialisation that developed in parallel, each contributing to the growth of R&D:

- ***Specialisation by discipline – the growth of science and engineering:*** In the nineteenth century, the emergence of specialised scientific disciplines like engineering, chemical engineering or electrical engineering was accelerated by the development of new powerful research tools and techniques as well as by increasing opportunities for application. Each of these developments reflected radical but uneven improvements in the underlying knowledge base, and in particular the emergence of technological paradigms (DOSI 1982).

82 "The enormous amount of knowledge of disciplines like macromolecular chemistry, physical chemistry, nuclear physics and electronics could never have emerged from casual observation, from craft skills or from trial and error in existing production systems (...)" (FREEMAN/SOETE 1997: 199).

83 For a very interesting historical overview of "Innovation through Time", the author would like to recommend reading the article of thus name by BRULAND/MOWERY (2005).

- ***Specialisation by corporate function – the growth of business R&D:*** Many of the initial industrial R&D laboratories just performed the task of controlling the quality and composition of materials in the increasingly large-scale production plants and thus of reducing costs. But management progressively recognised that the hitherto science-based technological opportunities were applicable to a whole range of existing and new products. Since then, corporate R&D departments have been devoted full time to invention and innovation and thus have provided an efficient and effective method to systematically create, combine, and coordinate increasingly specialised and complex knowledge and skills to develop and test prototypes, and to translate them into commercialised products or implemented production processes (PAVITT 2001b: 13221).

- ***Specialisation by institution – the growth of academic research:*** As described in the chapter on endogenous growth theory (chapter 2.4), the prospect of commercial returns greatly stimulates the development of innovative activities by firms in capitalist competition. But as has been mentioned before, public authorities would need to support complementary public research of a more fundamental nature, in order to avoid diminishing returns, to open up new opportunities, and to provide trained and highly skilled researchers. Besides the fact that firms directly profit from scientific output, like published papers, articles or books, they mainly benefit from basic research rather indirectly, in the form of young recruits. As PAVITT (2001b: 13222) notes, in all advanced countries, governments have become the main source of the funding for research activities, and related postgraduate training to provide industrial R&D laboratories with scientists and engineers who are aware of the latest research findings and methods (see also FREEMAN 1994b: 82).

FREEMAN and SOETE (2009: 586) emphasise that exactly those developments distinguish modern, industrial R&D, and justified the R&D focus in the OECD's Frascati Manual. These developments constituted a new scale, scientific content and extent of professional specialisation, because a much greater part of technological progress appeared attributable to R&D activities performed in specialised laboratories, departments or pilot plants by full-time qualified staff. "It was this sort of work that one wanted to get officially recorded in R&D statistics. It was totally impracticable to measure the part-time and amateur inventive work typical of the nineteenth century." (FREEMAN/SOETE 2009: 586). Briefly, present R&D statistics are really a measure of the professionalisation of this activity, the professional recognition of R&D activities as a separate activity within firms or of whole organisations carried out by professional researchers.

2.6.4 Determinants of firm level R&D

There are several characteristics of firm level R&D which are likely to result in considerable differences, particularly between smaller and larger firms regarding their R&D intensity (RAMMER ET AL. 2009: 37). Since the later writings of SCHUMPETER and his well-known hypothesis that large size is conducive to R&D investment (COHEN ET AL. 1987: 543; NOOTEBOOM 1999: 793f.), innovation scholars have been interested in the relationship between a firm's size and the rate at which it innovates (PLEHN-DUJOWICH 2007:1). Indeed, the relationship between firm size, R&D, and innovation is one of the most frequently analysed (BECHEIKH ET AL. 2006: 655f.). Various theoretical arguments have been put forth supporting Schumpeter's hypothesis that *large* firms should be more innovative (COHEN ET AL. 1987: 544f.; LEE 1995: 392; PLEHN-DUJOWICH 2007: 1; RAMMER ET AL. 2009: 37):

- First of all, R&D is associated with ***high entry cost***, for instance specific investment in laboratory or technical equipment and human capital in terms of highly skilled employees. Thus, there is some critical threshold of financial or personnel resources for firms to invest in R&D. With regard to small and medium-sized enterprises (SME) this means, that they have to invest a relatively higher share of their scarce resources than large firms.

- Secondly, firm level R&D requires a certain ***minimum project size*** to generate useful results because of technical indivisibilities. But in firms with smaller financial or human resources like SMEs, R&D expenditures prevent other necessary investments in non-R&D areas such as marketing or distribution. R&D minimum costs may even be so high that they prevent SMEs from investing in any R&D (GALBRAITH 1952: 92).

- Thirdly, R&D costs are largely ***fixed costs*** which in the case of SMEs must be borne by a smaller sales base, reducing either profitability or price competitiveness and restricting available cash flow to finance future R&D (COHEN/KLEPPER 1996b: 233f.).

- Fourthly, R&D activities are highly ***idiosyncratic***. Thus, it is difficult for external financers to assess and evaluate the success chances of a firm's R&D efforts without a firm-specific history of R&D success and failure. SMEs, and especially young firms, are less able to provide such a track record and may suffer restricted access to external finance.

- Fifthly, there is a considerable ***time-lag between R&D expenditures and, if any, the returns***. While R&D demands pre-financing, most R&D expenditures are on staff and materials and do not qualify as collaterals for debt financing. Lack of collaterals and information asymmetries on R&D project perspectives raise the costs of external funding (FREEL 2000a; CZARNITZKI 2006; TIWARI ET AL. 2007). Financing in-house R&D thus strongly rests on internal funding sources whose availability tends to be more restricted in SMEs, due to a smaller cash flow.

- Finally, large firms ***spread the risks*** of R&D by holding diversified portfolios, and so may invest in more risky projects, which typically earn a higher return (HOLMSTROM 1989). Moreover, particularly in the case of SMEs, the failure of a single R&D project may jeopardise the entire business if the R&D is financed through debt using non-R&D assets as collateral.

These features of R&D typically result in the fact that a firm's R&D intensity is positively associated with firm size (RAMMER ET AL. 2009: 37; SHEFER/FRENKEL 2005: 26), which was regularly revealed by various empirical studies based on data from innovation surveys covering a random sample of firms from different size classes (e.g. KLEINKNECHT 1989; SANTARELLI/STERLACCHINI 1990; DAMANPOUR 1992, 1996; MAJUMDAR 1995; NOOTEBOOM 1991; TSAI 2001; STOCK ET AL. 2002; LEE 2004).[84] Nevertheless, these results did not turn out to be very robust because several other studies could were not able to confirm a positive (e.g. COHEN ET AL. 1987; COHEN/KLEPPER 1996a; BOUND ET AL. 1984; ACS/AUDRETSCH 1988, 1991a; PAVITT ET AL. 1987; PLEHN-DUJOWICH 2007) or linear relationship between firm size and R&D (e.g. BERTSCHEK/ENTORF 1996). For example, ACS and AUDRETSCH (1991b) found that small firms contribute 2.4 times more innovations per employee than do large firms. Likewise, PLEHN and DUJOWICH show in their recent study (2007) that small firms are more innovative per dollar of R&D than large firms. Especially in the high-tech sectors of manufacturing industry, a large number of small firms can be encountered that engage in R&D activities to a great extent (ACS/AUDRETSCH 1993;

84 A detailed discussion of firm size and R&D intensity is provided by FREEMAN/SOETE (1997: 227ff.)

KLEINKNECHT 1989; SCHEIRER 1991).[85] Thus, it turned out that these contradictory findings can only be resolved by comparing the R&D propensity of small and larger firms under similar frame conditions, like similar sectors, market structures, or technological regimes (e.g. FISHER/TEMIN 1973; DOSI 1988a; ACS/AUDRETSCH 1987, 1988, 1991a, 1991b; VEUGELERS/CASSIMAN 1999). For instance, NOOTEBOOM (1991) and VOSSEN and NOOTEBOOM (1996) revealed that small firms participate less in R&D, which favours the 'mark II' hypothesis of SCHUMPETER. But when small firms do participate in R&D, they tend to do so more intensively than large firms. Both results hold true for all sectors identified by PAVITT (1984), except the science-based industries. There, large and small firms seem to be equal in both participation and intensity.

But as GRIFFITH and WEBSTER (2010: 472) point out, many of the studies found market structure and and/or firm size to be a determinant of R&D intensity have not controlled for the underlying conditions of opportunity and appropriability. Therefore, it appears that size and market structure are unlikely to be the dominant determinants of innovation, since the findings commonly depend on which control variables are included in the model. More consistent results were obtained by several studies that account for the opportunities provided by the scientific sector and the conditions under which firms can more or less easily appropriate their profits from R&D (e.g. COHEN ET AL. 1987; STERLACCHINI 1994; GRILICHES 1995; OLTRA/FLOR 2003).

Another, smaller strand of research focused on financial hurdles for firms to invest in highly uncertain, risky and collateral-free R&D projects (e.g. KAMIEN/SCHWARTZ 1982; HALL 2002; BLOCH 2005) and found that higher levels of retained earnings facilitate higher levels of R&D, ceteris paribus. The stream of management research followed a resource-based perspective (see chapter

---

85 According to FREEMAN/SOETE (1997: 230), small firms that do perform R&D tend to fall into three categories:

- Firms which have just begun to develop or exploit a new invention. In this case sales could be relatively low in relation to R&D and thus a very high R&D intensity could be expected.
- Highly specialised firms which have a particular expertise, sustained by an intensive research programme in a very narrow field.
- Firms struggling to survive in industries in which new product competition makes R&D increasingly necessary. Under these circumstances, a very varied management response may be expected, with some firms trying to scrape by with a sub-threshold R&D effort, others relying mainly on co-operative R&D, and still others taking high risks with an ambitious R&D programme.

3.3.3) by concentrating on the role of the firm-internal capabilities as a determinant of their likelihood to invest in R&D (e.g. NELSON/WINTER 1982; PAVITT 1991; SOUITARIS 2002; LEE 2002; MUROVEC/PRODAN 2009; CUERVO-CAZURRA/UN 2010) largely without focussing on the issues of firm size, market structure, scientific opportunities, or the conditions of appropriability. As the overview shows, their results differ and suggest that R&D is not equally important in all circumstances and for all kinds of knowledge (e.g. MUROVEC/PRODAN 2009: 861; also: GRUNFELD 2004; SCHMIDT 2005).

2.6.5 R&D as a firm level driver of innovation and economic success

As already discussed, endogenous economic growth models assume that firms invest in new technology through expenditure on R&D if they perceive the opportunity to appropriate the returns on R&D investment by above-average profit. As a consequence, the firms enlarge their market share and, with increasing returns to scale, they enjoy greater production efficiency and a higher rate of economic growth (SCHMOOKLER 1966: 240ff.; SEGERSTROM 1991: 807). Greater production efficiency enables industries to expand their domestic market share through import substitution, to increase local consumption and, at the same time, to penetrate new foreign markets and increase their export share (SHEFER/FRENKEL 2005: 25; PORTER, 1990).

In his book, BAUMOL states that "(...) virtually all of the economic growth that has occurred since the eighteenth century is ultimately attributable to innovation" (BAUMOL 2002: 13). A high technological level of an economy results from high innovation activities conducted by R&D which in turn result in a higher share of new goods and an extended use of new techniques in production (FAGERBERG 1987: 88). "There is ample empirical evidence supporting the hypothesis that R&D expenditures are a *sine-qua-non* for the firm's level of innovation activities" (SHEFER/FRENKEL 2005: 25).

Therefore, it is not very surprising that the economic impacts of R&D investments are a major subject of considerable interest to firm managers, entrepreneurs, policy-makers and economists in general. "The reason is obvious: investment in R&D and innovation is expensive and one would like to be sure that there is a positive return (...)" (HALL ET AL. 2010: 5). The outcome of R&D investment at the firm level is risky and uncertain, because it is not clear whether the firm will be able to transfer the expenditure in successful output, or whether competitors are doing research on similar issues and will produce results faster, which would make the R&D expenditures partially obsolete. But if the R&D process is successful and leads to superior new products or production methods, the firm would be able to generate innovation rents ("Schumpeterian rents") and increase its profits. Generally, BURR (2004: 33) names several reasons for firms to undertake and invest in R&D:

- ***Improvement of the firm's cost position*** by developing new, more efficient manufacturing processes and methods
- ***Development of new or improved products*** with superior characteristics in terms of quality, cost or range of functions compared to competitors
- ***Development of "technological capital"*** to increase the firm's attractiveness, e.g. for technology exchange with other firms (e.g. suppliers, customers, competitors) through collaboration or (cross) licensing
- ***Raise the firm's image*** as a strategic technological leader
- ***Increase the firm's absorptive capacity*** to recognise and adopt technological developments outside the firm as well as increase its ability to successfully implement such technologies within the own firm.

The motivation for firms to invest in formal R&D is not an end in itself. The ultimate reason is to improve or maintain a firm's performance, for example, by increasing demand or reducing costs while the decision to invest in R&D or not depends on the probability that the expected returns on R&D investment are positive (i.e. they positively contribute to the firm's economic objectives).

As can be seen from these previous suggestions, the relationship between R&D and a firm's economic success is moderated by the innovation dimension. In a first step, R&D, if successfully conducted, results in innovations like new products or manufacturing processes which positively affect the firm's economic ratios. Secondly, these positive effects caused by innovation enhance the long-term economic outcome of the firms (THORNHILL 2006: 690; FREEMAN/SOETE 1997: 221), for instance, by increased efficiency, employee growth or growth of profit (Figure 5). Thus, industrial R&D intentionally aims to generate a new commercial product, manufacturing process or method which helps the enterprise to overcome a specific problem or unsatisfactory situation which could not be solved by recourse to existing knowledge (GRUPP 2008: 87; DREHER 1997: 22) to achieve competitive advantage, and, finally, to maintain and enhance the firms economic performance (BROWN/EISENHARDT 1995: 352f.; ROSENTHAL 1992: 1; STOCK ET AL. 2002: 537f.; KIRNER ET AL. 2009a: 447; FREEMAN 1994b: 83; RAMMER ET AL. 2009: 35; RAYMOND/ST. PIERRE 2010: 48; BURR 2004: 33).

Additionally, as described by BURR, the performance and successful completion of R&D in terms of innovation and economic success have some positive side effects like enhancing the firm's technological capital and knowledge, its image as an innovative, technological leader or its absorptive capacity (COHEN/LEVINTHAL 1989, 1990) which in turn help to increase the firms R&D ability and to positively decide for future R&D investments (SHEFER/FRENKEL 2005: 26).

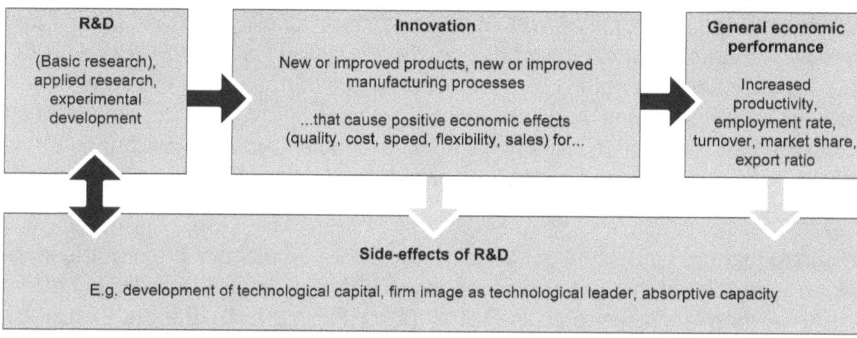

Figure 5: R&D as a driver of firms' innovativeness and economic success (Source: own illustration)

Unfortunately, empirical research does not always clearly follow this cause-and-effect chain by frequently not distinguishing between the R&D and the innovation dimension (LACHENMAIER 2007: 73; BARGE-GIL ET AL. 2008: 3; KIRNER ET AL. 2009a: 448). Thus, there are basically two perspectives of empirical studies which can be distinguished:

- Examining of the relationship between R&D expenditure and innovation
- Examining of the relationship between R&D and general economic outcome.

While the first perspective regards R&D expenditure as an input measure of innovation and tries to explain its effects on the innovation performance of firms, the second perspective uses R&D investment as an indicator of firms' general innovativeness (RAYMOND/ST-PIERRE 2010: 49). By taking R&D as a proxy for innovation, the firm's ability to innovate as such is equated with the mere existence of R&D investments in the second perspective. This perspective is thus mostly in the direct tradition to endogenous growth models and examines the relationship between firms' R&D intensity and their economic impact on productivity or firm growth. In contrast, the first perspective is mainly rooted in strategic management approaches which understand R&D expenditure as the determining management measure to enhance the firm's innovative performance.

2.6.5.1 Relationship between R&D intensity and innovation

In this context, R&D conducted in the private sector is commonly oriented towards concrete problems as well as the direct applicability of the solutions and the quick market exploitation of the newly generated knowledge (BURR 2004: 34). As the changing nature of competition places firms under increased pressure to rapidly develop and commercialise new innovations to ensure their survival, the successful release of new products undoubtedly brings firms plenty of profits and revenue growth (HUANG ET AL. 2009: 155; CAPUTO/PALUMBO 2005: 194).

In consequence, as BECHEIKH ET AL. (2006: 655f.) empirically proved, the innovation determinant that – besides the issue of firm size – received most attention from innovation research is R&D[86] (also: RAYMOND/ST-PIERRE 2010: 48), particularly in the context of SMEs (DE JONG/VERMEULEN 2006: 587f.; ROGERS 2004: 142f.). Moreover, nearly 80 percent of the studies revealed a significant positive relationship between R&D and innovation (BECHEIKH ET AL. 2006: 655). Investments in R&D are thus one of the most important mechanisms to determine the overall level of innovation in a given sector or industry (BALDWIN/HANEL 2003: 68ff.).

Empirical research in the context of the legendary SAPPHO project supported the view that specialised R&D has been increasingly important, both for the major radical innovations and for the minor improvement innovations of incremental nature, and thus into sustaining long-term success of industrial firms (FREEMAN/SOETE 1997: 221). The results showed conclusively that the growth of enterprises was closely related to their technological progress. This technological progress, however, was in turn strongly related to the performance of R&D and of specialised technical services. The comparisons of UK firm growth rates and R&D intensity conducted by the FEDERATION OF BRITISH INDUSTRIES

86 More than the half of the 108 innovation studies that were screened in the context of an excellent overview of empirical innovation research conducted by BECHEIKH ET AL. (2006) viewed R&D as explanatory variable of innovation.

(1947, 1961) also showed positive but weak correlations. In short, the results suggested that firms that perform little or no R&D in industries or do not introduce rapid technical change are likely to stagnate or even disappear. Instead, firms with high R&D intensity may sometimes achieve exceptionally high growth rates. But variations within the middle level of R&D intensity, however, did not show any statistical effect on firm growth.

It is the spirit and intrinsic purpose of industrial R&D investments to create and develop new or improved products and manufacturing processes and successfully implement them in the firm's market or production lines. As LEE (1995: 394) pointed out, this is particularly true for product or process innovations in new-technology settings, where it is very costly and difficult, even impossible, to acquire new technologies produced by competitors. A direct effect on the firm's innovativeness in terms of higher rates of new product and technical process development appears to be obvious and has been attested by various empirical studies. Thus, firms with significant R&D expenditures are more likely to be more innovative in terms of their development rates and sales of new products which are superior regarding their degree of novelty, range of functions, level of quality or outstanding technology (CASSIMAN/VEUGELERS 2002; BECKER/DIETZ 2004; GRAVES/LANGOWITZ 1996; HALL/BAGCHI-SEN 2002; KEIZER ET AL. 2002; LANDRY ET AL. 2002; LI/SIMERLY 2002; STERNBERG/ARNDT 2001; OERLEMANS ET AL. 1998; BIRCHALL ET AL. 1996; KLEINKNECHT 1996; DOSI 1988a; ACS/AUDRETSCH 1988, 1991a; KIRNER ET AL. 2009a), as well as in terms of the successful implementation of new technical manufacturing processes (GRAVES/LANGOWITZ 1996; KEIZER ET AL. 2002; LANDRY ET AL. 2002; LI/SIMERLY 2002; STERNBERG/ARNDT 2001).

Unfortunately, most studies dealing with R&D as a determinant of innovation have not explicitly differentiated between product and process innovation, respectively between R&D specifically devoted to the development of new products and R&D specifically devoted to the development of new manufacturing processes (RAYMOND/ST-PIERRE 2010: 49; GRUPP 1997: 84). This differentiation is appreciable as product and technical process innovation may not necessarily be linked to each other. For instance, a technical process innovation may be introduced not only in accordance with a new product that requires a new manufacturing method, but also to enhance the production quality, efficiency or speed of already existing products (LEE/PARK 2006: 1046; GRUPP

1997: 84; STONEMAN 1995: 3).[87] Thus, as recently proved by RAYMOND and ST-PIERRE (2010: 52ff.), product and process R&D may be assumed to be two types of activities which may differ with regard to their determinants and nature.[88]

Third, to assess the effects of R&D on the development of innovations new to the market (market novelties), SCHMOOKLER (1966) was among the first to systematically use patent statistics in his study of technical change and inspired many others to make use of this indicator for innovation output (e.g. FREEMAN ET AL. 1963, FREEMAN 1987a, 1987b; GRILICHES 1984, 1990; SIRILI 1987; ARCHIBUGI 1988a, 1988b; PATEL/SOETE 1988, GRUPP 1992).[89] Despite its various drawbacks, patent enables the measurement of an important intermediate output of innovation activities.[90] As BOUND ET AL. (1984: 51) conclude, some but not all of the firms which perform R&D also patent, and there is a strong relationship between the two activities. Interestingly, small firm that do R&D tend to patent more per invested R&D dollar than larger firms. Nevertheless, it is important to mention that patents frequently are not regarded as essential by innovators (MANSFIELD 1984: 145) which lead to rather unstable and highly context dependent results concerning the relationship between R&D and patenting.

87 LEE/PARK (2006: 1046) further suggests: "Because process innovation entails achieving process improvements or increases in production efficiencies, it is generally incremental, stepwise, and ubiquitous. On the other hand, product innovation, in general, has similarities with invention or the opening of a new market; it therefore is often radical in character, a "breakthrough" in the market." This view, however, may be questioned, as, on the one side, product innovations in one sector are process innovations for another sector (GRUPP 1997: 85), and, on the other side, product innovation can also be distinguished in to radical and incremental innovations (OECD 2005: 47). Most interestingly, the majority of all successful product innovations are incremental innovations (PLESCHAK/SABISCH 1996: 3).

88 For a more detailed theoretical discussion about the relationship between product and process innovation see GRUPP (1997: 84ff.).

89 For an overview of the various advantages of using patent data to measure innovation, see GRILICHES (1990).

90 For example: patents refer to the output of inventive efforts and thus do not measure commercialised innovations, not all innovations are patentable, not all patentable innovations are patented, the quality or innovative content of patented innovations may differ greatly among them, they depend on the background of varying national legal systems, they depend on varying propensities of industries or technological regimes to patent, non-consistent classification systems, etc. (see, among others, GRILICHES (1990), PAVITT (1982), GRUPP (1991, 1992).

Fourth, but nonetheless important, R&D activities indirectly work as a general enabler for innovation as such. They allow firms to better acquire, assimilate, transform, and exploit new technologies appearing outside the enterprise (COHEN/LEVINTHAL 1989, 1990; DEBACKERE ET AL. 1996; TYLER/STEENSMA 1998) by providing the necessary technical knowledge incorporated in R&D engineers (MOWERY 1983b; ARORA/GAMBARDELLA 1994). Thus, performing R&D creates an organisational climate that is propitious to questioning, thus favouring the flexibility of firms, their capacity to integrate new concepts and their adaptability to market changes (FREEL 2000b). This is also supported by BROUWER and KLEINKNECHT (1996), who suggest that the knowledge and experience gained from past R&D activities, as well as their existence on a permanent rather than sporadic basis, are considered to enhance the innovation ability of enterprises. RAYMOND and ST-PIERRE (2010) stress this finding by empirical evidence on the particular relevance of R&D activities for firms located in turbulent, highly dynamic market environments with short innovation cycles to maintain and increase competitiveness. Last but not least, firms' attractiveness to potential collaboration partners in the outside environment is increased by doing R&D (HALL/BAGCHI-SEN 2002; PISANO ET AL. 1988; VEUGELERS/CASSIMAN 1999; Colombo et al. 2006) which in turn increases the opportunities for inter-firm knowledge exchange and knowledge spillovers. Hence, the relationship between R&D and innovation can be characterised as self-referential or a closed-loop chain, as firms will decide to increase their current level of R&D investment in consequence of their past success in generating innovation, and so on (SHEFER/FRENKEL 2005: 26).

2.6.5.2 R&D expenditure and general economic outcome

While growth accounting tries to measure the effect of technological change at the level of industries or economics, microeconomic approaches try to measure the impact of R&D investment at the firm level. Economists have developed various methods to estimate the rate of long-term return to R&D investments. Regarding the indicators used to measure the economic outcome of R&D investments, the most frequent measures are those of productivity, sales, firm growth and export ratio.

For the most part, the literature has deployed the familiar growth accounting framework (production function) of neoclassical growth theory and its successor endogenous growth theory, underlaid with measures of R&D investment related to the growth of (labour or total factor) productivity (BRONWYN ET AL. 2010: 5; LACHENMAIER 2007: 71ff.). R&D can increase productivity by improving the quality or reducing the production costs of existing products, or simply by extending the firm's product portfolio or its available intermediate inputs. "Given good data, it [the production function – *amendment by the author*] can tell us something about average returns to R&D investments in the past and whether they appear to be changing over time" (GRILICHES 1995: 80). Applying the production function approach, much literature was devoted to firm level analyses trying to investigate the relationship between R&D investment and productivity.[91] As LÖÖF and HESHMATI (2006: 2) state, the existence of a strong association between R&D expenditures and subsequent benefits is no longer in doubt, as empirical research established an economically meaningful and statistically significant relationship between R&D investment and subsequent benefits, in the form of increased productivity, earnings and shareholder value. This relationship holds at the national, industry and firm levels. Nevertheless, HALL ET AL. (2009: 15), as well as LÖÖF and HESHMATI (2006: 2), note that the empirical literature still does not provide a unique answer in terms of the magnitude of this impact which may be due – as they suggest – to the variability and uncertainty inherent in innovation.

Starting from the original question of "Does technology create or destroy jobs?", recent research analysed the relationship between innovation and employment by looking at the impact of different types of innovation on the quantity of employment change (PIANTA 2005: 568).[92] As he argues, the evidence of the overall employment impact of innovation at the firm level tends to be positive. Firms that innovate in products and also in processes grow faster and are more likely to expand their workforce than non-innovative ones, regardless of industry affiliation, firm size or other characteristics (PIANTA 2005: 576). Thus, FREEMAN and SOETE (1997: 222) argue that the direct causality between R&D and firm growth is rather weak, and suggest instead that the real cause for firm

91 For a selection of empirical studies see: Griliches (1958, 1964, 1979, 1986, 1988, 1992, 1995), Mansfield (1961, 1962, 1965), Schmookler (1966), Griliches/Mairesse (1984), Cohen/Klepper (1996a), Patel/Soete (1988), mairesse (1990), Mairesse/Sassenou (1991), Harhoff (1998), Hall (1993), Hall/Mairesse (1995), Klette/Griliches (2000), Klette/Kortum (2004).

92 Besides analysing the mere quantitative effects of innovation on employment, quality aspects have received increasing attention (PIANTA 2005: 568). They deal with questions about the type of jobs that are created or destroyed by innovation, for example, how the composition of skills would change (e.g. BRESNAHAN ET AL. 2002; WOLFF 1996), or how wage structure would change under the impact of innovation (e.g. BLACK/LYNCH 2000; VAN REENEN 1996).

growth lies in successful innovation output.[93] On the one hand, economists like MANSFIELD (1968a, 1968b) have provided convincing evidence that successful technical innovation leads to the rapid growth of the firm. On the other hand, a high failure rate of innovation may lead to bankruptcy, despite a high intensity of R&D.

However, as R&D activities may aim to develop product or process innovation, there are arguments for both increasing and decreasing effects on employment growth. Negative effects can be expected as process innovation in terms of new machinery and equipment may substitute the production factor of labour by capital.[94] KATSOULACOS (1986) points out the importance of distinguishing between product and process innovation as he revealed different effects on employment for both types of innovation. The evidence shows that R&D in terms of product innovation has a generally positive employment impact, while process innovation tends rather to cause negative employment effects. A recent overview of empirical analyses of the effects of R&D on employment is provided by PIANTA (2005) or CHENNELLS and VAN REENEN (2002).[95] In summary, most empirical studies so far found positive effects for product innovations while the results for process innovations remain somewhat ambiguous so far (LACHENMAIER 2007: 132f.).

Considering the export dimension, its use as a measure of firms general economic performance does not seem as obvious as the case of productivity or employment. Firstly, theoretical arguments for the relevance of export ratios

---

93 "Although the statistical relationship between R&D intensity and subsequent growth by firms is not very strong, the association between successful innovation and subsequent growth of the firm is strong." (FREEMAN/SOETE 1997: 222)

94 For instance, the introduction of automated assembly lines reduces the number of employees. However, from a cross-sectoral perspective, such negative effects balance each other out, as the process innovation of one firm (e.g. introduction of industrial robots) may be the product innovation of another firm (e.g. manufacturers of industrial robots). Against this background, it is important to mention that these results briefly presented in the following focuse on the firm level. Nevertheless, after implementing a productivity-enhancing process innovation, the firms may be able to produce it products at lower manufacturing cost and thus with higher profit margins. However, lower manufacturing cost can also lead to decreased product prices which may in turn lead to an increased demand for the products, and which will finally result in higher sales and firm growth.

95 For a selection of empirical studies see: Entorf/Pohlmeier (1990), Zimmermann (1991), König et al. (1995), Brouwer et al. (1993), Smolny (1998), Doms et al. (1995), Peters (2004), van Reenen (1997), Greenan/Guellec (2000), Piva/Vivarelli (2004, 2005).

can however be drawn from international trade theory. An economy's performance is best visible when its actors directly compete with other economies in international markets (LEGLER/FRIETSCH 2007: 5). Following the theory of the "technological gap" (POSNER 1961; HIRSCH 1965; VERNON 1966; GOMULKA 1971; CORNWALL 1976, 1977)[96] which extends the approach of comparative economics of RICARDO (1951) and the HECKSCHER-OHLIN theorem[97], advances in the level of technology and quality of the goods and services offered in international competition can compensate the competitive cost advantages of other competing economies. In other words, goods and services that are characterised by a high level of technology could outperform other cheaper goods and services in international markets. The economy will be able to export these high-tech goods (its "flagships" among all produced goods) and services as long as other competing economies are not able to produce an identical product or service at equal or lower cost. In consequence, the level of technology (in terms of R&D) incorporated in an economy's good determines whether it is able to attain international competitive advantage by achieving superior

96 The main hypothesis of the technological gap approach to economic growth may be summarised as follows (FAGERBERG 1987: 88): a) Existence of a close relation between a country's economic and technological level of development; b) the rate of economic growth of a country is positively influenced by the rate of growth in its technological level; c) it is possible for a country facing a technological gap, e.g. a country at a lower technological level than the countries on the innovation frontline, to increase its rate of economic growth through imitation; d) the rate at which a country exploits the possibilities offered by the technological gap depends on its ability to mobilise resources for transforming social, institutional and economic structures.

For positive empirical testing of this theory see for example SOETE (1978); FAGERBERG (1987), or DOSI ET AL. (1990).

97 The HECKSCHER-OHLIN theorem has been named after the Swedish economic historian ELI F. HECKSCHER and the Swedish economist BERTIL OHLIN. HECKSCHER'S original article (1919) explains the impact of differences in factor endowments on inter-country income distribution and international specialization, and demonstrates that the resulting trade in mobile goods is a substitute for factor mobility in reducing factor-price differentials. His brilliant insights were propagated through the writings of his student OHLIN. OHLIN'S PhD thesis (1933) contains particularly the replacement of the century-old RICARDIAN labour-cost theory of comparative advantage with a multifactor general-equilibrium formulation.

According to HECKSCHER and OHLIN, economies vary in their endowment of the input factor labour and capital. Thus, economies would specialize in those products which correspond to their available input factor. For example, a capital-intensive economy would specialise its production on capital-intensive goods, while labour-intensive economies would specialise on labour-intensive goods. Thereby, an economy can exploit its comparative cost advantage regarding the availability of a certain production factor.

profit compared to other economies with less technology-intensive goods (LEGLER/FRIETSCH 2007: 5f.; GEHRKE ET AL. 2007: 13). These superior profits realised by technologically advantageous goods and services obviously correspond to Schumpeter rents, which are attained through the transient monopoly market position due to the competitive advantage of innovative new products.[98]

KLODT (1990: 61) distinguished innovative, technology-intensive industries from industries with lower technology-intensity. In the former case, he spoke of "Schumpeter industries" while the latter were named "Heckscher-Ohlin industries". Analogously, if the focus is on the level of goods instead of industries, the corresponding markets are named "Schumpeter markets" or "Schumpeter goods" and "Heckscher-Ohlin markets" or "Heckscher Ohlin-goods" (GRUPP 1997: 200).[99] Schumpeter goods or industries profit from the characteristics (highly developed stock of technological knowledge, high level of R&D expenditure, and high educational level of workforce) of highly developed economies (LEGLER/FRIETSCH 2007: 6). Vice versa, because Schumpeter goods attain competitive advantage on international markets which result in superior profits, they play a crucial role in attaining high levels of employment and economic growth within the home economy.

Secondly, the models of endogenous growth theory endogenise the rate of innovation and predict dynamic effects of international trade on innovation activity (e.g. GROSSMAN/HELPMAN 1990, 1991a, 1991b). The more intensive competition in international markets is, the more firms that want to extent their

---

98 This argumentation was taken up by the OECD (1994) on the basis of the Frascati Manual (OECD 2002) to classify the R&D intensity of manufacturing sectors and their goods by the labels of "low-", "medium-", and "high technology" sectors. The classification is based on the average share of sales spent on private business R&D within different industry sectors. Industrial sectors that spend less than 0.9 percent of their total sales are classified as "low-tech", sectors between 0.9% and 3% as "medium-low-tech", sectors between 3% and 5% as "medium-high-tech" and sectors that spend more than 5% of their total sales on R&D are labelled as "high-tech". Following this classification, the traditional metal, rubber, plastics, paper or textile industries are typical examples of so-called low-tech sectors. Due to structural and technological changes in economies, recent modifications of this typology (LEGLER/FRIETSCH 2007) differ in only using three categories of sectoral R&D intensity: less than 2.5% average sectoral expenditure for R&D in the case of low-tech sectors, between 2.5% and 7% for medium-tech sectors, and more than 7% for high-tech sectors.

99 As GRUPP (1997: 200) amends, the reason for preferring the distinction along sector- or industry-based data lies in the fact that there is hardly any administrative data available on the technological intensity at the market or goods level.

market abroad are forced to develop innovative products in order to become or remain competitive. Firms are unlikely to export goods and services unless they have sufficient capabilities in organisational learning and innovation to enter and compete in foreign markets where they lack experience. Exporters may need to adapt products to local market conditions, offer customised applications, and take advantage of new market opportunities through rapid new product development (LÓPEZ 2005; FILATOTCHEV/PIESSE 2009). Moreover, selling products abroad is also more costly than serving the domestic market. These costs predominantly include transportation, but also distribution, marketing, and foreign customer support that usually require additional skilled staff (HORVATH/JANGER 2008). GREENAWAY and KNELLER (2007) argue that differences in firm productivity are able to explain why not all firms export with the same intensity, which means that only the most productive firms select themselves into foreign markets. Hence, the positive relationship between R&D and export intensity is also mediated by the performance dimension of increased productivity. Empirical evidence strongly supports the relationship between high productivity and exporting for many countries (e.g. ARNOLD/HUSSINGER 2004; GIRMA ET AL. 2004; AW ET AL. 2005, 2007; CASSIMAN/GOLOVKO 2007; WAGNER 2007a, 2007b; SERTI/TOMASI 2008).

However, export activities can also directly influence how firms innovate through feedback effects. Exposure to a wider range of technologies than those available in the domestic market can give exporting firms a competitive edge over domestic rivals, encouraging them to invest in R&D activities (GIRMA ET AL. 2008; HARRIS/LI 2009). For instance, it is plausible that a firm that exports an innovative product will also experience an effect on production costs as due to the larger market scale economies will lower production costs. But for now, I would like to limit my view to the causal relationship ranging from R&D via innovation to productivity and export.

A considerable number of empirical studies provided evidence that the export propensity and export intensity of firms is positively affected by the innovativeness of firms. HIRSCH and BIJAOUI (1985) used R&D measures of Israeli firms and found positive effects on their export propensity. Likewise, STERLACCHINI (2001) shows that R&D investments particularly related to product innovation have a significant positive effect on exports for Italian firms. TOMIURA (2007) analysed a dataset of 118,300 Japanese manufacturing firms and found that exporters are on average more active in innovation than non-exporters, own more patents and report higher R&D expenditure. Other recent contributions explicitly explored the effects of process and product innovations on exports. All these studies find a positive relationship: product innovation is associated with better export performance of companies compared to firms that have not introduced novel products on the market. For instance, for Germany and the United Kingdom, ROPER and LOVE (2002), found out that the successful development of product innovations is positively correlated with the export probability of firms. Similar findings are available for Spain (CASSIMAN/GOLOVKO 2007; CASSIMAN/MARTINEZ-ROS 2007), Germany (BECKER/EGGER 2009), and the Netherlands (BROUWER/ KLEINKNECHT 1993, HESSELS 2007). However, like earlier studies (e.g. WILLMORE 1992, KUMAR/SIDDHARTHAN 1994, LEFEBVRE ET AL., 1998, WAKELIN 1998), the recent analysis conducted by LACHENMAIER (2007), did not confirm positive effects of innovation on exports. As he summarises, "this could suggest that either the causality runs from exports to innovation or that there are still unobservable characteristics that lead to a self-selection of successful firms in innovation and in exports simultaneously." (LACHENMAIER 2007: 205)

When we return to the causal model of R&D effects on innovation and firms' economic performance, there is broad consensus about the positive relationship between R&D intensity and innovativeness in terms of new product or technical processes. Further agreement is reached on the role of R&D as an enabler or amplifier of other dimensions related to innovation such as absorptive capacity, technological knowledge or collaborative attractiveness. Regarding the general economic performance dimensions, most agreement seems to be about the positive relationship between R&D investment and productivity. The positive relationship between R&D intensity and firm growth seems to be valid only in the case of R&D devoted to product innovation. In contrast, R&D activities that aim at process innovation are rather negatively correlated with employment growth. Last but not least, the findings about the theoretically assumed positive relationship between R&D and firms' export activities remain ambiguous so far on the basis of the recognised literature, and thus need further clarification.

2.6.6 Internal and external R&D – A complementary or supplementary relationship?

As described above, firms' R&D activities are considered as the critical determinant of innovativeness and competitive advantage. However, while in-house R&D has traditionally been an important source of technical know-how for firms, it is not the only possible source. As a consequence of the increasing complexity, risks and costs of innovation, firms more and more recognised that they should be open to external sources of innovation (e.g. BECKER/DIETZ 2004: 209; GALLIE/ROUX 2008: 2ff.; CHRISTENSEN ET AL. 2005; CHESBROUGH 2003; RIGBY/ZOOK 2002).[100] PISANO (1990: 153) argues that R&D procurement decisions become particularly relevant for established enterprises which are confronted with broad and rapid changes in their core technologies. During such phases of "creative destruction" (SCHUMPETER 2008b: 81ff.), established firms' in-house R&D may lack the relevant technological skills to perform R&D competitively against new entrants. In consequence, the research community extended its analytical focus from purely internal R&D activities to a broader view of R&D including inside and outside sources of R&D knowledge (ENKEL ET AL. 2009: 311). "Not all the smart people work for us. We need to work with smart people inside and outside our company" (CHESBROUGH 2003: xxvi). To assess such external R&D knowledge, firms have basically three opportunities, which are briefly described in the following.[101]

- To purchase or sub-contract R&D activities to external R&D organisations (e.g. SCHMIEDEBERG 2008; KUROKAWA 1997)
- To participate in R&D collaboration with external partners (e.g. BECKER/DIETZ 2004)
- To open up their innovation processes to reach out to all kinds of external parties for innovative ideas or R&D knowledge (e.g. CHESBROUGH 2003)

As can be seen from these possibilities, sourcing external R&D may cover the whole range from mandatory, contract-based purchase of R&D knowledge in the market, via in-between market and hierarchy modes of inter-firm collabora-

---

100 An extensive overview of empirical literature on technical learning from external sources can be found in FREEMAN (1994a: 469ff.)

101 Beyond that, GALLIE/ROUX (2008) first provided empirical insight into the heterogeneity of R&D relations between firms and its external partners by developing a typology on several characteristics: a) nature of the partnership (common research, sub-contracting, multi-partnership, management of a common structure), b) type of research (market-oriented, research-oriented, mix), c) its duration (short- vs. long-term), d) organisational arrangement (contract, specific investments).

Five classes of collaborations have been obtained from the particular combinations of these relational characteristics: a) long-term relationships for managing a common structure, b) multi-partnerships in upstream research, c) common market-oriented research, d) subcontracting relations, and e) contractual relationships based on strong collaboration and involvement.

tion through to loose, tentative arrangements between a firm and its external stakeholders (e.g. customers, consumers). Thereby the term of "open innovation" (CHESBROUGH 2003) covers both contracted and collaborative R&D activities (ENKEL ET AL. 2009) and the "open sourcing" of knowledge and ideas by including customers and lead-users into the innovation process (VON HIPPEL 1976, 1986, 2001, 2002). Within this context, I would like to restrict external R&D to R&D contracting and R&D cooperation, because the aspect of user-integration must not necessarily be bound to concrete R&D projects and may also be strongly focused on issues like design or product-related services. Thus, external or extramural R&D is assumed to be provided either by external, independent R&D organisations like universities (BENEITO 2006: 506; also BELDERBOS ET AL. 2004: 1478; FAEMS ET AL. 2005: 239f.) or other enterprises like suppliers, customers or competitors (BELDERBOS ET AL. 2004: 1479ff.; LHUILLERY/PFISTER 2009: 46ff.).

Cooperation in R&D is commonly defined as any form of voluntary collaboration of at least two economically and legally independent companies in product research and development. This definition can be extended by other characteristics, for example by a clear purpose of goals, the possibility to quit the collaboration at any time and benefits for all actors (win-win situation) (STRAUBE 1972: 65; EGGERS/ENGELBRECHT 2005: 3). In contrast to R&D cooperations which are based on a certain degree of commitment, trust, reciprocal interdependency and mutual flows of knowledge exchanges, contracted or purchased R&D is defined as mere market transactions by which R&D knowledge is provided in exchange for money.

Traditionally, the positive relationship between R&D and cooperation is based on COHEN and LEVINTHAL'S (1989, 1990) theory of absorptive capacity. It suggests that the benefit a firm can realise through cooperations highly depends on the firm's already existing stock of R&D knowledge. Subsequently, firms with higher internal R&D intensity will obtain higher profits from R&D cooperation. In contrast, other scholars argue that firms with an already high internal R&D intensity would not need to cooperate in the field of R&D as they already possess the relevant knowledge and do not want to let this asset knowledge diffuse to other firms or organisations (e.g. BAYONA ET AL. 2001; PITTAWAY ET AL. 2004). Moreover, firms with high R&D intensity and thus high absorptive capacity may more easily profit from general, freely available ("public") stocks of R&D knowledge, which in turn decreases their propensity to participate in R&D cooperations (ABRAMOVSKY ET AL. 2009). Thus, there are good arguments why either R&D-intensive or less R&D-intensive firms have the higher propensity to cooperate in the field of R&D (BARGE-GIL 2010: 198). However, as was described in the state of research on non-R&D-intensive firms, there is empirical evidence for both perspectives.

In the literature, several reasons are mentioned as to why firms participate in R&D cooperations with other enterprises, or with public research organisations, or decide to buy external R&D knowledge (e.g. BECKER/DIETZ 2004: 210f.; TETHER 2002: 953ff.; KUROKAWA 1997: 124ff.). In general, the benefits of R&D cooperation on innovation can be located at two levels (BECKER/DIETZ 2004: 211). On the side of innovation input, the adaption of additional external resources leads to an extension of firms' technological capabilities and knowledge about developing new products or production processes. On the other side of innovation output, R&D cooperations enable additional capacities to be implemented from the outside to raise the probability of successfully realising innovation projects.

The empirical literature emphasises that, in order to develop an innovation, firms require complementary intangible assets, basically tacit knowledge and know-how, which cannot be easily contracted and monitored through market-based transactions. R&D cooperations enable firms to internalise incoming knowledge spillovers and allow them to increase knowledge transfers voluntarily among the cooperating partners (e.g. PICOT ET AL. 2001; EGGERS/ENGELBRECHT 2005; KATSOULAKOS/ULPH 1998; CASSIMAN/VEUGELERS 2002; BELDERBOS ET AL. 2004).[102] Studies in management and technology policy stress the role of factors such as sharing risks and costs (e.g. DAS/TENG, 2000; TYLER/STEENSMA 1995), shortening innovation cycles (e.g. PISANO 1990), as well as exploiting economies of scale and scope in research and development (e.g. DAS/TENG 2000; KOGUT 1988). Last but not least, R&D cooperation may be undertaken to enable partners to increase market power (e.g. MARTIN 1994; GREENLEE/CASSIMAN 1999), for example, by learning through monitoring technologies and market developments (e.g. HAMEL 1991) or dealing with regulations, industry standards and government policies (e.g. NAKAMURA 2003). A large body of empirical literature found that engaging in R&D cooperation has positive impacts on innovation performance (e.g. BELDERBOS ET AL. 2004; JANZ ET AL. 2003; VAN LEEUWEN 2002; Faems et al.

102 Harking back to endogenous growth theory, these knowledge spillovers play the central role in economic growth as they serve as vehicles to diffuse the economy's accumulated stock of technological knowledge among its economic actors (see for example GRILICHES 1991). R&D spillovers are externalities where not only the innovator benefits, but also other parties involved (e.g. COHEN ET AL. 2002; SMOLNY 2000).

2005), sales growth (e.g. BELDERBOS ET AL. 2004; CINCERA ET AL. 2003), and productivity growth (e.g. HALL/MAIRESSE 1995). As can be seen from this short overview, firms can achieve superior innovation and economic performance not only by internal, in-house R&D, but also by cooperating with external R&D partners like other firms or dedicated research organisations, or by subcontracting R&D activities.

Besides these merits, however, external R&D has some disadvantages that are mainly caused by transaction costs (HUANG ET AL. 2009: 156; BECKER/DIETZ 2004: 211; PISANO 1990: 156ff.), especially for coordinating, managing, and controlling the activities of the external partners involved. According to BECKER and DIETZ (2004: 211) these transaction costs are mainly related to:

- Coordinating distinct organisational routines, cultures, processes
- Combinating complementary assets, resources, and knowledge
- Fixing transfer prices of intangible goods like information or know-how
- Regulating the exploitation (appropriation) of the results and the profits of the joint R&D activity.

These disadvantages are empirically reflected in a broad range of literature which emphasises that many external R&D agreements are inherently unstable and often produce disappointing outcomes (e.g. PORTER 1987; KOGUT 1989; BARKEMA ET AL. 1997; KALE ET AL. 2002; REUER/ZOLLO 2005; OKAMURO 2007; LHUILLERY/PFISTER 2009), due to incompatible partner characteristics or opportunistic behaviour of the R&D partners involved (KOGUT 1989; KALE ET AL. 2002; REUER/ZOLLO 2005).

Thus, external R&D is an efficient strategy for implementing external resources only if the cost– benefit relationship (trade-off) of joint R&D is positive, or can least be expected to be positive. If adaptation is cheaper than in-house R&D, inter-organisational arrangements in R&D are an efficient way to expand and optimize firms' innovation activities (BECKER/DIETZ 2004: 211). However, an evaluation of make-or-buy decisions about R&D may also depend on the degree to which firm-specific (idiosyncratic) knowledge is involved (the more the worse), or the degree to which the economic intentions of the involved partners are congruent (the more the worse). For instance, BENEITO (2006: 506) points out that the requirements a firm should accomplish for technically complex research projects are more likely to be met when R&D activities are internally organised. "The more complex and firm-specific the research project the more severe the need for interaction and information exchange among various functions within the firm. In-house organised R&D activities are the main source for those more significant innovations, whereas externally contracted R&D services are more productive in terms of incremental innovations." (BENEITO 2006: 513) Thus, contracted R&D is less likely to result in significant innovations as measured by patents, unless they are combined with in-house capabilities (BENEITO 2002, 2006). This leaves a broad range for firm-specific and complex innovation strategies of firms regarding the organisation of their R&D activities, often including both internal innovation activities and the involvement of external R&D partners (SCHMIEDEBERG 2008: 1492; NOOTEBOOM 1999: 795f.). Such make-or-buy decisions about whether to undertake external R&D activities therefore depend on multiple influences and thus may be of a highly firm-specific character.

Most interesting against the background of this thesis, however, is the question whether – in case of non-R&D-intensive firms – missing internal R&D activities can be compensated by external R&D or not. This leads to the assumption that internal and external R&D activities are not independent of each other as they could either substitute or complement each other (SCHMIEDEBERG 2008: 1493). On the one hand, a substitute would mean that firms can replace or compensate internal R&D activities by external R&D. On the other hand, a complement would mean that the more internal R&D it is already doing, the more a firm would engage in and profit from external R&D.

From a competence oriented view, it sounds reasonable that firms who do not undertake internal R&D may compensate for this by regularly or occasionally drawing upon external R&D, because R&D outsourcing is a better and quicker option than building up the required skills internally if suitable in-house capabilities are lacking (HUANG ET AL. 2009: 156; TIDD/TREWHELLA 1997: 359f.). Outsourcing R&D while giving up own research is argued to be a reasonable decision in order to exploit the partners' specialised knowledge and the economies of scale associated with specialisation (VEUGELERS/CASSIMAN 1999: 66). The exploitation of external R&D competences may thus compensate for technological competences that the firm does not possess or in which its existing knowledge is inadequate (Huang et al. 2009: 156). Additionally, some researchers stressed the point that R&D outsourcing would decrease the costs associated with new product development because external R&D organisations are often more efficient in running and managing R&D activities that the outsourcing firm (e.g. CHESBROUGH 2007, QUINN 2000).

However, as COHEN and LEVINTHAL and other scholars of absorptive capacity stressed, internal R&D may be a necessary precondition to participate in and profit from external R&D. Through the in-house innovation process, it is suggested that the involved individuals as well as the whole firm will gain experience with technologies and research methods which in turn enables them to understand the research results produced by others outside the firm. Absorptive capacity has two main implications in the context of external R&D. First, it facilitates the search for appropriate cooperation partners or R&D suppliers because it permits appraising the potential partners' quality more easily (SCHMIEDEBERG 2008: 1494; TYLER/STEENSMA 1998: 945). Secondly, absorptive capacity through in-house R&D raises the expected outcome of external R&D projects as firms will choose more profitable or promising R&D projects if external R&D is conducted to supplement in-house R&D (ARORA/GAMBARDELLA 1994: 102). Moreover, as absorptive capacity improves communication and coordination between internal and external R&D, the firms will be more likely to complete cooperative projects successfully (BOUGRAIN/HAUDEVILLE 2002: 743).[103]

Both possible relationships were examined by empirical research. LOVE and ROPER (1999) show that both forms may be substitutes rather than complements. However, they did not differentiate between R&D cooperation and R&D subcontracting. In their study of 2001, they confirmed these results for the UK and Ireland, without however finding a clear substitute or complementary relationship in Germany. Based on patent indicators and the measures of utility models, BENEITO (2006) found a positive effect of contracted R&D and internal R&D which points to the role of absorptive capacity (COHEN/LEVINTHAL 1989, 1990). As already mentioned above, BENEITO underlines the particular aspect of complementarity that internal R&D rather leads to more radical innovation, while contracted R&D is primarily used for incremental innovation. By testing the impact of university cooperations, BECKER and PETERS (2000) revealed two complementary effects: first, they found a positive influence of university cooperation on the intensity of in-house innovation. Second, R&D cooperation with universities is positively related to the regular conduct of R&D on patent production. However, R&D cooperation between firms and contracted R&D are

---

103 However, there is increasing criticism of the fact that in theory the firm's absorptive capacity is closely interlinked with the intensity of its R&D activities (COHEN/LEVINTHAL 1989, 1990). As recent studies (SCHMIDT 2005, SPITHOVEN ET AL. 2010, MUROVEC/PRODAN 2009) emphasise instead, absorptive capacity is a much more multidimensional concept which encompasses both firm-internal (i.e. routines, employee skills beyond the scope of R&D) and external resources (i.e. alliances, cooperations, or networks with other firms, stakeholders, scientific community). See also the paragraph on "dynamic capabilities" in chapter 3.3.3.3.

not included in their investigation. More recently, JIRJAHN and KRAFT (2006) analysed the degree to which a firm's R&D intensity and its research cooperations are complementary regarding the production of product and process innovations and patents. In summary, they interpret their findings in terms of a rather substitutive relationship between both dimensions. Most recently, the work of ARUNDEL ET AL. (2008), RAMMER ET AL. (2009), KIRNER ET AL. (2009b) and TSAI/WANG (2009) found empirical evidence that supports the complementary perspective. Firms with higher R&D intensity are thus aslo more likely to engage in external R&D.[104]

Unfortunately, the short literature review did not provide a clear picture, although the majority of empirical insights tends towards the complementary perspective (SCHMIEDEBERG 2008: 1493). Of course this topic needs to be further researched as obviously additional influencing factors arevimvolved which have not yet been addressed.

2.6.7 Summary and criticism

Both the currently dominant macro and micro perspective of R&D point out that the relationship between innovation and economic success is characterised as a steady-state growth pattern which can be adjusted relatively easily by "turning the knobs of the R&D process" (VERSPAGEN 2005: 504) and which supposes a linear relationship between input (R&D) and innovation output (VERSPAGEN 2005: 504; Hirsch-Kreinsen 2008b: 10f.).

As described before, the issue of professionalised R&D emerged with the social and economic developments caused by the industrial revolution in the late 19th and early 20th century (BRULAND/MOWERY 2005: 358ff.; FREEMAN/SOETE 2009: 585). Until then, innovation was rare and depended greatly on the randomly distributed individual ability of persons to recognise chances and possibilities to develop new products or processes. This historic dynamic is also reflected in SCHUMPETER'S shift of analytical focus from the heroic innovating entrepreneur to the specialised R&D departments of large corporations. Thereby, the nature of innovation was transformed into a systematic, science-based, institutionalised, and predictable activity of specialised functional departments

104 Nevertheless, KIRNER ET AL. (2009b) also found that, if non-R&D-intensive firms decided to cooperate in R&D with external partners they show higher rates of innovation success in terms of their share of sales with product innovations. Similar findings are also provided by KINKEL/SOM (2010) for non-R&D-intensive firms of the German mechanical engineering industry.

within these firms. This new mode of innovation was considered in itself to be a highly relevant institutional innovation and was described as "the heart of the modern capitalist engine" (NELSON 1990: 193; BARGE-GIL ET AL. 2008: 5).[105] Therewith, the influence of the "linear model of innovation" of science push emerged, which assumes that the innovation process follows a well-defined sequence from basic research via applied research to experimental development and, finally, to the stage of diffusion. Or as KLINE and ROSENBERG (1986: 285f.) trenchantly formulate: "One does research, research leads to development, development to production, and production to marketing. These events are implicitly visualised as flowing smoothly down a one-way street, much as if they were the "begats" of the Bible."[106]

This shift towards a science-driven mode of R&D-based innovation was also reflected in the design of innovation surveys and was established in the OECD's Frascati Manual (1963). Thereby, "over the last century formal R&D expenditures rose gradually to become the most widely used measure of technological performance of countries, of sectors, and of firms (...)." (FREEMAN/SOETE 2009: 585).

As a result, using R&D as an innovation indicator has fundamental advantages (SMITH 2005: 154; FREEMAN 1994a: 473):

- R&D data is available over long time periods
- R&D data covers detailed sub-classifications. Its disaggregation still offers rich and exploitable opportunities for research
- R&D data is available for many countries (e.g. the entirety of developed OECD countries)

105 "(...) what became distinctive about modern, industrial R&D (...) was its scale, its scientific content and the extent of its professional specialisation. A much greater part of technological progress appeared attributable to research and development work performed in specialised laboratories or pilot plants by full-time qualified staff." (FREEMAN/SOETE 2009: 586)

106 This strong reliance on scientific principles as a major source of innovativeness and economic success of the entire firm is also reflected in the Tayloristic management paradigm which emerged at the same time. Corresponding to the dichotomy of R&D (where the knowledge generation and learning activities are supposed to take place according to scientific principles) and production (where the mere production and distribution activities are supposed to take place following the economic principles), the management school of TAYLOR (2005) strongly differentiated between tasks of "head" (e.g. planning, controlling, learning) and "hand" (mere execution of explicit assembly steps).

- Good harmonisation of R&D data across countries.

Against this background, a long tradition of empirical innovation research investigated the role and relevance of firm level R&D at the interface between basic research and production as a driver of innovation and in consequence competitive advantage and superior economic performance. On the one hand, there is plenty of empirical evidence that firm level R&D activities positively influence the innovation performance of firms in terms of the successful development and market implementation of new products and technical processes. On the other hand, it was also shown that, mediated by increased innovation output, R&D also contributes towards enhancing firms' long-term economic performance in areas such as productivity, firm growth and export intensity. Hence, firm level R&D intensity is a very powerful determinant to explain differences in the innovation and economic performance of manufacturing firms.

However, in more recent innovation research literature, more and more scholars are becoming aware of the limitations of the R&D concept in explanatory power. I would like to highlight a few major shortcomings of the R&D concept and its established methods of measurement in empirical research from a strategic innovation research perspective at the firm level:

- ***R&D measures focus on innovation input:*** R&D data itself is always constrained as an innovation indicator by the fact that it measures an innovation input only (KLEINKNECHT ET AL. 2002: 110ff.; BECHEIKH ET AL. 2006: 649; FLOR/OLTRA 2004: 324; LACHENMAIER 2008: 17). Measures of R&D investment as well as the share of R&D personnel only provides information about the amount of financial or human resources that are spent for innovation purposes. It says nothing about how successful (in terms of effectiveness and efficiency) the firm has been with regardi to the value or usefulness of its innovation results. On the one side, there may thus be firms that are highly efficient in the use of their R&D resources, as they create highly marketable products with comparably low investments in R&D. On the other side, firms may spend vast amounts of R&D resources but fail to succeed with new products or processes that justify their expense.

- ***Role of existing knowledge:*** The way basic and applied research is defined in the Frascati Manual creates some problems, because performaning of these activities requires effort to obtain new knowledge. "The basic criterion for distinguishing R&D from related activities is the presence in R&D of an appreciable element of novelty and resolution of scientific and/or technological uncertainty, i.e. when the solution to a problem is not readily apparent to someone familiar with the basic stock of commonly used knowledge and techniques in the area concerned." (OECD 2002: 33f.). Thus, many other activities with a scientific or technological base (e.g. acquisition of products and licences, product design, trial production, training and tooling up, acquisition of machinery and equipment related to product or process innovation) as well as activities in the area of education and training in the context of innovation are kept distinct from R&D (SMITH 2005: 154). Moreover, the definition of experimental development creates substantial problems as many activities that are based on the use of already existing knowledge (e.g. engineering, design) are sometimes included as R&D and sometimes excluded (ARUNDEL ET AL. 2008: 6). This suggests that many innovative activities based on new uses of existing knowledge are not covered by the definition of R&D. But, for instance, purchasing manufacturing equipment gives the customer access to all the imbedded R&D that was necessary to produce it; an assumption which comes close to the idea of knowledge externalities or spillovers raised by the endogenous growth theory (SMITH 2005: 158). Therefore, any other, non-R&D paths of knowledge access which are related to the firm's innovation activities are not captured by R&D (BARGE-GIL ET AL. 2008: 6f.; BENDER/LAESTADIUS 2005: 131ff.).

- ***R&D measures not designed for research purposes:*** Some of the R&D indicators and classifications often originate from already existing statistical series that were originally not specifically designed for measuring R&D (GODIN 2001c: 3). For instance, the first edition of the OECD Frascati Manual suggests that the classification of R&D by economic sector corresponds in most respects to the definitions and classifications employed in other statistics of national income and expenditure (OECD 1963: 21). This suggestion aimed to facilitate comparisons with existing statistical series (e.g. gross national product, net output, investment in fixed assets etc.). Another reason for the simple adoption of already existing measurements rather than the development of new ones can be assumed in their availability (GODIN 2001c: 3f.).

- ***Social desirability:*** Due to the large historical background of theoretical and empirical research on the positive relationship between R&D and innovation respectively economic performance, R&D is popularly viewed as a "good" or "desirable" characteristic, which could encourage respondents to report their firm's innovative activities as R&D (ARUNDEL ET AL. 2008. 6f.).[107]

- ***R&D measures and SMEs:*** SMEs in particular could fail to report R&D because of the lack of clarity as to how R&D is defined (SMITH 2005: 154; ARUNDEL ET AL. 2008: 7). The definitions of R&D given in the Frascati Manual, particularly in the case of experimental development, leave substantial room for interpretation. Problems in determining what is and what is not R&D could be common among managers in firms that neither have a formal R&D department nor full-time researchers, and therefore do not show specialised R&D investments in their key figures[108] as the existence of an R&D department implies the formal allocation of resources and also the continuity of innovative efforts within the firm (FLOR/OLTRA 2004: 325). As a consequence, there may be a systematic undercounting of small-firms R&D activity (KLEINKNECHT ET AL. 2002: 111; LACHENMAIER 2008: 16f.; FLOR/OLTRA 2004: 325). Furthermore, because R&D represents one of the most costly and risky opportunities to achieve innovation success (RAMMER ET AL. 2009: 37), some firms, particularly those of small and medium size, face particular barriers while investing in in-house R&D and thereby refrain from undertaking continuous in-house activities. For example, high fixed costs and a high minimum size of R&D projects require great financial commitment, put pressure on profits, and imply high risk of firm survival in case an R&D project fails. As GODIN (2001a: 14) summarises "non-budgeted R&D is the rule in SME."[109]

107 To give an example, research by Statistics Canada shows that more firms report R&D activities in the Canadian innovation survey than claim an R&D tax credit (where evidence of R&D is required (ARUNDEL ET AL. 2008: 6f.).

108 This especially applies to firms in the services sector where many R&D activities could have a social science base (ROSA 2004; DJELLAL ET AL 2003). For the manufacturing sector, ROPER (1999) shows that only 20% of UK manufacturing firms with 20 to 49 employees have a formal R&D department. In contrast, 66% of German firms within this group report having an in-house R&D department. This suggests that many UK firms may not be aware that some of their innovative activities are indeed R&D (ARUNDEL ET AL. 2008: 7).

109 In his paper of 1987, KLEINKNECHT (1987) estimated that 33% of firms devoted less than one man/year to R&D. LIPSETT/LIPSEY (1995) confirmed these results using data on R&D tax credits.

2.7 Excursion - Analysis of the economic relevance of different R&D-intensive sectors in German manufacturing industry

By implication, recognising the superior effects of R&D leads to the question whether firms without any own R&D are even able to survive in market competition. With innovation being a generic process of firms to deal with increasing competition and market dynamics (TIDD/BESSANT 2009: 54) and R&D being such an important enabler for firms' innovativeness, it could be assumed that non-R&D-intensive firms and industries would rapidly lose economic relevance. For this reason, it is an interesting question whether non-R&D-performing firms are still a relevant group within German manufacturing industry or if they rather represent some kind of "phase-out model".

As, to the best of my knowledge, unfortunately no data exist which provide information about the macro-economic relevance (i.e. performance) of non-R&D-performing *firms*, this question has to be approached at the industry level. This corresponds to the proceedure in previous studies by SANDVEN ET AL. (2005), BENDER (2006), HIRSCH-KREINSEN ET AL. (2003), HAHN (2009), or HIRSCH-KREINSEN (2004). Of course, this approach is not satisfactory against the backdrop of an assumed inter-sectoral heterogeneity of non-R&D-performing firms, but so far it is the only possibility to access their macro-economic performance. Detailed analysis of the economic outcome with regard to these three dimensions on micro-economic firm level will be presented later (see chapter 5).

As was shown in the state of research, growth performance across countries is not correlated with shares of high tech sectors in manufacturing industry. Non-R&D-intensive sectors still account for the largest part of manufacturing output and employment in the OECD and European economies, and their preponderance is decreasing only at a very slow rate (SANDVEN ET AL. 2005; BENDER 2006; HIRSCH-KREINSEN ET AL. 2003; for Germany: GEHRKE/LEGLER 2010; SOM ET AL. 2010; HAHN 2009; HIRSCH-KREINSEN 2004). Thus, there has been till now no trade-driven "hollowing out" of low-tech sectors caused by moving their production in low-wage countries as predicted by some authors (e.g. LU 2007; HEIDENREICH 2009).

But as most of these findings are dated at least some years back, analyse an aggregated entity of OECD or European countries, or focus only on the comparison of low- and high-tech sectors, the author conducted some additional analyses to provide more current, tailor-made figures for the German manufacturing industry regarding the performance of R&D-intensive sectors (medium- and high-tech sectors) and non-R&D-intensive sectors. The detailed analysis can be found in chapter A.3 in the annex of this thesis. Corresponding to the state-of-the-art, these additional analyses were conducted with macroeconomic data of the OECD database for structural analysis STAN[110] which provides more than thirty indicators for production and employment structure, labour productivity and labour costs, investment, business research and development expenditures and international trade performance. For Germany, the OECD STAN data are very reliable because they are congruent with official statistical data from the German Federal Bureau of Statistics (GEHRKE/LEGLER 2010: 10). To answer the question about the economic relevance of non-R&D-intensive sectors in today's German manufacturing industry, this chapter draws on the three main performance dimensions of a) value added, b) employment rate, and c) export intensity that were already discussed before.

In short, the findings are:

- ***Value added:*** Despite their lower level, the picture also shows that non-R&D-intensive sectors still increasingly contribute to the overall value added of German manufacturing industry as no sector-specific dynamics can be recognised over time. But at least until 2006 the trend clearly showed a decreasing share of value added of non-R&D-intensive sectors which indicates a shrinking economic relevance of these sectors up to that point. However, it should be remarked that this decrease in value added of about 24% over the considered time period of 35 years is relatively small compared to what could have been expected on the basis of the growth theory's argumentation. Moreover, this decrease of non-R&D-intensive sectors' share of value added turns out to be weaker in the USA, Japan or the EU-10[111] than in Germany's manufacturing industry. In the case of the EU-14[112], the share of non-R&D-intensive sectors even increased between 1995 and 2007. Thus, because of the special role of medium-tech industries in Germany, the numbers presented on the value added of non-

110 More information about the OECD STAN database is available at http://www.oecd.org/document/54/0,3343,en_2649_34445_21573686_1_1_1_1,00.html (last accessed on 15th of July 2010).

111 EU-10 = "new" EU member states since May 2004 (Cyprus, Czech Republic, Estonia, Hungary, Lithuania, Latvia, Malta, Poland, Slovenia, Slovakia).

112 EU-14 = "old" EU member states without Germany (Austria, Belgium, Denmark, Spain, Finland, France, United Kingdom, Greece, Ireland, Italy, Luxemburg, Netherlands, Portugal, Sweden).

R&D-intensive sectors may represent the lower end of the scale in an international comparison.

- ***Employment:*** About 22% of the total working population in Germany were employed in the manufacturing industry in 2006. This share is almost equally distributed among R&D- and non-R&D-intensive sectors. Both shares have been relatively stable in the considered time period. The share of the non-R&D-intensive sectors slightly decreased from 13% in 1996 to about 11% in 2006, while the share of R&D-intensive sectors largely remained stable at 11%. This means that about half of all employees in manufacturing industry still work in non-R&D-intensive sectors. Thus, with regard to this indicator, there is no evidence of any drastic decreasing economic relevance of non-R&D-intensive sectors within the German manufacturing industry.

- ***Export intensity:*** As expected, the export ratio of R&D-intensive sectors is significantly higher than that of non-R&D-intensive sectors. Today, it is almost four times higher for the R&D-intensive sectors than in 1980, while the export volume of non-R&D-intensive sectors has approximately increased by the factor 3 within the same time. Nevertheless, with regard to the question of the economic relevance of non-R&D-intensive sectors in Germany, it seems remarkable that non-R&D-intensive sectors significantly increased their direct export volume, especially between the years 2001 and 2006. Thus, with respect to these absolute numbers, the export volume of non-R&D-intensive sectors is indeed below that of R&D-intensive sectors, but it cannot be stated that products and goods produced by non-R&D-intensive sectors are not competitive in international markets. And, in contrast to the reasoning of international trade theory, this is not necessarily due to their superior cost performance, but may also be rooted in the superior product quality (GEHRKE/LEGLER 2010: 24). Despite this positive trend in absolute numbers, the relative export share of non-R&D-intensive sectors declined between 1980 and 2007 from about 38% to 30%. Conversely, the export share of R&D-intensive sectors increased in the same period from about 63% to 70%. But since 2004, the export share of non-R&D-intensive sectors slightly increased about 2%. While for a long time R&D-intensive sectors were believed to be the main drivers of the economy and growth, the past years rather show a trend of consolidation and increasing importance of non-R&D-intensive sectors regarding the export performance of German manufacturing industry (GEHRKE/LEGLER 2010: 1).

To summarise, R&D-intensive sectors of German manufacturing industry indeed show a higher economic performance compared to non-R&D-intensive sectors. However, despite this difference in the comparison, in absolute numbers there is no justification for assuming a declining economic relevance of non-R&D-intensive sectors in Germany. Furthermore, if one considers the indirect contributions of non-R&D-intensive sectors in their roles of both customers and suppliers of R&D-intensive sectors, their contributions to economic growth, employment and international competitiveness are most probably considerably higher (SOM ET AL. 2010).

2.8 The shift from the R&D paradigm to a systemic innovation approach

R&D represents one of the most crucial resources of firms to generate new knowledge which flow into the successful development and market implementation of new products and technical processes to gain competitive advantage and economic success. The discussion of neoclassical growth theories revealed the theoretical roots of the prevailing R&D focus and the linear model of the innovation process within innovation literature.

Following the theoretical and empirical arguments outlined above, firms which do not undertake R&D cannot themselves generate the necessary knowledge for innovation. They can not make similar profit from knowledge-spillovers as R&D-performers do, because they have more difficulties in accessing the economy's stock of knowledge and thus face higher barriers, or are even unable to develop new products or production methods. Especially against the backdrop of export-oriented, high-wage economies like Germany, in which the input factor of labour is relatively expensive, non-R&D-intensive firms which have to compete on costs are likely to lose economic relevance over time, as they are either relocated abroad (in countries with lower labour costs) or go bankrupt as they are no longer able to compete with technology-intensive firms. Further, as they are less innovative, they can only appropriate lower innovation rents, if at all, which again results in less economic success. In consequence, it can be assumed that the economic underperformance of non-R&D-performing firms will cause them to disappear from the market over time.

But at least for the German manufacturing industry, this assumption has been partially contradicted by the short analysis of the macro-economic performance of non-R&D-intensive sectors. Although their shares of value-added and export intensity are lower in relation to R&D-intensive sectors, non-R&D-intensive sectors still show positive performance trends in absolute numbers. With regard to the employment rate, their share has remained almost unchanged since 1996.

Also against the backdrop of such "stylised facts", the view of R&D as the *sole* source of firms' innovativeness and competitiveness has waned more and more in academic discussion. Even the authors of the Frascati Manual have recognised these limitations of R&D measurements. In the very first edition of the Frascati Manual (1963), they already pointed to the need to measure a much wider range of scientific and technical information services, technical consultancy, training etc. (FREEMAN 1994a: 473). First doubts concerning the performance of R&D measures arose when patent statistics (e.g. SCHMOOKLER 1966) showed that professional R&D accounted only for about half of industrial inventions in the USA. Despite the fact that a large group of the residual inventions originated in governmental or university R&D, a significant number of inventions could still not be attributed to specialised professional R&D (FREEMAN/SOETE 1997: 220). Obviously, important inventions were still developed by production engineers, private inventors or by those actually operating the manufacturing processes. This was reflected by the fact that some firms set up "technical" or "engineering" departments which functioned as an intermediary between R&D and production, and which often contributed far more to the technical improvements of an existing process than the formal R&D department (FREEMAN/SOETE 2009: 586).[113] Hence, the use of R&D measures as a proxy for a wider range of technical and learning activities turned out to be unsatisfactory (FREEMAN 1994a: 473).

As a consequence, "innovation" began to attract more and more attention as the definition of industrial R&D increasingly turned out to be too restrictive. It was this dissatisfaction with R&D indicators that led to the development of a new set of "science, technology, and innovation" (STI) indicators within the framework of the OSLO Manual (1992).[114] They extend the then prevailing R&D focus of the Frascati Manual to the broader term of "innovation" and proposed harmonised guidelines for collecting and interpretating firms' innovation measures (FREEMAN/SOETE 2009: 586). Thus it is widely recognised today that R&D does not capture all aspects of innovation which often occur through other channels, and which could lead to a severe underestimation of the impact of innovation on productivity in non-R&D-performing or non-R&D-

113 See for instance, HOLLANDERS (1965) study of the sources of increased efficiency in Du Pont rayon plants.

114 Until today, the OSLO Manual has been published in three revisions (1992, 1997, 2005), each taking into account the progress made in understanding the innovation process and its economic impact.

intensive firms (HALL ET AL. 2009: 15f.; NELSON 2000: 13; DOSI 1988b: 222; KLINE/ROSENBERG 1986: 302; KLEINKNECHT ET AL. 2002: 110f.; PRAHALAD/HAMEL 1990: 83; FAGERBERG 2005: 16; SCHMIEDEBERG 2008: 1492; SANTAMARÍA ET AL. 2009: 508; BARGE-GIL ET AL. 2008: 6; SMITH 2005: 168; ARUNDEL 1997: 6).

But even if one accepts the view of R&D as being only one source of innovation among equals, it can not be ignored that R&D represents some kind of institutionalised answer of firms to their need to allow groups of people within the firm sufficient freedom to experiment with new technical opportunities, mobilise the knowledge base, and develop new solutions when confronted with problems and challenges (FAGERBERG 2005: 11; FORAY 2006: 54). As R&D is not subject to the same kinds of economic constraint as the regular production of goods and services, the ability to perform exploratory activities that would otherwise not be possible in everyday business life is a key factor supporting rapid knowledge advances in firms. Last but not least, R&D activities are also valuable because of their positive side effects, like an increased attractiveness for collaboration partners or a positive firm image as an innovation leader.

Before I proceed to the development of an evolutionary view on innovation, one important aspect should be kept in mind with regard to the following definition of non-R&D-performing firms. The literature review showed that internal and external R&D rather complement each other, which means that firms with no in-house activities in most cases do not undertake external R&D either. For this reason the definition of non-R&D-performing firms developed in chapter 4.3.1 does not necessarily need to distinguish between internal and external non-R&D-performers.

In theory, the decline of the R&D focus is on the one hand driven today by a shift of analytical focus towards innovation-related activities that go beyond the scope of formal R&D, and, on the other hand, by a changing understanding of the nature of the innovation process itself, for instance in the works of DAVID (1996), FORAY (1998), LUNDVALL and JOHNSON (1994) or EDQUIST and TEXIER (1998). The main developments are:

- A broader understanding of firms' internal innovation resources expressed by terms of "routines", "capabilities" or "competences" (NELSON/WINTER 1982; WINTER 1987; TEECE/PISANO 1994; PRAHALAD/HAMEL 1990; GRANT 1991), by generally referring to "knowledge" as the most important predictor for innovation (e.g. GRANT 1996a; SPENDER/GRANT 1996).

- The increased appreciation of engineering, design, production and distribution activities (KLINE/ROSENBERG 1986; VINCETTI 1990; WALSH 1996; FREEMAN/SOETE 1997; HANSEN/SERIN 1997; KOSCHATZKY ET AL. 2001) as well as investment in capital equipment related to innovation are further determinants of successful innovation (e.g. EVANGELISTA ET AL. 1998; EVANGELISTA 1999).

- The awareness that the firm's ability to exploit systematically the effects produced by new combinations and uses of components and practices in the existing stock of knowledge may be another crucial enabler for successful innovation (DAVID/FORAY 1995; KLINE/ROSENBERG 1986); for example frequently labelled as "architectural innovation" (HENDERSON/CLARK 1990: 12) or "innovation without research" (COWAN/VAN DE PAAL 2000: 3). As KLINE and ROSENBERG (1986: 302) argue, when firms are faced with having to innovate they first look into their existing stock of knowledge, and if the answer cannot be found there, they then consider whether it makes sense to spend on R&D or not.[115]

- The increasing recognition that firms are embedded in social systems of innovation (e.g. LUNDVALL 1992; EDQUIST 1997; NELSON 1993). This highlights the systemic nature of innovation processes, emphasising that firms do not normally innovate in isolation, but in collaboration and interdependence with other organisations (e.g. suppliers, customers, competitors, etc.), non-profit entities (e.g. universities, schools, government ministries), institutions (e.g. laws, rules, norms), and other social entities (e.g. local residents, consumers). Moreover, firms intentionally may actively make use of their surrounding external sources in terms of collaboration (e.g. DYER/SINGH 1998; NOOTEBOOM 1999), user-driven innovation (e.g. LUNDVALL 1985; VON HIPPEL 2004), or "open innovation" (e.g. CHESBROUGH 2003).

In parallel with the increasing dissatisfaction with the R&D focus in current innovation theory, the understanding of innovation processes has shifted from the linear, sequential and thus predictable nature towards complex and self-referential cycle models (KLINE/ROSENBERG 1986; ROTHWELL 2003; DODGSON 2000; TIDD/BESSANT 2009) which account for multiple recursive feedback loops and other sources of innovation knowledge. As ROSENBERG (1994: 139)

---

115 This is another quality than the "make or buy" decision examined by VEUGELERS/CASSIMAN (1999), because the "make" is here subdivided into "make through R&D" and "make through other means" (ARUNDEL ET AL. 2008: 5).

claims: "Everyone knows that the linear model of innovation is dead." Very few scholars of innovation research defend such a theoretical understanding of innovation anymore. Most often in theory, innovation is considered to result from cooperation in social *and* economic activities. Thereby, the innovation process links technical and economic considerations. It normally includes many kinds of interactions, and innovations do not have to be radical; on the contrary, they involve incremental social and organisational changes, as well as technological advance. Consequently, innovations are not just the results of scientific work in a laboratory-like environment. They are arrived at in networks where actors from different backgrounds are involved in the process setting new demand for innovativeness.[116]

As a result, the innovation process is understood as complex and variable. There is no one best way to innovate. Instead, against the background of modern, knowledge-based economies, the use of R&D as a proxy or surrogate measure for a wider range of innovation is no longer adequate (KLINE/ROSENBERG 1986: 302; FREEMAN 1994b: 83; ARUNDEL ET AL. 2008: 5; RAYMOND/ST-PIERRE 2010: 48) and the theoretical focus needs to shift "from R&D to learning processes" as all knowledge produced within a firm cannot be attributed to formal research activities (FORAY 2006: 7). Instead, any activity involving the production or use of a good (or service) can generate learning and hence knowledge production, which means that there is a complex of different ideas and solutions which are equally important for effective innovation. Thus, formal R&D remains only *one* among many other inputs and sources of innovation within the firm (SMITH 2005: 168; ARUNDEL 1997: 6; FREEMAN 1994b: 83; Nelson 2000: 13; SCHMIEDEBERG 2008: 1492).

In their recent article, FREEMAN and SOETE (2009: 587) speak about three shifts in the understanding of innovation:

- 19th century innovation paradigm: the inventor-entrepreneur

116 GRUPP (2008: 87) also campaigns for a rather functional instead of a determining understanding of the relationship between R&D and innovation. This means that R&D serves as a tool for solving problems that may occur in each stage of the innovation process (idea, theory, discovery / technical design / product design, innovation / imitation, improvement, diffusion, exploitation, disposal) which cannot be solved by drawing on the existing stock of knowledge and experiences alone. "But innovation is possible without R&D if the knowledge stock available to the firm and in the published science is sufficient."

- 20th century innovation paradigm: in-house corporate R&D with external communications
- 21st century innovation paradigm: networking collaborative systems of innovation with knowledge bases distributed all over the entire innovation system

Surprisingly, although innovation and economic theory widely acknowledge this theoretical shift, there are still only few empirical studies that are already based on this 21st century innovation paradigm and take into account innovation sources other than R&D, because R&D data still represents one of the most reliable and easily available measures to asses firms' innovation activities. Thus, the majority of empirical work still includes R&D as an explanatory variable of innovation (BECHEIKH ET AL. 2006: 655; BARGE-GIL ET AL. 2008: 8; ARUNDEL ET AL. 2008: 5).[117] As GODIN (2006a: 659f.) and ARUNDEL (2007: 50) argue, the linear model of R&D-based innovation is still attractive due to its very simplicity.

> "It is a thought figure that simplifies and affords administrators and agencies a sense of orientation when it comes to thinking about allocation of funding to R&D. However, official statistics are, in fact, more important in explaining the continued use of the linear model. By collecting numbers on research as defined by three components and presenting and discussing them one after the other within a linear framework, official statistics helped crystallize the model as early as the 1950s. In fact, statistics on the three components of research were for a long time (and still are, for many) the only available statistics allowing one to understand the internal organisation of research, particularly in firms. Furthermore, as innovation came to define the science-policy agenda, statistics on R&D were seen as a legitimate proxy for measuring technological innovation because they included development (of new products and processes). Having become entrenched in discourses and policies with the help of statistics and methodological rules, the model became a social fact." (GODIN 2006a: 559f.)

The mentioned efforts to overcome or modify this model were limited in their impact due to the complexity of threi analytical framework.[118] In particular, efforts to measure these new interactive models of the innovation process have

---

117 Most innovation research has focused on R&D-performing firms. Only about 5% of research papers based on the CIS, for example, specifically evaluated innovation by firms that do not perform R&D (ARUNDEL ET AL. 2007).

118 ARUNDEL (2007: 50): "Between 2004 and 2006, [researchers of – *author's note*] UNU-MERIT interviewed 67 members of the policy community - 55 from 15 European countries and 12 from Canada, Japan, Australia and New Zealand - on their use of and need for innovation indicators. R&D indicators were the most widely used and were considered to be the most valuable."

hardly been successful, at least in the official literature, because standardised statistics and indicators on flows of knowledge between economic sectors, performers and users of research, and types of related activities are still missing (GODIN 2006a: 660). "The countless announcements of the death of this model and its presumed replacement with 'systematic' models using Schumpeterian definitions or innovation are decidedly premature." (ARUNDEL 2007: 50). One serious consequence of this in many cases still prevailing focus on firms' R&D activities is that firms that do not perform R&D have largely been neglected or even ignored by policy-makers and academics in the past (ARUNDEL ET AL. 2008: 5; BARGE-GIL ET AL. 2008: 8).

3 Firms' heterogeneity and sources of innovation and economic success beyond the scope of R&D

This chaper deals with the previously raised points of criticism about the R&D paradigm of innovation by exploring heterogeneous innovation patterns of non-R&D-performing firms in German manufacturing industry. In doing so, it leaves the R&D-focused paradigm of the 20th century behind and explicitly refers to the systemic, "21st century" paradigm of innovation.

In order to answer the basic research question of this thesis of whether, in general, heterogeneous patterns of non-R&D-performing firms can be expected or not, it is necessary to develop a theoretical framework to a) explain why inter-firm heterogeneity of non-R&D-performing firms can be expected at all (structural dimension), and if yes, b) in which of the dimensions mentioned above this heterogeneity may manifest itself (substantive dimension).

The following chapters will depict and discuss theoretical approaches in recent innovation literature which aim to explain inter-firm heterogeneity. The first part (chapters 3.1 and 3.2.) adresses the structural explanation of firm heterogeneity by referring to a strand of theory which has been one of the origins of the 21st century innovation paradigm and which explicitly focuses on explaining inter-firm heterogeneity: evolutionary economic theory. Subsequently, the second part (chapters 3.3 and 3.4) enriches the abstract evolutionary concept of firms' heterogeneity by focussing on the firm level approach of strategic management theory called the resource-based theory of the firm. It aims to identify potential dimensions of non-R&D-performing firms' innovation behaviour for the later empirical analysis.

3.1 Evolutionary economic theory and the heterogeneity of firms

The origins of the evolutionary economic theory can be traced back to the works of MARSHALL (1920), VEBLEN (1898), GEORGESCU-ROEGEN (1971), VON HAYEK (1969) and of course SCHUMPETER (2006). Nevertheless the evolutionary approach did not establish itself as a self-contained strand of economic theory until the beginning of the 1980s with the most prominent contribution by NELSON and WINTER (1982) and their development of "An Evolutionary Theory of Economic Change". This milestone represents a synthesis of the existing unconnected evolutionary strands and shapes a theoretical approach to explain endogenous technological change and economic growth processes.

While the neoclassical approach consists of a relatively homogeneous set of interrelated sub-strands (models), the field of neo-SCHUMPETERIAN or evolutionary economics consists of a loosely connected set of different contributions. Both of these approaches agree on basic issues such as the importance of innovation and technology for economic growth, and the supporting role that can be played by government policy. However, they disagree about underlying behavioural basics, i.e. they have a different understanding of the nature of the economic system, technological progress and innovation. These differences are reflected in their respective analytical approach. While neoclassical theory sacrifices a significant amount of realism in terms of accounting for the actual complexity of innovation processes in return for a quantitative modelling approach that favours strong analytical consistency, evolutionary approaches embraces the micro complications of the innovation process and applies a more eclectic approach, (VERSPAGEN 2005: 492).

The evolutionary theory of innovation is often considered as "post- or neo-Schumpeterian" because SCHUMPETER also conceptualised an evolutionary economic process which is continually driven by all kinds of innovation (SCHUMPETER 2006: 161ff; RAHMEYER 1993: 260; SCHWITALLA 1993: 39; GRUPP 1997: 73). SCHUMPETER criticised the neoclassic paradigm of risk neutral, profit maximising behaviour of economic actors and described it as a very special case of economic processes. Instead, the everyday and most essential conditions of economic processes are characterised by dynamics, discontinuities and ongoing changes. "The competitive environment within which firms operate is one of struggle and motion." (NELSON/WINTER 1974: 890) Unforeseen challenges and unknown future developments rather demand courageous, risk-taking but also careful and prudent behaviour (SCHUMPETER 2006: 171ff). It is obvious that these characteristics are not equally shared by all economic subjects. This argument leads to the key arguments of evolutionary economics which point out that firms in the "real world" show considerable heterogeneity with regard to the behaviour and strategies they apply. NELSON (1991) and NELSON/WINTER (1982) argue that this is, among other things, due to different skills of their workers, different experiences, and differences in firm size and organisation structures. Empirical evidence for the heterogeneity in firms' innovative behaviour is further provided by economic and organisational analysis of innovation (e.g. SRHOLEC/VERSPAGEN 2008; LEIPONEN/DREJER 2007; JENSEN ET AL. 2007; CHRISTENSEN 2002; MASSINI ET AL. 2005; PAVITT 1984).

Despite the fact that the evolutionary theory of innovation is not a uniform theoretical concept (HODGSON 1996: 699, 1997: 12; FAGERBERG 2003: 127)[119], there is consensus among evolutionary theorists in rejecting the basic assumptions of profit maximisation and economic states of equilibrium. By describing economies in principle as dynamic systems, the neoclassical mechanistic explanation of economic growth and technological change is contradicted. "The essential forces of growth are innovation and selection, with augmentation of capital stocks more or less tied to these processes (NELSON/WINTER 1974: 890). Furthermore, evolutionary theorists criticised the neoclassical approach on the grounds that the underlying assumption of the perfectly rational behaviour of firms is unrealistic for the analysis of innovation. As mainstream optimisation posits only a single, optimal and most rational solution, there is not much room left for heterogeneity in actors' behaviour. Evolutionary economic theory is rooted in the view that firms are complex learning organisations which develop different ways to solve similar problems and eventually are selected by the environment through market mechanisms. Thereby, evolutionary theory allows for connected evolutionary processes at multiple levels of skills and jobs, of workgroups or communities of practice within organisations, of organisations within industries, and of industries within wider socio-economic systems (NOOTEBOOM 2009: 211). The change of individual units (ontogenetic development) like single firms for instance, as well as of species or populations (phylogenetic development) like industries is explained by the fundamental mechanisms of "mutation / variation", "selection" and "reproduction / retention / replication" (HERRMANN-PILLATH 2002: 203ff.; ALDRICH/RUEF 2006: 16ff.; VROMEN 1995: 107ff.; SCHWITALLA 1993: 40; NOOTEBOOM 2009: 207). Some authors explicitly connect these terms to Darwin's biological theory of evolution (e.g. ALCHIAN 1950; WINTER 1964; NELSON/WINTER 1982) arguing that firms are selected in the sense that they survive market competition.

But several authors state that a pure "natural" kind of selection mechanism runs the risk of tautology and needs therefore to be amended by additional assumptions (e.g. CAMPBELL 1994; HODGSON 1993; MARCH 1999; VROMEN 1995). Their view of economic Darwinism takes the concept of natural selec-

119 Even Schumpeter suggested some aspects which contradict the views of other representatives of evolutionary theory (see FREEMAN 1990: 22ff; GRUPP 1997: 73f.).

tion as an analogy[120] and implies that the actual process of economic natural selection is very similar to Darwinian selection in biology (KNUDSEN 2002: 443f.).[121] Despite different opinions about the compatibility and transferability of the biological and economic process of evolution, especially concerning the role of accident and intention in the creation of variation, it is possible to identify similarities in the epistemological structure of biology and economics (HODGSON 2004a: 57ff.). As both paradigms address evolving systems, biological and economic evolution share a common abstract level with regard to the general relevance of variation, selection, retention, and the universal force of competition for scarce resources.[122] "Establishing a Darwinian core for evolutionary economics does not mean reducing economics to biology, or ignoring important differences between evolutionary processes in their two domains. Neither does it mean that humans become automata, bereft of deliberation or self-reflection (HODGSON 2004a: 72)". Instead it means applying the general concept (i.e. "meta-theoretical framework" HODGSON/KNUDSEN 2006: 16) and approach of Darwinism to the real economic facts of firms' heterogeneity, competitive selection and retention of certain patterns of firms' (innovation) behaviour. As a whole, they are on the one hand subject to external market selection and on the other hand, are characterised by some kind of internal evolution in adapting or even interacting with changes in their external environment.

By paying attention to the dynamics of socio-economic systems, the evolutionary theory usually focuses on the element of change in terms of the endogenous, unpredictable innovation behaviour of enterprises. Last but not least, evolution itself is a dynamic process par excellence. But its founders SCHUMPETER and MARSHALL already conceptualised economic change as an interplay between change, on the one hand, and routine and stability on the other (RAHMEYER 2005: 2). It can be reasonably argued that stability is necessary for processes of evolution for several reasons. Firstly, the selection

---

120 Following COHEN (1993: 13), an analogy is an instrument of cognition and means the transition of theoretical elements or arrays between different disciplines of science to promote the solution of problems.

121 For a detailed discussion of the transferability of Darwin's biological concept of evolution into economic analysis, see HODGSON (1993, 1995, 1996, 2002), KNUDSEN (2002), WITT (1992a, 1992b, 1992c, 1996, 1997, 1999, 2003) and for an overview GEISENDORF (2004).

122 Darwin explicitly referred to ADAM SMITH'S (idea of specialisation and division of labour) and MALTHUS' work ("struggle for existence") (SCHWEBER 1985: 35f.; GOULD 2002: 122).

mechanism needs a certain constancy in the existence of variations to be ablte to select them. If a variation is not stable enough over a minimum period of time, it will disappear before it can prove itself in competition. Secondly, without stability and the preservation of once achieved quality characteristics, socio-economic systems will not be able to accumulate and secure stocks of knowledge over time (VON HAYEK 1969). The evolutionary theory therefore needs to explain both, stability and change (HERMANN-PILLATH 2002: 204). But this kind of stability is not to be confused with the static situation of the neoclassical equilibrium, which can be interpreted as stagnancy between cycles of change, while change is the permanent condition of evolutionary theory. The evolutionary concept of stability rather describes some kind of "regularity" or "order" within the permanent processes of change of needs, technology and environmental conditions (HERMANN-PILLATH 2002: 147f.).

The sub-division of evolutionary theory into an individualistic-behavioural and an institutional-historical branch by SCHWITALLA (1993: 40f.), GRUPP (1997: 76ff.) or VERSPAGEN (2005: 497ff.) corresponds to these elements. The individualistic-behavioural strand focuses on the dynamic element of change and the role of either individual (e.g. SCHUMPETER'S "entrepreneur") or collective, organisational patterns of behaviour (e.g. the "routines" of NELSON and WINTER). Within this strand, evolutionary ideas are also used to formulate formal models of economic growth and technology to push forward the mathematical formalisation of evolutionary theory.[123] On the other side, the institutional-historical strand analyses changes in the socio-economic or technological system from a historic perspective. Such historical analysis is often used by evolutionary scholars to develop or explore heuristic typologies, patterns, and classifications of technological change that can be used to describe and categorise evolutionary developments more generally in a rather non-formalised evolutionary theorising. Representatives of this evolutionary strand are e.g. the studies of FREEMAN (1982, 1988), PAVITT (1984), DOSI (1982, 1988a), or MALERBA/ORSENIGO (1995) at sectoral, and JENSEN ET AL. (2007), LEIPONEN/DREJER (2007), MARSILI/VERSPAGEN (2002), SRHOLEC/VERSPAGEN (2008), ARUNDEL ET AL. (2008) at firm level. As this thesis investigates heterogeneous patterns of innovation behaviour of non-R&D-intensive firms, and

123 Among others, such models were for example first developed by NELSON/WINTER 1982; CHIAROMONTE/DOSI (1993), SILVERBERG/VERSPAGEN (1994). A very recent strand within formalised evolutionary modelling is represented by the so-called "history-friendly models" which aim to bring evolutionary models closer to empirical reality by reproducing the historical evolution of a particular industry (e.g. MALERBA ET AL. 1999).

does not model evolutionary processes, it pursues the institutional-historical strand.

While most theorists (e.g. NELSON/WINTER 1982; NELSON 1991; PRAHALAD/HAMEL 1990) explain inter-firm heterogeneity by the firms' heterogeneous behaviour (e.g. in terms of individual routines, competences, abilities) they justify the heterogeneity of firms in terms of substantive dimensions which can be observed as ex post manifestations of heterogeneity. This means that firm-specific routines, competences etc. are the observable, heterogeneous result of evolutionary processes. In contrast, the following chapters will primarily rely on original evolutionary arguments to explain why inter-firm heterogeneity can be *principally* expected *ex ante* on the basis of the three main evolutionary core mechanisms of variation, selection and replication.

3.1.1 Variation

From an evolutionary perspective, the stimulus for enterprises to search for something new usually results from dissatisfaction with theirs current economic performance and the perceived need to adapt to changed or new external requirements (WINTER 1975: 105; WITT 1996: 712). By implementing such innovations, the firm seeks to unlock new opportunities for action and solutions for problems it is faced with. According to ALDRICH and RUEF (2006: 18ff.), any departure from existing practices of producing goods and services and problem-solving of a firm can be seen as a variation and corresponds to the concepts of mutation or recombination in nature (NELSON/WINTER 1982: 18). The evolutionary change is driven by the heterogeneity in the competences of individuals as well as in the routines of firms which result from the bounded rationality of actors (METCALFE 1995: 471; SRHOLEC/VERSPAGEN 2008: 5). Following the evolutionary understanding, enterprises are thus characterised as experience- and knowledge-based, learning organisations (WINTER 1988: 175).

Variation can be further subdivided into intentional and blind variation (ALDRICH/RUEF 2006: 18ff; GEISENDORF 2004: 87). Blind variations are independent of conscious planning. They result from accidents, mistakes, trial and error learning, luck, imitation, misunderstandings or even idle curiosities in the everyday fulfillment of operational processes within the firm (BRUNSSON 1985; MARCH 1981). Other sources of blind variation are spontaneous reactions of organisations to unexpected shocks or discontinuities in their environment. In contrast, intentional variation means the purposeful attempt to generate alternatives and seek solutions to problems. Intentional variations result from conscious responses to difficult situations, challenges, planning sessions, or advice from external partners, for example. Sources of intentional variation within firms include formal programmes of experimentation and imitation (e.g. R&D), direct and indirect incentives offered to employees or the encouragement of unfocused variation or "playfulness" (ALDRICH/RUEF 2006: 24f.). It is important to mention that, in economics, a great deal of variation is indeed caused by the firms' perceived need and intention to adapt to external challenges and requirements and their intentional, targeted search for new problem solutions. But, for instance, as some enterprises or organisations intentionally promote kinds of "intelligent failure" (SITKIN 1992: 239) as a method of constructive experimentation, the boundary between both types of variation appears to be fluid. Moreover, elements of blind variation presumably influence the intentional search for variation at any stage. For example, a singular, random lucky mistake must be intentionally reproduced with an assignable and manageable degree of probability to unfold its economic benefits over time. As NELSON and WINTER (1982: 11) conclude: "both elements are involved and difficult to disentangle".

Economic evolutionary theory is often criticised for overemphasing the role of blind variation (WITT 1992a, 1992b, 1992c, 1997, 2003; FOSTER 2000). But the need for variation does not explain the solution (CZIKO 1995: 288ff.). The selection of variations follows from their effects and consequences, not from the intentions of the actors who generated the variations. "Thus, blind variations can be as effective as intentional ones." (ALDRICH/RUEF 2006: 18). But in general, and that is a crucial element of evolutionary theory, each variation or mutation is "blind" in the sense that there is – in contrast to neoclassical theory – no ex ante way of telling whether or not it will improve the performance of the firm. Hence, from the evolutionary perspective, particular outcomes are not necessarily prefigured or predicted in advance (VERSPAGEN 2005: 496; HODGSON/KNUDSEN 2006: 11; NOOTEBOOM 2009: 218).

The selection mechanism is neutral with respect to the type or the source of variation as it just selects from a population of variations (HERRMANN-PILLATH 2002: 211). But variation itself is indeed a necessary, but not sufficient condition for heterogeneity. On the one hand, variation may only take place with regard to a single dimension. On the other hand, variation can be wiped out by the second evolutionary mechanism: selection. Thus, the mechanism of selection is assumed to play a decisive role in creating, respectively approving heterogeneity.

3.1.2 Selection

3.1.2.1 The functioning of selection

While evolutionary change produces heterogeneity through variation, on the one hand, it simultaneously reduces this heterogeneity by its second basic mechanism of selection, on the other hand. Selection describes all kind of forces that differentially select or selectively eliminate certain types of variations (ALDRICH/RUEF 2006: 21). Selection will reward strategies that produce high competitiveness and punish strategies that imply low competitiveness in the struggle for scarce resources. "There are winners and losers in Schumpeter's process of 'creative destruction', and these are not determined mainly in ex ante calculations, but largely in ex post actual contest" (NELSON 1991: 66). In consequence, selection changes the relative frequency of entities in the population by positive selection of entities that "fit" the selection criteria.[124] Formation, adaption, and diffusion of variation change the population which is being selected. Over time, populations are more likely to be characterised by the attributes of surviving organisations than by the attributes of those that were eliminated. Firms with superior adaption processes (e.g. showing superior efficiency) dominate the competition, parallel to increases of internal capacity and productivity. This in turn enables them to undertake new variation activities which at last intensify the process of selection for less adapted enterprises. As selection needs a certain degree of variation between its objects of selection to be effective, a well-functioning market competition over a sufficient period of time is a necessary prerequisite (WITT 2001: 66).

But social entities – in contrast to mere biological organisms - are able to consciously react and intentionally interact with their environment. They are even capable of anticipating changes in their environments. Hence, as several authors note, they are supposed to be much less exposed to environmental selection than biological organisms. Thus it is assumed that variations in human action or social behaviour are subject to a kind of "internal" selection than to the external selection pressure (e.g. WITT 1997: 7, 1999: 295).[125,126] The con-

124 For a discussion of the fundamental theorem of natural selection see METCALFE (1995, 1998).

125 FOSTER (2000: 314ff.) shares this view by subordinating the mechanism of selection to the superior principle of "self-organisation" (ibid.: 314; also WITT 1996: 709, 2001: 49). For the relationship between self-organisation and selection, see DEPEW/WEBER (1995: 479ff, 1996) and HODGSON (2002).

cept of "dynamic capabilities" may be interpreted as a kind of firm-internal selection mechanism, as these "meta-routines" are responsible for the identification, selection, exploitation, modification, and refusal of the firms' problem-solving and adaptive behaviour (e.g. TEECE ET AL. 1997; EISENHARDT/MARTIN 2000; ZOLLO/WINTER 2002; HELFAT/PETERAF 2003).

On the one hand, the argument of internal selection challenges the relevance of the external selection mechanism in its basic tenets. On the other hand, selection does indeed take place, but is rather considered to be endogenous to the subjects of selection. This criticism is countered by the argument that internal selection follows the same principles as external selection.[127] For instance, selection criteria are set through the operation of market forces, competitive pressures or the logic of internal organisational and strategical structuring (ALDRICH/RUEF 2006: 21). Although human respectively economic actors can intentionally adapt themselves to external requirements or even anticipate future changes, in the end their ideas have to prove themselves under the external conditions. After the internal "pre-selection" (GEISENDORF 2004: 90), their ideas, strategies and patterns of behaviour have to come to terms with those of the other market competitors. Moreover, due to their bounded rationality, social actors are not able to completely capture the whole range of possible actions. Finally, the difference between internal and external selection criteria is only rooted in their different levels of reference. Each level of selection has its own relevant "environment" which provides a set of non-parallel aligned test criteria.[128] Nevertheless, it is the purpose of both natural and socio-economical selection to prevent well-adapted subjects from extinction and to eliminate the non-adapted. Thus, if one accepts selection as a universal, not only biological, principle (HODGSON 2002), it is reasonable to assume that selection in social or economic systems is characterised by a temporal sequence of various interdependent processes of internal and external selection. Moreover, according to the system approaches of innovation (LUNDVALL 1988; EDQUIST 1997), it is reasonable to assume the existence of subordinated

126 DAWKINS (1976) also proposed a second level of selection operating at the human level – the "meme".

127 GEISENDORF (2004: 86ff.) provides a good overview of the counter arguments.

128 A biological example of this multidimensionality of selection is the difference between "natural" and "sexual" selection. The struggle for sexual reproduction may create variations (e.g. extra long tail feathers) which may be dysfunctional to the general adaptive capability. But in turn, such reproductive variations could also mean that the organism is so perfectly adapted to its environment that it can "afford" this variation.

non-economic levels of external selection, which consist of institutionalised political, social or ethical norms of selection (ALDRICH/RUEF 2006: 21).

3.1.2.2 Selection and heterogeneity

Many popular contributions to evolutionary theory simplify the adage "survival of the fittest" by arguing that, as a result of selection, only limited heterogeneity is left. For instance, FRIEDMAN (1953) states that economic evolution leads to the survival of firms that homogeneously maximise profits only and hence that the traditional profit-maximising theory is adequate. But also the approach of national and sectoral innovation systems is implicitly based on the assumption that such national or sectoral systems constitute similar criteria of selection (e.g. importance of technological know-how, competitive pressure, availability of resources) which lead to relatively homogeneous innovation behaviour of the embedded firms (SRHOLEC/VERSPAGEN 2008: 9f.). Following this assumption, sectorally-based studies on innovation have taxonomised sectors and the firms within them into larger groups, each characterised by its typical innovation strategy (e.g. PAVITT 1984; MALERBA/ORSENIGO 1995, 1996; MARSILI 2001; ARCHIBUGI 2001; PENEDER 2003).[129] But referring to the discussion about different levels of selection, reducing selection to a process which unequivocally leads to homogeneity is insufficient (SRHOLEC/VERSPAGEN 2008: 8).[130] To date, the economic evolutionary theory has unfortunately not paid much attention to the question of how selection impacts heterogeneity. But analogousy to biological evolution, there are good reasons why significant heterogeneity may be left even after selection has been working over time.

First, external selection criteria as a part of the environment can also be highly variable. If the selection environment is completely stable for a long time, the evolutionary process would come to rest in a kind of steady state where all niches for survival are occupied by highly specialised and evolved species. But as RICHERSON ET AL. (2005) stated, the environment is a "restless world" which results in the fact that selection criteria keep continuously changing irregularly

---

129 PORTER'S market-based approach (1980, 1985) can also be classified as such, as firms' opportunities to gain competitive success are determined by the external market structure in which they are embedded. Nevertheless, the original approach of PORTER is obviously not rooted in an evolutionary tradition, as he presupposes a fully rational behavior of firms.

130 See also ESSLETZBICHLER/RIGBY (2005) for a brief discussion of this and the presentation of evidence on heterogeneity in plants' productivity.

over time and thereby create new niches, thus stimulating new variation, and itself increases heterogeneity.

The second argument relies on the so-called neutral theory of evolution (KIMURA 1983). It is closely linked to the point mentioned above, that selection works on the tangible effects and consequences of variations, not their nature. At the latest, the emergence of the situative or contingency approach of organisational theory (e.g. LAWRENCE/LORSCH 1967; KHANDWALLA 1977) rejected the "one best way" paradigm of organisational structure. Strategic goals and organisational targets can be reached by various ways, methods and instruments. Thus, even an assumed homogeneity in firms' goals can theoretically be pursued by a broad range of different operational strategies. In the biological analogy, this means that certain parts of the variation are not subject to selection. SRHOLEC/VERSPAGEN (2008: 8) exemplify this argument by the external and internal research and development of firms (R&D). In certain selection environments, innovations resulting from internal and external R&D result in equally competitive products. Hence, external selection would not favour any particular firms with respect to their internal or external R&D activities. More generally, this means that when the selection environment contains many local niches rather than one global market, heterogeneous behaviour tends to be affected relatively less by market selection. Firms which occupy such local niches, and therefore appear as heterogeneous in terms of their behaviour, need to show large differences in their observed economic performance.

The third reason is based in evolutionary game theory and relies on the argument that a mixed strategy is an evolutionary stable strategy within a given population (BERGSTROM/GODFREY-SMITH 1998; HERMANN-PILLATH 2002). But this argument can also be traced back to the work of ADAM SMITH and his concept of the division of labour, as it addresses the issue of functional specialisation and differentiation. Again, SRHOLEC/VERSPAGEN (2008: 9) explain this reason by referring to the example of the well-known game of doves and hawks. The point is that in a population of either only doves or only hawks, the entry of the other strategy would be evolutionary stable.[131] Evolutionary stable strategies are obviously determined by their external selection criteria which impact the competition of both groups. Whether an aggressive (hawk) or a passive

---

[131] „Starting with only doves, the first hawk to enter could obviously survive, since it will always 'win'. But with only hawks, two doves entering would also survive, since they would share the resource on the (rare) occasion that they meet, and then have a higher pay-off than the hawk-hawk meeting (SRHOLEC/VERSPAGEN 2008: 9)

(dove) strategy is good or bad, results from the functionality of behaviour, which in turn has to adapt to environmental frame conditions (HERRMANN-PILLATH 2002: 113). It can be assumed, that the evolutionary stable composition of the system lies in a certain proportion of both strategy types. But this proportion itself is not stable and changes similarly to the environmental conditions of selection.

None of these arguments in favour of a certain degree of substantial heterogeneity remaining after economic selection are part of the original economic evolution theory so far (SRHOLEC/VERSPAGEN 2008: 9). But they provide good reasons for the assumption that heterogeneity in firms' behaviour can be expected, even within similar frame conditions (in the sense of similar selection criteria). The thesis thus follows the approach of DOSI (1988a) who argues that, despite common inter-sectoral frame conditions of firms (like technological opportunities, possibilities of technological acquisition, or public support), a "large, unexplained residual" (GRUPP 1997: 81) of intra-sectoral, institution-specific innovation behaviour remains which results from differences in firm-specific structures, proportions and innovation strategies.[132] Insofar, the neoclassical assumption of the "representative enterprise" has to be abandoned (IBID.: 82).

3.1.3 Replication / retention

The third evolutionary mechanism involves operating of a retention mechanism to maintain positively selected variations. The retention process preserves duplicates or otherwise reproduces variations (ALDRICH/RUEF 2006: 23ff.). Positively selected variations (e.g. patterns of behaviour, skills, competencies ...) are individually conserved and duplicated, for example, by internal growth of existing or the entry of new enterprises or organisations (RAHMEYER 2005: 17f.) GODFREY-SMITH (2000: 413) defines replication as the process which explains the heritability of variation, requiring similarity in relevant aspects. AUNGER (2002) generally argues that replication is a relationship between a

132 The possibility to bridge the gap between the micro- and macroeconomic level is outlined by DOSI/ORSENIGO (1994).

copy and some source exhibiting certain characteristics.[133] Hence, the process of retention represents the element of stability in the evolutionary process, because it accumulates and preserves once achieved quality characteristics, and thus secures stocks of knowledge over time on which future generations can rely to produce new variations.[134] Besides its dynamic elements, evolution encompasses also features of coordination and stability. Thus, in addition to the element of "exploration", the second important source of evolution is the "exploitation" of already existing, retained knowledge and experiences (MARCH 1991: 71f.; NOOTEBOOM 2000: 303).

The meaning of retention or "replication" as coined by DARWIN and its consequences on socio-economic evolution have been broadly discussed in the literature.[135] In DARWIN'S original theory, replicators (i.e. genes or DNA) generate, in ontogenesis, the characteristics of the interactors (individuals, firms, species) carrying them. In contrast, socio-economic (inter-)actors are also able to shape their replicators by cognitive construction (NOOTEBOOM 2009: 212). Thus, the "universal" approach of Darwinism[136] extends the mechanism of retention to other contexts and systems, especially as a key feature of human socio-economic systems (HODGSON 2004b: 56). Due to the controversial and difficult debate about memetics (BECKER 2004: 663), some authors propose the two alternative and mutually related concepts of "habits" and "routines" as units of replication (HODGSON 2004b: 61ff.).[137] Hodgson defines habits as individual dispositions of social or economic actors to engage in previously adopted or acquired behaviour, triggered by specific stimuli. At the level of organisations or firms, routines correspond to habits as they are defined as organisational dispositions to energise conditional patterns of behaviour within an organised group of individuals, involving sequential response to cues

133 These aspects are: *causation* (the source must be causally involved in the production oft he copy), *similarity* (the copy must be like its source in relevant aspects), *information transfer* (the process that generates the copy must obtain the information that makes the copy similar to its source from that same source), *and duplication* (during the process, one entity gives rise to two or more others).

134 This is one of the basic arguments of VON HAYEK'S concept of cultural evolution (1969). See also VANBERG (1994: 5f.)

135 See, for example, the contributions of HULL (1988), GODFREY-SMITH (2000), SPERBER (2000), AUNGER (2002), WITT (2005), HODGSON/KNUDSEN (2006) or NOOTEBOOM (2009).

136 See also DAWKINS (1983), CAMPBELL (1965) or CZIKO (1995).

137 Winter as one of the pioneers of evolutionary economics has also pursued the issue of the replication of routines (e.g. WINTER 1990, 1995).

(HODGSON 2004b: 70-71). From early on, the term "routine" has been linked with the notion of "patterns" (Becker 2004: 644ff.) The term "patterns" emphasizes the regularity that the concept stands for.[138]

But habits and routines consisting of knowledge, qualification, skills, competencies, and memorised experiences cannot simply be transferred from generation to generation like biological characteristics. Instead, they need to be actively acquired and absorbed by other individuals or organisations. Hence, as ontogenetically produced replicators may be adopted by others, this means that the process of socio-economic evolution must at least contain some elements of LAMARCKIAN evolution (NELSON/WINTER 1982: 11; SAVIOTTI/METCALFE 1991: 36; KNUDSEN 2001: 122f.; NOOTEBOOM 2009: 212). LAMARCKIAN evolution as the "inheritance of acquired characteristics" is characterised as an evolution through transformation (HUMPHREYS 1996: 295; MAYR 1972: 55; MAYR 1991: 388). It emphasises the continuous development of an organism, behavioural routines or technical artefacts towards higher complexity and perfection by environmental adaptation, preserving once achieved quality characteristics by shaping its replicators and thereby transmitting them to the next generation ("transformative evolution") (RAHMEYER 2005: 20). In Darwin's theory, evolutionary change is independent of the environment, as the process of retention or replication operates directly at the genotype level ("explorative evolution"). In contrast, LAMARCKISM suggests that the changes in the environment trigger changes in the habits and routines of its inhabitants (phenotypes), which in

138 WINTER (1964: 263) defined a routine as "patterns of behaviour that is followed repeatedly, but is subject to change if conditions change". For detailed discussion of the concepts of routines in literature see BECKER (2001, 2004).

turn lead to changes in the organisms themselves.[139] Against the interplay of evolutionary mechanisms, this means that – in contrast to mere biology – under the condition of LAMARCKIAN evolution the mechanism of variation "learns" how to generate "smarter" mutations which have a higher propensity to be positively selected by market selection (VERSPAGEN 2005: 496).

Due to the combination of explorative and transformative elements in economic retention processes, it can finally be assumed that heterogeneity produced by variation and selection is also preserved in subsequent generations and populations of firms. As they are of high individuality and specificity, it is suggested that the LAMARCKIAN elements in retention thereby contribute specially to heterogeneity in firms' behaviour.

3.1.4 Summary and criticism

By rejecting the basic assumptions of profit maximisation and economic states of equilibrium, evolutionary economic theory represents the antithesis of the mechanistic neoclassical explanation of economic growth. Its approach is based on the axiom that both individuals and social entities (i.e. firms, organisations) are unable to cope in a fully maximising way with the complexities of their environments in general, and technology in particular. Economic actors thus operate under a scheme of bounded rationality in which relatively simple

139 There was a long ongoing debate in biology between supporters and opponents of LAMARCKIAN evolution (MAYR 1972:90; BALTER 2000: 38). Today, LAMARCK is remembered mostly for the discredited theory that evolution occurs when parent organisms pass on to their offspring characteristics they have acquired during their lifetimes. But nevertheless, the French naturalist, who lived from 1744 to 1829, was one of the great scientists of his age. He was the first to systematically study invertebrate animals, and he was an early founder of the idea that evolution rather than divine intervention was responsible for changes in plants and animals over time. But by the early 20th century, LAMARCK'S concept of evolution was superseded by DARWIN'S theory of natural selection and the genetic laws of GREGOR MENDEL. "(...) the recognition that DNA does not directly participate in the making of the phenotype and that the phenotype, in turn, does not control the composition of the DNA, represents the ultimate invalidation of all theories involving the inheritance of acquired characters" MAYR (1972: 55). Recently, however, the ideas of LAMARCK were rediscovered by scientific literature, especially in the context of socio-economic evolution to explain how individual or organisational habits and routines that are generated by the interactor can be transmitted to future generations of individuals or organisations. In biology, this corresponds to the growing number of studies that analyses changes in genetic expression that are not linked to alterations in DNA sequences ("epigenetics") (BALTER 2000: 38).

and occasionally adaptive behavioural rules ("rules of thumb" or "routines") which can change over time, especially under the influence of environmental challenges, are used to make decisions (VERSPAGEN 2005: 496). Due to the impossibility of perfectly managing or foreseeing the complexity, dynamics and thus uncertainty of external environments, economic actors indeed still show profit-seeking behaviour, but each output or outcome can only be satisfying or unsatisfactory in the context of what the actor knows about his environment.

In evolutionary economics, the explanation of aggregate economic performance mainly relies on the two forces of variation and selection. Variation is the evolutionary mechanism that constantly generates novelty through any deviation from existing habits, actions, processes or ways of problem-solving. Over time, this variety of novelty present in the system is reduced by the mechanism of selection in terms of favouring those entities that are better adapted to circumstances to grow, and sanctioning those that are not. Thus, evolution is the outcome of the constant interplay between the mechanisms of variation and selection. Innovation is a novelty-generating process whose novelties are selected by the market as the most viable selection mechanism in the context of economic evolution. While the generation of novelty is entirely blind in biology, socio-economic actors and decision-makers are able to plan their actions to generate potentially successful innovations in a process that closely resembles LAMARCKIAN evolution.

For this reason, economic evolutionary theory is more than the mere transmission of biological analogies to socio-economic systems. Instead, the evolutionary theory and its mechanisms of variation, selection and retention should rather be understood as "universal characteristics" of socio-economic systems that are applied to explain the evolution of economic subjects and contexts. The evolutionary theory in its broad variations provides good arguments that a) heterogeneity in firms' innovation behaviour is produced by continuous adaptive, more or less blind generation of variation to fit the challenges and requirements of their external environments; b) that selection is a multilevel and multidimensional mechanism which does not select along static and homogeneous criteria; c) thus, even under similar external frame conditions of markets, technology or sectors, heterogeneity is not weeded out by selection; and d) that the mechanism of retention and replication preserves not only features at the level of replicators, but also cumulates and transfers once achieved stocks of experiences and lessons learned by the carrier of the replicators. Based on the mechanisms of variation and selection, such stocks of cumulative experiences and knowledge are likely characterised by a high level of individuality. The process of replication therefore amplifies and preserves heterogeneity over time.

However, in the context of this thesis, economic evolutionary theory leaves some questions unanswered:

- ***What exactly is being selected?*** In the context of selection arising from competition in markets firms can suffer bankruptcy, take-over, break-up or management buy-out. But even in the case of their complete failure, at least some of their capabilities may be adopted by other firms which consider them useful. Hence, it remains unclear whether selection means that the replicators or the interactors disappear from the market (NOOTEBOOM 2009: 214ff.).

- ***Impact of selection on market heterogeneity:*** It has not yet been analysed, how selection impacts on firm level heterogeneity in the real world. As a result, we know relatively little about how much heterogeneity can be expected after firms have brought their innovation strategies to the market (SRHOLEC/VERSPAGEN 2008: 8). How much heterogeneity does the market require to ensure that it functions? How, and by which level of heterogeneity is the market's evolutionary stable mix of strategies characterised?

- ***Focus on incremental innovation:*** In its basic understanding, evolution is usually considered to proceed in small, incremental changes and by the adaptation of individuals or entities to successively changing environments in the long run. With regard to this point, even SCHUMPETER would have had some conflicts with evolutionary theory, as he focused strongly on radical or breakthrough innovation by speaking of new combinations of production factors. In evolutionary theory, variation ranges always between the presently given standard criteria of selection. But what is it about such radical, groundbreaking variations that go beyond the scope of existing internal or external economic, ethical, moral or political selection criteria? Thinking along evolutionary lines, this would be only possible if the selection criteria themselves undergo a radical change or breakdown (e.g. financial crisis, war). Thus we still know little about the dynamics of the selection mechanisms itself. Against the background of the current economic depression, it can be reasoned that the selective pressure on economic actors is increasing. For example, existing selection criteria are either tightened or shifted, or new selection criteria arrive on the scene. Thus, an enhanced understanding of the dynamics of the selection mechanism would probably benefit firms' anticipation capabilities.

- ***Meaning of interactors and replicators:*** Biological evolutionary theory basically distinguishes between replicators (i.e. genes) and interactors (i.e. the carriers of genes which are in turn shaped by them).

While in biological evolution, the survival of the replicators is inevitably bound to the survival of the corresponding interactor, this it not necessarily the case in socio-economic evolution, because routines, competences or knowledge are able to either diffuse among other interactors or may be adopted long after the interactor has gone (e.g. if the employee leaves the firm). But if replicators may be disembodied from interactors, does the notion still make sense? (NOOTEBOOM 2009: 212; also: NELSON 2006). This problem gets worse even when one recognises the possibility of LAMARCKIAN evolution which means that the interactors are able to modify and shape their replicators. Built on subjective perceptions and experiences of a changed environment, this shaping of replicators would (if working completely and fast) reliably reflect any shift or variety of selection environment without being proved by selection. As a consequence, survival would not be an indicator of success any more, as perhaps many worthless or deleterious traits would be replicated along with favourable ones (NOOTEBOOM 2009: 212; also: HODGSON/KNUDSEN 2006). Finally, this could cause evolution to break down. Thus although the idea of LAMARCKIAN evolution is very attractive for socio-economic evolution, for evolution to work there must be some kind of isolating mechanism which protects the replicators from influence by interactors (HANNAN/FREEMAN 1984: 150f.)

– ***Differences between biological and socio-economic mechanisms of replication:*** In biology, the mechanism of replication works automatically (as only those species survive which are best adapted) and completely (the whole set of genes is copied and transmitted to future generations). In contrast, replication in socio-economic systems entails reproduction and imitation of knowledge and competencies at different levels and occurs through observation, communication, and apprenticeship. All these elements comprise linguistic processes of expression, sense and reference, and cognitive processes of assimilation into mental schemata or mental models. Thus, replication of "genes" in socio-economic systems is at least partly voluntary and subject to choice (one adopts what is perceived as "successful"), partial (one may adopt only part of a bundle), and subject to decay, distortion, reduction, extension, and transformation. To conclude, there are good reasons to assume that in socio-economic systems the mechanism of replication itself at the same time entails a considerable degree of variety generation (NOOTEBOOM 2009: 216f.).[140]

- ***Structural explanation and description of heterogeneity:*** Evolutionary theory as such represents a framework of assumptions concerning the nature of individuals and socio-economic entities and their behavioural conditions. Based on the interplay of variation, selection and replication, the evolutionary theory explains how heterogeneity in economic behaviour is generated. Thus, evolutionary theory is able to explain why heterogeneous innovation patterns of firms can be expected, but says nothing about the dimensions in which this heterogeneity is reflected. As a consequence, evolutionary theory today mostly limits itself to outlining the complexity of interactions and unpredictable dynamics of the economic environment. Some authors (e.g. VERSPAGEN 2005: 509) therefore see one of the major challenges for evolutionary theory in developing models that are relevant for practice, for instance, with regard to specifying policy advice.

3.2 Structural heterogeneity of non-R&D-performing firms

At the beginning of this thesis, two main assumptions underlying the existing empirical research on non-R&D-intensive or non-R&D-performing firms were identified: a) non-R&D-intensive industries and sectors are characterised by similar criteria of market selection and thus consist of a homogeneous population of non-R&D-intensive firms, and b) for the reason that non-R&D-performing firms all lack the same important innovation resource of R&D, they will of necessity display a similar, rather homogeneous innovation pattern.

To overcome these assumptions, the previous chapters discussed why inter-firm heterogeneity can be expected even if they are embedded in similar market environments or are all characterised by the lack of R&D as an important source for innovation. The argumentation went beyond the reasoning why inter-firm heterogeneity of non-R&D-performing firms may be visible in *ex post* evolutionary processes (e.g. in terms of firm-specific routines, competences, capabilities) by focussing on reasons why this heterogeneity might be particularly rooted *ex ante* in the interplay of the mechanisms of variation and selec-

140 The pro-active role of socio-economic actors by individually selecting and choosing those elements which are perceived as useful or successful leads to considerable diversity, for instance, in the ways firms handle technical innovations (see for example: DREHER/LAY 1992).

tion. Following biological evolutionary theory, three good reasons were discussed as to why significant heterogeneity may survive even after (sectoral) selection has been in place over a long time:

- ***External selection criteria as a part of the environment may also be highly variable:*** The selection environment itself is a "restless world". Thus, selection criteria keep continuously changing irregularly over time and thereby create new niches, stimulate new variation, and thus itself increase heterogeneity.

- ***"Neutral theory of evolution":*** Selection only works on the tangible effects and consequences of variations, not on their nature. Innovation goals can be reached by various ways, methods and instruments. Thus, even an assumed homogeneity in firms' goals can theoretically be pursued by a broad range of different operational strategies.

- ***Evolutionary stability of mixed strategies:*** Only heterogeneous strategies are evolutionary stable within a given population. If, in the struggle for scarce resources, all firms compete by using identical competitive strategies, their payoff decreases over time until none of them remains alive. Instead, the first alternative competition strategy will receive a higher payoff. Thus, it can reasonably be assumed that the evolutionary stable composition of a socio-economic system lies in a certain proportion of different strategy types. But this proportion itself is not stable and changes similarly to the environmental conditions of selection.

These evolutionary arguments strongly support the assumption that even within similar frame conditions (in the sense of similar selection criteria or the common lack of the important innovation resource R&D), significant heterogeneity in non-R&D-performing firms' innovation patterns can be expected. Thus, if we hark back to one of the basic research questions of this thesis about whether heterogeneous innovation patterns can be expected among non-R&D-performing firms or not, this result derived from evolutionary theory provides a first positive answer: yes, non-R&D-performing firms in manufacturing industry can be expected to show heterogeneous innovation patterns.

As was already mentioned in the points of criticism above, the explanations derived from evolutionary theory are, however, neither able to explain how and in which dimensions this heterogeneity of non-R&D-performing firms manifests itself, nor to predict to what extent heterogeneous innovation patterns among these firms can be expected. With regard to the latter aspect, the original research question

> *Can inter-firm heterogeneity in non-R&D-performing firms' innovation patterns generally be expected? If so, why?*

can be re-phrased as follows:

> *To what extent do heterogeneous patterns of non-R&D-performing firms exist in German manufacturing industry?*

This research question can only be assessed if we have an idea about the theoretical scope of innovation dimensions along which inter-firm heterogeneity can occur. It is therefore now important to theoretically identify and narrow down the dimensions within which firms' heterogeneity in innovation behaviour could probably take place. Otherwise, the explorative empirical analysis would just be of an arbitrary character. Therefore, the following chapter develops an evolutionary-based concept of possible dimensions of non-R&D-performing firms' sources of innovation and competitive advantage by drawing on firm level approaches from strategic management theory.

3.3 Sources of firms' competitive advantage in strategic management

For the purpose of further theoretically investigating the ability of non-R&D-performing firms to obtain competitive advantage and economic success, the field of strategic management theory is explicitly focused on the question why some firms persistently outperform other firms (BARNEY/CLARK 2007: 3; WELGE/AL-LAHAM 2008: 43). This question, however, should not be so interpreted that these performance differences between firms constantly exist. Instead, this questions points to the fact that, in some situations and some of the time, persistent performance differences do exist between firms. A firm is said to be successful, when it succeeds in reaching sustainable competitive advantage compared to its competitors.

In the past, these questions were addressed at the broadest level by two distinctive approaches in the literature. The first, and older one, was originally founded by MICHAEL PORTER (1979, 1980, 1985) and relied strongly on the "structure-conduct-performance" (SCP) paradigm of industrial organisation economics (BAIN 1956, 1968), which attempted to develop a concept for the optimal economic market structure at the industrial level.[141] Unlike this market-based perspective, the second major approach of strategic management explains persistent differences in firm performance and sustainable competitive advantage by firm-specific sets or bundles of strategically relevant resources and resource combinations. Correspondingly, this approach was named the "resource-based view" or "resource-based theory" and refers to the works of WERNERFELT (1984), RUMELT (1984) and BARNEY (1986a). In the following chapters, both approaches will be discussed and analysed to determine whether they can explain how non-R&D-performing firms gain competitive success.

3.3.1 The market-based approach of strategic management

MICHAEL PORTER early recognised the analytical potential of BAIN'S industrial economics approach of the SCP paradigm for the business management level (PORTER 1981: 609ff.). Starting from the criticism that the market behaviour of actors had been frequently disregarded in empirical studies of the SCP approach, PORTER instead suggested that the economic performance of firms is also determined by their market behaviour, which is shaped by their competitive strategies. On the one hand, he argued that differences in the economic performance of firms within the same industrial sector result from different strategic behaviour and are not directly determined by the industrial market structure (PORTER 1981: 616). But, on the other hand, PORTER still emphasised the high relevance of the market structure by suggesting that only the firm's thorough analysis of the market structure can guarantee the right choice of competitive strategy and thus economic success (PORTER 1999: 33ff.). Finally, he came to the conclusion that a firm's competitive advantage is determined by two basic aspects: the nature of the industry which the firm is embedded in, and the strategic decisions by which it positions itself within this industry (HOSKISSON ET AL. 1999: 426.).

According to PORTER, the competitive advantage of firms is determined by their relative product-market position, which can be strategically managed by adapting to the predominant sector- and market-specific competitive forces

---

141 According to the SCP paradigm of industrial economics, the characteristics of industrial sectors (= structure, e.g. product differentiation or the degree of concentration) influence the strategic behaviour of buyers and sellers (= conduct, e.g. price policy or advertising efforts) which in turn determines the economic success of an enterprise (= performance, e.g. profitability, efficiency) (BAIN 1968: 112ff.).

and choosing the optimal competitive strategy. This analysis of the sector's competitive market structure should thereby not be limited to the degree of concentration of existing competitors. Instead, the firms should also focus on the characteristics of customers, suppliers, the probability of entrance of new competitors, or the risk of its own products being substituted by products of competitors (PORTER 1999: 33ff.). Due to this line of reasoning, PORTER'S approach was has been named the "market-based approach".

PORTER derived five universal competitive forces of economic markets for which he claimed universal validity, regardless of their national or international, product- or service-oriented character. These forces are (PORTER 1999: 33ff.):

- the conditions for the entry of new competitors
- the threat from substituting products
- the bargaining power of buyers
- the bargaining power of suppliers
- other forces that are released by the concurrency of already existing competitors.

This "five forces framework" describes competitive imperfections in product markets and how they can develop their effects on the firms' possibilities to make profit by raising prices above a competitive level.

The most important strategic goal of managers should thus be to identify and pursue the optimal market positioning of the enterprise in such a way that it can prevail against these influencing market forces to the best possible extent. Starting from these market forces, PORTER developed three basic types of competitive strategy for firms to adapt to their respective competitive frame conditions (PORTER 1999: 70ff.):

- the strategy of differentiation
- the strategy of cost-leadership, and
- the niche-strategy

PORTER explicitly underlined the necessity for each firm to make a clear, consistent decision in favour of one of these strategies to profit from the respective economic profits of each strategy (PORTER 1999: 78ff.). If a firm becomes "stuck in the middle", which means that it does not manage to make a consistent decision for one of these strategies and instead tries to pursues more than one strategy simultaneously, it will subsequently lose both groups of customers, those who demand highly differentiated products as well as those looking for low-cost products. Additionally, an inconclusive concentration on sector or niche markets aggravates this problem even more.[142] The market-based view assumes that firms are originally endowed with similar sets of strategic resources and that these resources are to a large part mobile across firms within common markets (PORTER 1991: 108; DE WIT/MEYER 2004: 250ff.). "Resources are not valuable in and of themselves, but because they allow firms to perform activities that create advantages in particular markets" (Porter 1991: 108). Hence, according to the market-based perspective, resources themselves could never be the starting point for strategy formulation. Instead, the development and accumulation of resources necessarily needs to follow the demands of market and industry.

3.3.2 Summary and criticism

To summarise, the market-based approach explains the persistent economic success of firms by external characteristics of industry structure, rather than the firm's internal managerial, technical, marketing, and other resources. This perspective emphasises the need for a "strategic fit" between the firm's competitive strategies and the external frame conditions it is confronted with. From an evolutionary understanding, this means focussing on the selection mechanisms of the product market which in the end decide whether a product is bought or not. For PORTER it is thus not so important how the firms realise their competitive strategies, but that they succeed in doing so, which means that they manage to achieve the intended performance effects. For this purpose, PORTER and other representatives of the market-based view of industrial eco-

142 In his contribution in the Financial Times, JOHN KAY (1996) showed that firms do not become 'stuck in the middle' as predicted by PORTER. Compared to the competition, firms with medium costs and medium quality have achieved higher returns on investment than those with either low-low or high-high strategies. Furthermore, some firms that pursue a combination of high quality and low-cost compared to its competitors also attain high financial returns.

Due to continuous criticism, PORTER himself more and more weakened this radical position in the last years. In a publication in 1990 (PORTER 1990: 39f.), PORTER already points to the necessity for firms to simultaneously attach value to both positions of cost and quality to attain persistent economic performance.

nomics provide well operationalised and practicable models and tools of competitive analysis (ZANKER 2011: 83).[143]

But as a consequence, the competitive forces approach to the theory of the firm says little about the limits, if any, to a particular firm's ability to enter new markets or lines of business. The following points of criticism are frequently raised against the strategic management approach of industrial economics:

- ***Technological change:*** By tracing the competitive success of firms back to the simple choice of the right or wrong, adequate or inadequate competitive strategy which best fits the sector-specific existing market forces, PORTER tends to overestimate the scope of action of business management and its heuristics of strategic decision-making. In the contrary, he underestimates the power of technological change to transform and override industrial structures (TIDD ET AL. 2009: 173). Consequently, PORTER proceeds from the assumption of a unidirectional relationship between competitive strategy and technology, as the choice of a competitive strategy determines the deployment of technology. Thus, PORTER disregards the fact that the availability of certain existing or future technology may also shape the competitive strategy of firms. For example, technological advances (e.g. IT solutions, new manufacturing equipment) can increase opportunities for economic success in so-called mature or traditional sectors, for instance, like the textile industry or metal processing industry.

- ***Effects of technology:*** This argument is closely connected with the last point. In the market-based view, the use of technology is always instrumentally related to improving a certain dimension of firm performance (SKINNER 1984: 121). The expected effects are exclusively limited to the attainment of higher product quality, higher productivity or a broader product mix. But it is reasonable to assume that a higher quality of manufacturing processes would also result in lower manufacturing cost. Hence, the possibility that new manufacturing technology may overcome the conflicting goals of quality and cost optimisation is not considered due to the monocausal modelling of performance outputs.

- ***Focus on existing product markets:*** PORTER'S framework is only applicable to existing markets or industries with considerable and relatively stable structures. It provides no answers on how to choose

143 See also Rytter et al. (2007).

the "right" competitive strategy in emerging markets which are not yet characterised by stable structures, not to mention the problem of strategically managing the prospective entry of a firm in future markets by introducing a radical product innovation. Moreover, PORTER overlooks the fact that firms and technology must not necessarily fit perfectly into industrial structures, as some firms are typically active in more than one product market.

- ***Perfectly informed and fully rational behaving actors:*** A fully rational choice of competitive strategy by strategic management as assumed by PORTER would reasonably imply perfect and full information of the firm and its actors about their current and future competitive situation, the type and strength of prevailing market forces and the optimal strategic positioning of the firm in its market. BURR (2004: 114f.) points to the increasing dynamics and complexity of competitive frame conditions and the resulting discontinuities and structural changes with which the firms have to cope in the age of internationally interlinked economies and economic processes. As SCHUMPETER already stated, change and mutation are the natural characteristics of the economic system (SCHUMPETER 2008: 82). Hence, the assumption that a firm could assess all relevant market information at one time or even predict all future developments as a reliable basis for its strategic decision-making appears unrealistic. Thinking this aspect of fully rational behaviour through to the end would result in similar market and competitive power and economic success of (fully rationally behaving) firms which have chosen the same optimal competitive strategies according to their identical competitive conditions. Thus, the market-based approach is not capable of explaining immanent variance and heterogeneity in firms' competitive strategies and fails as well to explain why firms differ in their economic and competitive success, despite the fact that they are operating under similar competition conditions and are thus expected to pursue similar competitive strategies.

- ***Exclusiveness of strategic options:*** Another important point of criticism refers to PORTER'S assumption that firms can constantly choose between all three types of competitive strategy. Firms that face different markets will behave and perform differently, but if the external market conditions were reversed, so also firm behaviours. Where the market-based view admits product differentiation, different firms will produce different products, but it assumes that any firm can choose any niche. Therefore, there are differences between firms but there is no essential, autonomous quality to them (NELSON 1991: 64). Instead, followers of the evolutionary economic theory argue that the initial choice of a competitive strategy results in some kind of technological "path-dependency" or "trajectories" (DOSI 1982; NELSON/WINTER 1982) in terms of the technologies deployed. The decision for a certain product market influences, for example, the development of organisational processes and structure as well as the development of competencies within the firm, or may be accompanied by great investment cost.[144] Due to such firm-specific technological trajectories and the firm-specific technological and organisational competencies to exploit them, a radical change between PORTER'S basic types of competitive strategy is not possible for firms at any time (TIDD ET AL. 2009: 173).

Against the background of these criticisms, it becomes obvious that the market-based approach of strategic management contributes only little to the underlying research question of how non-R&D-performing firms gain competitive advantage. This is not to say that the market-based approach is of little explanatory value in general. But as it focuses on the role of the industrial, sectoral or market environment of firms as the most important determinants of firms' competitive success, it is incompatible with the prevailing research perspective which explicitly addresses evolutionary heterogeneity at firm level. Moreover, by focussing solely on the role of markets, the market-based approach emphasises the role of market selection. But, as argued before, the broadest level of inter-firm heterogeneity can be expected at the level of variation, as some elements of variations are neutral to the market. Hence, analysing only at the market or industry level would a priori constrain the possible heterogeneity in non-R&D-performing innovation patterns. Last but not least, the incompatibility of the market-based approach with the underlying evolutionary framework is also reflected in their contradicting assumptions. Like neoclassical models of growth, the market-based approach starts from the axiom of decision-makers behaving fully rational who are always able to recognise the optimal strategic option to position their firm in its markets. Evolutionary theory explicitly neglects the perfectly rational behaviour of actors. The

144 TIDD ET AL. (2009: 191) give some examples of persistent differences amongst industrial sectors in the sources and directions of technological trajectories such as a) size of innovating firms, b) type of product made, c) objectives of innovation, d) sources of innovation, and e) locus of own innovation.

following chapters will therefore concentrate on the internal, resource perspective of firms as the major source of competitive success.[145]

3.3.3 The resource-based theory of the firm

Similar to the theoretical strands of industrial economics, the dependent variable of the resource-based theory of the firm is the attainment of persistent economic success of an enterprise through competitive advantage. But in contradiction to the preceding approaches of the market-based perspective, the resource-based theory of the firm proceeds from the assumption that the economic success and competitive advantage of firms cannot only be determined by their market environment, but rather by their internal assets and resources (SCHENDEL 1996: 3).

The terms "resource-based view" and "resource-based theory" serve as an umbrella for all approaches and theories in strategic management which explain persistent differences in firm performance and sustainable competitive advantage by firm-specific sets or bundles of strategically relevant resources and resource combinations (WELGE/AL-LAHAM 2008: 87). An enterprise is only assumed to be successful, if it either possesses a unique and superior set of resources, or is able to exploit them more effectively or efficiently than its competitors (MAHONEY/PANDIAN 1992: 364f.; PETERAF 1993: 180). Thus differences in the competitive advantage and economic performance of firms are explained by analysing the differences between their resource setting and their processes of combination or reconfiguration of resources. In contrast to the market-based view of strategic management, resource-based theory does direct its primary analytical focus to the external market environment, but rather the internal and firm-specific resources and competencies. While the external market environment and competitive framework conditions cannot, to a large extent, be influenced by firms, the development and acquisitioning of resources are subject to firms' strategic decision-making. Thus, the resources developed, held and controlled by firms are assumed to provide a much more stable and reliable basis for business strategies in the long run than cross-

---

145 It should be added at this point, that, in his more recent work, PORTER relativised this strict market focus by admitting the important role of firm-specific resources and competences (PORTER 1996: 124) for realisating of competitive strategies in the sense of a “strategic fit” between resources and competitive strategy. However, according to PORTER, resources are only the necessary, enabling condition of competitive strategies, not the sufficient one (PORTER 1996: 129).

sectional analyses of market and competitive framework conditions (PRAHALAD/HAMEL 1990: 81; BURR 2004: 114; WELGE/AL-LAHAM 2008: 87).

To date, the resource-based theory has gained remarkable importance and can probably be considered the "most influential framework" (BARNEY ET AL. 2001: 625) or the "dominating approach" (MOLDASCHL/FISCHER 2004: 122; also FOSS 1997a: 5, 1997b: 7) in strategic management research. Until today, a vast amount of scientific literature dealing with the resource-based theory has been published, especially the 1990s can be characterised as a "wildfire of interest in resources, capabilities, and competences" (FOSS 1997a: 5).[146] Moreover, an overview of key empirical tests of resource-based theory within strategic management and other management disciplines, which would far exceed the scope of this discussion, provides a comprehensive body of empirical evidence of resource-based theory.[147]

The following paragraphs outline on this approach's conceptual development from an earlier resource-based "view" to a later resource-based "theory" and discuss some of its core concepts to explain the competitive advantage of firms. Furthermore, a brief overview in the theoretical antecedents of resource-based theory can be found in the annex (A.1).

---

146 See for example: rumelt (1991, 1994), Wernerfelt (1995), Wernerfelt/Montgomery (1986), Hansen/Wernerfelt (1989), Collis/Montgomery (1995, 1998), Barney (1986, 1991, 1997), Dierickx/Cool (1989), Mahoney/Pandian (1992), Peteraf (1993), Prahalad/Hamel (1990), Campbell et al. (1995), Amit/Shoemaker (1993); Mahoney (1993), Connor (1991), rumelt (1991), Connor/Prahalad (1996), Kogut/Zander (1996), (Grant 1996), Liebeskind (1996), Sanchez et al. (1996), Sanchez (1997), Sanchez/heene (1996), Foss (1997a, 1997b). For a selection of German contributions, see: Nolte (1998), Krüger/Homp (1997), Thiele (1997), Bamberger/Wrona (1996a, 1996b), Rasche (1994); Hammann/Freiling (2000), Freiling (2001).

For a comprehensive overview on the various thematic fields in which resource-based research work in the 1990s was undertaken, see FREILING (2001: 9f).

147 This overview of recent, resource-based empirical research (e.g. in the fields of the relationship between resources and firm performance, corporate strategy, international strategy, firm alliances) as well as resource-based research in other management disciplines (e.g. human resources, marketing, entrepreneurship, management information systems, operations management, technology innovation management) can be found in chapter #11 of BARNEY/CLARK (2007: 221ff.).

3.3.3.1 From resource-based view to resource-based theory

Having mainly focussed on external, respectively market-driven influence factors of competitive advantage in the 1980's, the research approach of the resource-based theory of the firm drew attention to firm-internal conditions and resources as the explanatory variable for the competitive and economic success of enterprises. To date, it gained remarkable relevance and can probably be considered the "most influential framework" (BARNEY ET AL. 2001: 625) or the "dominating approach" (MOLDASCHL/FISCHER 2004: 122) in strategic management research.

As BARNEY and CLARK (2007: 14) illustrate, the first resource-based theory publication in the field of strategic management was written by WERNERFELT (1984) and attempted to form a theory of competitive advantage based on the resources a firm develops or acquires to implement its product market strategy. Thus WERNERFELT'S contribution conceptualised a complementary or alternative perspective to PORTER'S approach (1980, 1985) of competitive advantage based on the firm's market position.[148] WERNERFELT'S approach suggests that the bundle of product market positions that a firm possesses is reflected in the bundle of resources it controls. Thus, WERNERFELT interpreted PORTER'S competition on the best product market position just in terms of the competition among resource positions held by firms. By specifying a resource profile for a firm, he argued, it is possible to find the optimal product market activities (WERNERFELT 1984: 171). For every concept that enables a firm's product market position to be assessed (e.g. barriers to entry), a complementary concept should exist that enables the level of resources controlled by a firm to be analysed (e.g. barriers to resource imitation). WERNERFELT'S main achievement was to recognise that the resource profile of firms and the competition for resources among firms have important implications for the ability to occupy competitive product market positions. "In this way, WERNERFELT anticipated some of the critical elements of resource-based theory as it developed in the 1990s" (BARNEY/CLARK 2007: 15).

---

148 "This is why Wernerfelt called his ideas the resource-based 'view' - since he was simply viewing the same competitive problem described by Porter from the perspective of the resources a firm controls." (BARNEY/CLARK 2007: 14).

"For the firm, resources and products are two sides of the same coin." (WERNERFELT 1984: 171)

While WERNERFELT tried to develop a theory of firm performance differences that result from the resources controlled by a firm, the second important contribution to resource-based theory, published by RUMELT (1984), described a strategic theory of the firm that looked at the ability of firms to generate economic rents more efficiently than other forms of economic organisation, including markets. Instead of primarily existing to reduce transaction costs, as suggested by the transaction cost theory (WILLIAMSON 1975, 1979), RUMELT argued that firms represent the most efficient form of organisation to generate and appropriate economic rents. This anticipated linkage of rent generation, transaction cost, and governance were later realised in the contributions of CONNER and PRAHALAD (1996), LIEBESKIND (1996), GRANT (1996), KOGUT and ZANDER (1996), and especially TEECE (1980). By defining firms as a bundle of productive resources (RUMELT 1984: 557-558) and suggesting that the economic value of these resources will vary, depending on the context in which they are applied, and that the imitability of these resources depends on the extent to which they are protected by "isolating mechanisms" (IBID.: 567), RUMELT developed many ideas that can be discovered in the resource-based theory.

Finally, the third resource-based article in the history of strategic management was written by BARNEY (1986a). Like his predecessors, he was also convinced about the idea that persistent superior firm performance can be explained by the characteristics of the resources a firm controls. However, the work of BARNEY went beyond the previous approaches, as he argued that a resource-based theory can have very different implications from theories of competitive advantage which are based on product market positions (BARNEY 1986a: 1239f.). By introducing the concept of strategic factor markets, where firms acquire the resources they need to implement their product market strategies, BARNEY shows that if strategic factor markets are perfectly competitive, the acquisition of resources in those markets will anticipate the performance those resources will create when used to implement product market strategies. But the fact that strategic factor markets can be perfectly competitive implies that theories of imperfect product market competition (like PORTER'S approach) are not sufficient for the development of a theory that explains persistent performance differences between firms (BARNEY/CLARK 2007: 16f.). At its extreme, BARNEY'S argument suggests that it is not possible for firms to attain economic rents if strategic factor markets are always perfectly competitive.

Apart from the fact that firms can accidentally be lucky, the reason that factor markets are not always perfectly competitive stems from the fact that particular firms have unusual, specialised insights about the future value of resources they are acquiring or developing in a strategic factor market (BARNEY 1986a).

BARNEY concludes that the resources a firm already controls are more likely to be sources of economic rents for firms than resources that it acquires from external sources (BARNEY 1986a: 1239). BARNEY cites the example of a low-cost manufacturing strategy. He suggests that firms without any special skills in low-cost manufacturing can only expect normal returns from imitating the low-cost manufacturing strategies of other firms. Other firms with cultural or other advantages or skills in low-cost manufacturing can exploit them to obtain above normal returns from implementing a low-cost strategy (BARNEY 1986a: 1239f.). Put differently, the choice of superior product market strategies has to fit the set of resources a firm can acquire or develop in its factor markets. Only firms that manage to *create* such imperfect product market positions will enjoy above normal returns. In that sense, BARNEY has shifted the focus from the development of a merely complementary resource-based "view" of market-based approaches in strategic management towards an independent resource-based "theory" which is opposed to the ideas of PORTER.

In 1989, DIERICKX and COOL extended BARNEY'S argumentation by introducing attributes of a firm's resources that make them less likely to be subject to the strategic factor market competition, for instance resources that are subject to time compression diseconomies, that are causally ambiguous, that are characterised by interconnected asset stocks, or that are characterised by asset mass efficiencies.

Subsequent to these seminal contributions, other contributions followed and further advanced and consolidated the resource-based approach. To mention just a few, in his further contributions, BARNEY extended his resource-based logic of explanation to organisational culture (1986b), to mergers and acquisitions (1988), and refined the analysis of the potential of several firm resources to generate sustained competitive advantage (1991a) and their characteristics (1991b). Further advances in resource-based theory building were achieved, for example, by GRANT'S (1991) "resource-based theory" , HUNT'S (2000) "resource advantage theory", the more industrial economics related "resource dependence approach" of MAHONEY (1995), the analysis of the relationship between resource-based theory and other microeconomic schools of thought by CONNOR (1991) respectively theories of competitive advantage by MAHONEY (1993), or the paper of CASTANIAS and HELFAT (1991) that showed how the creation and appropriation of economic rents aligns the interests of managers and equity holders. Moreover, in her seminal paper, PETERAF (1993) thoroughly grounded resource-based argumentation in microeconomics (see chapter A.1 in the annex).

More empirical work was done by RUMELT (1991) who provided empirical evidence that firm level effects explained more variance in firm performance than either corporate or industry level effects. Likewise, HANSEN and WERNERFELT (1989) demonstrated that the characteristics of a firm's organisational culture had a more significant impact on its economic performance than the attributes of the industry within which it operated. These results are consistently in line with the resource-based argumentation and contradicted earlier findings that industry effects were a more important determinant of firm performance than firm effects.

As the resource-based theory was developing, it did not last long until researchers from other research traditions were also developing theories of competitive advantage that followed the argumentation of the resource-based perspective. According to BARNEY and CLARK (2007: 19), two of the most important of these parallel streams were the theory of invisible assets by ITAMI (1987) and the work of PRAHALAD and BETTIS (1986), respectively PRAHALAD and HAMEL (1990), on a competence-based theory of corporate diversification. Besides numerous similarities, these approaches differ significantly in their analytical accent by arguing that a firm needs distinct competencies to effectively generate competitive advantage on the basis of their available resources. Moreover, dynamics and changes in the external environment and the ability of firms to adapt to them increasingly became the focus of the resource-based perspective, and led to the development of the strand of "dynamic capabilities" mainly founded by TEECE ET AL. (1997).

Other, more recent progress in resource-based approaches single out specific types of resources which are assumed to be of particular importance for firms' economic success. Starting from a differentiation between tangible resources (e.g. personnel, technology, and financial resources) and rather intangible, knowledge-based resources (DEMSETZ 1988; KOGUT/ZANDER 1992; NONAKA 1994; NONAKA/TAKEUCHI 1995; CONNER/PRAHALAD 1996; TEECE ET AL. 1997; GALUNIC/RODAN 1998; NOOTEBOOM 1992; GRANT 1996a; SPENDER 1996a, 1996b) the knowledge-based approach emerged. It argues that the firm-specific stock of knowledge as well as the firm's ability to modify and adapt this stock of knowledge through processes of learning represent the central influencing factor of firms' heterogeneity and differences in economic performance.[149]

Another strand of resource-based theory focuses instead on the ability of firms to participate in organisational networks and alliances by having the "right" structural and social status in their network of organisational alliances. This strand of the "relational view" (DYER/SINGH 1998) encompasses the "competence to cooperate" (DYER 1996; SIMONIN 1999; DYER/NOBEOKA 2000) in inter-firm cooperations as well as the stock of "social capital" as a determining resource for a persistently superior economic performance of firms (BURT 2003; KOKA/PRESCOTT 2001; TSAI 2000) and argues that competitive advantage derives to a large part from strategic alliances between firms.

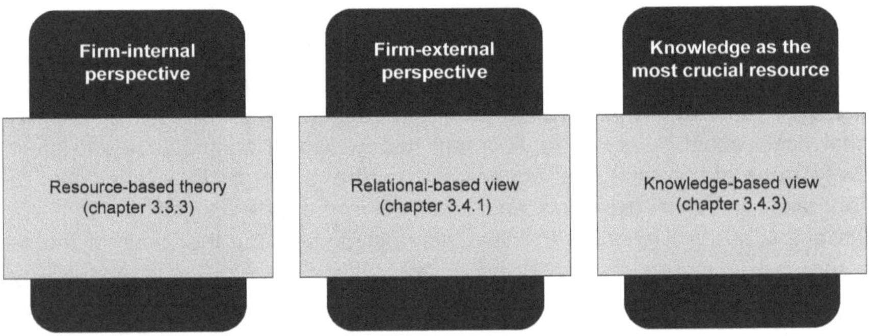

Figure 6: Overview of resource-based theory and its modifications of the relational- and knowledge-based view
(Source: own illustration)

In order to identify the scope of possible innovation sources, the following chapters will describe the original resource-based theory as well as its two recent modifications of the relational- and knowledge-based perspective in detail (Figure 6) and discuss how each of them may contribute to explaining sources of non-R&D-performing firms' competitive advantage.

149 The idea that knowledge nowadays constitutes the most crucial asset, and hence the ability to develop and employ knowledge is the most crucial organisational capability, was very recently taken up and pursued in depth by the two commendable books by FORAY ("The Economics of Knowledge" (2006)) and NOOTEBOOM ("A Cognitive Theory of the Firm" (2009).

3.3.3.2 Main elements of the resource-based theory of the firm

According to BARNEY (1991: 105; also: FOSS 1997a: 7; WELGE/AL-LAHAM 2008: 88; WITT 2008: 48f.; WOLF 2008: 571), two core assumptions underly the argumentation of the resource-based theory:

- all resource-based approaches rely on the existence of systematic, empirically observable differences in the resource configuration of firms
- due to the restricted mobility, transferability and shortage of these resources in the market, the heterogeneity in the firm-specific sets of resources is relatively stable.[150]

Taking these two core assumptions as his starting point, BARNEY (1991: 105f.) explains that if a resource can be expected to contribute to a firm's sustainable competitive advantage, it must have four attributes (Figure 7):

BARNEY defines resources as being valuable when they help to grasp an opportunity in the firm's environment or when they help to cancel out external threats or at least shield the firm against these threats. Thus, valuable means, that resource contains a relevant strategic value for the firm. Considering the rareness of resources, BARNEY seems to have a simple counting sense in mind (FOSS/KNUDSEN 2003: 293): if a million firms control a certain resource, it is not likely to be rare (even if a billion firms badly need the relevant resource).

From an economic perspective, rare resources are always valuable. Thus, it can also be argued from the strategic perspective that a rare resource should be considered as a valuable resource. A firm that controls valuable and rare resources obtains competitive advantage. The condition of non-inimitability pays attention to the circumstance whether a competitor can acquire or accumulate resources with similar attributes and levels of attributes to the desired resource. On the contrary, the condition of non-substitutability directs attention to the fact of whether a competitive firm can assess different resources that will allow them to implement identical strategies because they cause similar effects to the originally desired resource.

---

150 This assumption is inevitably necessary against the background of strategic management. If the differences in firms' resources are not considered to be stable, they would not provide a solid, reliable basis for developing and formulating business strategies. "Since strategists are primarily concerned with rents over a longer term, the condition of heterogeneity must be relatively durable to add value." (PETERAF 1993: 182)

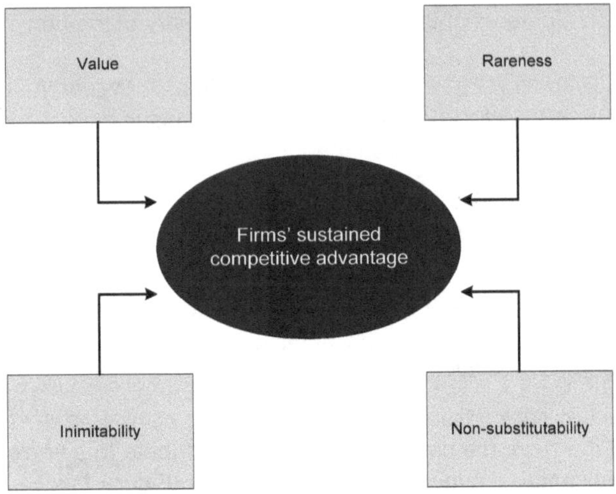

Figure 7: Necessary attributes of resources to sustain a firm's competitive advantage (Source: own illustration, adapted from FOSS/KNUDSEN 2003: 294).

PETERAF'S (1993) analysis of the conditions for sustainable competitive advantage is different from BARNEY'S (1991). She more explicitly draws on price theory (economics of rents) and takes individual resources as the relevant level of analysis (not strategies). Starting from the same two core assumptions as BARNEY (1991), she adds the conditions of "ex post limits to competition" and "ex ante limits to competition" (PETERAF 1993: 182ff.), and thereby derives the "cornerstones of competitive advantage"[151] - all of which conditions must be met to enable firms to achieve persistent competitive advantage through the protection, control and utilisation of their internal resources (Figure 8).

Heterogeneity describes the condition that each firm is endowed with a unique bundle of internal resources, which was historically developed through evolutionary processes (WOLF 2008: 571). These firm-specific pools of resources

151 In 1993, this was the title of a paper published by MARGARET A. PETERAF in the Strategic Management Journal and represents "the perhaps most systematic exposition of a resource-based perspective on the conditions for sustained competitive advantage" (FOSS 1997a: 8). Unfortunately, PETERAF is not entirely clear about whether these conditions constitute the minimum set of jointly necessary conditions, or whether they are individually necessary conditions, or whether they are merely collectively sufficient for a firm's sustainable competitive advantage (FOSS/KNUDSEN 2003: 295).

constitute a broad range of resource heterogeneity which distinguishes each firm from its competitors (WERNERFELT 1984; BARNEY 1991). Based on this axiom, the resource-based theory seeks to explain the heterogeneity in firms' economic performance and attainment of competitive advantage (MAHONEY/PANDIAN 1992: 365; PETERAF 1993: 186; BURR 2004: 114f.). Based on the role of SCHUMPETER'S entrepreneur, RUMELT (1984: 566) mentions technological change, changes in consumer demands or the change of governmental regulation as possible sources of profits realised by resource heterogeneity. The uncertainty that firms (respectively, theirs entrepreneurs) face with future developments outside the firm, unpredictable future effects of different resources and resource combinations involved, as well as the existence of exclusive resources (like patents, firm reputation or exclusive information) represents the "isolating mechanisms" by which heterogeneity in resources is created.

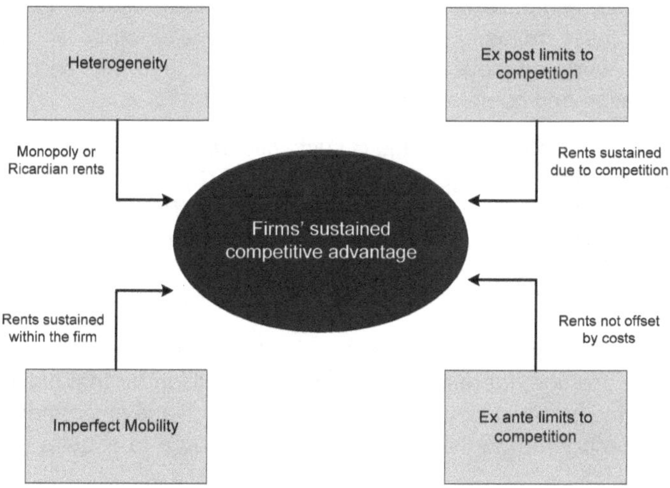

Figure 8: Cornerstones of competitive advantage
(Source: own illustration, originally referring to Peteraf 1993: 186, modifycations adapted from Witt 2008: 50; Al-Laham 2003: 115).

The type of competitive advantage connected to this aspect can be described by the characteristics of RICARDIAN or monopoly rents ("BAIN rents").[152] These rents result either from the control of scarce resources (e.g. patents, locational advantage) or from the possession of firm-specific, unique, and heterogeneous bundles of resources (WELGE/AL-LAHAM 2008: 89; AL-LAHAM 2003: 122f.).

The second condition of *imperfect mobility* refers to the assumption that firm-internal resources responsible for competitive advantage can neither be traded, nor transferred easily between firms (DIERICKX/COOL 1989: 1505f.), or lose their value for competitiveness when deployed outside the firm context of purpose (PETERAF 1993: 183; RASCHE 1994: 82ff.)[153], for example highly specialised production machinery, highly efficient work teams or innovation culture. Such resources cannot be acquired in factor markets and have to be developed within the firm over time. Competitive advantage is thus created by the circumstance that these resources are not accessible or imitable for other competitors because of their high context dependency. Another case of such non-tradable or immobile resources are "cospecialised assets" (TEECE 1986: 289), which must be used in conjunction with one another or which have higher value when employed together, due to their high degree of organisational integration and complementarity within the firm.

In this case, competitive advantage is sustained within the firm and can be described by so-called "quasi-rents" (KLEIN ET AL. 1978) which are also known as "PARETO- or MARSHILIAN rents". They describe the difference between the rent which is achieved if the resource is used in the most effective and efficient way and the rent which is attained with applying only the second-best solution (WELGE/AL-LAHAM 2008: 89f.).

Ex ante limits to competition mean that, prior to any firm's establishing a superior resource position, there must be limited competition for that position. A superior resource position could only be a source of above normal economic success if some firm had the foresight or good fortune to acquire it in the ab-

152 The concept of BAIN rents is usually connected with rents from a monopolistic position in product markets due to barriers of market entry like economies of scale and product differentiation (BAIN 1959). In contrast, the resource-based theory explains monopoly rents by the existence of firm-specific, unique, and heterogeneous bundles of resources instead of superior products.

153 Such resources can be defined as "idiosyncratic" resources as their effect relies on strong interdependencies with other firm resources (WELGE/AL-LAHAM 2008: 93; WOLF 2008: 573).

sence of competition (PETERAF 1993: 185). As BARNEY (1986: 1233ff.) points out in his concept of "strategic factor markets", all kind of resources are principally tradable in factor markets, even immobile and imperfectly mobile resources, as both DIERICKX and COOL (1989) and BARNEY (1989) noted. However, assuming perfect competition in these factor markets, the price of a resource would correspond to the expected profits that result from its usage and, subsequently, firms would not be able to gain above normal rents. Thus, the economic performance of firms depends not only on the returns from their strategies, but also on the cost of implementing those strategies. Only if the difference between the ex post value of a necessary resource is greater than the ex ante cost of acquisitioning, would the firm attain economic profit. From this it follows that only imperfect competition in factor markets caused, for instance, by information asymmetries between sellers and buyers, can explain the strategic relevance of resources for individual firm performance (DIERICKX/COOL 1989: 1505ff.; also: RUMELT 1987).

Competitive advantage of firms emerge if the buyer of a certain resource has superior information about its actual value, respectively the value potential, which he can exploit by using this resource more efficiently and effectively than his competitors (WELGE/AL-LAHAM 2008: 93f.). As RASCHE (1994: 37f.) adds, due to their intransparency and complexity, the value potentials which resources are able to unfold when deployed by a firm are very difficult to assess by competitors ex ante. Thus competitive advantage, respectively rents, result from the differences between the acquisition cost of resources and their economic value.

Finally, the fourth condition of *ex post limits to competition* mentioned by PETERAF deals with probably the most crucial issue of resource-based theory of how heterogeneity in resources can be sustained and contributes to persistent superior performance of firms (WELGE/AL-LAHAM 2008: 94f.). To explain superior competitive advantage and the economic success of firms, heterogeneity must be relatively durable. But, according to PETERAF (1993: 182f.), this will only be the case if forces exist which limit competition for those rents achieved by superior firm performance. Otherwise, market competition will dissipate those rents by increasing the supply of scarce resources. BARNEY (1991) and DIERICKX/COOL (1989) focused on two factors which limit ex post competition: imperfect imitability and imperfect substitutability. The latter one thereby refers to one of PORTER'S "five forces" (1980).[154] Resources relevant for competitive advantage may be substituted by competitors, either in terms of deploying similar resources with similar performance effects, or in terms of completely different resources which constitute an alternative strategy for realisating of the intended economic effects. Technology based resources in particular are continuously threatened by substitution due to technological change (e.g. the substitution of manual lathes by computerised numerically controlled (CNC) lathes).

By coining the term "isolating mechanisms", RUMELT (1984) pointed to other phenomena which protect firms from resource imitation and thus preserve their superior economic performance. These isolating mechanisms include property rights to scarce resources and other aspects like time lags, information asymmetries, and frictions which impede imitative competition (RUMELT 1984: 567ff.). For example, "causal ambiguities" (LIPPMAN/RUMELT 1982: 420) about the cause and effect chains of efficiency differences among firms prevent possible imitators from knowing exactly what or how to imitate.[155] In many cases, the contribution of single resources to the economic performance is not ex post separable. DIERICKX and COOL (1989: 1507) emphasise the role of dynamic developments of the firms' resource bundles which are, in the sense of a historical path-dependency[156], unique to each firm and can thus not be imitated. The same applies to firms' individual stocks of knowledge which some-

154 But in contrast to PORTER, the resource-based theory understands substitutability not as substitution of products but with respect to the substitution of (scarce) resources.

155 BARNEY (1991: 108), however, emphasises that both the competitors as well the firm that controls the resource, are necessarily confronted with this causal ambiguity. Otherwise, if internal causalities are unambiguous to the members of a firm, they could be transferred or betrayed to other firms for example by the move of employees to other firms. REED/DE FILIPPI (1990: 100f.) point to the dilemma that, even if the management of a firm is confronted with this causal ambiguity about the effects of resources, this would imply that the firm itself is not always able to fully exploit its resources in the most effective and efficient way. In other words, the central dependent variable "attainment of persistent competitive advantage" would solely rely on random, unpredictable, and uncontrollable resource combinations (PETERAF 1993: 187; SCHREYÖGG 1999: 402). This aspect was investigated empirically by KING/ZEITHAML (2001). They analysed the relationship between "causal ambiguity" and "resource imitability" and found out that, on the one hand, causal ambiguity of resources leads to increased firm performance. But, on the other hand, causal ambiguity can also simultaneously have negative effects on firm performance if managers are no longer able to figure out the complexity of resource effects in their firm (KING/ZEITHAML 2001: 94f.).

156 For the role of such path dependencies for the future development of enterprises, see, for example, the contributions of RASCHE (1994: 71) and DOSI ET AL. (1998: 20).

times were accumulated over decades (RASCHE 1994: 70). To summarise, the strategic value of resources can also be assessed by the probability with which they can be imitated by competitors. The higher the degree of non-substitutability and inimitability, the higher too is the strategic potential and value of a certain resource.[157]

Rents resulting from such ex post limits to competition can be described as "Schumpeter rents" (SCHUMPETER 2008b: 87ff.), as they persist until imitators manage to catch up. While SCHUMPETER, however, regarded such rents to be of a temporary nature in the dynamic processes of competition, the resource-based approach emphasis the stabilisation of such first-mover advantages, for example by pointing to the concept of the isolating mechanisms (WELGE/AL-LAHAM 2008:89).

Considering the basic conditions under which a firm's resources can develop their positive effects on the firm's competitiveness and economic success, it becomes clear that the term "resources" used in resource-based theory goes far beyond the notion of resources (= production factors of labour, capital and land) in traditional economics. To take a closer look at the various concepts which are linked with the term "resources", the following sub-chapter gives a brief overview of different dimensions, definitions and understandings of firm-specific resources.

3.3.3.3 Terms and definitions of the resource-based theory to describe factors of production controlled by firms

Against the background of its theoretical origins and historical development briefly described above, it is clearly recognisable that the resource-based theory does not form a homogeneous theory of strategic management. The different theoretical concepts which have gathered under the umbrella of the resource-based approach results in "ongoing confusion" (BARNEY/CLARK 2007: 22), "vagueness" (MOLDASCHL/FISCHER 2004: 127), a "tangle" (MOLDASCHL 2006: 9) or a "terminological soup" (FOSS 1997a: 11) about the terms used to describe factors of production controlled by a firm.

Various authors tried to develop typologies of tangible and intangible assets to suggest that different types of assets can have different competitive effects for

157 For a tabular overview of the variety of isolating mechanisms within resource-based theory, see FREILING (2001: 105).

firms. One of the most comprehensive definitions is proposed by BARNEY (1991: 101) who defines resources as "(...) all assets, capabilities, organisational processes, firm attributes, information, knowledge etc. controlled by a firm that enable the firm to conceive of and implement strategies that improve its efficiency and effectiveness". It should be remarked, that this definition stresses three important points: firstly, it emphasises the interdependency of resources and strategies. Secondly, it finally determines that a positive impact of firm performance defines what is a resource or not.[158] And thirdly, the wording "all assets (...) controlled by a firm (...) includes the possibility that a resource can also be assessed and controlled by a firm through collaborations and strategic alliances with external partners.[159]

Other authors like WERNERFELT (1984: 172) speak of resources in terms of "(...) anything which could be thought of as a strength or weakness of a given firm", or define resources as "(...) stocks of available factors that are owned or controlled by a firm" (AMIT/SHOEMAKER (1993: 35). MONTGOMERY (1995: 257) remains even vaguer when suggesting that resources are "(...) something that can be used for support or help; an available supply that can be drawn on when needed". Last but not least, the definition of GRANT (1991: 118f.), that resources "(...) are inputs into the production process (...)", remains close to the terminology of traditional economics.[160]

Depending on the particular analytical focus, "resources" may cover a wide range starting from physical, tangible resources up to abstract and intangible meta-resources like dynamic capabilities. Thus, in favour of a common basis of understanding, it is necessary to pay attention to some terminological definitions. I shall to proceed from the more general level of ubiquitous input factors to the particular concept of dynamic capabilities by loosely following the analytical differentiation used by TEECE ET AL. (1997: 516). Based upon the observations of these authors, (1) input factors, (2) resources, (3) organisational

---

158 This second aspect causes some serious problems for the explanatory value of the resource-based theory and is discussed later in the paragraph on shortcomings and criticism on the resource-based approach (see chapter 3.3.4).

159 This aspect will be later discussed in the chapter dealing with the "relational-based view" (see chapter 3.4.1)

160 Such a definition of resources has frequently claimed that resources (in a resource-based sense) are similar to the "input factors" or "factors of production" of the traditional firms' production function, a circumstance which should be absolutely avoided, because of the different qualitative dimension of "resources" in the understanding of the resource-based theory (FREILING 2001: 12f.).

routines and (core) competencies, and (4) meta-competencies as dynamic capabilities are distinguished and described.

Factors of production

TEECE ET AL. (1997: 516) define factors of production as "(...) 'undifferentiated' inputs available in disaggregate form in factor markets." As they point out, "undifferentiated" means that these inputs lack a firm-specific component, like unskilled labour or financial capital for instance. GERSCH ET AL. (2005: 44f.) as well as FREILING (2004: 30) consider input goods as homogeneous, basically marketable, internal or external factors which form the starting point for further activities of refinement or utilisation within value-added processes. WELGE and AL-LAHAM (2008: 88) subsume these factors of production under the term of "fundamental financial, physical resources". Examples are the geographical location, physical capital resources, access to raw materials, buildings, physical technology, or the machinery and plant equipment of a firm (BARNEY/CLARK 2007: 24).

Those entities characterised as "input factors" or "factors of production" are mainly synonymous with the "assets" in the sense of DIERICKX and COOL (1989: 1505f.). They argue that, due to their free availability in the factor markets as well as their tradable character, such input factors cannot be considered as native sources of competitive advantage since they are usually freely accessible to each firm.

Resources

The notion of resources represents the next level of complexity as they can be understood as the "result of successful asset refinement processes" (FREILING 2004: 30; GERSCH ET AL. 2005: 45f.) of the above described input factors, which produce sustainable heterogeneity of the owning firm in competition and enable the firm to withstand competitive forces (FREILING 2004: 30). Refined processes encompass all qualitative modifications of input factors in the context of current or future market demands and are ideally aligned with corresponding strategic goals (GERSCH ET AL. 2005: 46).

As TEECE ET AL. (1997: 516) point out, resources need to be considered as "(...) firm-specific assets that are difficult if not impossible to imitate" - like trade secrets, certain specialised product facilities or engineering experience. Such assets are difficult to transfer among enterprises because of their transaction and transfer costs, and because they may contain tacit and implicit knowledge. Resources are thus characterised by a certain level of firm-specificity and -individuality (HELFAT/PETERAF 2003: 997; GERSCH ET AL. 2005: 45f.). Thus specificity and individuality work as an "isolating element" (RUMELT 1984) to secure the resources' unique character and to prevent imitation or substitution by competitors, and finally, to allow the sustainable generation of competitive advantage in terms of quasi-rents (PETERAF 1993: 183; GERSCH ET AL. 2005: 46).

Resources themselves can be further differentiated and systematised into material or immaterial resources (COLLIS/MONTGOMERY 1995: 123; MOLDASCHL/FISCHER 2004: 138f.), limited, regenerative, and generative resources (MOLDASCHL/FISCHER 2004: 138f.), resources which are either processed under the participation of external partners of the firm ("integrative") or which are refined autonomously by the firm ("autonomous") (GERSCH 1998: 139ff.). According to KNAESE (1996: 17), immaterial resources could be furthermore divided into resources that are bound to persons (knowledge, competencies) and those that are independent of single individuals (e.g. patents, treaties). Another differentiation was made by WERNERFELT (1984: 172ff.) who distinguishes between "fixed long run capacity (fixed assets)", "practically unlimited capacity (blueprints)", and "limited short-run, but unlimited long-run capacity (cultures)".

As can be seen from this short overview, the term “resources” deployed in literature encompasses both a broad qualitative range from "high-ordered" material resources to the intangible and invisible competencies of the firm. To keep the terminology as clear as possible, I shall adopt the argumentation of GERSCH ET AL. (1995: 47f.) who understand resources as the basic rudiments of further business processes. The mere existence of resources in the sense of a refined input factor at a given point in time is not sufficient to gain competitive advantage. Instead, the firm needs the relevant competences and capabilities to exploit these resources in order to attain economic success.

Routines and competences

At this point, the concepts of "routines" and "competences" come into play. To generate a strategic advantage, firms need the ability to deploy, combine and coordinate their input factors and especially their resources more effectively and efficiently than their competitors (GRANT 1991: 116ff.; FREILING 2001: 22f.) This view is also supported by AMIT and SHOEMAKER (1993: 35) who define competences as a firm's organisational ability to make use of their resources to achieve their intended strategic goals. Similarly, AAKER (1989: 93f.) distinguishes between "assets" and "skills". While assets describe the ownership of resources, skills refer to the competence to exploit them. Sometimes, as in HILL and JONES (1992), HITT ET AL. (1997), and MAKADOK (2001), the literature also distinguishes between resources and capabilities. The argumentation remains the same: resources represent a firm's fundamental financial, physical, individual, and organisational capital attributes, while capabilities are those attributes of a firm that enable it to exploit its resources in implementing strategies or to improve the productivity of other resources (BARNEY/CLARK 2007: 22f.). According to FREILING ET AL. (2006: 57), competences encompass both the individual competences bounded to single individuals and competences at the collective level of organisations or networks of organisations. They propose that the term “competences” should be used at the organisational level, while the individual competences of employees, managers or organisation members in general should be referred to as "capabilities".[161] The differentiation between resources and competences is often regarded as one of the most important achievements within the evolution of the resource-based theory and has closed a gap in the explanation of persistent performance differences between firms (FREILING 2001: 23).

The so-called "competence-based view" of strategic management is considerably younger than the resource-based approach. Nevertheless, its origins can be traced back to SELZNICK'S (1957) notion of "distinctive competences" and the early discussion of SNOW and HREBINIAK (1980), respectively HREBINIAK and SNOW (1980). But the breakthrough of the competence-based view in strategic management was made in the seminal paper of PRAHALAD and HAMEL (1990) who coined the term "core competences" to describe those internal factors of a firm that decide on its competitiveness and economic performance (PRAHALAD/HAMEL 1990: 81f.).[162] Following the competence-based view, the sustainable competitiveness of firms originates in their ability to form, strengthen and expand their core competences to develop and produce new products more successfully. As PRAHALAD and HAMEL (1990: 82) define, "core competences are the collective learning in the organisation, especially how to coordinate diverse production skills and integrate multiple streams of tech-

161 These "capabilities" in the sense of individual competences should not be confused with the concept of "dynamic capabilities" which focuses on the organisation's or firm's ability to learn and evolve over time (TEECE ET AL. 1997).

162 For a historical review of the competence perspective, see for example KNUDSEN (1996).

nologies". As example, they adduce Sony's capacity to miniaturise, Canon's know-how in optical technologies, or Philip's's optical-media expertise. The firm is depicted as a widespread tree whose nourishing roots represent its core competences which enable to firm to successfully develop and implement new products.[163]

But core competences are not only about harmonising streams of technology. They are also about organisating work and delivering value, about communication, involvement, and the commitment to working across organisational boundaries (PRAHALAD/HAMEL 1990: 82). Thus, core competences are to be considered as immaterial resources which are constituted by different skills of individuals in- or outside the firm and are enriched within the organisation over time through its "organisational memory". They are characterised by their high value, high specificity, their rareness, their difficulty of imitation and their non-substitutability, their wide variety of potential application, their incremental nature, as well as the fact that they will get lost if they are not used. But most importantly, "unlike physical assets, competences do not deteriorate as they are applied and shared. They grow." (PRAHALAD/HAMEL 1990: 82).

Today, the competence-based view represents an autonomous strand of research within the research-based theory. While the resource-based approach identifies ex post those resources which account for economic performance, the competence-oriented perspective tries a priori to understand which resources (competences) are necessary for future competitive advantage. The competence-based view thus takes a far more process-oriented perspective than the original resource-based approach.

This process-oriented perspective of a firm's competence is even more strongly expressed in the concept of "organisational routines". TEECE ET AL. (1997: 516) define competences as organisational routines and describe them as follows: "When firm-specific assets are assembled in integrated clusters spanning individuals and groups so that they enable distinctive activities to be performed, these activities constitute organisational routines and processes." They "(...) are typically viable across multiple product lines, and may extend outside the firm to embrace alliance partners." By using the term of "routines", TEECE ET AL. refer their definition of competences to the concept of "organisa-

163 In more detail, HAMEL/PRAHALAD (1990: 81ff.) distinguish between "core competences", "core products" and "end products". Core products represent the physical or material manifestation of core competences and are located between core competences and end products at an intermediate level.

tional routines" of NELSON and WINTER (1982: 98) which is central to their evolutionary theory of the firm. According to them, a routine encompasses various knowledge-based abilities which are fixed in the firm's organisational guidelines or in the employees' minds, such as repetitive patterns of action as well as individual skills, or the effectiveness of an organisational or individual performance. Organisational routines describe "... [the – *author's note*] way we do things around here" (TIDD/BESSANT 2009: 70) in answer to the same generic questions (e.g. how to manage quality, how to manage innovation) every enterprise is recurrently confronted with. Thus, routines are supposed to have considerable impact on firms' performance and economic success, as they ensure that available resources are being successfully transformed into desired strategic or innovative outcomes

Routines represent firm-specific, repetitious patterns of action and therefore are supposed to have a considerable impact on the innovation ability and economic success of an enterprise (NELSON/WINTER 1982; NONAKA 1991; TIDD ET AL. 2009; BECKER/ZIRPOLI 2006), as they ensure that available resources are being successfully transformed into the desired innovative outcomes.

Despite the fact that there is still little consensus concerning the nature of organisational routines, most authors agree that nothing can be a routine, and thus a competence, without occurring repeatedly (BECKER 2005: 249f., 255). The second important qualification of routines and competences lies in the notion of "recurring patterns" or "repetitive patterns" (NELSON/WINTER 1982: 98; GRANT 1996a: 115; TEECE/PISANO 1994: 541; TEECE ET AL. 1997: 518; DYER/SINGH 1998: 665; SANCHEZ ET AL. 1996: 7; PENTLAND ET AL. 2010: 917), which means that a firm or an organisation must be able to repeat the actions of resource use, combination, and exploitation over time. Such repetitions of action build sequences over time, which in turn form patterns of action. Consequently, BECKER proposes a definition of routines which are understood as "recurrent interaction patterns" (BECKER 2005: 256).[164] Conversely, if they are not used frequently once achieved competences and routines can disappear in the course of time. In this context, RASCHE (1994) speaks of "asset erosion".

Because routines are usually located at the organisational level and describe a collective pattern of action or particularly, collective action potentials (BECKER

164 For a more detailed elaboration of organisational routines, see the contributions of BECKER (2004, 2005). Besides a comprehensive review of the literature on organisational routines (2004), he also discusses measures to clarify terminology as well as fruitful areas of future research (2005).

2004: 644ff.; FREILING ET AL. 2006: 57) and they have to be distinguished from "individual skills" that are located at the individual level.

Routines are not to be confused with "brilliant improvisation" or mechanisms of single problem solving, because there is neither a general purpose routine nor is problem-solving to be considered as highly patterned and repetitious (WINTER 2003: 991). Such activities could be rather seen as intended rational actions which are usually triggered by an external or at least singular event evoking a singular problem-solving reaction. Such activities are not characterised by a recurrent nature, in contrast to the concept of routines.

It becomes apparent that the notion of competencies and routines are often used synonymously. This is a visible expression of the conceptual diffuseness which resource-based approaches are often blamed for. In my view, following MARCH and SIMON (1958), the concept of competencies seems more likely to be linked to specific, outstanding abilities, knowledge stocks which a firm possesses and which have a certain qualitative value. The concept of routines instead rather stresses the aspect of the recurrent and repetitive nature of behavioural patterns (with less regard to their content dimension) by which these competences are generated, accumulated, and exploited.

Authors like STALK ET AL. (1992) and COCKBURN ET AL. (2000) early emphasised the point that competences and routines cannot be assumed to be stable over time. They argue that firms need to adjust and reconfigure their existing competences, or must even be able to develop new competences and routines in reaction to future market changes and discontinuities, if they do not want to lose their competitive position (STALK ET AL. 1992: 62, 66).[165] To maintain their competitive advantage, firms must be able to identify external signals in their environment and adapt their competence portfolio to them (COCKBURN ET AL. 2000: 28ff.). These ideas form the starting point for the conceptual development of "dynamic capabilities".

---

165 They identified five of the competences a) "speed" (of reaction to changes in the market structure, demand structure, or technologies), b) "consistency" (with customer needs), c) "acuity" (to correctly identify external changes), d) "agility" (to be simultaneously active in different competitive environments), and e) "innovativeness" (to generate and (re-)combine new or existing sources of competitive advantage) (STALK ET AL. 1992: 63).

Dynamic capabilities

In this way, the work of STALK ET AL. (1992) and COCKBURN ET AL. (2000) laid the foundation for further research work on the dynamic development and evolution of firm competences and routines. In times of increased dynamics in technology and competition, firms are constantly faced with new challenges. Product and factor markets are becoming increasingly complex and more integrated, new technologies shape new forms and rules of competition. For these reasons, firms are permanently forced to adapt to changes in their competitive environment to take advantage of them. Moreover, firms have to identify ways and possibilities to help shape this change, for example, by developing corresponding technological or non-technological innovations (HELFAT ET AL. 2007: 1).

The concept of dynamic capabilities emerged in the mid 1990s (AMIT/SHOEMAKER 1993; TEECE ET AL. 1997; for one of the few German contributions, see: CANTNER ET AL. 1996) and deals with the central question of how firms can sustain competitive advantage responding to and shaping environmental change, e.g. by using their strategic resources in response to the dynamic and changing market and external environments (TEECE ET AL. 1997: 510; TEECE 2007: 1319ff.; HELFAT ET AL. 2007: 1; HELFAT/PETERAF 2009: 94f.; FRIED 2007: 202;). Within the present framework of the resource-based theory, competitive advantage can flow only at a point in time due to the ownership or control of scarce, relevant and difficult-to-imitate assets and know-how. However, in fast moving business environments open to global competition, characterised by spatial dispersion of sources of innovation and manufacturing, persistent and sustainable competitive advantage and economic success need more than the selective control of certain resources and competences (TEECE 2007: 1319; HELFAT/PETERAF 2003: 998f.; MOLDASCHL/FISCHER 2004: 128; SCHREYÖGG/KLIESCH 2005a: 4).

The concept of dynamic capabilities thus goes beyond the mere resource-based perspective by explaining competitive advantage:

- by a firm's internal mechanisms or routines to adapt its resource basis to external changes
- by a firm's ability to recognise the necessities and opportunities for change early and to choose the appropriate measures for action (TEECE 2000: 3; HELFAT ET AL. 2007: 2).

Further questions deal with the impact of technology and the market environment on firms' ability to adapt their performance potential over time, and why firms which succeed in the same sector show similar, as well as completely different shapes of dynamic capabilities (DE TONI/TONCHIA 2003: 953f.).

According to the authors (TEECE ET AL. 1997; and with a different connotation: LEI ET AL. 1996; EISENHARDT/MARTIN 2000; ZOLLO/WINTER 2002; HELFAT/PETERAF 2003) of the dynamic capabilities approach, the answer to these questions lies in highly firm-specific (heterogeneous) capabilities which enable the firm to continuously (re-)configure and adapt its path-dependable, firm-specific, intangible resources, competences, and know-how to dynamic environments (TEECE 2000: 3). "We define dynamic capabilities as the firm's ability to integrate, build, and reconfigure internal and external competences to address rapidly changing environments (TEECE ET AL. 1997: 516). More recently, HELFAT ET AL. (2007: 4) define them as "(...) the capacity of an organisation to purposefully create, extend, or modify its resource base". Other definitions point to the continuous character ("on-going processes" - MEDCOF 2000: 14), their dependence on the formal and the informal structure of the firm (TEECE 2000: 3), or the fact that dynamic capabilities result from learning processes (ZOLLO/WINTER 2002: 340; WU 2006: 448) and thus have to be distinguished from operational capabilities. Because dynamic capabilities aim at processes of renewal and reconfiguration, they particularly promote the attainment of Schumpeterian rents. They support necessary changes in the resource basis of the firm which favour the development of corresponding innovations (TEECE ET AL. 1997: 527).

The term "dynamic capabilities" represents the two conceptual pillars on which it is based. "Dynamic" refers to the firm's capability to continuously renew and adjust its existing resources and competences corresponding to the dynamic nature of the world outside the firm. Thus, the firm-external environment is designed as a SCHUMPETERIAN world of dynamic innovation-based competition, and the creative destruction of existing competences. As a result of the ongoing renewal, firm resources and competences are subject to some kind of firm-internal processes of evolution, because existing resources and competences are constantly subject to variation, selection and retention. Those resource configurations that best fit the currently prevailing external challenges are positively selected, while dysfunctional resources or competences are either modified or even wiped out. In consequence, authors like TEECE ET AL. (1997: 509f.) closely refer to the evolutionary approaches of SCHUMPETER (2006) and NELSON and WINTER (1982), who were among the first to integrate individual and organisational knowledge and capabilities with the notion of dynamic competitive environments.

The second part of "capabilities" emphasises the key role of strategic management to integrate, configure, modify, and exploit firm-internal and -external resources and competences according to challenges and demands of its environment (HELFAT ET AL. 2007: 121; COLLIS 1994: 144ff.)[166], as well as the organisational processes to achieve these goals (TEECE ET AL. 1997: 524; EISENHARDT/MARTIN 2000: 1105ff.; GALUNIC/EISENHARDT 2001: 1229). TEECE ET AL. (1997: 516) closely stick to a competence-based perspective by differentiating explicitly between factors of production (input factors) which can be acquired in factor markets and firm-specific assets like unique, hard-to-imitate, tacit bundles of resources, routines, or (core) competences. These firm-specific assets are generated through regular, recurring processes that are deeply grounded in the organisation and shaped through evolutionary paths of development (TEECE ET AL. 1997: 518ff.). In their very recent contribution, TEECE ET AL. (2000: 339) dissociate themselves to some extent from this original definition by emphasising that "dynamic capabilities (...) reflect an organisation's ability to achieve new and innovative forms of competitive advantage *despite* [*italics by the author*] path dependencies and core rigidities in the firm's organisational and technical processes". In that sense, dynamic capabilities even represent a mechanism to overcome evolutionary path dependencies and lock-in effects. The changes in resource configurations may thus not only be limited to incremental steps, but also to more basic changes in the resource structure of a firm.[167]

COHEN and LEVINTHAL (1989, 1990) developed one of the most influential conceptualisations of dyanamic capabilities. Their seminal contribution highlights the fact that firms cannot benefit from external knowledge flows simply by being exposed to them. Instead, firms must develop the ability to recognise the value of new, external information, assimilate it and apply it to commercial ends. Considering this definition, it can be stated that the absorptive capacity of a firm can be distinguished by two interdependent dimensions: a) the capability to search and acquire new, external information about technological trends, and b) the capability to adapt internal processes and resource configu-

166 "Dynamic managerial capability refers to the capacity of managers to create, extend or modify the resource base of an organisation" (Helfat Et Al. 2007: 121).

167 As Teece et al. (1997) assumed, firms indeed are always in possession of dynamic capabilities, however, they play a subordinate role for their competitiveness in organisational environments with just low dynamics of change. In contrast to them, Zollo/Winter (2002: 340) assume that dynamic capabilities are always crucial for a firm's competitiveness, even "in environments subject to lower rates of changes".

rations in such a way that their competitive potential is fully exploited (ESCRIBANO ET AL. 2009: 98; also: ZAHRA/GEORGE 2002). Originally, the absorptive capacity of a firm was closely interlinked with the intensity of its R&D activities (COHEN/LEVINTHAL 1989). However, there this R&D-based, firm-internal operationalisation of absorptive capacity was increasingly criticised. Thus, as recent studies (SPITHOVEN ET AL. 2010: 132, MUROVEC/PRODAN 2009: 860ff.) emphasise, absorptive capacity is a multidimensional concept which encompasses both firm-internal (e.g. routines, learning processes, employee skills) and external resources (e.g. alliances, cooperations, or networks with other firms, stakeholders, the scientific community).

Due to their complexity and their implicit, highly firm-specific nature, dynamic capabilities are almost not replicable by other firms. "Some sources of competitive advantage are so complex that the firm itself, let alone its competitors, does not understand them" (TEECE ET AL. 1997: 525). For this reason, they remain inimitable, even if the firms' employees are headhunted by competitors. On the other side, this causes long periods of initial training for new employees of a firm.

This means that the firm's competitiveness depends on its past strategic decisions and limits the possibilities of future options for strategic development. For this reason, they are often conceptualised as routines of practice and learning (TEECE ET AL. 1997: 518; EISENHARDT/MARTIN 2000: 1107). Thus, the ability of an enterprise to permanently reconfigure its resources and competences is an acquired and learnable organisational capability. The more often this capability is needed, the more easily it can be learned and applied (TEECE ET AL. 1997: 521; BURR 2004: 133). Dynamic capabilities are supposed to be based upon explicit and implicit collectively held knowledge which has been obtained by overcoming previous difficulties and challenges. Such experience-based knowledge is assumed to allow faster adaptation to similar situations in the future.[168] Thus, a crucial aspect of the concept of dynamic capabilities is the perception of external and internal changes as well as the self-reflection of those changing processes (HELFAT ET AL. 2007: 2).

Although the dynamic capability approach in its basic assumptions is strongly based on the resource-based perspective, the concept of dynamic capabilities

---

168 This point clarifies the close connection to theories of organisational learning (CYERT/MARCH 1992) when it comes to the micro-foundation of dynamic capabilities (HELFAT ET AL. 2007: 19; also: DIERKES ET AL. 2001; GAVETTI 2005; TEECE 2007).

should be principally differentiated from other types of intangible resources like competences or routines (DOSI ET AL. 2000: 3ff.). First, "ordinary" organisational routines only require limited knowledge and do not leave much room for learning processes through trial and error, as they are highly specialised and static. On the contrary, dynamic capabilities need profound and broad knowledge about technology, market, product, distribution, or leadership. Secondly, dynamic capabilities reflect some degree of intent, even if not fully explicit. Therefore, dynamic capabilities should be distinguished from organisational routines, which consist of rote organisational activities that lack intent (HELFAT ET AL. 2007: 5). Moreover, they require open-minded and courageous ways of managerial thinking (VOLBERDA 1998: 108ff.). For instance, ZOLLO and WINTER (2002: 340) suggest that "a dynamic capability is a learned and stable pattern of collective activity through which the organisation systematically generates and modifies its operating routines in pursuit of improved effectiveness". WU (2006: 448) underlines the aspect that dynamic capabilities are antecedent to organisational and strategic routines by which managers transform their resource base to create new value-creating strategies. WINTER (2003) suggests a hierarchical relation between organisational routines and dynamic capabilities and further distinguishes between ordinary capabilities (which can be translated into "resources") and dynamic capabilities. The latter operate to extend, modify or create ordinary capabilities while the so-called "ordinary capabilities" only secure the firm's survival in the short term. Dynamic capabilities therefore can be considered as a kind of "meta routine" which fulfils a coordinating function. In this context, TEECE ET AL. (1997: 542) explicitly use the term "coordinative routines", while WINTER (2003: 994) and SANCHEZ ET AL. (1996: 9) speak of "higher-order capabilities".

From this point of view, dynamic capabilities are a direct, autonomous source of firms' competitive advantage (COLLIS 1991: 50ff.) because the ability to learn faster and adapt better to external changes constitutes the only tenable competitive factor in the long-run (FOSS 1997c: 350; DE GEUS 1988: 71). Consequently, the ability to learn itself becomes one of the crucial dynamic capabilities of the firm. Organisations must constantly learn, and those which learn most proficiently will be most able to enhance and sustain their competitive advantage (HELLELOID/SIMONIN 1994: 213).

However, authors like EISENHARDT and MARTIN (2000) and MAKADOK (2001) who focus on the dynamic character of the firm's market environment and its impact on the development of dynamic capabilities, have questioned this view. Dynamic capabilities are "(...) the organisational and strategic routines by which firms achieve new resource configurations as markets emerge, collide, split, evolve, and die" (EISENHARDT/MARTIN 2000: 1107). Thus, in their understanding, dynamic capabilities do not represent an autonomous source of competitive advantage, because they only serve as an enabler for new combinations of resources which finally cause competitive advantage.

They distinguish between "moderately dynamic markets" and "high velocity markets" (EISENHARDT/MARTIN 2000: 1111ff.) In moderately dynamic markets, they assume that dynamic capabilities show a homogeneous and equivalent character between firms. They are focused on the evolutionary, incremental (and thus relatively stable) development and expansion of the firms' stock of knowledge (EISENHARDT/MARTIN 2000: 1116f.). In contrast, in high velocity markets changes become non-linear and less predictable. Effective dynamic capabilities in such markets are assumed to consist of simple, not too complicated routines which keep managers focused on broadly important issues without locking them into specific behaviours, or the use of past experience that may be inappropriate in the context of a particular situation.[169] Thus, highly dynamic markets favour the development of dynamic capabilities which allow the rapid creation of new, situation-specific knowledge by engaging in experiential actions in order to learn quickly and thereby compensate for limited, relevant existing knowledge. They are less stable and their effects are hardly predictable due to their high level of context dependency.

MAKADOK (2001), who also distinguishes between moderate and highly dynamic markets, describes two causalities of how firms achieve economic rents: a "resource-picking" and a "capability-building" mechanism. While the first one argues that a firm is more effective in its choice of resources, the second means that a firm is more effective in the use of its resources than its competitors (MAKADOK 2001: 396f.). Depending on the market context, these two rent-generating ways could either complement or substitute each other.

In that sense, the consideration of routines as an explanatory model for heterogeneity and sustainable competitive advantage is shifted to another, even more abstract level of analysis which draws mainly on the dynamic process of reconfiguring organisational resources and routines. The source for persistent superior competitiveness does not lie primarily in the existence of dynamic capabilities, but in the new, adjusted resource combinations which can be realised. Dynamic capabilities are thus only a means to an end which ensures that

169 "Often these routines consist of a few rules that specify boundary conditions on the actions of managers or indicate priorities, important in fast-moving markets where attention is in short supply" (EISENHARDT/MARTIN 2000: 1111).

the firm adapts its resource development and use faster and more intelligently to external changes than its competitors (EISENHARDT/MARTIN 2000: 1110f.). As they suggest, dynamic capabilities are a set of specific processes (like product development, the practice of strategic decision-making, cooperation) that show similarities in their main characteristics amongst all firms and are therefore assumed to be at least partially imitable und substitutable. The authors thereby explicitly weaken the argument in favour of the heterogeneous, unique character of dynamic capabilities.

As can be seen from the above, the fundamental characteristic of the concept of dynamic capabilities is that the firm's rapidly changing environment is taken into account. Facing (increasing) market dynamics, the concept of dynamic capabilities overcomes the static and asset-oriented perspective of the traditional resource-based view and tries to incorporate adaptability and change into the concept of organisational competence. In contrast to the traditional resource-based approach, the concept of dynamic capabilities thus responds to the often mentioned criticism that (core) competences and routines can fast change into "core rigidities" (LEONARD-BARTON 1992). This aspect represents the dysfunctional flip side of competences which can lead to path-dependency or structural inertia caused by economic and cognitive-emotional processes in dynamic environments (SCHREYÖGG/KLIESCH 2005b). The dynamic capability approach tries to offer solutions for the phenomenon of "organisational path-dependency" (TEECE ET AL. 1997: 522f.; SCHREYÖGG/KLIESCH 2005a: 14) as it is aware of the significance of this problem and declares it to be the starting point of its contributions.

As the analysis focus on the processes of resource combination and resource development rather than the characteristics of the resources themselves, it is controversially discussed whether the concept of dynamic capabilities can be considered as a new approach within the resource-based theory at all, or if this approach even constitutes an independent research paradigm in itself (MOLDASCHL/FISCHER 2004: 127f.). For instance, HELFAT and PETERAF (2003) labelled their approach as the "dynamic-resource-based view". TEECE ET AL. (1997: 510) even call their concept an independent paradigm of strategic management research. "As a result, dynamic capabilities can be seen as an emerging and potentially integrative approach to understanding the newer sources of competitive advantages." Others who emphasise the impact of market contexts (EISENHARDT/MARTIN 2000; MAKADOK 2001) position their work as an integrating perspective between the market-based view and resource-based view. Following BURR (2004: 132), I suggest understanding the concept of dynamic capability as one research strand under the subset of the broad resource-based paradigm because on the one hand, it strictly relies on the fundamental assumptions of the resource-based theory and, on the other hand, describes a super-ordinate subset of the amount of firm-specific resources that serve as "meta-routines" for (re)configuring and coordinating routines and resources.

3.3.4 Summary and criticism

The following illustration (Figure 9) based upon BURR (2004: 132) summarises the concepts to explain competitive advantage under their common umbrella of the resource-based theory. For this purpose, the concepts were arranged with respect to their quantitative share in the total resource portfolio of a firm, their level of complexity, intangibleness, lack of transparency, their underlying time frames, as well as their interrelationships in strategic planning.

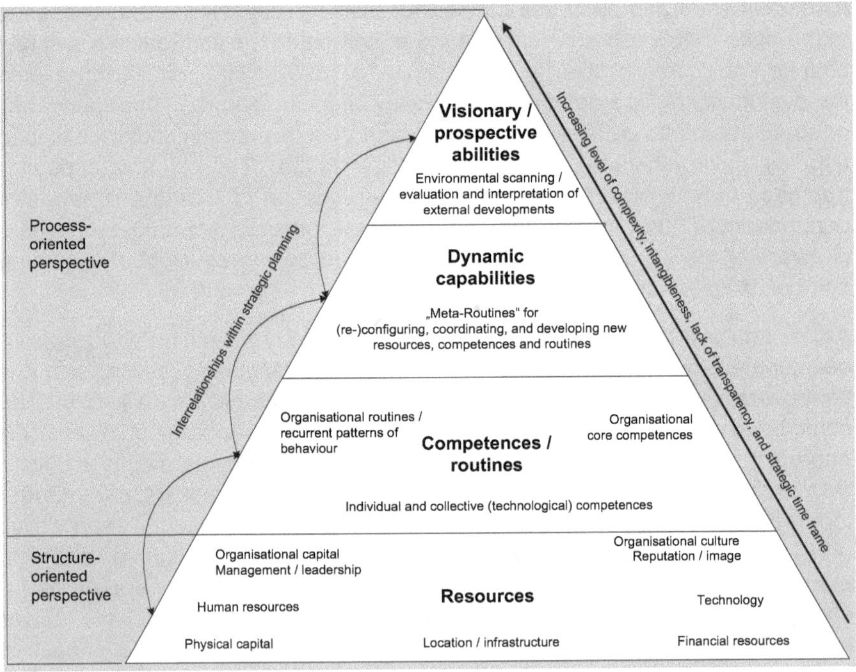

Figure 9: Overview of different resource concepts under the umbrella of resource-based theory
(Source: based on BURR 2004: 132; own modifications)

Obviously, such a typology (like most typologies) can never be absolutely consistent in itself and always remains arbitrary to some extent. Despite the fact that numerous typologies of resources are available in the literature, the followers of the resource-based approach did not manage to develop a consensual categorisation of a firm's resources. But, nevertheless, and that is why I decided to suggest such a - rough - typology, this model may help to structure, understand, and interrelate the various concepts of resource-based theory that are used to explain persistent competitive advantage of firms.

As can be seen, the figure contains another dimension at the top called "visionary and prospective abilities". As BURR (2004: 134f.) and Sanchez (1993: 269f.) correctly allude, these firm's capabilities should be considered as an implicit part of the resource-based theory because they dispose and form the development of the other levels, especially the level of dynamic capabilities.[170] Thus, visionary-prospective abilities are necessary to monitor changes and developments in the market, in the technological or political environment of the enterprise and to link the perceived changes to internal competencies and capabilities. It could be argued that this type of resource is very similar to the concept of dynamic capabilities. However, while dynamic capabilities primarily aim at the ability to (re)configure and (re)coordinate internal competences and resources, visionary-prospective abilities represent the competence of observing and forecasting external change in order to adjust internal resources by using dynamic capabilities. The assumption of an existing strategic vision of a firm as an internal resource with high strategic relevance finally helps to explain the visionary-prospective capabilities of a firm endogenously and thus completes the structure of resource-based argumentation (BURR 2004: 136).

I am aware that classifying organisational culture or leadership in the structural resource category appears to be questionable. However, I decided to do so, because these resource variables were originally introduced by early proponents of the resource-based theory like BARNEY (1986b) or PENROSE (1959) who focused in their empirical research rather on the observable structural

---

170 "In this regard, the resources and capabilities view seems to make an implicit assumption that capable strategic managers will somehow be able to develop superior insights into the most likely outcomes in dynamic markets and will therefore be able to lead the firm to internalise the right set of resources" (SANCHEZ 1993: 269).

BRUMAGIM (1994: 97) describes them as the "imagination of some future desired state" - a definition that fits closely with those of a firm's strategic vision (BURR 2004: 134).

manifestation of the organisational culture than on its intangible constitutive processes.[171]

Consistent with central contributions to the resource-based approach (such as DIERICKX/COOL 1989; PRAHALAD/HAMEL 1990; GRANT 1991; AMIT/SHOEMAKER 1993; TEECE ET AL. 1997), the typology distinguishes between "resources" and "competences/routines". While resources can be both tangible (like physical capital) and intangible (like human capital), competences and routines are always intangible. Furthermore, competences and routines emerge from the interaction of multiple actors and exist in a sense independent of individual persons. The economic distinction between resources and competences or routines is that a particular part of resources is tradable, while competences are always non-tradable. However, non-tradable competences and routines are built up from tradable resources, because they represent the various ways in which a firm uses, exploits, develops and configures its resource pool. Competences and routines are in turn shaped and administered by dynamic capabilities which in turn refer to the visionary-prospective capabilities of the firm. Thus, multiple interrelationships between the different levels of concepts can be assumed.

During the last years, the distinction between resources and competences caused an increasing split between a "structural school" and a "process school" within the resource-based theory (MOLDASCHL/FISCHER 2004: 123ff; similarly FOSS ET AL. 1995: 9ff.; FOSS 1997b: 14ff.).[172] The structural school follows the traditional neoclassical economic paradigm by presuming that markets inherently pursue a largely stable state of equilibrium distribution. Any disturbance of this equilibrium is regarded as a quickly passing exception caused by external changes. For this reason, efficiency rents are not taken into account since it is expected that the speedy introduction of efficiency improvements (and thus the restoration of equilibrium) causes such rents to disappear. The main interest of research within the structural school applies to those resources that result in Ricardian rents which can be achieved even in the market equilibrium (DEMSETZ 1973; LIPPMAN/RUMELT 1982; BARNEY 1986a). Hence,

---

171 For the theoretical derivation of the dualism of action and structure, see GIDDENS (1984).

172 FOSS ET AL. (1995: 10) entitled their discussion "Equilibrium or process?". With reference to the Schumpeter-hypotheses I & II, FOSS (1997b: 13) distinguishes two versions of the resource-based perspective ("RBP"): the "Mark I" (structural, static) and the "Mark II" (process, evolution), and traces back their respective roots to the work of DEMSETZ (for Mark I) and PENROSE (for Mark II) as the two central precursors of these schools.

this school of resource-based theory starts by clarifying and examining the conditions that must prevail in order for resources to yield rents in equilibrium. Representative contributions of this microeconomic school are, for example, those of WERNERFELT (1984), RUMELT (1984), BARNEY (1986a, 1991), DIERICKX and COOL (1989), and PETERAF (1993).[173]

By contrast, the process school of resource-based theory starts from the assumption of dynamic markets. The firms' efforts are not conceived merely as exercises in adaptation to the externally pre-set market equilibrium, but must also be explained by endogenous, dynamic factors like innovation, organisational learning, resource-accumulation, competence-building, or the development of mental models by the management team (FOSS 1997b: 13; MOLDASCHL/FISCHER 2004: 124). Under these framework conditions, rents can be achieved by firms if they manage to better utilise existing resource combinations (efficiency rates) or creatively generate new resource combinations (Schumpeterian rents). For the process-oriented perspective, competitive advantage does not only arise through the possession of idiosyncratic resources, but also and primarily through the processes of resource combination and coordination, through organisational competences and routines, and through dynamic capabilities which flexibly adapt them to changes in the external environment. As the process school of resource-based theory cannot be traced back to a singular theoretical basis (with the exception of its rejection of traditional, neoclassical economic thinking), it mixes elements of other theoretical traditions like behavioural science, theories of learning, and especially the approach of evolutionary economics by NELSON and WINTER (1982). Important authors of this process- and strategy-oriented strand are for example TEECE ET AL. (1997), PRAHALAD and HAMEL (1990), or ZOLLO and WINTER (2002).

For this reason, the impression could probably be conveyed that the various concepts within the resource-based theory conflict or even contradict each other (GERSCH ET AL. 2005: 12). This is especially the case if the various concepts are treated and interpreted as autonomous, partial theories of strategic management. In fact, the sole focus on - for instance - dynamic capabilities would imply that individual competences and skills are replaced by the firm's

173 The linkage between resource-based argumentation to neoclassical economic theories (e.g. neoclassical micro-economics (e.g. FOSS ET AL. 1995), transaction cost theory (e.g. FOSS/FOSS 2004), industrial economics (e.g. MAHONEY/PANDIAN 1992), or property rights theory (e.g. BAMBERGER/WRONA 1996a)) is blisteringly criticised by FREILING (2001: 62ff.) and GERSCH ET AL. (2005 62ff.), because both theoretical paradigms show principal inconsistencies and incompatibilities in their basic assumptions and premises.

mere ability to change and learn. To put it bluntly, this would mean that the concrete content of resources and competences does not matter anymore. Instead, the mere ability to acquire and learn relevant knowledge is a sufficient precondition for a firm's competitive advantage; "this sounds absurd" (MOLDASCHL 2006: 6). Furthermore, narrowing down resource-based analyses to the distributive characteristics of valuable resources in factor markets certainly does not explain why firms differ in their competitive performance although they are using similar resources.

By quoting this illustration, it should be emphasised that resource-based theory should rather be understood as an umbrella built on homogeneous, common premises and basic assumptions. Under this umbrella, the various conceptual branches equally contribute to explanating firms' sustainable competitive advantages as they shed light on different, but complementary aspects of strategic resource and competence management.

To briefly summarise, the resource-based theory...

- ...seeks to explain a firm's sustainable competitive advantage by the more effective and more efficient use of firm-internal, valuable, heterogeneous, immobile, specific, non-imitable and non-substitutable resources and competences compared to its competitors. But, BARNEY'S (1991: 101) wording "all assets (...) controlled by a firm (...)" leaves room for the interpretation that such resources need not necessarily be owned by a firm. This includes the possibility that a resource can also be assessed and controlled by a firm through collaborations and strategic alliances with external partners.

- ...defines competitive advantage as the attainment of economic rents, especially in the form of Ricardian rents, Schumpeterian rents, or firm-internally generated quasi-rents.

- ...can be differentiated into different conceptual strands that, on the one hand, distinguish between resources (structure-oriented perspective) and their corresponding competences and processes (process-oriented perspective) of usage, and, on the other hand, into strands of static (resources, competences, routines) and of a dynamic, evolutionary (dynamic capabilities, visionary-prospective abilities) consideration of these firm assets.

- ...focuses on medium- up to long-term strategic planning, as supposed by the formulation of the dependent variable as "sustainable competitive advantage" or "persistent differences in firms' performance" that suggest a certain stability of resources relevant for competitiveness. Moreover, the process-oriented strand of the resource-based theory points to the fact that both inimitable and non-substitutable (core) competences, organisational (meta-) routines or institutionalised patterns of behaviour do not emerge ad hoc, and thus cannot be planned or reconfigured in short time cycles (HAMEL/PRAHALAD 1997: 304).

- ...characterises the firms' external (market) environment as a strong, well-functioning competition which forces the firms to constantly enhance their levels of efficiency and which aggravates the persistence of monopoly rents. This corresponds with the assumption of constant returns to scale (explicitly: no economies of scale) and economies of scope. Thus, small enterprises do not show differences in terms of efficiency compared to larger firms and thus the market is not dominated by a few big corporations (no oligopolistic competition) (Penrose 1985: 10). In addition, the concept of dynamic capabilities explicitly assumes that the firms' environment is characterised by dynamic processes of (technological) change which cause a high degree of uncertainty and risk.

Based on these arguments, the resource-based theory has to be acclaimed for its insight that persistent differences in firms' competitive performance are based on heterogeneous, highly firm-specific bundles of resources, competences and capabilities. Thus, it stresses that these resources should be treated and analysed uniformly, true to the motto "one size fits all" (MOLDASCHL/FISCHER 2004: 146). Instead, especially the process-oriented school of the resource-based perspective points out that the resources' high level of individuality, specificity, and idiosyncracy should been taken into account. Regarding the process of strategy development, the popular approach of proclaiming best practice examples as simple rules of thumb or "management fashions" (TIDD/BESSANT 2009: 71f.) for competitive advantage and applying them incautiously and unreflectively to other firms may turn out to be a dead end.

Another benefit of the resource-based approach is the notion of firms as natural, open operative systems. Thus they are not constrained to only passively react to external changes ("outside-in perspective"), but are able to actively influence their environmental framework conditions themselves. Moreover, this aspect requires paying attention to questions of how firms manage to simultaneously handle externally existing resources and self-generated, internal resources.

Last but not least, one needs to particularly appreciate the fact that the strand of dynamic capabilities has clearly overcome the traditional equilibrium model of mainstream economics by accentuating the dynamic and disruptive character of the firms' competitive environments. Thus, a balanced relationship between a firm's internally and externally controlled or developed resources, competences and routines is seen as an exception in the competitive process. Instead, the firm has to continually adapt to the dynamic changes in its competitive environment to sustain its competitive advantages.

Besides these assets, it was already suggested that the resource-based theory also has some severe shortcomings and was thus subjected to frequent criticism. In the following, I will refer to the most frequent points of criticism (see for instance FOSS/FOSS 2004; FOSS/KNUDSEN 2003; FOSS 1997a, 1998; Moldaschl/FISCHER 2004; MOLDASCHL 2006; FREILING 2001; FREILING ET AL. 2006; GERSCH ET AL. 2005; PRIEM/BUTLER 2001a, 2001b; PORTER 1991, 1994, 1996a).

- ***Eclecticism:*** Despite these efforts, however, it must be stated that the resource-based theory today does not represent a homogeneous, conceptually closed theoretical construct within strategic management. Apart from the common idea of the heterogeneity and uniqueness of firms to explain differences in their competitive performance, there is neither consensus about basic assumptions nor can any progress in theoretical development be spotted (FREILING ET AL. 2006: 41). This is shown, for example, by the fact that many authors combine resource-based assumptions with basic elements of other mainstream economic theories. Especially the linkage between resource-based argumentation to neoclassical economic theories (e.g. neoclassical micro-economics (e.g. FOSS ET AL. 1995), transaction cost theory (e.g. FOSS/FOSS 2004), industrial economics (e.g. MAHONEY/PANDIAN 1992), property rights theory (e.g. BAMBERGER/WRONA 1996a)) is blisteringly criticised by FREILING (2001: 62ff.) and GERSCH ET AL. (2005 62ff.) because both theoretical paradigms show principal inconsistencies and incompatibilities in their basic assumptions and premises. While some authors claim for a draw-back of resource-based theory to traditional economics, others even extend the resource-based perspective by including behaviouristic or sociological perspectives and approaches (FREILING ET AL. 2006: 42). Moreover, there is no integration of different analytical levels. For instance, SANCHEZ and others (SANCHEZ ET AL. 1996, SANCHEZ/HEENE 1996; SANCHEZ 1997) prefer a holistic perspective of competences at the system level. On the contrary, FREILING (2001) and FREILING ET AL. (2006) argue on the basis of a methodological individualism.

- ***Terminological confusion:*** The situation described above results in what several authors have named "ongoing confusion" (BARNEY/CLARK 2007: 22), "vagueness" (MOLDASCHL/FISCHER 2004: 127; also: SANCHEZ ET AL. 1996: 2), a "tangle" (MOLDASCHL 2006: 9), "all-inclusive resources" (PRIEM/BUTLER 2001b: 32), or a "terminological soup" (FOSS 1997a: 11) about the terms used to describe factors of production controlled by a firm. Similar types of resources or concepts are assigned a "flood of different terms and definitions"[174] and similar terms are in turn used for different contents. But this problem was predestined by its founding fathers. Definitions like "[...] anything which could be thought of as a strength or weakness of a given firm." introduced by WERNERFELT (1984: 172) early opened the doors to uncontrolled and random growth of terms and definitions. As it lacks any conceptual focus, it neither dismisses trivial nor irrelevant issues. The same accounts for weaknesses as resources of a firm. It remains fully open in what sense a non-existence of a resource can also be regarded as a resource. Another example of such vagueness and arbitrariness is given by the definition used by BARNEY (1991: 101) "[...] all assets, capabilities, organisational processes, firm attributes, information, knowledge etc. [...] that enable the firm to conceive of and implement strategies that improve its efficiency and effectiveness." Likewise, the mere enumeration of possible firm resources appears not very helpful, as the inductive path from a particular to a general level will never result in a general, clear, theory-driven, deductive definition. However, BARNEY has recently discounted (BARNEY/CLARK 2007: 248ff.) this issue by arguing that changes in the labels would not change the central propositions of the resource-based theory.

- ***Circular reasoning / tautology:*** The definition of resources as the independent variable according to BARNEY (1991: 101; likewise e.g.: THIELE 1997; SANCHEZ ET AL. 1996) causes another serious problem, because they are linked by definition to the dependent variable of economic success. Thus, resources are defined a priori by their positive relationship to the firm's competitiveness. The resource-based approach states that firms need valuable resources and competences

174 Especially for the cases of "competences", "capabilities" or "competitiveness" this "flood" is impressively demonstrated by MOLDASCHL (2006: 4f.). Furthermore, FREILING (2001: 11ff.) extensively discusses the various problems of different types of resource definitions.

to succeed in market competition. But the definition says that firms succeed in competition *because* of their valuable resources and competences. Thus both statements constitute a circular reasoning as they can neither logically be falsified, as they explain everything or nothing (PRIEM/BUTLER 2001a: 60, 2001b: 27ff.).[175]

- ***Missing explanation of the emergence and controllability of valuable resources:*** The definitional linkage between resources and their impact on the firm's competitive advantage reveals another problem. The value of certain resources in terms of their contribution to the firm's economic performance can be only assessed ex post, because the effect is temporarily subordinate to its cause. Thus strategic management from a resource-based perspective seems like driving by looking in the rear-view mirror. As FOSS (2002: 156) and GOSHAL ET AL. (2002: 291f.) stated, most of resource-based literature focuses not on the endogenous creation of new resources as it takes the existence of heterogeneous firm resources and competences as given.[176] Especially the structure-oriented, resource-based perspective of firm rents in equilibrium says very little about how the heterogeneous conditions underlying differential rents arise.

  But a resource-based theory of strategic management should include a more general explanation of resource-building processes (BARNEY/CLARK 2007: 257f.). In the context of strategic management, it is not only interesting how existing resources can be exploited, but also how future developments of existing resources can be predicted, how new valuable resources can be generated within the firm, how this generation can be controlled, and what problems and limits thereby may thereby occur (NOOTEBOOM 2009: 2). Although the process-oriented school of resource-based theory does address dynamic issues by concepts of dynamic capabilities or organisational learning, it remains to date in rather diffuse and incoherent terms. Moreover, even dynamic capabilities concepts are static within themselves. This means that the ability that enables firms to develop new resources or

---

175 The complete discussion about tautology in resource-based theory between PRIEM/BUTLER (2001a, 2001b) and BARNEY (2001) can be found in the 26th Volume of the Management Academy Review. Again, BARNEY has recently negated this complaint of tautology: "Can resource-based theory be stated as if it was tautological? Yes. Is resource-based theory tautological? No." (BARNEY/CLARK 2007: 253).

176 The same criticism that PENROSE directed against the neoclassical theory of the firm is also applicable to the resource-based theory: there is "... no notion of an internal process of development leading to cumulative movements in any one direction" (1959: 1).

to adapt existing ones to external requirements is also assumed to be fixed. Thus, the concept of dynamic capabilities is actually no more dynamic than other versions of the resource-based theory (BARNEY/CLARK 2007: 259f.).

Moreover, regarding the central output dimension of efficiency of the resource approach, generating and accumulating valuable resources only within the firm is not always the most efficient strategy due to cost and time involved. "For some skills [...] the cost of internal development may be almost infinite." (HAMEL 1991: 99) Moreover, it can be argued that a firm does not need to have all resources relevant for competition available at all times, while some valuable resources may even only emerge or can only be assessed after the firm collaborates with external partners. But despite the fact that authors of the competence-based view (HAMEL 1991; LEONHARD-BARTON 1995) already pointed to the relevance of the cooperative assessment of complementary resources, resource-based theory did not paid much attention to the fact that some resources may not be generated or controlled until a firm collaborates with external partners or networks (DUSCHEK/SYDOW 2002: 3). This shortcoming is explicitly addressed by the relational-based view, which will be discussed in the following chapter (3.4.1).

- ***Missing linkage between strategy and resources:*** The missing provision for external context variables is often criticised, because the economic value of resources is ultimately influenced only by exogenous factors, such as market conditions and can be thus evaluated only in relation to external market requirements (PRIEM/BUTLER 2001: 62f.; PORTER 1991: 108; BOWMAN/AMBROSINI 2000: 2ff.). As PORTER (1991: 108) put it, resources - in contrast to one of the basic assumptions of the resource-based theory - cannot be valuable in themselves, because they allow firms to perform activities that create advantages in particular markets. Once firm-specific resources prove valuable in the market, however, the rarity and inimitability associated with firm-specific resources enable such resources to play a more strategic role than general resources. However, it should be mentioned that the proponents of the resource-based theory are aware of this problem (e.g. BARNEY 2001: 42f.). Resources are thus only meaningful in the context of performing certain activities to achieve competitive advantages. For this reason, the resource-based theory remains too vague in its performance definition in terms of mere effectiveness and efficiency. It is likely to overlook that a higher level of effectiveness and efficiency of internal processes cannot be equated automatically with competitive advantages, as effectiveness and efficiency have to fit the strategic market positioning of the firm (PORTER

1996: 114f.). From an evolutionary perspective, market selection works on the effects and product characteristics (e.g. quality, price) the firm is able to offer. However, it can be assumed that this selection mechanism impacts the internal deployment and evaluation of a firm's resources.

This perspective is already included to some extent in the concept of dynamic capabilities, but has to be strengthened by further investigations of the relationship between the specific resources and the competitive strategy of firms. Valuable resources help to build strong strategies; strong strategies help to build up valuable resources (WITT 2008: 67). From my point of view, the character of the causal relationship between the firm's set of resources and their economic impact also remains unanswered. As it seems, the resource-based theory implicitly assumes a linear causal relationship between the employment of resources relevant for competitive advantage and the observed effects on the firm's competitive success. To put it simply, the increase of the amount of a resource by one unit leads to an increase of competitive performance by one unit. I think it is reasonable to question this assumption, as it neglects both the general issue of marginal utility of resources and the specific character and complexities of innovation processes within the firm. Thus it can be suggested that the character of the causalities between some resources and economic effects are rather (inversely) U-shaped, as they lose or even reverse their positive effects. This may be especially the case the more tangible the resource is. Instead, intangible resources may show rather linear causalities until they approximate to a marginal value. On the other hand, the strength of the economic impact of resources may vary between firms due to their specific internal processes and routines. For instance, higher investment in R&D would only show positive effects if the additional funds are transformed effectively into new products by corresponding processes. Thus, different dimensions of tangible and intangible resources may interfere with each other by enhancing or weakening their effects.

- ***Difficulty of measuring and operationalising the theoretical concepts:*** Finally, some of the concepts of the resource-based view are difficult, if not impossible, to operationalise for empirical analysis. While tangible, physical resources or competences that are based on explicit knowledge can be measured relatively easily, more abstract, intangible or complex types of resources like routines, implicit competences, or dynamic capabilities can hardly be observed as they are latent and intangible variables. As BARNEY and CLARK (2007: 251) admit, the empirical testing and measuring of resource-based concepts produced "(...) much better theories about how *not* to measure re-

> sources than (...) about how to measure resources". Considering the fact that a common terminological framework does not exist so far, it is not surprising that no harmonised sets of indicators for measurement have been developed either. And even if latent and intangible variables like routines or dynamic capabilities can be successfully operationalised through indicators, researchers have to cope with the circumstance that it is almost impossible to empirically isolate resources (due to their idiosyncracy) in a discrete manner. Thus, researchers have to deal with inter- and intra-correlations between different indicators which represent such variables. Secondly, constructing adequate indicators for such variables is often a trade-off between simplicity, and thus feasibility, on the one side, and an adequate provision for the necessary degree of complexity, on the other side. Thus, by measuring the addressed concepts only at a very superficial and simplified level, such indicators offer only limited options for interpretation. Especially against the backdrop of strategic management, these difficulties in measuring complex and intangible resources cause serious problems for firms. As the old management adage "you can't manage what you can't measure" argues, unless something is measured one does not know if it is getting better or worse. It is impossible to manage for resource improvement if it is not possible to determine if the efforts have been fruitful or not.

Based on these criticism, some authors like FREILING (2001, 2004), FREILING ET AL. (2006), and GERSCH ET AL. (2005) recently tried to integrate the different strands and concepts within the existing resource-based perspective into a comprehensive "competence-based theory of the firm" by building upon the common analytical core of the different concepts and by establishing a common set of clear conceptual definitions. Others, like MOLDASCHL (2006), seek to disentangle the numerous competence- and capability dimensions by suggesting "institutional reflexivity" as an alternative to grasp a firm's "changeability".

To incorporate this criticism into the context of this thesis, three points are of particular relevance: the problem of tautology which is closely connected to the missing link between the resource constructs and strategy, and the difficulty of measuring and operationalising all these resources. The first two aspects are encountered by the fact that the non-R&D-performing firms which use such resources should not assumed a priori to show competitive advantage in terms of innovation output and economic success. Instead, these resources only represent *possible* sources of their economic success. Thus empirical analysis needs to contrast the use of different resources with measurements of the actual innovation output and economic performance of the analysed firms. The problem of measurability and operationalisation, and how this thesis deals with this problem, will be explicitly addressed in chapter 4.5.

3.4 Extending the firm-internal resource-based theory

3.4.1 The relational-based view of the firm

In answering the question why firms differ in their performance, strategic management typically choose to view firms as autonomous entities which search for competitive advantage, either through market or industry conditions or internal resources, competences, and capabilities. Scholars of the young relational-based view assume that these approaches overlook the important fact that the advantages of individual firms are often linked to the advantages of the social networks of relationships in which the firms are embedded (DYER/SINGH 1998: 660; GULATI ET AL. 2000: 204).

3.4.1.1 Development and demarcation

This view arose from the increasing phenomenon of strong inter-firm networks and the increasing intensity of competition between these alliances (CONTRACTOR/LORANGE 2002: 498; D'AVENI 1995: 199ff.). According to OELSNITZ and GRAF (2006: 85), the analysis of firm alliances and networks may be able to address some of the current changes in the market environment of firms, such as:

- an increasing complexity of industry- and consumer goods which requires high financial efforts for product development with simultaneous higher levels of uncertainty and risk
- technological convergence which leads to an increased demand for system solutions
- increased need for firms to attain economies of scale and scope forced by the necessity to cover international markets
- shortening of product life-cycles results in shorter periods of pay-off and increases the need for first-mover advantages through collaboration
- existing barriers of trade and barriers to market entry require collaborative strategies of market entrance

The relational-based view is thus focused on dyadic and network relationships and cooperations between enterprises without being exclusively restricted to specific types of inter-firm collaboration.

From an inter-organisational perspective, the acquisition of valuable resources can be coordinated through market mechanisms, hierarchical or cooperative forms of organisation. As already discussed in the previous chapter, the acquisition of resources in factor markets does not play a very important role as resources which can be traded on markets are not characterised by specificity, idiosyncracy or inimitability since they have to be internally refined and modified by the firm to develop their full impact on its competitive advantage. Similarly, hierarchical, only firm-internal modes of resource acquisition are not always the most efficient strategy to generate and accumulate valuable resources relevant for competition due to issues of cost and time (RASCHE 1994: 230; HAMEL 1991: 99). For these reasons, the acquisition and control of missing, valuable and intangible resources through collaboration with external partners is a promising and efficient strategy for firms.

Despite its analytical potential, strategic management research has neglected the systematic examination of inter-organisational rent-generating processes (DYER/SINGH 1998: 661). As HAMEL (1991: 99f.) clarifies, the acquisition of valuable resources through inter-firm cooperation is just "a half-way house on the road from market to hierarchy" (IBID.: 99), because such cooperations only represent a strategic option to appropriate missing resources and competences by learning ("outlearning") from external partners. Once they have been appropriated, the cooperation becomes obsolete. Thus the hierarchical integration of those resources acquired through cooperation is the only way to attain *sustainable* competitive advantage. The firm (respectively the firm-specific benefits from cooperation) remains the analytical unit as a carrier of unique and valuable resources. Consequently, researchers were more interested in how firms collaborate to generate economic rents and tended to focus on particular benefits which result from collaboration activities such as, for instance, lower transaction costs, increased knowledge transfers, pooling of resources, or shared risks of investment (see for example: STUART 2000: 792; DAS/TENG 2003: 297; KOZA/LEWIN 2000: 147; AHUJA 2000: 317).

3.4.1.2 Competitive advantage through relational rents

The relational-based view however argues that sustainable competitive advantage *directly* results from, or even may *inevitably* be bound to collaborations, alliances, and networks (DYER/SINGH 1998: 660; GULATI ET AL. 2000: 203). The central goal of the relational-based view is to explain how the competitive advantage of firms is earned and preserved through network and collaboration routines and processes. DYER and SINGH (1998: 661) defined inter-organisational competitive advantage by coining the term of "relational rents" which means above-normal profit "[...] jointly generated in an exchange relationship that cannot be generated by either firm in isolation and can only be created through the joint idiosyncratic contributions of the specific alliance partners" (DYER/SINGH 1998: 662; likewise: DUSCHEK 2002: 258).[177]

Because conventional, "arm's length" (DYER/SINGH 1998: 661) market relationships are usually characterised by non-specific asset investments, minimal information exchange (prices as coordinating information devices), separable technological and functional systems within each firm, and low transaction costs and minimal investment in governance mechanisms, it is relatively easy for firms to substitute trading partners with little penalty or opportunity cost.

In order to transform such conventional forms of inter-firm market relationships into relational rents-generating, "high ordered" forms of inter-firm cooperation, collaborating firms have to meet four necessary conditions (Figure 10) (DYER/SINGH 1998: 662ff.; GAITANIDES 2007: 289ff.).

While within resource-based theory, assets have to be rare, valuable, immobile and inimitable, the relational-based view is more focused on the creation of *assets that are specialised in conjunction with the assets of an alliance partner* (DYER/SINGH 1998: 662). Following WILLIAMSON (1985), DYER and SINGH (1998: 662) distinguish three types of relation-specific assets:

- site specificity
- physical asset specificity
- human asset specificity

177 This definition closely corresponds to PETERAF'S (1993) definition of "quasi-rents" as relational rents enable firms to achieve returns that exceed a factor's short-run opportunity cost and are in excess of the returns to a factor in its next best use. Thus, the collaborating partners in an inter-organisational network constantly generate quasi-rents as they extend their internal routines and processes beyond their organisational boundaries. However, the authors avoid the term "quasi-rents" because it "suggests that the rents are not permanent in nature" (DYER/SINGH 1998: 661, footnote #2).

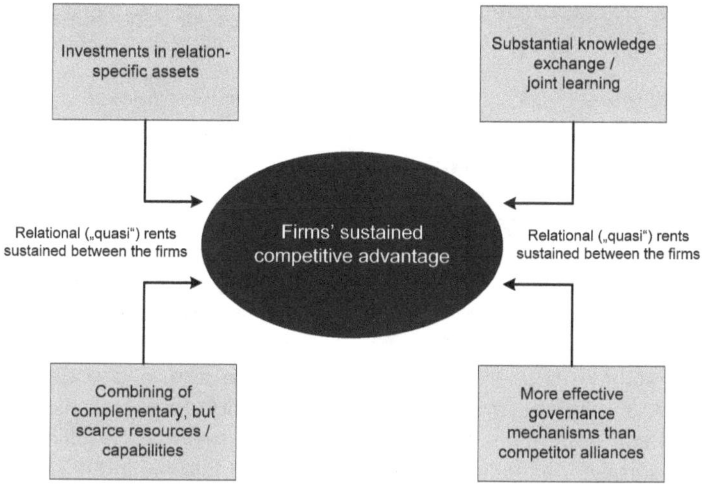

Figure 10: Determinants of firm's competitive advantage achieved by relational rents (Source: own illustration)

Site specificity refers to the geographical proximity of the collaboration partners. Thereby, successive production processes that are immobile by nature are located close to each other and help to reduce inventory or transaction costs, as well as coordinating activities. Physical asset specificity instead corresponds to transaction-specific capital investments (like cooperation-specifically designed or modified machine tools, machinery equipment or software-interfaces for enterprise resource planning systems). Finally, human asset specificity describes the stock of knowledge and know-how which arises through the joint (long-term) collaboration processes (e.g. through the training of employees who were trained for the individual particularities in the systems and procedures of the cooperation partner).

With regard to these relation-specific assets, two important sub-conditions positively influence the ability of partners to generate relational rents (DYER/SINGH 1998: 664). First, given the durable and fixed-cost nature of many, especially technological investments (e.g. machinery), cooperation partners are more likely to make such durable and costly relation-specific investments when safeguards for these investments are provided for a certain period of time. Secondly, there is a positive relationship between the total volume of alliance transaction and the potential to generate relational rents through relation-specific assets, because alliance partners are able to increase the efficiency associated with inter-firm exchanges as they increase the volume and scope of transactions between them.

Considering the second condition of *substantial knowledge exchange* between the collaborating firms, the relational-based view suggests that a firm's cooperation partners are often the most important source of new ideas and information that result in performance-enhancing technology and innovations.[178] Thus, collaborating firms can generate relational rents by developing superior inter-firm knowledge-sharing routines in terms of a "regular pattern of inter-firm interactions that permits the transfer, recombination, or creation of specialised knowledge" (DYER/SINGH 1998: 665). To create and establish such routines of inter-firm knowledge-sharing, firms have, on the one hand, to develop partner-specific absorptive capacities.[179] They enable the firm to decode, recognise and assimilate different forms of valuable knowledge (e.g. tacit as well as explicit knowledge) from its particular alliance partners. In addition, partner-specific absorptive capacity is enhanced as individuals within the alliance partners get to know each other well enough to know who knows what and where critical expertise resides within each firm. The ability of cooperation partners to attain relational rents through knowledge-sharing also depends on the incentives that encourage the partners to make their knowledge and information transparent and to allow for knowledge transfer, as well to prevent that a firm is free-riding on the knowledge and information acquired from its partner (DYER/SINGH 1998: 666).

The third way firms can generate relational rents is by *combining complementary resource endowments*. Ideally, the alliance partners bring distinctive resources into the cooperation which, when combined with the resources of their partner, result in synergistic effects, whereby the combined resource endowments are more valuable, rare and difficult to imitate than they had been before they were combined and complemented by each other (DYER/SINGH 1998: 667). The more the distinctive resources are complementary to each other, the higher the potential to attain relational rents. However, the ability of firms to

---

178 To support this argument, DYER/SINGH (1998) refer to scholars and findings in the field of organisational learning (e.g. MARCH 1958) and open innovation (e.g. VON HIPPEL 1988).

179 Although DYER/SINGH (1998) refer to the concept of "absorptive capacity" developed by the famous contribution of COHEN/LEVINTHAL (1990), they underline that having absorptive capacity does not imply being equally capable of learning from all other organisations outside the firm. Instead, firms adapt their absorptive capacity to the specific context of their collaborations to attain relational rents (DYER/SINGH 1998: 665).

synergise their complementary resources increases with the firm's prior alliance expertise, the investment in internal search and evaluation capability (e.g. by establishing an internal department of strategic alliances), and the ability to occupy an information-rich position in its social and economic networks (DYER/SINGH 1998: 668). Moreover, this ability also relies on a certain degree of proximity or compatibility between the cooperation partners regarding their organisational systems, processes, and organisational cultures.

The fourth condition for attaining relational rents through inter-firm cooperation is related to establishing *governance mechanisms within the firm alliance that are superior in their effectiveness compared to competitor alliances.* In general, governance within inter-firm alliances can be realised through third-party enforcement of agreements (e.g. legal contracts or a legitimate organisation authority) or through self-enforcing agreements (e.g. informal mechanisms like trust or goodwill, as well as formal safeguards like reciprocal interdependency in the field of finance or investment) (DYER/SINGH 1998: 669).[180] Following the argumentation of transaction cost theory, DYER and SINGH (1998: 669f.; also DYER 1997; HANSEN ET AL. 1997) state that cooperating firms have to efficiently and effectively align the transactions with governance structures to minimise transaction cost and to increase the incentives for partners to engage in value-creating initiatives, and thus to finally obtain relational rents. Regarding minimising transaction costs, self-enforcing mechanisms are more effective than third-party enforcement mechanisms because of their lower contracting cost, lower monitoring costs, lower adaption costs, lower recontracting costs, and superior incentives for value-creation initiatives (DYER/SINGH 1998: 671). Within such self-enforcing mechanism of alliance governance, informal mechanisms like mutual trust, goodwill, or reputation have a higher potential to obtain relational rents as they cause lower marginal costs and are more difficult to imitate, due to their social complexity and idiosyncracy (DYER/SINGH 1998: 671). Moreover, not only does the joint coordination of collaboration require effective governance, but distributing the obtained relational rents does also (see DYER ET AL. 2008).

3.4.1.3 Isolating mechanisms

Like the isolating mechanisms of RUMELT (1984) that preserve and maintain the value of firm-internal resources, the relational-based view also postulates

180 See also TELSER, L. G. (1980).

isolating mechanisms that work on preserving relational rents (DYER/SINGH 1998: 671ff.). Some of them are adopted from the resource-based literature like causal ambiguity and time compression diseconomies (BARNEY 1991; DIERICKX/COOL 1989; LIPPMAN/RUMELT 1982; REED/DE FILLIPPI 1990). For example, trust and goodwill between cooperation partners are both subject to considerable causal ambiguity, due to their social complexity and situational specificity, and subject to time compression diseconomies, as they cannot be instantly developed, nor can they be bought or sold on markets (DYER/SINGH 1998: 672).

In addition to these isolating mechanisms already known from resource-based theory, the relational-based view suggests four relational-specific mechanisms that preserve the competitive advantage of inter-firm collaborations (DYER/SINGH 1998: 672ff.; DUSCHEK/SYDOW 2002: 429f.):

- ***Inter-organisational asset interconnectedness:*** Initial investments in relation-specific assets entail additional, specialised investments in the future. The result is a cumulative process of investments in relational assets that are interconnected and based on each other. The key strategic implication is that alliance partners may need to make related relation-specific investments in order to realise the full potential of relational rents.

- ***Partner scarcity:*** Relational rents may be difficult to imitate because potential alliance partners with the necessary complementary resources and relational capability are rare. The key strategic implication is that there are strong first-mover advantages for those firms that develop a capability of quickly identifying and allying with adequate, matching partners.

- ***Resource indivisibility:*** Partners may combine resources or jointly develop resources in such a way that they are path-dependent, idiosyncratic, and indivisible as they cannot be dissected ex post by competitors. The strategic implication is that the partners' resources and capabilities may co-evolve and change over time, thereby restricting each firm's ability to control and redeploy the resources.

- ***Institutional environment:*** An institutional environment that encourages or fosters trust among trading partners, for instance, by effective institutional rules or social controls for enforcing agreements, may facilitate the creation of relational rents. Indeed, these arguments can be extended to consider the issue of national or country advantage. The strategic implication of this isolating mechanism is that firms may need to locate operations in particular institutional environments in order to realise the benefits associated with extra-hybrid institutions.

But despite the fact that more and more enterprises have recognised the importance and potential of inter-firm collaboration and alliances to improve their competitiveness, the success and benefits of inter-firm collaborations and networks are often limited by, or even fail, due to problems like mutual distrust (ZAHEER ET AL. 1998: 141ff.; ENDRES/WEHNER 1995: 9), communication problems (LOOSE/SYDOW 1994: 164), different practices of conflict-solving (SCHUH ET AL. 2005: 31), different abilities to learn (D'AVENI 1995: 117; DYER/SINGH 1998: 664ff.), or missing management competences (DYER/SINGH 1998: 669; KALE ET AL. 2002: 759ff.). Thus it seems that the analytical consideration of inter-firm cooperation is based solely on economic aspects, like joining togehter valuable, complementary resources or minimising governance costs, is insufficient. Instead, inter-firm collaboration and alliances are also social entities (DAS/TENG 2002, 447f.; GULATI 1998: 95f.) which are - besides economic transactions - equally based on immaterial, social processes. In this context, variables like the socio-structural network of firms, reputation, political influence, and the ability to appropriate the economic outcomes of these alliances come into play and are bundled under approaches of the firm's "social capital" (e.g. BURT 2003; TSAI 2000; KOKA/PRESCOTT 2001).

3.4.1.4 Cooperative competences

Against this background, the relational-based perspective of firms' sustainable competitive advantage has - similar to the resource-based theory - originated a process-oriented strand of research which focuses on the firm's "relational capability" (LORENZONI/LIPPARINI 1999), "collaborative know-how" (SIMONIN 1997), "cooperation ability" (EISELE 1995), "cooperative core competences" (DUSCHEK 1998), or "relational and network competence" (STAHL 1996; PEITZ 2002).[181] The logic of argumentation thereby also follows the resource-based perspective, as the cooperative competence of a firm determines whether it is able to attain relational rents or whether the cooperation is unsuccessful. "Understanding of collaborative mechanisms and transfer processes favours knowledge absorption by eliminating many of the unnecessary tasks and disruptive noise of cooperation." (SIMONIN 1999: 603) The term "collaborative compe-

181 For a more detailed overview of different concepts of the firm's cooperative competences see OELSNITZ/GRAF (2006: 88ff.)

tences" thus represents in some way a melting pot for any of a firm's competences to maintain and optimally exploit its collaboration or network activities. Thereby, cooperative competences are a multidimensional construct which encompasses personnel as well as organisational skills or characteristics (OELSNITZ/GRAF 2006: 91).

According to their literature overview, OELSNITZ and GRAF (2006: 93) distinguish between three major dimensions of a firm's collaborative competence:

- functional, technical core competences and overarching management know-how as a meta-resource
- compatible systems of management, organisation and communication
- organisational reputation, social status, and social capital.

In order to be attractive to potential alliance partners, firms on the one hand need to have specific technical ("general purpose") resources and competences (e.g. in the fields of technology, R&D, marketing, manufacturing) as a constitutional element of the collaboration, as well as certain general, specific, and immaterial resources which can be contributed to the collaboration to enhance its effectiveness or efficiency ("alliance capability") (INKPEN/CURRALL 2004: 592; OELSNITZ/GRAF 2006: 93; likewise: KALE ET AL. 2002; LORENZONI/LIPPARINI 1999; HAMEL/PRAHALAD 1997).

On the other hand, a firm has to be compatible with the processes of external collaboration partners (see, for instance ROBSON 2002; DYER/SINGH 1998). This does not necessarily imply that cooperation partners should conform entirely in their operative or communicative processes, technological backgrounds or social characteristics, like their organisational culture.[182] More important is that the firm can handle the similarities or dissimilarities of their external alliance partner. This can be realised through the design and arrange-

182 BOSCHMA (2005) argues that other forms of proximity may frequently substitute for geographical proximity. He distinguishes between five forms of proximity – geographical, cognitive, organisational, social and institutional. BOSCHMA defines cognitive proximity as a function of the similarity between organisations' knowledge bases. Simply, organisations are cognitively proximate where they posses similar market and technological competences. And, building on shared experiences and understandings, cognitive proximity facilitates effective communication and collaboration.

For the aspect of cognitive distance, see also NOOTEBOOM (2009: 95ff.).

ment of organisational interfaces and by individual skills of persons ("boundary spanners") who work in these interfaces (KIRNER 2005).

Last but not least, a firm's social embeddedness turned out to be of crucial importance for the collaborative ability and attractiveness of a firm (GULATI 1998: 300). A firm's social embeddedness is thereby composed of both its structural positioning and its social status within its social network. The particular importance of this social dimension relies on the fact that inter-firm cooperations are not seen as a punctual phenomenon but as continuous and path-dependent activities. The behaviour and performance of a firm are stored in the collaborative memory of its alliance partners in the network and form - together with the structural positioning - the social capital of a firm (OELSNITZ/GRAF 2006: 95). Subsequently, the social capital of a firm serves as a predictor of its collaborative ability and represents an intangible, inimitable, and non-substitutable resource of the firm (GULATI 1998: 298ff.; also: KALE ET AL. 2000; LORENZONI/LIPPARINI 1999; BURT 2003; KOKA/PRESCOTT 2001; TSAI 2000). As a consequence, if a firm possesses social capital, it is able to enhance its social positioning and reputation within its network by enhancing its collaborative attractiveness for top-level partners and by obtaining information advantages about opportunities for further alliances. "[...] firms that had more (positive - *author's note*) prior alliances, were more centrally situated in the alliance network, or had more focused networks, were more likely to enter new alliances and did so with greater frequency" (GULATI 1998: 300f.).

3.4.2 Summary and criticism

To date, a number of empirical studies underline the potential of inter-firm collaborations and alliances for attaining above-normal economic rents (GULATI 1998: 309f.) and the specific sources of relational rents and the barriers of imitation as elaborated by DYER and SINGH (see, for instance: DYER 1996, 1997; ZAHEER ET AL. 1998, MÖLLER 2006; DYER/NOBEOKA 2000; DUSCHEK/SYDOW 2002; DYER/HATCH 2006). As the relational-based view strongly relies on the assumptions and premises of the resource-based theory as well as referring to the concept of a firm's (core) competences, it is usually seen as an advance in, rather than an autonomous strand of, strategic management (DUSCHEK 2002: 256ff.; DUSCHEK/SYDOW 2002: 10; LAVIE 2006). Nevertheless, some authors argue that the relational-based view could also be subsumed under the market-based perspective of strategic management, because the chance to obtain relational rents through inter-firm collaboration strongly depends on the structure and characteristics of the firm's external market environment (BECK 1998: 55). Following this line of argument, strategic alliances between firms increase the availability of complementary resources, support the implementation of competitive strategies, and thus increase the firms' market power as well as barriers of market entry for other competitors (BECK 1998: 52ff.; PORTER 2000: 619). Despite the fact that this argumentation sounds reasonable, it leaves no room for aspects of isolating mechanisms, like causal ambiguity or inimitability of relational specific resources which are central to the theory of DYER and SINGH (1998) and others.

The relational-based view explicitly overcomes the prevailing intra-firm perspective of the resource-based theory by accounting for valuable resource that are located either outside the firm at other external partners or that are created unless firms combine their complementary resources to build even more valuable, unique, and inimitable resources or competences through collaboration. Thus, inter-firm collaboration represents another strategic option for firms to enhance the effectiveness and efficiency of their resource deployment as they enable them to assess resources relevant for competition, either without having them internally available, or despite not being able to develop the required resources themselves. Against this background, the relational-based view is the long overdue and compatible extension of the resource-based theory (DUSCHEK/SYDOW 2002: 10).

Another important achievement of the relational-based view is the process-oriented extension of the structure-oriented, economic perspective by considering both social conditions (like social positioning, social reputation, social capital) and social competences, like overarching collaborative competences or general management abilities for cooperation at the individual and organisational level which are necessary for firms to realise relational rents. Thereby, the original concept of absorptive capacity (COHEN/LEVINTHAL 1990), which was frequently criticised due to its empirical linkage to R&D activities, is broadened and supplemented by the general ability of firms to identify collaboration partners, establish strategic alliances, and organise and exploit their collaborations in such a way that competitive advantage is obtained.

However, the approach of the relational-based view suffers from some shortcomings which should be briefly mentioned here:

- ***Missing explanation of rent-generating processes:*** Similar to the resource-based theory of strategic management, the relational-based view has failed so far to theoretically explain the processes of relational rent generating. The actual character of value-generating processes and the modalities of the evolution of valuable resources within inter-firm collaboration and strategic alliances remain unanswered (DUSCHEK/SYDOW 2002: 11; MORAN/GOSHAL 1999: 409; in detail:

DUSCHEK 2002). However, this aspect would be of great importance in strategic planning and organisation of such inter-firm collaborations.

- ***Neglect of firm-internal resources:*** Although authors writing about the relational-based view explicitly stress the importance of the availability of certain valuable *firm-internal* resources as the necessary fundament for achieving *inter-firm* relational rents, they constantly run the risk of focussing only on the collaborative or network dimension in the analysis of relational rents (DUSCHEK/SYDOW 2002: 11. However, an adequate extension of resource-based approaches towards an inter-firm dimension of rent generating should necessarily consider both the firm-internal as well as the inter-firm dimension and the recursive and interdependent relationships between these two levels.

- **Knowledge resources with high complexity, specificity, and high strategic value are more likely to be accessed through acquisitions than by inter-firm alliances:** As recent findings shown, firms are more likely to exploit complex knowledge by expanding through wholly owned subsidies rather than engaging in alliances (CARAYANNOPOULOS/AUSTER 2010: 256). As KOGUT and ZANDER (1993: 633) suggest, the more complex a manufacturing technology, the more difficult it should be to transfer. This is reasoned by the assumption that complex knowledge is comprised of a great number of tacit interdependencies (across individuals, departments, or skills) making the totality of the knowledge difficult to understand or integrate for other firms, and which is difficult to transmit across firm boundaries as they are deeply embedded in the firm's social context (e.g. SORENSON ET AL. 2006; VAN WIJK ET AL. 2008; SIMONIN 1999: REED/DEFILIPPI 1990). The same applies to the specificity of knowledge (e.g. PARUCHURI ET AL. 2006), and knowledge resources that contain high profit-generating potential (CARAYANNOPOULOS/AUSTER 2010: 257; INKPEN 2000) as firms are not likely to disclose their unique, very specific knowledge assets or to share above-normal rents with external partners when they have the possibility to obtain them by themselves. These arguments show that the condition of extensive knowledge exchange may cause severe problems for the success of inter-firm collaborations, as the firms' willingness to exchange knowledge tends to decrease with the level of its strategic relevance.

- ***Inconsistent terminology of collaborative competences:*** Regarding the more process-oriented strand of the relational-based view which accounts for the social embeddedness of firms and their necessary social, collaborative competences, it must be stated that no common, consistent terminology about (core) competences for inter-firm collaboration exists (OELSNITZ/GRAF 2006: 87ff.). Instead, the various concepts remain at a rather superficial analytical level and are often characterised by anecdotal evidence (in most cases resulting from case studies of big Japanese corporations) depending on their respective analytical focus. For this reason, one of the most urgent tasks of relational-based research is to develop a consistent and structured terminology. Furthermore, the focus on the firm's general collaborative ability shifts the level of analysis of relational rents back to the intra-firm perspective, because the likelihood of obtaining relational rents is not primarily located in the collaboration itself, but in the relevant collaborative competences within the firm which can also be reasoned within the conceptual framework of competence-based or dynamic capability-based perspectives. Thus, with respect this point of criticism, future terminologies of collaborative competences have to account for this by underlining their instrumental character and to building an argumentative bridge between the attainment of relational rents at the inter-firm level and firm-internal preconditions.

Besides these main criticisms, other authors mention that the relational-based view does not distinguish between different modes of inter-firm collaboration, and thus does not account for the attractiveness of specialised types of vertical collaboration like supply chains and their specific barriers of imitation (e.g. SCHMIDT 2009: 135). Last but not least, only few studies to date deal with questions of the appropriating and distributing the obtained relational rents between the involved firm partners. As the literature suggests, designing mechanisms to appropriate relational rents may be one of the central reasons for firms to participated in inter-firm collaboration and strategic alliances (for a first approach see for instance DYER ET AL. 2008).

The relational-based view's shortcoming in neglecting the role of firm-internal resources is of particular relevance in this thesis. Basing the empirical analysis only on a relational-based perspective would underestimate the importance of non-R&D-performing firms' valuable, in-house resources. Thus, care should be taken that the relational-based view is supplemented by the resource-based perspective to account for both, internal and external innovation resources. Moreover, the point that knowledge resources with high complexity, specificity, and high strategic value are more likely to be accessed through acquisitions than by inter-firm alliances underlines the necessity of employing a broader focus on external sources for competitive advantage. Ideally, the empirical analysis should include a variable to measure the firm's expenditures for external acquisitions (e.g. acquisition of machinery, equipment, licences, software etc.).

3.4.3 The knowledge-based view of the firm

The concept of knowledge as a strategic resource of firms gained in interest since industrialised economies experienced a shift in importance from tangible resources towards intellectual, intangible assets (NOOTEBOOM 2009: 11). This shift is commonly described by the term "knowledge-based economy" (FORAY 2006: IX) and is reflected in an ever-increasing proliferation of the production, processing, and transfer of knowledge and information. This evolution spread across the whole economic system and highlighted the importance of knowledge as a key driver of economic growth and thus mainly contributed to the development of a knowledge-based theory of the firm (HANSEN ET AL. 2009: 106; CZARNITZKI/WASTYN 2009: 1).

3.4.3.1 Development and demarcation

The "knowledge-based view of the firm" (SPENDER/GRANT 1996; GRANT 2002) encompasses recent approaches of strategic management research which focus on the firm's specific knowledge bases and its ability to evolve its stock of knowledge through learning processes as the major, persistent source of firms' competitive advantage in the context of dynamic market environments (EISENHARDT/SANTOS 2002: 139; SPENDER/GRANT 1996: 5f; DECAROLIS/DEEDS 1999: 953ff.; WELGE/AL-LAHAM 2008: 98; AL-LAHAM 2004: 406).

The knowledge-based view of the firm is composed of a broad range of ideas and strands of research, including the resource-based theory (BARNEY 1986A; WERNERFELT 1984; RUMELT 1984), especially drawing on its competence- and capability-based analysis (PRAHALAD/HAMEL 1990; GRANT 1991; AMIT/SHOEMAKER 1993), epistemology contributions (POLANYI 1958, 1966; VON HAYEK 1945), as well as organisational learning (LEVITT/MARCH 1988; HUBER 1991; ARGYRIS/SCHÖN 1978). Key contributions to the knowledge-based view of the firm were made for example by DEMSETZ (1988, 1991) who analysed the knowledge-based boundaries of firms, the examination of knowledge-based organisations (BROWN/DUGUID 1991), KOGUT and ZANDER'S (1992) view of the firm as a knowledge processing institution, and NONAKA'S (1994) analysis of knowledge creation within the firm. With regard to theory construction, GRANT (1996a, 1996b), LIEBESKIND (1996), SZULANSKI (1996), and SPENDER (1996a, 1996b) were among the first who tried to reconcile and integrate some of these contributions and thereby contributed significant evidence to the importance of knowledge as a process, and how the coordination of knowledge is necessary to achieve strategic competitive advantage.[183] The knowledge-based view of the firm emerged in the context of the rapid upcoming and enormous competitive success of Japanese corporations like Honda, Canon, Matsushita, NEC, or Sharp in the 1990s, and their ability to respond quickly to customers, create new markets, rapidly develop new products, and dominate emergent technologies. As the secret of their competitive success was seen, among others, in their unique approach to managing the creation and coordination of new, especially tacit, knowledge (NONAKA 1991: 96f.), this paved the way for the development of the knowledge-based view of the firm.

The knowledge-based view of the firm is not so much a new theory of the firm as a number of explorations into different aspects and into its organisation of production that are unified by their focus on the role of knowledge as a factor of production (GRANT 1996a: 110; GRANT 2002: 133).[184] In this sense, the knowledge-based view primarily represents an advance or a re-labelling of the resource-based theory by focussing on knowledge as the most relevant strategic resource of firms, and its related dynamic processes of identification, creation, modification, accumulation, transformation, transmission and communication within, as well as between firms (WELGE/AL-LAHAM 2008: 98). However, other scholars see the knowledge-based view either as a useful extension of organisational learning to strategy and organisational functioning (e.g. KOGUT/ZANDER 1992, 1996)[185] or weakening the strategic resource char-

183 For more contributions on the knowledge-based view, see for example: Leonard-Barton (1992, 1995), Kogut/Zander (1993, 1996), Nonaka (1991), Nonaka/Takeuchi (1995), Nonaka/Konno (1998), Nonaka/Toyama (2003), Sveiby (1997), Spender/grant (1996), Tsoukas (1996), Conner (1991), Conner/Prahalad (1996), von Krogh et al. (1998), Cook/Brown (1999), Cowan/Foray (1997), Simonin (1999).

184 WELGE/AL-LAHAM (2008: 100ff.) provide a good systematisation of the various disciplinary strands within the knowledge-based paradigm.

185 For a detailed debate of the question whether the knowledge-based view represents the emergence of a new theory of strategy or not, see the contribution of EISENHARDT/SANTOS (2002).

acter of knowledge by emphasising the ongoing social processes of knowledge construction (e.g. SPENDER 1996a).[186]

What is common to all knowledge-based contributions is that firms are conceptualised as heterogeneous, knowledge-bearing entities which cannot be interpreted only by looking at their contractual constitution (KOGUT/ZANDER 1992: 383).[187] The proponents of the knowledge-based view underline the functional dimension of firms as organisational entities which are repositories of distinct productive (technological and organisational) knowledge and can learn and grow on the basis of this knowledge (FOSS 1996: 470). Thus, following the argumentation of organisational theory, the analysis of firms' competitiveness should be grounded in the understanding of what they know how to do.

As a basic premise, knowledge is subject to economies of scale and scope (GRANT 2002: 136). Its initial creation is more costly than its subsequent replication. In contrast to physical resources, it does not decrease with use. Thus, economies of scale in knowledge, together with the complementarities of different types of knowledge imply increasing returns in knowledge-intensive contexts. To the extent that knowledge is not specific to the production of a specific good, economies of scale translate into economies of scope.

3.4.3.2 Structural and processual dimensions of firms' knowledge

As EISENHARDT and SANTOS (2002: 140f.) note, researchers in the strategy field traditionally used a concept of knowledge that is focused on the explicit nature of knowledge. Knowledge of firms was modelled as an unambiguous,

---

186 SPENDER (1996a) argues that a dynamic theory of the firm based on knowledge should be conceptually different from the resource-based approach, because knowledge does not represent an observable and transferable commodity. Instead, the knowledge which a firm possesses derives from ongoing learning processes and from the social interactions between the individual and organisational identity. Knowledge thus becomes the basis of human meaning and communication (EISENHARDT/SANTOS 2002: 144).

187 "Our view differs radically from that of the firm as a bundle of contracts that serves to allocate efficiently property rights. (...) Rather, we suggest that organisations are social communities in which individual and social expertise is transformed into economically useful products and services by the application of a set of higher-order organising principles. Firms exist because they provide a social community of voluntaristic action structured by organising principles that are reduceable to individuals (KOGUT/ZANDER 1992: 384). This view has for instance criticised by FOSS (1996: 474) who argues that one should not oversee the fruitful complementarities that exist between a contractual and a knowledge perspective of the firm.

reducible and easily transferable construct, while knowing was associated with processing information.[188] In contrast to this was the traditional conception which emphasises the absolute, static, and non-human nature of knowledge (e.g. expressed in propositional forms in formal logic). According to SPENDER (1996a: 47ff.), the knowledge-based view primarily focuses on two main dimensions of knowledge that account for the competitive advantage of firms.

The first conceptualisation of knowledge is based on a "positivistic-functionalistic" (AL-LAHAM 2004: 408) perspective which defines knowledge as "(...) knowledge as a dynamic human process of justifying personal beliefs as part of an aspiration for the "truth" "(NONAKA 1994: 15). In this context, the firm's knowledge represents the objective and explicable aggregate of all individual cognitions and capabilities which are generated through individual learning processes. Against the background of strategic management, such objectifiable stocks of knowledge can be comprehended as a static resource input for production which has to meet the conditions of high strategic value, rareness, inimitability, and non-substitutability as formulated by the resource-based theory (BARNEY 1991: 106ff.).

The second conceptualisation is more strongly linked to "pluralistic-interpretative" (AL-LAHAM 2004: 409) epistemology and moves away from a universally valid definition of knowledge. Instead, knowledge is seen as a complex, social construct which is closely embedded and intertwined in its social context. Due to its context dependency it can hardly be objectified or explicated. Thus, following this constructivist approach, the analysis of knowledge emphasises the contextual understanding of knowledge and particularly the social interactions by which knowledge is generated, shared, used, and imitated.

---

188 This "has given rise to several theories that suggest a machine-like functioning of organisations. For example, scientific management theories [like Taylorism - *author's note*] posit that the organisation of work should be entirely determined by codified knowledge, and that the knowledge of the firm is held by a select number of individuals [=management - *author's note*]. Similarly, the information-processing perspective treats organisations as machines that use rules and routines to address the individual information processing requirements caused by interdependent work and environmental uncertainty" (EISENHARDT/SANTOS 2002: 140).

This conceptual distinction is reflected in the basic considerations of NONAKA (1994: 15f.) and KOGUT and ZANDER (1992: 386f.) who basically categorise the firm's knowledge into "information" and "know-how":

- By ***information***, they describe all kinds of knowledge which can be transmitted without loss of integrity once the syntactical rules required for deciphering it are known. Thus, information includes facts, data, axiomatic propositions, and symbols. For public dissemination, information is standardised in order to be understood at minimal cost. Knowledge in terms of information implies knowing *what* something means.

- Drawing on the definition by VON HIPPEL (1988), ***know-how*** describes the accumulated practical skill or expertise that allows one to do something smoothly and efficiently. The notion of "accumulation" thereby stresses the assumption that know-how must be learned and acquired. Knowledge in terms of know-how implies knowing *how* to do something. Such procedural knowledge is a description of what defines current practice inside a firm (e.g. regularity of the structuring of work, the interactions of employees).

Following this basic conceptualisation, the perhaps most fundamental differentiation of the knowledge-based view is that between *the general, basic stock of knowledge ("knowledge stocks")* of the firm and *the processes of learning ("knowledge flows")* which aim to modify and renew the knowledge stock. The knowledge-based view thus comprises a static-structural as well a dynamic-processual dimension of firms' knowledge (WELGE/AL-LAHAM 2008: 105; AL-LAHAM 2004: 409).[189]

3.4.3.2.1 Knowledge stocks - The structural dimension

As an empirical matter of fact, firms persistently differ in their relative competitive performance. Using a knowledge-based view, this means that firms persistently differ in their performance because they differ in their information and know-how, their knowledge stocks and their knowledge flows. Thus, the persistence of differentials in firm performance lies in the joint problem of the diffi-

---

189 As AL-LAHAM (2004: 409) amends, the differentiation between stock and flow dimensions of competitive resources was first introduced in strategic management literature by DIERICKX/COOL (1989).

culty of transferring and imitating knowledge in terms of information and know-how (KOGUT/ZANDER 1992: 387). For this reason, the knowledge-based view has to progress beyond the mere classification of knowledge into information and know-how by reasoning *why,* due to which of its characteristics is knowledge not easily transmitted and replicated between firms.

For the purpose of developing a knowledge-based view of the firm, KOGUT and ZANDER (1992: 387ff.), GRANT (1996: 111f.), and NONAKA (1994: 16ff.) identified several characteristics explaining why knowledge may not be easily transferable and replicable between firms. These are:

- the type of knowledge (implicit and explicit knowledge)
- the carriers of knowledge (individuals and organisation)
- the content of knowledge (specificity, social complexity)

The first dimension differentiates between *tacit and explicit knowledge,* a conceptualisation which can be traced back to the work of POLANYI (1958, 1966) and was further made prominent by NONAKA (1994: 16f.) and NONAKA and TAKEUCHI (1995: 8f.). Explicit or codified knowledge refers to knowledge that can be assigned and transmitted by formal, systematic language. Explicit knowledge can be structured into a set of identifiable rules, mental models and relationships that can be easily communicated using language, and is thus alienable from the individual person by codification (e.g. like cooking recipes, user manuals, text books). In contrary, tacit (or implicit) knowledge has a strong personal quality which makes it hard to formalise and communicate. "Though I cannot say clearly how I ride a bicycle (...), yet this will not prevent me from saying that I know how to ride a bicycle (...)." (POLANYI 1958: 88) Tacitness is something personal, an ability or skill to do something or to resolve a problem that is based, in part, on one's own experiences and learning. For these reasons, tacit knowledge is a knowledge resource that is difficult to make explicit for transfer and reproduction (GALUNIC/RODAN 1998: 1195ff.). As NONAKA (1994: 19), NONAKA and TOYAMA (2003: 5), and NONAKA and KONNO (1998: 43) argue, firms need to establish mechanisms to convert tacit into explicit knowledge (for the purpose of knowledge transfer and reproduction), and - vice versa - to convert explicit knowledge into tacit knowledge (for the purpose of preserving and storing their relevant knowledge in terms of tacit knowledge). Moreover, the reproduction and transfer of tacit knowledge which cannot be made explicit at all requires time-consuming processes of socialisation.

Within the firm, tacit knowledge is created through idiosyncratic, experience-based learning processes. Thus, it is imbued with a high level of firm specificity which means that it can be used outside the firm only with significant deterioration (SPENDER 1996a: 52, also: SPENDER 1996b). Tacit knowledge is deeply rooted in action, commitment, and involvement in a specific context. It cannot be expressed outside the action of the person who has it. In general, individuals are not even aware of the fact that they possess such tacit knowledge (FORAY 2006: 71). Therefore, tacit knowledge is closely bounded to its firm-specific context within which it emerged and is used.

Obviously, these characteristics of tacit knowledge fit perfectly with the conditions for valuable competitive resources postulated by the resource theory. Because tacit knowledge cannot be easily explicated and transferred between individuals, is closely bounded to the specific social context of the firm, and thus is almost inimitable for competitors, it builds the main competitive resource for firms. While economies of scale and scope are great for explicit knowledge, information in particular, which is "costly to produce, but cheap to reproduce" (SHAPIRO/VARIAN 1999: 3), tacit knowledge tends to be significantly more costly to replicate. But nevertheless, these costs are probably lower than those incurred in its original creation (GRANT 2002: 136).

However, it is important to mention that POLANYI - contrary to the widespread interpretation and citation[190] - did *not* mean to establish a dichotomy between tacit and explicit knowledge (NEUWEG 1999: 47ff.; GRANT 2007: 175f.). For him, there is not an either/or between tacit and explicit knowledge as *all* knowledge has a tacit component which results from the individual heuristics, experiences and practices in making use of explicit knowledge (POLANYI 1958: 87; likewise: KOGUT/ZANDER 1992: 389).[191] Opposed to the mainstream opinion, Polanyi does not say that tacit knowledge cannot be transferred *per se*

190 See for example NONAKA (1994: 16) or NONAKA/TAKEUCHI (1995: 8) who made these two knowledge dimensions popular. Moreover, while KOGUT/ZANDER (1992), like NONAKA, believe that tacit knowledge can be made explicit through processes of codification, COOK/BROWN (1999) see tacit and explicit knowledge as completely distinct and complementary types of knowledge.

191 "The tacit cooperates with the explicit, the personal with the formal" (POLANYI 1958: 87). And further: "For just as, owing to the ultimately tacit nature of all our knowledge, we remain ever unable to say all that we know, so also, in view of the tacit character of meaning, we can never quite know what is implied in what we say" (POLANYI 1958: 95). This is rather more complex that the extensively used quote from "The Tacit Dimension" (POLANYI 1966: 4), which says "I shall reconsider human knowledge by starting from the fact that we know more than we can tell."

between individuals. Rather he suggests that, appropriately using a common language, most tacit knowledge can be shared between individuals. However, as tacit knowledge cannot be explicitly specified by codes and rules, it always remains closely interlinked to the specific context of action. The skill of bicycling is only transferable by observation, imitation and practice. It cannot be specified in detail or be transferred by prescription, since no prescription for it exists. "It can be passed on only by example from master to apprentice. This restricts the range of diffusion to that of personal contacts." (POLANYI: 1958: 53)[192] Thus, tacit knowledge should also not be equated with practical knowledge, as practical knowledge consists in large parts of explicit elements. Instead, it seems more adequate to interpret the tacit knowledge dimension in terms of "unconscious" knowledge.

Building upon this major differentiation, other proponents of the knowledge-based perspective suggested more detailed classifications of knowledge, because the distinction between tacit and explicit knowledge is not so much the result of the difficulty or complexity of knowledge, but rather how easy or difficult it is to express (TIDD/BESSANT 2009: 543). For example, WINTER (1987: 170ff.) distinguishes between the four dimensions of tacitness, complexity, systems dependence, and observability. The tacit dimension is further divided into teachability and codifiability. SANCHEZ (2001: 10f.) speaks about "know-how" (learning by doing), "know-why" (learning by analysing), and "know-what" (learning what is relevant to know). BLACKLER (1995: 1023ff.) even developed a finer typology of five types of knowledge: *embrained knowledge, embodied knowledge, encultured knowledge, embedded knowledge,* and *encoded knowledge.* But since these dimensions are closely tied to the tacit-explicit dimension and the individual-organisation dimension, I shall not further expand on them.

The second dimension of knowledge is located at the level of social interaction and deals with the differentiation between the knowledge carriers of *individuals and of the whole organisation.*[193] Considered as an entity, both levels form the

192 "By watching the master and emulating his efforts in the presence of his example, the apprentice unconsciously picks up the rules of the art, including those which are not explicitly known to the master himself. These hidden rules can be assimilated only by a person who surrenders himself to that extent uncritically to the imitation of another." (POLANYI 1958: 53)

193 The distinction between the levels of individuals, groups, and organisation (see e.g. REBER 1992; WIEGAND 1996) which is commonly used in the German literature is not popular in the Anglo-American literature (AL-LAHAM 2004: 428, amendment #15).

total stock of knowledge of firms. Individuals are the primary carrier of explicit and - most importantly - tacit knowledge. As was shown above, individuals are the sole instance within an organisation that can create tacit knowledge (NONAKA 1994: 21).

However, the knowledge-based view postulates that only collective, organisational knowledge serves as a source of competitive advantage. If individual knowledge stays implicit and is not being transferred and integrated among the members of the organisation or aggregated into some kind of organisational knowledge, it remains more or less valueless. Therefore, individual knowledge is primarily seen as the necessary precondition in which organisational, firm-specific, and tacit knowledge can emerge and develop (VAN DEN HOOF ET AL. 2004: 13; HAAS/HANSEN 2005: 2). Firms do not attain persistent competitive advantage until they manage to develop, exploit, and evolve a common stock of shared, "sticky" (VON HIPPEL 1994) knowledge.[194] Hence, firms possess a common stock of "organisational" knowledge which is shared by all its organisational members (PAUTZKE 1989: 78).

Important types of such organisational knowledge are (GRANT 1996: 116):

- ***Common language:*** Common rules of verbal communication between the individuals of the firm.
- ***Other common forms of symbolic communication:*** Literacy, numeracy, and familiarity with the same working appliances (e.g. computer software).
- ***Commonality of specialised knowledge:*** A certain basic stock of specialised knowledge or understanding in the context of the technologies, products, services applied or produced by the firm.
- ***Shared meaning:*** Common cognitive schema and frameworks, metaphor and analogy, and stories.

194 For example, HAAS/HANSEN (2005: 2) see organisational knowledge as "[...] a property of the overall firm, rather than of individual members or task units". Likewise, DYER/HATCH (2004: 1158) stresses the social dimension of organisational knowledge as they argue that individual explicit or tacit knowledge does not become relevant for competitive performance until it is "[...] embedded in the firm's routines, human skills, and relationships". In contrast, BERMAN ET AL. (2002: 15) suggest that the organisational knowledge required to perform a complex task is diffused among the individuals. Each individual possesses only a part of the whole organisation's knowledge.

- ***Recognition of individual knowledge domains:*** Recognition of the abilities or expert knowledge of other firm members.

Interestingly, organisational knowledge permits individuals to share and integrate aspects of knowledge which are not common between them, like highly specialised expert knowledge, and their tacit knowledge (GRANT 1996: 115; NONAKA/TAKEUCHI 1995: 80f.; GRANT 2002: 139f.). On the one side, the common stock of firms' knowledge thereby serves as an enabler of further knowledge integration by providing a communicative framework and communicative rules. On the other side, the firm's collective knowledge is generated and modified by ongoing integrative processes of individuals' knowledge, and thereby relies on knowledge-generating processes within firms.

As BECKER ET AL. (2006: 213) point out, such types of organisational knowledge represent, however, just the idea of what GIDDENS (1976: 107; 1984: 375) called "mutual knowledge".[195] It is the fundamental precondition which enables individuals to interact within social contexts.

According to BECKER ET AL. (2006: 213ff.) this is not sufficient to describe tacit organisational knowledge as a source of competitive advantage. Therefore, they add the categories of distinctive *areas of unshared, specific expert knowledge* within organisations and *"knowledge about knowledge"* (BECKER ET AL. 2006: 215). The first category describes expert knowledge which is not shared among all organisational members. It can be typically found at the level of departments, work groups or teams. As they argue, such expert knowledge at the intermediate organisational level - despite the fact that it is not integrated and shared within the whole organisation - is highly relevant because of its strongly idiosyncratic character. The second category focuses on the ability to access (knowing where the relevant knowledge is located within the firm) and to validate knowledge and sources of knowledge.

Finally, by starting from the dimension of content, the third dimension deals with the *specificity of knowledge.* The higher the specificity of knowledge for the firm, the lower the possibility to use this knowledge outside the firm or within other competing firms (COLLIS 1996: 151). As the level of specificity depends on the level of contextual embeddedness, this accounts particularly for the body of tacit knowledge (AL-LAHAM 2004: 411). Another source of specific-

---

195 "I use the term 'mutual knowledge' to refer generally to taken-for-granted 'knowledge' which actors assume others possess, if they are 'competent' members of society, and which is drawn upon to sustain communication in interaction" (GIDDENS 1976: 107).

ity lies in the synergetic integration of individual and organisational knowledge through long-term learning processes. The resulting complexity, context dependency, and social embeddedness increases the potential for value-added.

3.4.3.2.2 Knowledge flows - the processual dimension

As the knowledge-based view explicitly presumes a dynamic understanding of knowledge and is deeply influenced by streams of organisational learning and cognitive psychology, particular attention is paid to the process of knowing instead of knowledge as a or less transferable resource (EISENHARDT/SANTOS 2002: 141ff.). A firm's knowledge is considered to be socially constructed and being created through ongoing social interactions grounded in working practices (COOK/BROWN 1999: 387), and the specifics of the social and cultural setting of the firm (GALUNIC/RODAN 1998: 1198ff.).

With regard to the knowledge-generating process, the knowledge-based view refers to individual learning theories (WELGE/AL-LAHAM 2008: 106; for learning theories see REBER 1992; WIEGAND 1996: 336ff.). NOOTEBOOM (2009: 184ff.) distinguishes three types of learning: a) generation of new knowledge by adopting existing ideas, insights or skills from others through communication or imitation, b) learning through collaboration by recognising new ideas, insights or skills from outside the firm, and c) learning in the sense of creative discovery from experience. However, the particular focus of the knowledge-based perspective lies on creating organisational knowledge through integrating the scattered individual stocks of knowledge at the organisational level, as it is assumed that the firm's competitive advantage arises through the successful integration and building of a specific stock of knowledge at the organisational level.

Most fundamentally, knowledge is created and carried by individuals either through their formal qualification and expert knowledge (rather explicit knowledge) or through direct "hands-on" experience and practice (rather tacit knowledge). As only individuals can create tacit knowledge, they are the "prime mover in the process of organisational knowledge creation (...)" (NONAKA 1994: 21), and ideally supported by an organisational context which promotes individuals to create tacit knowledge.[196] Thus, the process of knowledge creation within the firm is initiated by enlarging its individuals' knowledge. Individual, explicit and especially tacit knowledge remain personal as long as they are articulated through social interaction between the firm's individuals.

While explicit knowledge can be easily transferred between different individuals, tacit knowledge cannot. Thus, NONAKA (1994: 18ff.) particularly stressed the need to converse tacit into explicit knowledge between organisational members through "externalisation". He argues that tacit knowledge can be transformed into explicit knowledge through a process of conceptualisation and crystallisation which makes tacit knowledge explicit and codified in some persistent form.[197] But as GRANT (1996: 114) argues, the mere conversion of tacit into explicit knowledge is not a sufficient condition for integrating individual knowledge at the organisational level. Instead, he suggests that is more efficient to establish modes of interaction so that the tacit knowledge of individuals is integrated without inefficient cross-learning processes of conversion.[198]

GRANT (1996a: 114f., 2002: 138ff.) proposes several mechanisms to integrate and transfer individual (tacit) knowledge, like rules and directives, sequencing,

---

196 NONAKA (1994: 21f.) mentions two central factors of work organisation: a) the "variety" and "quality" of an individual's experience. For example: if this experience is limited to routine operations, the amount of tacit knowledge will decrease over time because routine tasks do not promote creative thinking and the formation of new knowledge. b) the second factor is the coupling of knowledge and "bodily" (and not abstract) experience, for example in the way that firms encourage the use of judgement and knowledge formed through interaction with customers rather than by "objective", scientific conceptualisation.

197 The other types of conversion of individual knowledge into organisational knowledge proposed by NONAKA (1994) are: *socialisation* - tacit to tacit knowledge through personal exchange between individuals or within groups; *combination* - explicit to explicit knowledge through defined processes or technological systems; *internalisation* - explicit to tacit knowledge through organisational learning. Together with the dimension of *explication*, which is described in the main text above, these types of knowledge conversion build the "SECI-model" that can be seen as the birth of practical knowledge management (MIKIC ET AL. 2009: 26)? The SECI-model has incurred numerous criticisms due to practical and epistemological reasons (see, for instance GOURLAY 2006; GRIFFIN ET AL, 2001). Nevertheless, it still forms a central part of knowledge management (MIKIC ET AL. 2009: 28).

198 "As I have already argued, the key to efficiency on knowledge integration is to create mechanisms that reconcile the efficiency in knowledge creation (which requires specialisation) with efficiency in knowledge deployment (which requires integrating many types of knowledge)" (GRANT 2002: 138).

or techniques of group problem-solving and decision-making. The knowledge-based view draws thereby upon the concept of organisational routines (GRANT 1996: 115; KOGUT/ZANDER 1992: 388) coined by NELSON and WINTER (1982), which was mentioned earlier. Organisational routines constitute repetitive and recurrent sequences of complex patterns of autonomous interactions between individuals in the absence of rules, directives, or even significant verbal communication and are thus predestined for the integration of tacit forms of knowledge. Organisational routines thereby build the basis of collective learning in organisations and somehow represent a manifestation of the organisation's memory as they encode inferences from history, and guide individual and group behaviour (EISENHARDT/SANTOS 2002: 141).

By these mechanisms, the firm establishes and maintains a stock of collective organisational knowledge which comprises all elements of knowledge shared by and available for all its members. Such organisational routines represent the essential links between the different levels of a firm's knowledge (WELGE/AL-LAHAM 2008: 107). From an evolutionary perspective, it could be assumed that these mechanisms may fulfil the organisational task of knowledge retention and replication.

Few scholars have stressed the primacy of the individual dimension of knowledge as the nucleus for organisational knowledge (GRANT 1996: 112; SIMON 1991: 125). For them, an organisation is only able to learn either by the learning of its existing members or by ingesting new members who have knowledge the organisation previously did not have. Consequently, they rather underline those organisational processes by which firms access and utilise the knowledge possessed by their members. Others instead argue that the integration of individual knowledge is more than their mere aggregation, because this integration is a fundamental social phenomenon of complex social interactions, and which is therefore inseparable from the social context of the organisation (NAHAPIET/GHOSHAL 1998: 246; NELSON/WINTER 1982: 63; KOGUT/ZANDER 1992: 384, 1996: 503; SPENDER 1996a: 51; BROWN/DUGUID 2001: 200ff.).[199]

---

199 "(...) the possession of technical "knowledge" is an attribute of the firm as a whole, as an organized entity, and is not reducible to what any single individual knows, or even to any simple aggregation of competencies and capabilities of all the various individuals, equipments and installations of the firm (NELSON/WINTER 1982: 63).

For a detailed discussion of the individual or organisational locus of knowledge, see the paper by FELIN/HESTERLY (2007).

3.4.3.3 Combinative mechanisms

Whatever presumption one follows, both perspectives presume that firms as organisations exist as institutions for producing goods and services because they can create conditions under which multiple individuals can integrate their specialist knowledge (GRANT 1996: 112) and thereby - depending on the perspective - enable the creation of new (organisational) knowledge by replication and learning (KOGUT/ZANDER 1992: 384). In this sense, organisations - and so firms - represent the most effective and efficient coordinative framework conditions for the integration, replication and creation of knowledge relevant for competitive success.

But if it is argued that the knowledge of the firm causes competitive advantage, learning cannot be characterised as independent of the current capabilities. The firm-specific character and the uniqueness as necessary preconditions for knowledge to have effects on the firm's competitive performance do not emerge until the interaction between the firm's knowledge stocks and knowledge flows takes place (AL-LAHAM 2004: 411). By introducing the term of "combinative capabilities", KOGUT and ZANDER (1992: 391) point to the close relationship between the knowledge-based view and dynamic concepts of the resource-based theory of strategic management.[200]

Combinative capabilities mean the intersection of the firm's capabilities to exploit its existing knowledge and to explore the potential of new or recombined knowledge (KOGUT/ZANDER 1992: 391f.).[201] To describe this interplay of knowledge stocks and knowledge flows, the knowledge-based literature mainly refers to three concepts (AL-LAHAM 2004: 411f.):

- absorptive capacity

---

200 Against this background, organisational learning is perceived as an adaptive change process to external stimuli that is influenced by past experience, focused on developing and modifying routines, and supported by organisational memory (NONAKA/TAKEUCHI 1995: 115; EISENHARDT/SANTOS 2002: 141). NELSON/WINTER (1982) were among the first to integrate organisational knowledge and routines in the context of dynamic competitive environments. In their approach to evolutionary economics, the firm is understood as a repository of knowledge which is deposited by organisational routines that guide organisational action.

201 This resembles the notion by SCHUMPETER (2008a: 94f.) that the production of other things or the same things by a different method means combining these materials and forces differently. Development in his sense is then defined by carrying out new combinations.

- path-dependency
- causal ambiguity

The first concept of the firm's *absorptive capacity* (COHEN/LEVINTHAL 1989, 1990) is rooted in the theory of organisational learning (CYERT/MARCH 1992) and was briefly discussed in chapter 4.2.4.4. It describes the firm's ability to identify and absorb strategic relevant knowledge from its external environment and to implement this external knowledge in terms of concrete improvements or new developments, which increase its competitive performance. In the knowledge-based view, the firm's absorptive capacity is a competitive advantage in itself, because this ability cannot be acquired externally. The absorptive capacity of a firm is shaped by highly firm-specific, time-consuming, context-specific, implicit learning routines at the individual and the organisational level (NELSON/WINTER 1982: 75ff.). Thus, absorptive capacity describes the firm's ability to add new knowledge to existing knowledge by establishing processual heuristics of search, cognition, absorption, and implementation (GRANT 1996a: 111; EISENHARDT/SANTOS 2002: 141f.; GRANT 1996b). It is based upon explicit and implicit collectively held knowledge which was obtained by overcoming previous difficulties and challenges in the past. For this reason, absorptive capacity is a self-referential concept. By having recourse to the concept of absorptive capabilities, it becomes apparent how strongly the knowledge-based view, in its processual form, relates to other dynamic concepts of the resource-based theory (e.g. TEECE ET AL. 1997). In addition, this makes the point that knowledge can also be integrated externally through relational networks that span organisational boundaries (see, for instance GRANT 1996b; KOGUT/ZANDER 1996).

The second concept *path-dependency* originates in the evolutionary economics approach of NELSON and WINTER (1982). It describes the circumstance that the firm's ability to identify, recognise, absorb, and implement new knowledge is determined by its past trajectories due to the historical-evolutionary and idiosyncratic nature of the corresponding learning processes (AL-LAHAM 2004: 412). However, it is not clear, how the concepts of absorptive capacity (as a dynamic capability) and path-dependency depends on or even contradict each other, because dynamic capabilities may particularly contribute to overcoming the problem of path-dependency and cognitive lock-in effects (TEECE ET AL. 1997: 522f.; SCHREYÖGG/KLIESCH 2005a: 14).

The third concept of *causal ambiguity* addresses the observation that the causal relationship between the existing knowledge stock and the processes of knowledge generation, integration, and modification are characterised by high levels of intransparency and ambiguity (REED/DE FILLIPPI 1990; SIMONIN 1999). The level of intransparency and ambiguity increases a) the more knowledge which is relevant for competition is of a tacit nature, b) the more specific the knowledge is, and c) the more complex the relationship is between knowledge stocks and knowledge flows (REED/DEFILIPPI 1990: 91ff.). As described by scholars of the resource-based theory (LIPPMAN/RUMELT 1982; DIERICKX/COOL 1989), causal ambiguity constructs barriers to imitation because competitors are not able to reconstruct or reproduce such firm-specific knowledge systems.

3.4.4 Summary and criticism

As the previous remarks show, the knowledge-based view of the firm represents a current theoretical approach within the resource-based strategic management literature which can be seen as quite promising. It takes knowledge as the central determining factor to explain the heterogeneity among firms and thus to explain the persistent differences in their competitive performance. The heterogeneity of firms is thereby reflected both in their already existing stocks of knowledge *and* their ability to generate new knowledge or modify existing knowledge in terms of learning processes. Firm-internal routines and mechanisms to coordinate and integrate "knowledge stocks" and "knowledge flows" thereby cause and preserve this heterogeneity and firm-specificity, and thus finally lead to the firm's persistent competitive advantage (Figure 11).

"(...) organisations know more than what their contracts can say" (KOGUT/ZANDER 1992: 383). By this paraphrasing of POLANYI'S (1966: 4) famous dictum "(...) we know more than we can tell" means that a certain part of the firm's stock of knowledge which determines its competitiveness and economic success, is "tacit" and "social". Knowledge is produced and reproduced in social settings as well as being inseparably embedded in these settings and this not fully reducible to the firm's individual members (FOSS 1996: 471).

Within this argument, the knowledge-based view integrates elements and streams of epistemology, organisational learning, and evolutionary theory into the theoretical paradigm of the resource-based theory. But in contrast to the traditional resource-based perspective, the hitherto fuzzy and diffuse concept of firm-specific resources has been much more refined and narrowed to the competitive impact of a firm's stock of knowledge and its ability to generate and modify this stock of knowledge.

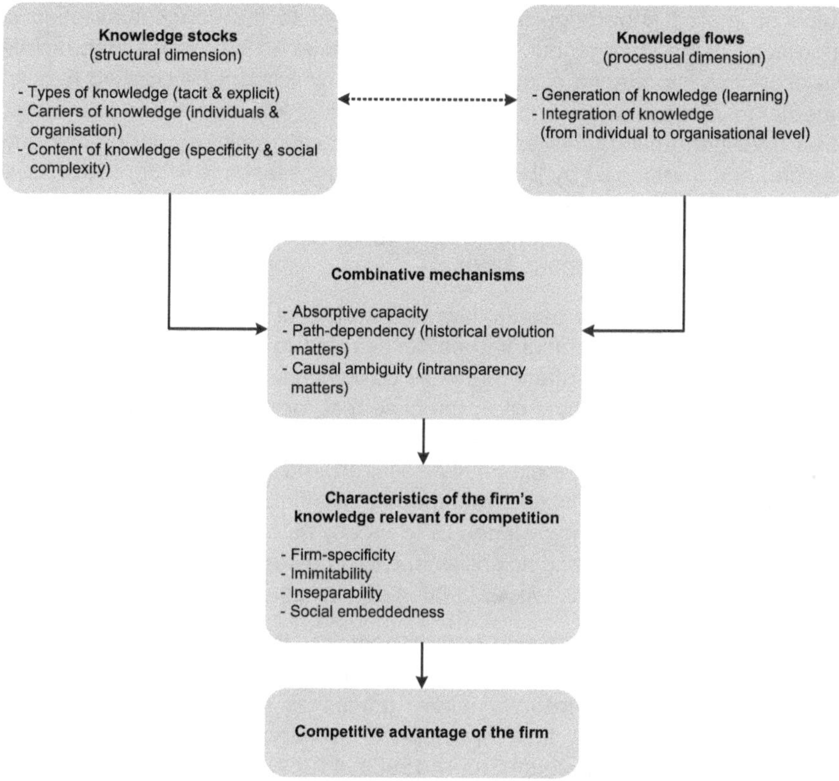

Figure 11: The knowledge-based view of the firm
(Source: own illustration, based on WELGE/AL-LAHAM 2008: 106; AL-LAHAM 2004: 410)

While firms were earlier described as a more or less blurred bundle of resources and competences, they are now described as, in essence, a "body of knowledge" (SPENDER 1989: 185) or a "distributed knowledge system" (TSOUKAS 1996). By elaborating and specifying the strategic characteristics of knowledge, the knowledge-based view firstly reveals those framework conditions under which the firm's knowledge stock can cause persistent competitive advantage and how this knowledge can be protected against imitation. In terms of strategy, both explicit and tacit knowledge can be very important. It is generally argued, however, that tacit knowledge is more strategically important as it is embedded in individuals and extremely difficult for competitors to replicate. In resource-based view terminology, it is inimitable. The particular importance of implicit knowledge is thereby opposed to the conceptualisation of knowledge in the endogenous growth theory (ROMER 1990) which emphasises the economic importance of explicit knowledge which can be codified (e.g. by patents) to increase to the economy's total stock of knowledge and from which other R&D-performing firms are profiting.[202] Instead, tacit knowledge can only be used and transferred by the individual who possesses it. Thus, tacit knowledge can in terms of SCHUMPETER be understood as "entrepreneurial knowledge". If it is not used or transferred, it remains irrelevant for the economic system (RÖPKE/STILLER 2006: XXXI).

Secondly, the knowledge-based view adds a distinct processual dimension by focussing on firm-internal learning processes and to the extent to which their characteristics positively affect the firm's competitive advantage. This ability to learn gains particular relevance in the case that the firm's existing knowledge stock becomes obsolete because of changes in the external environment; and new knowledge has to be externally acquired and learned. Thus, as knowledge is one of the most flexible resources that a firm has access to, its ability to quickly generate new knowledge or to modify its existing knowledge is vital for strategic adaptability and flexibility, and can be seen as a concrete example of "dynamic capability" (AL-LAHAM 2004: 425f.; WELGE/AL-LAHAM 2008: 98).

Third, in contrast to transaction cost economics, the knowledge-based view explicitly points out that firms (or organisations in general) provide a conditional framework which institutionalises processes by which individuals can integrate and coordinate their specialist knowledge more efficiently than it would be possible in markets (GRANT 1996a: 112).[203] Markets are not able to generate immobile, tacit knowledge, nor can they protect knowledge which has been made explicit from imitation.

For these reasons, the economic resource-based theory has tended more and more towards a competence- and knowledge-based view of the firm during the last years (VON KROGH/GRAND 2002: 163; BECKER ET AL. 2006: 204; also: VON KROGH/GRAND 2004). Subsequently, the issues of a firm's organisational competences, its organisational stock of knowledge, and the "collective learning in

---

202 Likewise, the very recent contributions of VANDENBUSSCHE ET AL. (2006) and AGHION/HOWITT (2006) within endogenous growth theory particularly focus on the role of explicit knowledge gained through higher education as the major driver of economic growth.

203 See also TSOUKAS (1996).

the organisation" (PRAHALAD/HAMEL 1990: 82) take centre stage in explaining persistent competitive advantage. As EISENHARDT and SANTOS (2002: 145ff.) show in their literature review on empirical research within the knowledge-based view, there have been numerous studies in the fields of knowledge sourcing, internal and external knowledge transfer as well as knowledge integration which support the basic line of theoretical argumentation. "Our view is that KBV *[the knowledge-based view - amendment by the author]* offers enormously useful theoretical insights, well grounded in empirical findings that address the multi-level social processes through which knowledge is sourced, transferred and integrated within and across organisations" (EISENHARDT/SANTOS 2002: 161). More recent empirical work was done, for example, by CZARNITZKI and WASTYN (2009), BHATT (2001), DARROCH (2005), CHEN ET AL. (2009), and ZHEN ET AL. (2009).

However, the knowledge-based view gives rise to several points of criticism. These points are:

- ***No consistent knowledge taxonomies:*** Like almost all theoretical strands within the resource-based paradigm, the knowledge-based view of the firm suffers from substantial inconsistencies in how knowledge is conceptualised (EISENHARDT/SANTOS 2002: 159f.). Even the most commonly accepted distinction between tacit and explicit knowledge is controversially interpreted. While some scholars (e.g. COOK/BROWN 1999) see tacit and explicit knowledge as two distinct and complementary forms of knowledge, KOGUT and ZANDER (1992) as well as NONAKA (1994) and NONAKA and TAKEUCHI (1995) believe that tacit knowledge can (or even should) be made more explicit through processes of codification or externalisation. Regarding further classifications of knowledge, some authors (e.g. ZANDER/KOGUT 1995) deploy the detailed taxonomy of WINTER (1987), while others draw on more aggregate constructs (e.g. SZULANSKI 1996). Even more confusing is the fact that some authors tend to equate tacit knowledge with procedural knowledge, a consideration that clearly contradicts the concept of POLANYI (1958) who suggested that all types of knowledge contain tacit and explicit elements.[204] As correctly noted by EISENHARDT and SANTOS (2002: 160), the knowledge-based view should centre on developing a consistent classification of knowledge which enables comparing empirical findings across different studies

204 "I identify knowing how with tacit knowledge and knowing about facts and theories with explicit knowledge" (GRANT 1996a: 111).

and improving the accretion of understanding about the knowledge-based view.

- ***Divergence on the meaning of knowledge between the theoretical and the empirical literature:*** As described above, the theoretical perspective of the knowledge-based view contains several conceptions of knowledge like thinking in terms of different kinds of knowing, linking mechanisms between individual and organisational knowledge, or the emphasis on the emergence and identity of knowledge. However, the empirical literature - especially the strategy literature - so far largely focused on one, relatively simplistic, functionalistic conception of knowledge as a "resource" (EISENHARDT/SANTOS 2002: 158; AL-LAHAM 2004: 419f.). Investigations of the processual perspective of learning processes and combinative mechanisms have been scarce to date. With regard to the concept of absorptive capacity, first progress was made by recent studies (e.g. ZAHRA/GEORGE 2002; ARBUSSA/COENDERS 2007; FOSFURI/TRIBO 2008; MUROVEC/PRODAN 2009) which stress the broad, dynamic perspective instead of cutting down the firm's absorptive capacity to its stock of R&D-generated knowledge. Thus, one of the major challenges for the knowledge-based view lies in the development of adequate indicators for empirically measuring of processes related to organisational learning.

- ***Missing link between the value of knowledge and contextual variables:*** In its basic argumentation, the knowledge-based view suggests a "the more the merrier" perspective of knowledge and knowledge-related processes. While this assumption may have surface appeal, it is hard to believe that more knowledge sourcing, transfer, and even integration are necessarily advantageous (EISENHARDT/SANTOS 2002: 159). Instead, it seems to be crucial to consider the respective strategic value of knowledge and learning abilities which may also be moderated by variables of the firm's environment (e.g. EISENHARDT/GALUNIC 2000; GUPTA/GOVINDARAJAN 2000). It can be reasoned that the strategic value of knowledge or processes of knowledge generation also depends, for instance, on the firm's external market dynamics or technological life cycles. Moreover, as already mentioned in the summary of the resource-based theory, the firm's knowledge and learning processes also have to align with tactical activities.

- ***Unclear relationship between individual and organisational knowledge as determinant of competitive advantage:*** There is still confusion within the knowledge-based view whether it is primarily individual or organisational knowledge that leads to the firm's competitive advantage (FELIN/HESTERLY 2007). Some scholars argued that individual stocks of (tacit) knowledge play a crucial role for competitive success (GRANT 1996; SIMON 1991) and thus vote for an actor-oriented view of methodological individualism. In contrast, other authors follow a collectivistic perspective by stressing the fact that only collective, organisational knowledge can cause competitive advantage, because the heterogeneity between the firms' stock of knowledge results from the heterogeneous, firm-specific processes of knowledge integration and their embeddedness in the social context of the firm (KOGUT/ZANDER 1992, 1996; NAHAPIET/GOSHAL 1998; NELSON/WINTER 1982; SPENDER 1996a; TSOUKAS 1996; BROWN/DUGUID 2001). But as BECKER ET AL. (2006: 205f.) state, such an exclusive collectivistic approach underestimates the importance of unshared knowledge at the individual level in terms of distributed expert knowledge which particularly may contribute to the generation of competitive advantage. Moreover, it neglects the increasing tendencies of organisations to decentralise or split off specialised functional areas to increase their flexibility and adaptability to external changes. HAAS and HANSEN (2005: 4f.), for instance, additionally point out that the management of collectively integrated, organisational knowledge may even cause competitive disadvantages because of the related cost of search and transfer. Last but not least, the knowledge-based view systematically underestimates political aspects, egoisms, and opportunism within the firm that may inhibit knowledge integration or even render it impossible (DUSCHEK 2002: 139ff.; TSAI 2002: 179). Thus, the knowledge-based view, on the one hand, needs to consider such aspects and, on the other hand, needs to develop adequate theoretical solutions to better integrate the competitive effect of both individual and collective stocks of knowledge within firms as "distributed knowledge systems" (TSOUKAS 1996).

- ***Equation of tacit knowledge and knowledge relevant for competitive success:*** It has been shown above that, due to its causal ambiguity, tacit knowledge is said to be of particular relevance for the firm's persistent competitive success, as it cannot be easily imitated by competitors, nor can it be deployed efficiently outside its original social context. But within strategic management, this argumentation causes serious problems because the concept of causal ambiguity impedes any intentional and purposeful strategic activity to generate or control this kind of knowledge which is most important for the firms' competitive success. Vice versa, if thought through to the end, the central dependent variable "attainment of persistent competitive advantage" would thus solely rely on random, unpredictable, unconscious and thus uncontrollable processes of tacit knowledge creation (BECKER ET AL. 2006: 206f.; likewise for the resource-based approach: REED/DE FILIPPI 1990: 100f.; PETERAF 1993: 187; SCHREYÖGG 1999:

402). However, if the knowledge-based view claims to be an approach within strategic management, this conceptualisation seems therefore not very satisfactory.

- ***Are hierarchical mechanisms of knowledge coordination really more efficient than those of markets?*** Authors like GRANT (1996a: 112) explicitly argue that firms are more efficient in providing adequate conditions for knowledge exchange and integration than markets are. In doing so, the knowledge-based literature seeks a theory of the firm that is independent of transaction cost logic and its behavioural assumption of opportunism. While others generally questioned the claim that a firm can be seen independently of opportunism (e.g. FOSS 1996: 473ff.), NICKERSON and ZENGER (2004: 617) note that the two fundamental arguments within the literature that support the efficiency of firms in knowledge exchange relative to markets are completely contradictory. While one argument claims that hierarchies exist to essentially avoid knowledge transfer (DEMSETZ 1988, CONNER 1991, CONNER/PRAHALAD 1996), emphasising the firm's capacity to exercise authority in directing others' actions; the other view claims instead that hierarchies exist to facilitate knowledge transfer (KOGUT/ZANDER 1992, 1996; NAHAPIET/GHOSHAL 1998), emphasising the firm's capacity to support the formation of shared language and identity. Thus the knowledge-based view needs to overcome its a priori assumption that firms are superior in terms of knowledge transfer and it should provide theoretical arguments under which conditions the hierarchy coordination mechanisms of the firm are preferred to those of markets, and vice versa.

  The work of NICKERSON and ZENGER (2004) first articulates the knowledge-based advantages and disadvantages of both markets and firm hierarchies and matches these to desired types of knowledge formation. NOOTEBOOM (2009: 72ff.) recently contributed some general theoretical arguments to that issue. He suggests that the firm's advantage relative to market lies in the circumstance that they provide more cohesion in exploitating of existing resources and knowledge while markets provide more variety for exploration.[205]

Nevertheless, the knowledge-based view so far represents a contemporary research programme which offers a theory of competitive advantage that, de-

205 "The duality of market and firm yields a solution to the problem of combining exploitation and exploration (NOOTEBOOM 2008: 121).

spite its shortcomings, has high explanatory potential (AL-LAHAM 2004: 425), and which offers a surprisingly consistent body of empirical results on a pluralistic understanding of knowledge, and a view of organisations as complex adaptive systems of knowledge (EISENHARDT and SANTOS 2002: 161).

As this thesis does not focuse on a deep analysis of different types of knowledge, but rather deploys this approach to round off the scope of non-R&D-performing firms' potential innovation resources, most of the criticisms above do not play a crucial role herein. What should be recognised, however, is that equating of tacit knowledge with knowledge that is relevant for competitive advantage should be avoided. Therefore, it would be preferable to focus not only on tacit knowledge, but also take sources of explicit knowledge into consideration.

3.5 R&D as only one resource among equals to explain firms' competitive advantage

After the discussion of the resource-based theory and its extensions in the form of the relational-based and the knowledge-based view of the firm, we can conclude that each provides a fruitful basis for analysing the innovation patterns of firms in general, and non-R&D-performing firms in particular. As BARNEY and CLARK (2007: 221ff.) show, numerous empirical tests so far have provided "(...) an impressive and comprehensive body of empirical evidence of resource-based theory. (...) the results demonstrate highly consistent findings across the domain of strategy and all of these other management disciplines" (IBID.: 245).

Table 1 compares the three approaches within the resource-based paradigm which have were discussed in this chapter. For instance, the resource-based theory focuses on both tangible and intangible resources which already exist within the firm. It encompasses for instance technological, financial or human capital as well as organisational capabilities like routines, competences, and even meta-routines to renew and modify the firm's resource base. In contrast, the relational-based perspective focuses on inter-firm collaboration and firms' strategic alliances as a source of competitive advantage by either combining already existing complementary resources or generating new valuable resources.

The knowledge-based view points to knowledge as a firm's most crucial competitive resource. Most interestingly, it argues that tacit knowledge represents the most valuable resource for competitive advantage; a position which contradicts an exclusive concentration of R&D, because R&D per definition seeks to overcome the tacit dimension by grounding solutions to inter-subjectively replicable methods. Besides, the knowledge-based view particularly grasps the dynamic perspective (that was already explained under the concept of 'dynamic capabilities' in the resource-based theory) and stresses the aspect of organisational learning as a prerequisite for the ongoing renewal, expansion, and modification of the firm's knowledge stock.

Nevertheless, it was also shown, that all of these approaches despite their numerous strengths also suffer from individual shortcomings. Therefore, for this thesis, it seems practical to integrate these three partial approaches of the resource-based perspective into an overall resource-based framework of firms' competitive advantage, because they may complement each other as each of them emphasises a particular aspect. The approaches within the resource-based paradigm, particularly those focussing on the processual dimension of resource generation and modification, are closely related to the evolutionary perspective presented in chapter three. They explicitly recognise differences and heterogeneity of resource endowments across firms and their individual, socially embedded heuristics of how they generate, combine, deploy, and modify their stock of resources to achieve competitive advantage (FOSS ET AL. 1995: 3). Integration these three approaches seems widely unproblematical, because they share a set of common premises and characteristics (isolating mechanisms) according to which the respective resources become valuable and relevant for competitive success. As the table shows, causal ambiguity, idiosyncracy, inimitability, and inter- or context dependency build common conditions for the competitive value of resources. Even the relational-based view, which expands the potential sources of competitive advantage primarily to external sources and thus has to deal with issues of transaction cost, remains compatible in its basic assumptions.[206]

206 As already discussed, all of these approaches lack a clear and consistent terminology to define what a "resource" means in their particular context. However, it will not be possible to solve this problem in this thesis. Nonetheless, it is important to come up with an adequate working definition of the term "resource" in the present context. I will address this issue in the following chapter when conceptualising the resource-based framework for the purpose of empirical analysis.

Table 1: Overview of resource-based approaches within strategic management (Source: own illustration)[207]

	Resource-based theory	Relational-based view	Knowledge-based view
Unit of analysis	The firm	Pair, groups or networks of firms	Firms and organisations as social entities
Primary sources of above-normal rent / competitive advantage	Rare, valuable - physical, technological or financial resources - human resources - individual and organisational competences - routines, repetitive patterns of behaviour and processes - Ability to renew and modify the resource base through dynamic capabilities	- Relation-specific investments - Routines for inter-firm knowledge exchange - Complementary resource endowments - Effective mechanisms of governance	- Stock of implicit knowledge which is generated through individual experience and held by individuals or the whole organisation - Ability of the firm to renew and modify the stock of knowledge through processes of learning - (Stock of explicit knowledge)
Isolating mechanisms	- Idiosyncracy - Causal ambiguity - Immobility - Inimitability	- Causal ambiguity - Interdependencies - Complementarities of interorganisational assets - Scarce choice of adequate partners - Indivisibility of common resources	- Causal ambiguity - Inimitability - Context dependency - Social embeddedness
Ownership/control of valuable resource	Individual firm	Collaborating partners	- Individuals (implicit and explicit knowledge) - Firm (shared explicit and implicit knowledge)
Origin of valuable resource	Internal	External	Internal (external)

207 Please note that the table was structurally adapted from DYER/SINGH (1998: 674) who compare the industry structure view, the resource-based view, and the relational-based view of strategic management.

As this study explores the heterogeneity of non-R&D-performing firms' innovation patterns, integrating all three approaches allows this very question to be taken into account, as the analytical framework is not narrowed down to either only the firm-internal or external dimension or exclusively limited to only one specific type of innovation input. Hence, the approaches broaden the range of what could serve as a firm's competitive advantage far beyond the issue of mere R&D, by explicitly emphasising heterogeneity (i.e. specificity, individuality, non-imitability) of firms' innovation resources and thus fitting into the previous argumentation based on evolutionary theory.

Of course, as shown in the literature review in chapter 2.6, R&D, in terms of R&D expenditure or personnel in R&D, is nevertheless an important resource for firms' innovativeness and competitive advantage. R&D represents a major source of explicit knowledge, as it strongly relies on the systematic and intentional deployment of scientific methods which can be documented and intersubjectively transferred between individuals. This kind of expert knowledge accumulated through R&D can be made visible for others (e.g. by publications, patents) in terms of unique characteristics, and may in turn increase the firm's attractiveness for cooperation partners or new customers.

Moreover, R&D may also fit the conditions of valuable resources in the sense of resource-based theory, because the process, routines and competencies necessary for R&D activities themselves can prove difficult for other firms to imitate. Thus, the firm-specific nature of the R&D process impedes imitation: when the R&D process has an important tacit or intangible element, it is difficult for others outside the firm to replicate the process (HELFAT 1994: 175). Hence, this thesis will not question the relevance or importance of R&D as a firm level driver of innovation.

However, starting from the resource-based approaches discussed above, it should be pointed out that firm level R&D is just one (although important) resource of a broad variety of other important sources of firms' competitive advantage. Therefore, the assumption that non-R&D-performing firms - *by definition* - have a systematically lower innovation performance, or are less competitive than R&D-performing firms due to their missing R&D resource, cannot longer be sustained in view of the resource-based paradigm.

Hence, the most important result of this chapter is that, despite (or because of) the fact that they lack the specific R&D resource, non-R&D-performing firms can basically draw on many other types of innovation input to enhance and maintain their innovativeness and competitive advantage: technical, non-technical, physical, immaterial, firm-internal, firm-external innovation resources as well as different types of knowledge. Thus, the heterogeneity of non-R&D-performing firms' innovation patterns can be assumed to manifest itself primarily via these different sources of competitive advantage. Hence, the question of *which innovation resources contribute towards enhancing and maintaining non-R&D-performing firms' competitive advantage* cannot be ultimately answered by resource-based theory. But with regard to the underlying resource question, the results from resource-based theory provide a good framework in which possible dimensions of non-R&D-performing firms' sources of competitive advantage can be located, and from which empirical analysis can start.

As the critics of the market-based view already argued, resources themselves may not turn out to be valuable until they result in economic effects that contribute to the firm's competitive advantage. As this is indeed a very serious problem of the resource-based theory, I will address this shortcoming by contrasting the resource-dimension with measurements of innovation output and economic performance in order to control whether these innovation resources indeed contribute towards enhancing the firms' economic success or not. For this reason, I will supplement the original research question "which innovation resources contribute towards enhancing and maintaining non-R&D-performing firms' competitive advantage?" by an additional output dimension:

> *If firm level innovation is ultimately related to attaining competitive advantage, what innovation output and economic performance is achieved by non-R&D-performing firms on the basis of their innovation resources?*

These questions will be first addressed in a descriptive firm level analysis (chapter 5.3) before they undergo cluster analysis (chapter 5.4). These questions should be kept in mind as they will be further developed in the following chapter, which integrates the theoretical backgrounds of evolutionary and resource-based theory into one conceptual and methodological framework for empirical analysis.

4 Development of a guiding conceptual framework for the empirical analysis

Basically, this chapter serves as a bridge between the theory and the following empirical analysis. The first step reexamines the research questions of this thesis, clarifying how they have been developed during the previous theoretical discussion, on the one hand, and how they will be addressed by the empirical analysis, on the other hand. It integrates the theoretical considerations of evolutionary and resource-based theory about the heterogeneous innovation patterns of non-R&D-performing firms into a conceptual framework for empirical analysis. Based on this theoretical framework, the key concepts will be defined and some important implications for the design and methodology choice of the empirical analysis will be derived. This chapter closes by erecting some guiding theses for the explorative empirical analysis. However, the function of these theses should not be confused with the role of hypotheses in the context of testing causal relationships. Instead, they should serve as a kind of verification instrument to assure that the later results are in line with the theoretical framework and that they are not arbitrarily detached from the previous theoretical suggestions.

4.1 The research questions revisited

First of all, I want to briefly revisit the basic research questions which have been discussed in the previous chapters (Figure 12). The starting point was this thesis' basic research question:

> *Do heterogeneous innovation patterns of non-R&D-performing firms exist in the German manufacturing industry?*

To answer this research question, it was divided into two aspects for the subsequent theoretical discussion. The first addressed whether structural heterogeneity of non-R&D-performing firms can be expected at all (chapter 3.1). The underlying research question was formulated as follows:

> *To what extent do heterogeneous innovation patterns of non-R&D-performing firms exist in the German manufacturing industry?*

Following the reasoning of evolutionary economic theory, chapter 3.1 argued that heterogeneous innovation patterns of non-R&D-performing firms can be expected, even if these are all located in similar industries and despite the fact that they all lack the same innovation resource of R&D.

Based on this, the second aspect dealt with, in which dimensions of innovation resource variation this heterogeneity may be visible and thus analysable (chapter 3.3 and 3.4), and addressed the following research question:

> *If not formal R&D, which firm-internal and firm-external innovation resources are deployed by non-R&D-performing firms in the German manufacturing industry?*

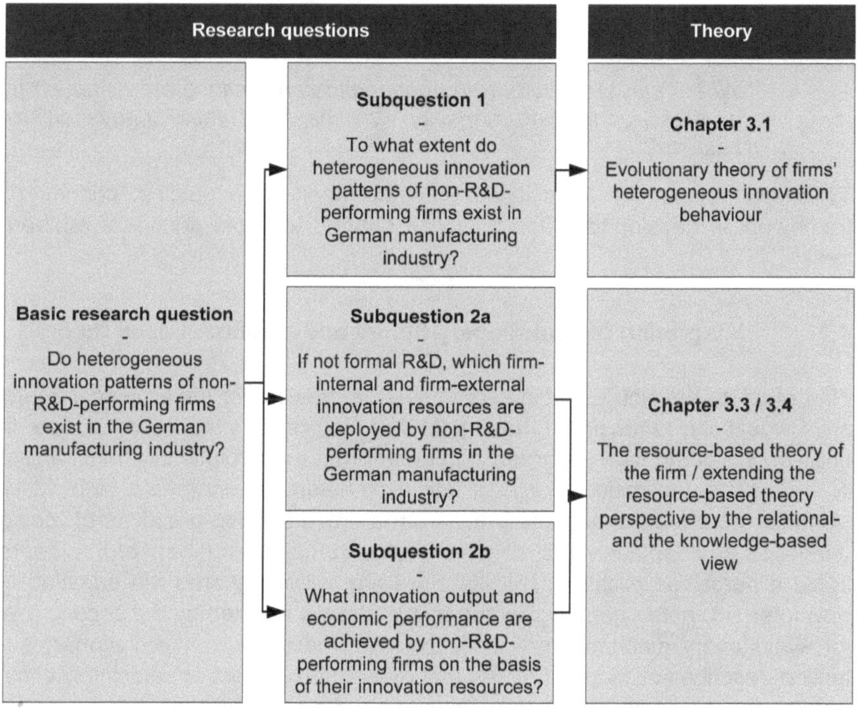

Figure 12: Overview of research questions to be addressed by the empirical analysis (Source: own illustration)

In line with the previously developed evolutionary framework, it has been argued that the evolutionary mechanism of variation which generates hetereogneity mainly corresponds to the input dimension of innovation, namely the use of different resources for innovation by non-R&D-performing firms. As they do not perform formal R&D, non-R&D-performing firms have to rely on alternative firm-internal and external resources to enhance and maintain their competitive advantage. Based on the resource-based theory of strategic management and its recent extensions by the relational- and knowledge-based perspective, it has been argued that – beyond formal R&D – there are multiple potential dimensions through which firms can obtain a competitive advantage.

However, the second evolutionary mechanism of selection, which corresponds to the dimension of innovation output and economic performance which is how the firm becomes visible on its market, is widely neglected by the existing resource-based approaches. In order to include this output dimension in the analysis of non-R&D-performing firms' innovation patterns, the following subquestion has been added:

> *What innovation output and economic performance are achieved by non-R&D-performing firms on the basis of their innovation resources?*

To answer these questions in an empirical analysis, a guiding, conceptual framework is developed in the following chapter to avoid analytical arbitrariness.

4.2 Integration of evolutionary theory and resource-based theory

First of all, to develop a guiding conceptual framework for the empirical analysis, I would like to recall the major points of the previous theoretical debate in chapter 3. Today, there is broad agreement that the evolutionary mechanism of variation is the major source of inter-firm heterogeneity (see chapter 3.1). Variation here means any departure from a firm's existing practices of "doing things" as an adaptation to changes in their external environment or in order to solve a perceived problem. But literature has widely ignored the question of how inter-firm heterogeneity created by variation is affected by the second major evolutionary mechanism of selection. The mechanism of evolutionary selection describes all kinds of forces that differentially select or selectively eliminate certain types of variations by rewarding strategies that result in the firm's enhanced competitive advantage and punishing those resulting in lower competitiveness in the struggle for scarce resources. As "selection destroys the measure of variety on which it depends, so that (...) the variance of behaviour is driven to zero by selection" (METCALFE 1994: 330), it may be assumed that the force of selection works against firms' heterogeneity, particularly in the case of relatively homogeneous selection criteria.[208]

This thought is taken up by sectoral approaches to innovation systems which – to some extent - are implicitly based on the assumption that innovation systems constitute similar selection criteria which lead to relatively homogeneous innovation behaviour on the part of the embedded firms (SRHOLEC/VERSPAGEN 2008: 9f.). Following this assumption, sectoral-based studies of innovation have taxonomised sectors and the firms within them into larger groups, each characterised by a typical innovation strategy (e.g. PAVITT 1984; MALERBA/ORSENIGO 1995, 1996; EVANGELISTA 2000; MARSILI 2001; ARCHIBUGI 2001; PENEDER 2001, 2003; DE JONG/MARSILI 2006; CASTELLACCI 2008b). Applied to the present research question, this implies that low-tech industries and sectors are characterised by similar criteria of market selection and thus consist of a homogeneous population of non-R&D-intensive firms.

But in analogy to biological evolution, three arguments were presented in chapter 3.1.2.2 for why significant heterogeneity may remain even after (sectoral) selection has taken place over time:

- External selection criteria as part of the environment may also be highly variable.

208 Selection here is understood to mean "market selection" in the economic sense. Market selection focuses on the economic output or competitive performance that is achieved by new variations, not on the nature of the variations itself. But firms as social entities and their variations in human action or social behaviour are also subject to a kind of "internal" selection (e.g. WITT 1997, 1999) that – besides mere economic selection criteria – may also be driven by normative or value-oriented selection mechanisms. For instance, an economic purpose can be achieved either through participative forms of work organisation or – perhaps even more efficiently (and thus favoured by pure economic selection criteria) - by exploitative forms of work organisation like slavery. But the latter is of course to be negatively selected by management if the firms and its members are embedded in a liberal-democratic social system. Thus, if one accepts selection as a universal, not only biological, principle (HODGSON 2002), it is reasonable to assume that selection in social or economic systems is characterised by a temporal sequence of various, interdependent processes of internal and external selection. Moreover, according to the system approaches of innovation (LUNDVALL 1988; EDQUIST 1997), it is reasonable to assume the existence of subordinated non-economic levels of external selection, which consist of institutionalised political, social or ethical norms of selection (ALDRICH/RUEF 2006). Thus, non-economic levels of selection may also work on the characteristics of variations, while economic market selection works on the economic effects that are achieved through variation. I consider this by assuming that variations which deliver their economic effects up to market selection have already passed through firm-internal normative and ethical selection mechanisms.

- "Neutral theory of evolution" (KIMURA 1983) which stresses that selection works on the tangible effects and consequences of variations, not on their nature.
- The evolutionary stability of mixed strategies based on evolutionary game theory argues that a mixed strategy can be an evolutionary stable strategy within a given population (BERGSTROM/GODFREY-SMITH 1998; HERMANN-PILLATH 2002).

These arguments provide good reasons for the assumption that even within similar frame conditions (in the sense of similar selection criteria), heterogeneity in firms' behaviour can be expected. This thesis thus follows the approach of DOSI (1988a) who argues that, despite common inter-sectoral frame conditions of firms (like technological opportunities, possibilities of technology acquisition, or public support), a "large, unexplained residual" (GRUPP 1997: 81) of intra-sectoral, institution-specific innovation behaviour remains which results from differences in firm-specific structures, proportions and innovation strategies. To this extent, the neoclassical assumption of the "representative enterprise" has to be abandoned (GRUPP 1997: 82). The increasing availability of firm data from national innovation surveys has resulted in several papers which are very critical of the presumed sectoral regularities in innovation patterns. Instead of classifying industries or sectors, they focus on the heterogeneous innovation types observed at the firm level such as CESARATTO and MANGANO (1993), HOLLENSTEIN (2003), ARUNDEL and HOLLANDERS (2004), LEIPONEN and DREJER (2007), or SRHOLEC and VERSPAGEN (2008).[209]

Thus, the assumption of heterogeneous innovation patterns of non-R&D-performing firms can be reasonably derived from evolutionary theory. Nevertheless, heterogeneity itself does not yet provide a substantive explanation of their possible characteristics and peculiarities. Therefore, it is necessary to put some flesh on the bones of heterogeneity. This was achieved by referring to the integrated resource-based theory of the firm.

209 Just to clarify: it is *not* the intention of this thesis to suggest that the sector level has no explanatory value for firms' innovation behaviour at all. It argues that - within the context of evolutionary theory - there are good reasons for an explorative analysis of the patterns of firms' innovation strategies starting at the firm level in order to adequately account for the maximum level of heterogeneity. Once such firm level patterns have been identified, further analyses could clarify to what extent they can be described or explained by sector variables. Another promising strategy was very recently presented by PENEDER (2010), who attempts to integrate both dimensions in a joint classification of firms and industries.

As the competitive advantage of non-R&D-performing firms cannot easily be explained by the existence of internal R&D, it is neccessary to look elsewhere for the sources of their innovativeness and competitive advantage. In evolutionary terms, this means focussing on other resource variations that achieve similar innovation success and allows firms to to survive market competition. Thus, it is necessary to choose an approach which allows for multiple, heterogeneous innovation input beyond mere R&D. Within the evolutionary theory, authors like NELSON and WINTER (1982), DOSI and MARENGO (1994), or TEECE ET AL. (1997) have integrated this perspective by referring to the works of PENROSE (1959) and streams of organisational learning like CYERT and MARCH (1992) or SIMON (1991). Thus, within the evolutionary paradigm of this study, the resource-based theory of the firm (BARNEY 1986a, 1991a, 1991b; WERNERFELT 1984; RUMELT 1984; PETERAF 1993) as well as its recent modifications by the relational-based view (DYER 1996; DYER/SINGH 1998) and the knowledge-based view of the firm (KOGUT/ZANDER 1992; GRANT 1996a; SPENDER 1996a; NONAKA 1994; NONAKA/TAKEUCHI 1995) seem to provide an adequate framework to analyse the innovation patterns of non-R&D-performing firms (see chapter 3.3 and 3.4).

Because the three streams within resource-based theory complement each other by each emphasising a particular aspect, it seems reasonable to integrate them into an overall resource-based framework of the innovation patterns of non-R&D-performing firms. They explicitly recognise the differences and heterogeneity of resource endowments across firms and the individual, socially embedded heuristics of how firms generate, combine, deploy, and modify their stock of resources to gain a competitive advantage. Additionally, it seems quire straightforward to combine these three approaches, because they share a set of common premises and characteristics (isolating mechanisms) under which the respective resources becomes valuable and relevant for competitive success.

Finally, the conceptual framework can be derived through the composition of evolutionary theory and resouce-based theory (Figure 13). Evolutionary economic theory is used to explain the structural dimension of innovation patterns by explaining why heterogeneous innovation patterns can be theoretically expected among firms without formal R&D. To substantiate the dimension of structural heterogeneity, the integrated resource-based perspective serves to identify possible innovation resources of non-R&D-performing firms beyond formal R&D. Hence, both theoretical aspects together provide a suitable framework for analysing the heterogeneous innovation patterns of non-R&D-performing firms. This makes it possible to modify the basic research question of this thesis as follows:

To what extent are the heterogeneous innovation patterns of non-R&D-performing firms in the German manufacturing industry reflected in their individual use of different innovation resources? How can these heterogeneous innovation patterns be characterised regarding their innovation output and economic performance?

As mentioned above, both the resource-based and the evolutionary approach share a common doctrinal basis represented by the seminal contribution of PENROSE (1959). Briefly, one cen see evolutionary theory as having developed the dynamic aspects laid out by PENROSE'S view of firm growth as a continuous and cumulative "unfolding" process (PENROSE 1959:1), while the resource-based view is more focussed on the analysis of the resources themselves. Moreover, both approaches are rooted in the central concept of firm heterogeneity caused by firm-specific, highly-individual cognitive heuristics and types of behaviour. Especially the concept of dynamic capabilities is strongly related to evolutionary aspects.

Despite this, there are also some major differences between evolutionary and resource-based theory which could not be resolved in this context.[210] For instance, as already mentioned, the earlier, structure-oriented perspective of the resource-based view is very different from evolutionary theory, as it is based on an equilibrium understanding of factor markets. Additionally, while evolutionary theory focuses on routines as "genes" of organisations via which evolutionary progress is replicated and stored, the resource-based perspective views them as one type of (intangible) resource for the firm's competitive advantage, among others.

Integrating both strands into a conceptual framework, therefore, does *not* mean that they are merged into one single, encompassing theory built on a common basis of assumptions or a common level of analysis. Instead, the two approaches are combined only for the instrumental purpose of addressing the two basic elements of structure and substancy in order to characterise non-R&D-performing firms' innovation patterns.

210 A very good overview of these commonalities and differences is provided in the book by MONTGOMERY (1995 – "Resource-Based and Evolutionary Theories of the Firm: Towards a Synthesis") and particularly in the contribution by FOSS ET AL. within this book (pp. 1-18).

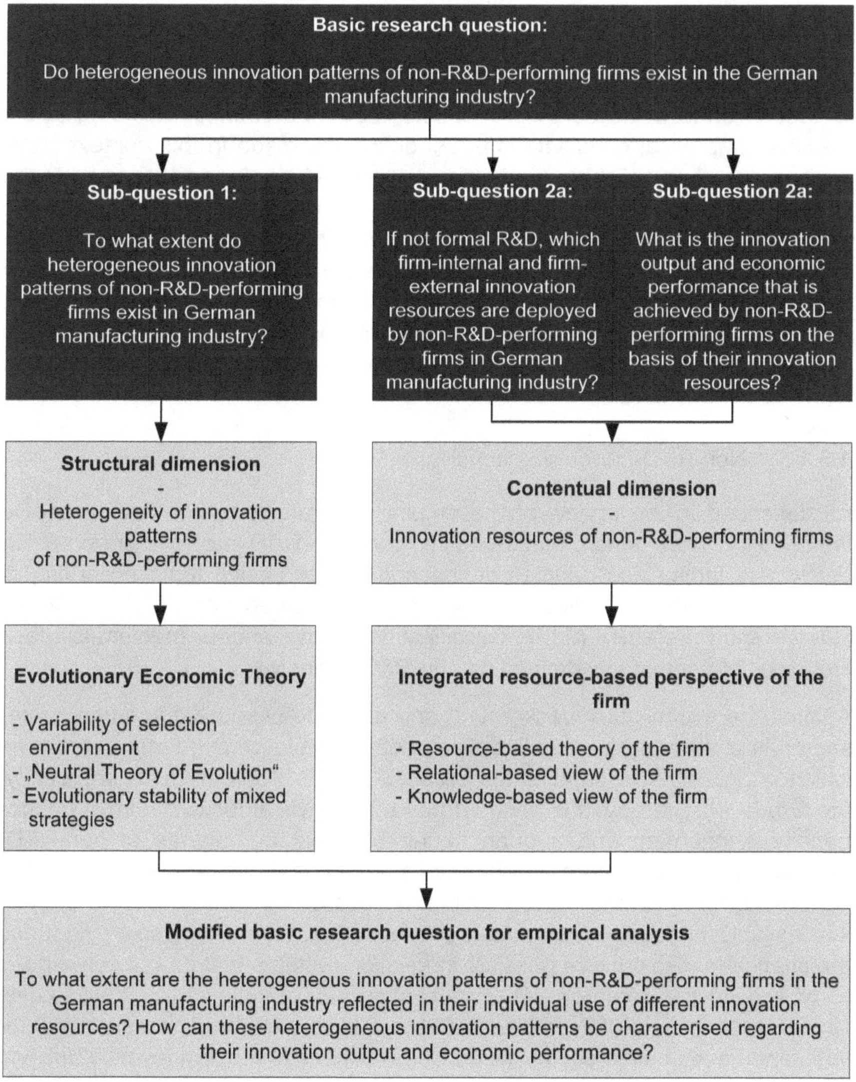

Figure 13: Analytical framework
(Source: own illustration)

4.3 Defining the key concepts

Based on these cornerstones of the conceptual framework, some clarification is needed on how the central constructs should be defined according to the previous theoretical considerations. All definitions made in this context serve the purpose of establishing a common understanding of the terms used in the subsequent analysis. Thus, the definitions applied here are nominal. This means, that no definition contains any empirical information. Instead, the mean of the definiendum (the term to be defined) is equated with the mean of one or a set of already known terms in a tautological manner (KROMREY 1998: 149). Therefore, nominal definitions cannot be considered true or false. They can only be justified or rejected by their appropriateness or inappropriateness to describe the definiendum.

4.3.1 Non-R&D-performing firms

As described in the review of the empirical literature on non-R&D-intensive firms, so far, there is no common definition of non-R&D-intensive or non-R&D-performing firms. Most studies define non-R&D-intensive firms according to their sector affiliation, which means that firms located in OECD's (1994) or LEGLER and FRIETSCH'S (2007) classification of low-, and/-or medium-low-tech industries are summarily defined as non-R&D-intensive.

Against the evolutionary background and resource-based argumentation presented in chapter 3, this thesis defines non-R&D-performing firms at the level which is best able to capture the broadest range of inter-firm heterogeneity: *the firm level*. Thereby, it is ensured that the analytical focus on firms' heterogeneity is not narrowed a priori to the pertinent aggregates of non-R&D-intensive, low-tech-sectors. For the same reason, the subsample of non-R&D-performing firms should *not* be limited in advance to firms that have already successfully introduced a product or process innovation. Doing so would forfeit the chance to also capture possible innovation patterns that do not correspond to or exceed the common set of innovation output indicators. Other studies (e.g. KIRNER ET AL. 2007, 2009a, 2009b) have also followed these considerations and have applied sector-based classifications to the firm level by building a firm level taxonomy based on the firms' individual share of R&D expenditure in total sales.

However, in the sectoral definition as well as in the firm-level definition based on sectoral classifications, non-R&D-intensive firms may still have R&D expenditure. But it can be questioned whether it is really reasonable to assume that a firm that spends less than 2.5% on R&D is less R&D-intensive than a firm with 3% R&D expenditure. This is also supported by the discussion in chapter 2.6.4, which argued that some of the major determinants of firm level R&D are high entry costs, a certain minimum size of R&D projects, and the high risk and uncertainty associated with R&D projects. Therefore, it can be assumed that the landmark decision for firms about R&D is not so much the actual R&D intensity (e.g. if 2.5% or 3% of total sales are invested in R&D), but whether to engage in formal, in-house R&D at all (yes/no). Accordingly, authors like RAMMER ET AL. (2009), BARGE-GIL ET AL. (2008), ARUNDEL ET AL. (2008) or HUANG ET AL. (2010) decided to select their subsamples of non-R&D-performers by applying a zero-level condition of R&D intensity at firm level. In doing so, the following types of non-R&D-performing firms can be distinguished:

- Firms that do not perform in-house R&D but may contract R&D performed by external firms or organisations
- Firm that neither perform in-house R&D nor external/contracted R&D.

However, regarding the issue of internal and external R&D expenditure, chapter 2.6.6 showed that, in most cases, internal and external R&D expenditures complement each other. "The majority of innovative firms without permanent in-house R&D refrain from any kind of R&D activity." (RAMMER ET AL. 2009: 38). To account for this circumstance, I do not distinguish between internal or external R&D expenditure when defining non-R&D-performers. Of course, R&D cooperations with external partners or obtaining knowledge impulses from external R&D organisations that go beyond the mere purchase of ready-to-use R&D solutions will be considered in the empirical analysis.

To counter these problems, this thesis defines of non-R&D-performers very strictly as follows:

> ***Definition:*** *Non-R&D-performing firms in the manufacturing industry are firms that do not perform any internal or externally contracted R&D in terms of R&D expenditure.*

4.3.2 Innovation

In capitalist market economies, competition is the basic coordinative principle underlying the economic order and shaping the social structures (BURR 2004: 58). Through competition, the emergence and distribution of profit and income is inseparably tied to performance. Economic actors who perform better will receive greater reward than those who do not perform well. This primarily performance-related distribution is also the basis for a secondary redistribution mechanism that follows the principle of need and is usually conducted by governments and public law/authorities. The guiding principle of competition is therefore closely intermeshed with general social and moral concepts of freedom, democracy, state of law, efficiency, social welfare, and peace. Pressured by competition, economic actors are forced to develop new products, services, manufacturing processes, and organisational concepts to maintain and increase their competitive ability. Thus, market competition drives firms to detect and exploit efficiency potentials and to choose the optimal (re-) combination of input factors and resources. In capitalist market economies, technological change and innovation are thus mainly driven by competition:

> "That innovation is the crucial source of effective competition, of economic development and the transformation of society has become very widely accepted." FREEMAN (2003: 4f.)

> "In the nineteenth century, de Tocqueville foresaw that the dynamics of capitalist competition would greatly stimulate the development of innovative activities that showed the prospect of a commercial return." (PAVITT 2001b: 13221)

> "(...) virtually all of the economic growth that has occurred since the eighteenth century is ultimately attributable to innovation." BAUMOL (2002: 13)

Innovation can occur in multiple ways. Most prominently, new products or service offers with new functionality, design, lower price, improved customisation, or superior quality allow firms to capture and retain market shares and to increase sales in those markets. However, process innovation is just as important as product innovation. This describes the ability of a firm to produce its goods in a way no-one else can (e.g. precision, flexibility), or in a way that is better than anyone else (e.g. price, quality, customisation). Based on the mechanism of market competition, innovation activities thus become a "generic core business process" (TIDD/BESSANT 2009: 54ff.) of firms and companies and leave firms with no option but to innovate in order to survive in capitalist economies.

Although most authors acknowledge innovation to be more complex than first appears, several authors attempted to define innovation (e.g. ROGERS 1998; READ 2000; TETHER 2003).[211] Despite the fact that a lot of research on innova-

---

211 "It is a serious mistake to treat an innovation as if it were a well-defined homogeneous thing that could be identified as entering the economy at a precise date." (KLINE/ROSENBERG 1986: 283)

tion has taken place to date in many different scientific disciplines, no single approach has succeeded in uniting the fragmented thinking under one consistent umbrella theory able to provide commonly agreed definitions and theoretical concepts. Since this thesis is also not able to develop a universal understanding of innovation, it follows the strategy of other existing studies which – depending on their specific research interests – define innovation on the basis of numerous different dimensions (see e.g. VAHS/BURMESTER 2002: 73ff.). In this context, it has to be clarified how the term "innovation" corresponds to the evolutionary framework of this thesis and what dimensions of innovation are to be covered by the subsequent empirical analysis.

Given the research objective of this thesis, the concept of innovation has to meet the following requirements. Firstly it has to fit into the evolutionary framework, secondly, it should explicitly address the resource-based perspective which has been chosen to analyse the innovation patterns of non-R&D-performing firms, and thirdly, it should be broad enough to capture the theoretically expected heterogeneity of the innovation activities of non-R&D-performing firms.

Etymologically, the term "innovation" means "new", "novelty" or "renewal" (DISSELKAMP 2005: 16). As a consequence, innovation is frequently equated with invention, which means the novelty aspects represents the prevailing element of innovation (Kirner et al. 2006: 4). But as originally introduced by SCHUMPETER (2006: 164; 2008: 91ff.), economic research should strictly distinguish between these two terms, because inventions themselves do not refer to economics.[212] An invention does not become an innovation until its successful market implementation and its economic exploitation (SCHUMPETER 2006: 158f.). He suggested that innovation comprises two dimensions, and has to be distinguished from general creative activity in the arts, philosophy or basic re-

212 The distinction between inventions (which focus on the "new") and innovation (which focuses on economic exploitation) made by SCHUMPETER, is rooted in his early conceptualisation of the entrepreneur as the driving promoter of innovation. This was influenced by the economic situation in the late 19th and early 20th centuries that was characterised by legendary entrepreneurs who adopted the ideas of inventors and made them commercially successful. Thus, for SCHUMPETER it must have been obvious that the process of invention does not need to follow the laws of economics. It is the genuine ability of the entrepreneur to recognise the commercial potential of these inventions, and to transform them into economically successful products or processes.

For a more detailed discussion of the differentiation between invention and innovation, see also BURR (2004: 25ff.).

search (IBID.: 142f.). The first element of invention refers to the dimension of novelty, which means developing or creating something new without planning its commercial application. Therefore, a mere novelty may have little if any economic value. The second element of innovation is linked to the successful commercial application of a novelty which helps to increase the firm's competitive advantage and economic performance as a result (SCHUMPETER 2006: 143).

Hence, innovation encompasses invention plus economic exploitation: the invention process creates a new idea and gets it to work; the exploitation process develops and disseminates the innovation commercially in such a way that it creates value for the enterprise (ROBERTS 1988: 13; DAMANPOUR/WISCHNEVSKY 2006: 272; VAHS/BURMESTER 2002: 44; ROGERS 1998: 9). Innovations describe the purposeful development and implementation of new, value-creating solutions to problems that aim to meet institutional and corporate targets by new means. The OSLO Manual (2005: 29) states that firms innovate for the ultimate reason of improving their performance, whereas it is of no importance if the innovation is new to the market. In doing so, the OSLO Manual follows a "subjective" view on innovation and notes, that "new" can also mean "new to the firm" (IBID.: 48f.). This aspect in the OECD definition underlines that novelty is a necessary but not sufficient condition for innovation by firms.

However, with the rising institutionalisation of formal R&D activities within firms, SCHUMPETER'S original distinction between invention and innovation has become more and more blurred, as both elements are closely intertwined by applied industrial R&D. But even without formal R&D it can be assumed that creating novel products or processes at the firm level is usually connected with solving a specific economic problem or task. Therefore, this thesis will not distinguish between invention and innovation.

Nevertheless, innovation is not an end in itself, and it is thus important to mention that it should not be solely reduced to the dimension of doing something new. For this reason, this thesis would like to shift SCHUMPETER'S original distinction between invention and innovation to a more general evolutionary level in order to separate the two dimensions of novelty and the impact on the economic performance of the firm and to link them to the previously developed resource-based perspective.

The later work of SCHUMPETER already presented a more general argumentation by drewing an analogy between innovation and production systems. In the same way as a production system, which describes how different resources or input factors are combined to produce a certain good, innovation describes the novel combination or re-combination of resources or inputs in such a way that either a new product or a new production process is developed or modified (SCHUMPETER 2008: 95ff.). But while the production function only describes changes in the amounts of given resources or input factors, innovation fundamentally changes the functional relationship between the resources or input factors (SCHUMPETER 2008: 94f.).

Against the evolutionary backdrop of this thesis described in chapter three, such novel combinations of the firm's resources in the innovation process are represented by the evolutionary mechanism of variation. If we recall the discussion in 3.1.1, variation means any departure from the firms' existing practices of producing goods (ALRDRICH/RUEF 2006: 18ff.) in order to unlock new opportunities or solve problems. Evolutionary variation is thus expected to impact the resources and the ways they are combined. In contrast, the environmental selection mechanism primarily impacts the economic effects rather than the underlying variations (3.1.2). In other words, market selection does not affect the novelty- or variation dimension of resources or resource-combinations, but their economic effects that become visible on the market in terms of certain performance characteristics that distinguish the firm from its competitors. Thereby, starting from SCHUMPETER, both dimensions of innovation - novelty and economic impact - can be captured by an evolutionary conceptualisation of variation and selection. Finally, this view implies a conceptual distinction between a firm's inputs *to* innovative activity and the outputs *of* innovative activity (see Figure 14) (ROGERS 1998: 9; KIRNER ET AL. 2006: 18ff.; TETHER 2003: 4ff.; OECD 2005: 16ff.).

Moreover, such an evolutionary understanding of innovation addresses the resource-based theory's shortcoming of being too focused on the firm's internal perspective (see chapter 3.3.4) by linking new combinations of a firm's internal resources to their economic effects which are subject to market selection mechanisms.

Because the creation of new products or processes is not considered innovation until it has been productively incorporated in the firm's activities, innovation is not something that can be thought of as separate to a firm's core activities. In line with the basic assumption of evolutionary theory that the surrounding external environment is dynamic and continuously changing, firms are forced to *continuously* adapt in order to survive. Firms are usually stimulated to try something new becasue of dissatisfaction with their current economic performance or becasue of the perceived need to adapt to changed external requirements (WINTER 1975: 105; WITT 1996: 712). For this reason, TIDD ET AL.

(2009: 54f.) point to the procedural and "generic" character of innovation.[213] Innovation should not be regarded as something that occurs selectively at a single point in time, but rather as a kind of "core business process" which assures the firm's survival and economic success. Innovation in this sense means the process of managing of different activities to realise and implement new and useful ideas which comply with socio-economic frame conditions.

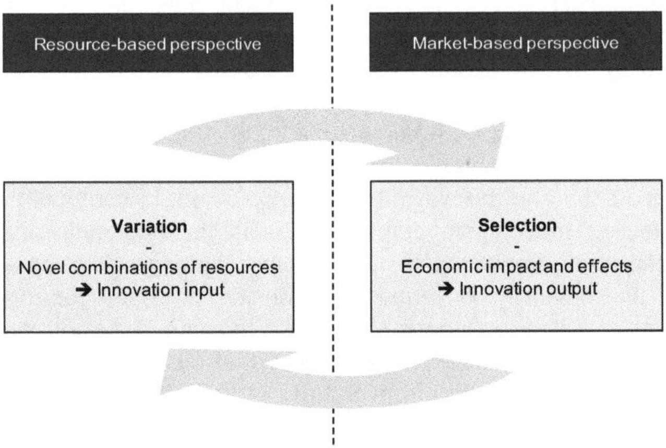

Figure 14: Evolutionary conceptualisation of innovation
(Source: own illustration)

Against this evolutionary background, a firm can be considered innovative if it has managed to survive market competition by successfully adapting itself, at least to a minimum degree, to the challenges of its dynamic environment. In the context of the empirical analysis which is based on a selection of innovation input indicators, the innovation activity of firms is thus defined as making use of at least one of these innovation resources or combinations of them and reporting one or more of the corresponding innovation output and economic performance indicators.

Understanding innovation to be a "generic", continuous process of firms' adaptation to external conditions thus represents the "natural" perspective of evolu-

213 This procedural view is shared by several authors (e.g. KLINE/ROSENBERG 1986; SOMMERLATTE 1997; DODGSON ET AL. 2005)

tionary theory which stresses the idea of continuous change. Therefore, analysing the innovation patterns of non-R&D-performing firms seems contradictory to the evolutionary perspective at first glance, because the term "pattern" usually refers to something "stable" or "static" as it describes some kind of structural regularity. This view is originally rooted in traditional or neoclassical economic theory, which understands structure to be a *result* of processes that lead to a state of equilibrium described by a few parameters.[214] This structural minimalism limits the analysis to processes that either support equilibrium or distort it (HERRMANN-PILLATH 2002: 149). In this sense, processes are temporary phenomena which only occur until the economic system reaches a state which none of the economic actors want to change. This equilibrium state creates structure and patterns.

In contrast, evolutionary economics has a completely different understanding of processes. It does not regard processes as an expedient to overcome economic disequilibria but as representing the *natural condition* of all types of systems resulting from their ongoing adjustment to dynamic environmental conditions. This means that structure or patterns are created through the continuous, repetitive character of viable adaptation processes.[215] As long as processes are considered viable for evolutionary adaptation to dynamic external frame conditions by the actors, they will be used and repeated. Thus, the aspect of viability covers both states of economic equilibrium and disequilibrium, because even disequilibrium can be viable. In other words, while static systems are interpreted under the circumstance of equilibrium, dynamic systems are interpreted under the condition of viability (HERMANN-PILLATH 2002: 150). While, in the neoclassical sense, static means the absence of change or any disturbance, the evolutionary approach understands stability or structure as a result of the ongoing and recurrent processes of evolution. To conclude, the innovation patterns of firms in an evolutionary understanding describe regulari-

---

214 Mainstream economics basically deals with three types of processes (HERMANN-PILLATH 2002: 148): a) *economic growth* in terms of an increase in the satisfaction of needs, b) *allocation* in terms of an achievement of the most efficient status of resource distribution, and c) *variations and disturbances* of economic activity which hinder the economic system in reaching its equilibrium. In all cases, these processes describe the problem of optimisation either at the level of firms or the whole economy.

215 The term "viability" goes back to the work of the radical constructivist GLASERSFELD (1987) and describes "practicable", "appropriate", "suitable" or "functional" ways of individual problem solving by recognising that there are multiple other possibilities.

ties or recurrent structures of the firms' viable innovation processes for surviving market competition.

Having derived an innovation concept that fits the evolutionary resource-based perspective, it is then necessary to identify which areas of firms' innovation behaviour should be covered by the empirical analysis in order to sufficiently account for heterogeneity.

In general, as elaborated above, the innovation processes of firms encompass all the activities aiming to improve competitive performance. As already described (see chapter 2.6.5), the basis characterisation used in the prevailing R&D focus of past empirical innovation research, distinguishes product and technical process innovation (RAYMOND/ST-PIERRE 2010: 48). The OSLO Manual defines product innovation as the introduction of a product whose either technological characteristics or intended uses differ significantly from those of previously produced products, or whose performance has been significantly enhanced or upgraded, whereas process innovation describes the adoption of technologically new or significantly improved production methods (OECD 2005: 32). However, as derived from the integrated resource-based approach, the innovation patterns of non-R&D-performing firms may not only aim to compensate for missing R&D intensity, but also be targeted towards other types of "non-technical" innovation (KIRNER ET AL. 2009a: 448; RAMMER ET AL. 2009), for example innovation in product-related services or in the use of organisational concepts (KINKEL ET AL. 2004).). For instance, the famous Toyota production system which was first analysed by WOMACK ET AL. (1990) underlines the point that superior process performance can also be based on the use of innovative concepts of firm-internal work and production organisation or human resource management.

In order to explore the heterogeneous innovation patterns of non-R&D-performing firms, it is therefore necessary to apply a broader, holistic definition of innovation, which does not only account for the technical (and thus largely R&D-oriented) dimensions of innovation. SCHUMPETER (2006: 159) already paved the way for such a holistic understanding, which also considers non-technological innovations to be a source of competitive advantages and economic success. Since then, this understanding has become widely established in the theoretical debate (DAMANPOUR/EVAN 1984; BESSANT/CAFFYN 1997; PIVA/VIVARELLI 2002; TOTTERDELL ET AL. 2002; TIDD/BESSANT 2009; LAM 2005;

HIPP/GRUPP 2005).[216] The latest edition of the OSLO-Manual (OECD 2005: 29, 46) has also adopted this view and defines innovation as "...the implementation of a new or significantly improved product (good or service), or process, a new marketing method, or a new organisational method in business practice, workplace organisation or external relations (IBID.: 46)". However, this definition causes some analytical problems. It remains open whether the "process" dimension of innovation also encompasses non-technical process innovations, or whether they have to be considered part of organisational "methods". Moreover, the term "product" here places the development of new material goods or products within the development of immaterial service products. This conceptualisation can be questioned because of significant differences in the causalities of how they emerge and their effects (e.g. PREISSL 2000; HIPP 2000; NIJSSEN ET AL. 2006; MILES 2007). For example, while R&D plays an important role in the development of new products, new services are often triggered by other departments within the firm, or by customers' feedback.[217]

For these reasons, this thesis refers to the innovation typology proposed by KINKEL ET AL. (2004: 11), DREHER ET AL. (2005a: 25) and KIRNER ET AL. (2009a: 448), which has been specially conceptualised for manufacturing firms (see Figure 15). Like the OSLO Manual's definition, it is also based on a holistic view of innovation and provides a slightly differentiated typology of firm level innovation on the basis of SCHUMPETER'S ideas. It explicitly distinguishes between the product and process dimension of innovation, on the one hand, and the technical (i.e. material) and non-technical (i.e. immaterial) dimension of innovation, on the other.

216 SCHUMPETER distinguished five types of "new combinations": a) production of a *new product* which is either new to the market or features a novel quality of function, b) implementation of a *new manufacturing method*, which is new to the industry and is not necessarily based on R&D, c) opening up *new markets*, d) exploitation of *new sources of supply*, and e) implementation of *new organisational practices and concepts*.

217 See also Tether/Howells (2007), Hipp/Grupp (2005), Leiponen (2005), Tether (2004)

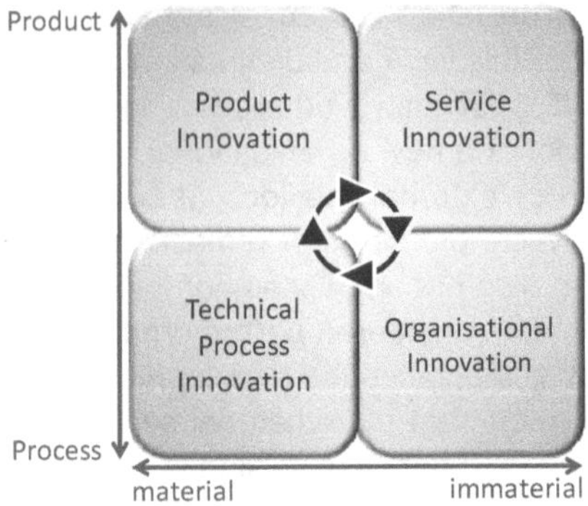

Figure 15: Typology of manufacturing firms' innovation activities
(Source: adopted from KINKEL ET AL. 2004: 11; DREHER ET AL. 2005a: 25)

Regarding manufacturing firms, product innovations can consist of either technical or non-technical products, and process innovations can involve technical and organisational aspects. They represent both the physical and the intangible aspects of process innovations. Each of these innovation types can be a distinct source of competitive advantage in itself. Depending on their individual frame conditions, available resources and competencies, firms can follow different innovation strategies.[218] Moreover, the figure underlines the interconnection between these fields as, in the complexity of the real world, an innovation activity in one field often triggers further innovation activities in adjacent fields, or is not always clearly distinguishable as product and process innovations. Because it postulates no "one best way" and thus leaves room for heterogeneity, this innovation typology meets the requirements of an evolutionary innovation concept.

218 Given the focus on manufacturing firms, innovation in the area of marketing is not considered an independent field of innovation. On the one hand, regarding the business-to-business context of investment goods, it can be reasoned that marketing as an autonomous field of innovation does not play such an important role as in business-to-customer markets. On the other hand, marketing contains dimensions of all innovation fields (i.e. it is related to product or service characteristics and involves certain organisational processes like the establishment and operating of a marketing department) and thus extends across this typology. For this reason, the scheme above should not be interpreted as a final typology of innovation, but rather as an ideal abstract along two dimensions which can be used to describe the innovation activities of manufacturing firms.

All the necessary elements of an appropriate definition of innovation in the context of this thesis have now been elaborated and innovation is defined as follows:

> ***Definition:*** *Innovation is defined as a firm's continuous and repetitive activities of developing new resource or input combinations to increase its economic performance or output and thereby to ensure its survival by successfulyl adapting to dynamic environmental frame conditions. These activities encompass technical and non-technical, as well as product- and process-related fields of innovation.*

> ***Definition:*** *Innovation patterns describe recurrent structures or regularities of viable elements within the continuous, generic innovation processes of firms.*

4.4 Implications for the empirical analysis of innovation patterns of non-R&D-performing firms

With these theoretical arguments in mind, two main requirements can be derived for the empirical analysis of non-R&D-performing firms' innovation patterns:

- Maximum consideration of possible heterogeneity in non-R&D-performing firms' innovation strategies
- Maximum consideration of different firm-internal and -external resources for innovation

Hence, it is precisely the variety in non-R&D-performing firms' behaviour, which motivates the search for regularities and systematic empirical patterns through the creation of classifications. As defined here, the term of "patterns" in an evolutionary context specifically refers to the occurrence of recurrent good, viable practice in the ongoing process of evolution and environmental adaptation. Thus, identifying innovation patterns allows inter-firm heterogeneity to be taken into account, while simultaneously requiring the selection of relevant dimensions in which this heterogeneity may occur.[219] In order to ensure the maximum consideration of non-R&D-performing firms' heterogeneity, the firms should not be pre-selected or pre-grouped using any structural variables like their sector affiliation or size.

One possible method would be to gather relevant indicators and then attempt to connect the dots between them using some kind of descriptive analysis. Such an approach is feasible if only a small number of variables is being taken into account. But since the dots that need to be connected tend to grow exponentially with an increasing number of variables, descriptive approaches suffer from several limitations. Instead, the analysis conducted here takes into account a large number of variables in order to provide robust empirical evidence of the difference facets of non-R&D-intensive firms' innovation behaviour.

In order to take up the idea of firms' evolutionary heterogeneity using empirical methods in line with the recent studies on innovation strategy at firm level of LEIPONEN and DREJER (2007), ARUNDEL and HOLLANDERS (2004), HOLLENSTEIN (2003), SRHOLEC and VERSPAGEN (2008) or PENEDER (2010), this thesis argues for the application of explorative cluster analysis, which is described as "the art of finding groups in data" such that the degree of natural association is high among members within the same group and low between members of different groups (KAUFMANN/ROUSSEUW 1990: 1; BACHER 2002: 2ff.; KOZELKA 1982: 6). Firstly, to identify the underlying structure of the data, a hierarchical cluster analysis seems appropriate as this starts at the broadest level of heterogeneity by assigning each observation to a single cluster before they are then successively grouped into bigger clusters (BACHER 2002: 297). Then, in a second step, the results can be sharpened by a non-hierarchical cluster analysis to increase the similarity between the elements within a cluster and to maximise the diversity between different clusters. This methodological approach increases the chance that all the dimensions of innovation resources are well represented in the final results.

The variables deployed in cluster analysis should reflect a broad set of non-R&D-performing firms' innovation resources, as the resource level represents

219 Innovation research always tried to link high diversity at the micro-level with a strong interest in taxonomic work (PENEDER 2010: 324). The most prominent example goes back to PAVITT (1984) whose classification proved extremely influential and motivated numerous extensions and further refinements. Among them, the most important taxonomies are provided by MARSILI (2001) and CASTELLACCI (2008b). Other notable examples are to be found in EVANGELISTA (2000) or DE JONG/MARSILI (2006).

the level at which the highest level of heterogeneity can be expected (variation), and the level able to adequately capture firms' multiple innovation resources beyond formal R&D. Nevertheless, as the dicussion against the background of the innovation definition underlying this thesis made clear (see chapter 4.3.2), innovation cannot be thought of without including the dimension of its economic effects on firm performance. Thus, any analysis of firms' innovation patterns should account for both dimensions.

However, since the understanding of innovation processes has shifted from a linear, sequential and thus predicable progression towards more complex and self-referential cycle models (see chapter 2.6.5) it can no longer be assumed either that a higher level of innovation input automatically leads to higher innovation output in general, or that a higher input of a particular innovation resource directly results in a higher level of a particular, corresponding economic output. Due to the complexity of cause and effects between innovation input and economic effects, it is obvious that any single innovation activity will impact other performance aspects to a greater or lesser extent and may be positively reinforced by prior performance as well (RICHARD ET AL. 2009: 726f.; MARCH/SUTTON 1997: 700) (see Figure 16).[220]

As a consequence, organisational performance is not a one-dimensional theoretical construct, nor can it be characterised by a single operational measure. Therefore it is necessary to access it using a broader set of performance measures covering different dimensions of firms' performance (RICHARD ET AL. 2009: 722).

As has also been illustrated, hardly any change in a firm's innovation resource is directly reflected in a measurable change in its economic performance as there is a certain time lag between the stimulus and effect. This time lag may vary across different kinds of innovation resources (MEADOW CONSORTIUM 2010: 119). For instance, the acquisition of ready-to-use machinery should cause positive economic effects in a shorter space of time than the accumulation of an experienced-based stock of knowledge through concepts of knowl-

---

220 This is especially valid for the field of organisational innovation. Organisational innovation encompasses many different concepts of how to change the present structure or procedures as various business strategies are fostered and triggered by different innovative organisational concepts. Hence, indicators of organisational innovation are not able to explain the specific performance impacts of single concepts. While, for example, product innovations have a relatively direct impact on sales, it is not reasonable to postulate such a direct performance relation for organisational innovations.

edge management or the participation in innovation cooperation. Moreover, in addition to the general problem of time-lags between innovation input and innovation output, the short-term effects of innovation activities may differ from their long-term outcomes (MARCH/SUTTON 1997: 700f.).[221]

At the moment, there is no real solution to these problems as research on these issues is still in its infancy. The use of panel data may be one solution, but suitable data on firms' innovation behaviour are scarce (SCHNELL ET AL. 1995: 231f.). However, these aspects and restrictions should be taken into account when analysing firms' innovation performance using cross-sectional data.

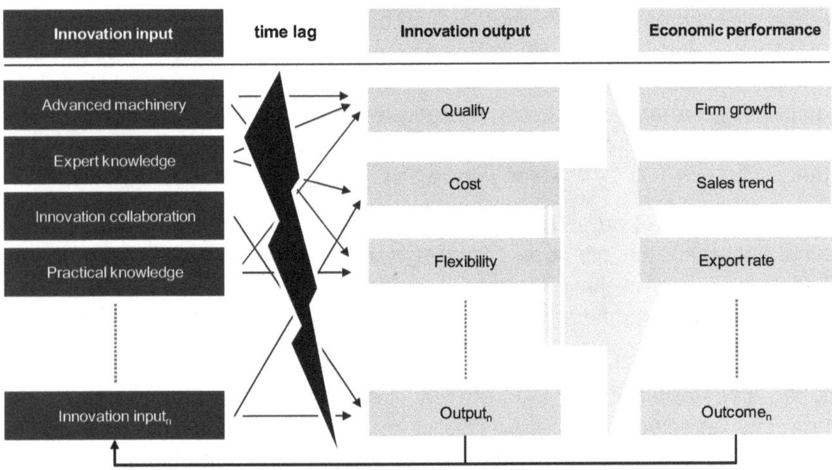

Figure 16: Contingency and time lags between innovation input and innovation output (Source: own illustration)

Based on these considerations, this thesis conducts a cluster analysis of the innovation patterns of non-R&D-performing firms which only considers the input dimension of innovation, and then cross-tabulates these with their corresponding performance output. As a consequence, the non-R&D-performing

221 See also MARCH (1994).

firms will also not be pre-grouped using any innovation performance or economic output variables.

This approach is preferable for three reasons:

- It enables the identification of innovation patterns which are more closely connected to recent actual innovation behaviour rather than primarily focusing on time-lag-biased performance measures or mingling resource and performance dimensions as though these occur at the same point in time.
- It allows for a broader range of heterogeneity, as it leaves enough room for the contingency that different innovation inputs may lead to similar economic performance.
- It addresses the limitation of resourced-based theory of neglecting the market or the performance dimension of resources and emphasises the aspect that resources are only valuable if they enable the firm to achieve competitive advantage in terms of economic success (PORTER 1996; BOWMAN/AMBROSINI 2000; PRIEM/BUTLER 2001).

The analytical model used to identify the innovation patterns of non-R&D-performing firms which is summarised by Figure 17.

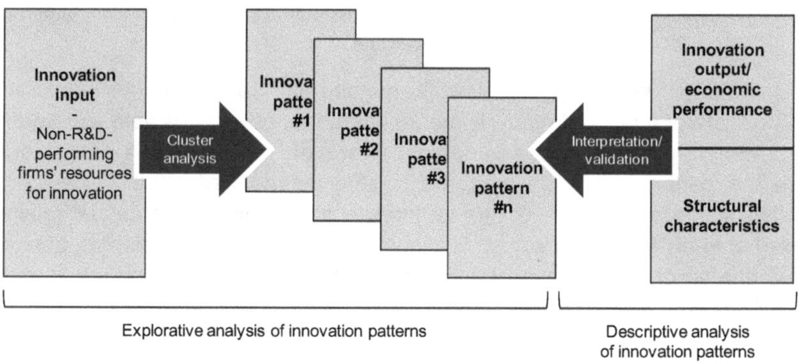

Figure 17: Analytical model of empirical analysis
(Source: own illustration)

Based on this analytical model, the conceptual framework for the following empirical analysis can easily be developed in the usual three-part basic structure (CHILD 1972: 2; PUGH ET AL. 1963: 297f.):

- ***Variables of configuration:*** This dimension encompasses all constructs, elements or measures which form the core of the analysis. Here, the variables of configuration are represented by variables describing non-R&D-performing firms' resources for innovation. These are based on the firms' individual strategic decisions and are used in the cluster analysis.

- ***Variables of context:*** This dimension describes those measures which can be reasonably assumed to influence the variables of configuration, or should at least be considered in order to ensure that the findings concerning the variables of configuration are interpreted correctly. In this analysis, this dimension is captures by the entity of variables describing the structural characteristics of the firms which are used to describe, interpret and validate the identified innovation patterns.[222]

- ***Variables of success:*** This dimension comprises all measure of success whose characteristics can be assumed to be either linked to the variables of configuration or the interrelationship of contextual and configuration variables. In this analysis, this dimension is represented by both the variables of innovation output and the firm's overall economic performance (outcome) which will also be used for cluster description, interpretation and validation.

WOLFF (2008: 39f.) suggests that a conceptual framework for explorative analysis should also contain some estimations about possible cause-and-effect chains between these variables. However, as this thesis seeks to explore the general structural patterns underlying the innovation behaviour of non-R&D-performing firms, it has no interest in identifying the causal relationships between these variables. Obviously, the principal relationship between the variables of configuration (such as the innovation input) and success (innovation output and economic performance) is already described by the defini-

222 As WOLF (2008: 38) states, sometime it is not always easy to clearly distinguish between the dimensions of configuration and context. Hence, the major point of differentiation refers to the question to what extent a variable can be influenced by the decision-makers within the analysis. If a variable cannot be strongly influenced by intentional decisions, it can be treated as a contextual variable; otherwise it should be treated as a variable of configuration.

tion of innovation underlying this thesis. Moreover, as this thesis follows a strictly evolutionary, resource-based approach, the influence of context variables (structural characteristics) should not be considered a priori, but analysed by ex post examination.

4.5 Quantitative measurement of firms' innovation resources

Due to their intangibility, complexity and latent character, some of the concepts of the integrated resource-based perspective are difficult if not impossible to operationalise for empirical analysis. Only the lowest level of resource complexity, the level of concrete physical resources or explicit competences, is easy to survey as this can be directly measured. But more abstract and complex concepts like routines, organisational culture or dynamic capabilities cannot be directly observed as they are latent and intangible variables. Even their operationalisation through indicators appears to be problematic. Firstly, due to their complexity, theoretical constructs like routines or behavioural patterns cannot be empirically isolated in a discrete way. Thus, researchers have to deal with inter- and intra-relations between the different indicators supoosed to represent such variables. Secondly, the construction of adequate indicators for such variables is often a trade-off between simplicity, and thus feasibility on the one hand, and an adequate provision of the necessary degree of complexity on the other. Thus, such indicators offer only limited possibilities for interpretation because they measure the addressed concepts at a very superficial and simplified level.

When discussing the measurement possibilities, it should be kept in mind that the resource-based theory itself does not represent a ready-to-operationalise theoretical concept. Due to its general assumptions regarding the micro-economic origin of the driving forces behind the firms' competitive and economic success, it can instead be characterised as a kind of meta-theoretical paradigm which paves the ground for firm level analysis.

As already mentioned, different core concepts of the resource-based approach vary distinctly in their direct empirical measurability depending on their level of complexity and tangibility. Empirical surveys (e.g. European Innovation Scoreboard (EIS), Community Innovation Survey (CIS)) therefore prefer to concentrate on directly measurable financial variables or human capital variables of the resource level such as:

- (R&D-expenditures)
- Innovation expenditures (i.e. all financial expenditure in the context of innovation projects)

- Early-stage venture capital
- Expenditures on information and communication technologies
- Expenditures for the modernisation of manufacturing systems
- Technology-related expenditures
- Share of highly trained employees
- Expenditures for employee training and qualification
- etc.

Although such measurements are one of the easiest and most reliable ways of empirically analysing firms' resources, they represent only a small part of resource-based theory. In order to access more latent resources (e.g. management, organisational capital or organisational culture) and routines, these have to be operationalised through indicators that allow for conclusions to be drawn about their condition. Three different strategies of questioning can thereby be observed here. The first asks for several rather isolated variables that represent different elements of underlying theoretical concepts like the external orientation of the enterprise, organisational culture or structure. Examples of such indicators are (e.g. Danish Research Unit for Industrial Dynamics (DRUID), CIS):

- Number of hierarchical levels
- Formal organisation, structural traits
- Customer / supplier relationships
- Participation in cooperation and collaboration
- Internal cooperation between departments
- Sources of information and knowledge
- Management responsibilities
- Content of work / work organisation
- Characteristic of decision-making
- etc.

The second draws on the use of concrete management or production-related concepts as these are assumed to represent the visible and measurable dimension of underlying routines, behavioural patterns and capabilities (e.g. the Project Danish Innovation System in a Comparative Perspective (DISCO), European/German Manufacturing Survey (EMS/GMS)), for instance:

- Concepts of knowledge management
- Concepts of total quality management
- Simultaneous engineering
- Supply chain management
- Just-in-time
- etc.

The third strategy questions which routines, resources or capabilities do not exist (e.g. DRUID, DISCO, CIS). This means that the interviewees are asked which are the main constraints or barriers to economic success or innovation ability. The variables mentioned can then be supposed to be closely cognitively related to innovation ability or economic success, and it is assumed that their non-existence contributes to the failure to meet economic targets. Another advantage of this strategy is that open questions can be designed which do not limit the answers a priori. DREHER ET AL. (2005b) chose a similar method in their study of the "time drivers" of innovation. These time drivers are supposed to reflect how well product innovation routines are functioning. These time drivers were measured by asking the firms to assess the relevance of various innovation routines in light of the delay which had occurred during the latest innovation process. Thus, the frequency of occurrence and the degree of delay caused by the dysfunctionality of certain routines serves as an indicator of the importance of these routines. Based on theoretically-derived product innovation routines they identified and surveyed the following potential delaying factors:

- Unclear objectives at the beginning of the project
- Unclear or changing product specifications
- Difficulties in the application of new technologies
- Lacking skills of workforce
- etc.

The above described indicators or questioning strategies only represent a short overview that could be supplemented by various other examples. Despite the progress made recently by empirical resource-based research, it should be acknowledged that the more intangible the concepts to be analysed, the more difficult it is to operationalise and measure them. This is especially valid in standardised written surveys as such indicators are usually only able to scratch the surface of this theoretical construct.

Because the available database of the *German Manufacturing Survey* 2009, which is described in more detail in chapter 5.1, does not contain resource measures in terms of non-existent resources or barriers to innovation, this thesis will follow the first and second measurement strategy described above. When necessary, the intangible, processual dimension of non-R&D-performing firms' innovation resources is estimated by using proxy variables or different organisational concepts.

4.6 Guiding theses for the empirical analysis

Because the research design of this thesis is explorative, it is not necessary to postulate hypotheses to be tested a priori as is the case in a confirmatory analysis. Nevertheless, in addition to the criteria of statistical model quality and interpretability of the cluster solution, several authors (BACHER 2002: 17; DENZ 1982: 134ff.; HOLM 1970: 710; BAGOZZI 1982: 14) point to the necessity of postulating advance hypotheses about which clusters or patterns are expected to be identified by the cluster analysis. These hypotheses will serve to test the construct validity of the identified clusters ex post.

Firstly, since innovation has been defined as a firm's continuous and repetitive "core" activity of developing new resource or input combinations to increase its economic performance or output and thereby to ensure the firm's survival by successful adaptation to dynamic environmental frame conditions, non-R&D-performing firms can be generally expected to be innovative if they make use of the considered innovation resources to realise innovation and economic success. Hence, from an evolutionary perspective all the firms existing at the time of analysis can be regarded as more or less innovative as they have obiously managed to survive (i.e. be positively selected) market competition. Furthermore, it can thus be expected that **the innovation patterns of the non-R&D-performing firms to be analysed will reflect the different kinds of innovativeness** by different levels of resource usage and different levels of innovation and economic success.

Secondly, it is assumed that the dimensions of the possible innovation resources of non-R&D-performers identified by the resource-based perspective above will serve as the elements or modules which make up their innovation patterns. This aussumption is strongly supported by the empirical findings so far (see chapters 1.1 and 1.2) which prove the relevance of several non-R&D innovation resources:

- ***Use of firm-internal innovation resources beyond formal R&D:*** The role of technological and non-technological concepts to enhance process speed, quality or efficiency as well as the importance of human resource management (BENDER 2006; NASCIA/PERANI 2002; HEIDENREICH 2009; KIRNER ET AL. 2009a, 2009b, RAMMER ET AL. 2009; BARGE-GIL ET AL. 2008).

- ***Access to and exploitation of firm-external innovation resources:*** The acquisition and purchase of machinery and equipment, the role of absorptive capacity as well as the participation in inter-firm collaboration and cooperation with external R&D partners (e.g. BENDER 2006;

ARUNDEL ET AL. 2008; SANTAMARÍA ET AL. 2009; TSAI/WANG 2009; HEIDENREICH 2009; BENDER/LAESTADIUS 2005; BENDER 2008; HIRSCH-KREINSEN 2004, 2008a).

- ***Use and exploitation of different kinds of knowledge:*** Customer-driven, practical, experience-based knowledge, „learning by doing", „learning by using" (e.g. HEIDENREICH 2009; VON TUNZELMANN/ACHA 2005; HIRSCH-KREINSEN 2004, 2008a, 2008b; SUNDBO 1996; ARUNDEL ET AL. 2008).

Thirdly, while there is broad agreement on some findings about non-R&D-intensive firms (e.g. less importance of product development, higher importance of external sources of knowledge), there are also many contradictory findings (e.g. higher vs. lower propensity to cooperate, technical process specialists vs. non-technical innovators). When viewed from the perspective of evolutionary theory, these **existing empirical findings do not necessarily have to contradict each other, as they may represent different elements or modules of distinct, heterogeneous innovation patterns.** Hence, another assumption is that all of these elements or modules can be found again in the distinct innovation patterns of non-R&D-performing firms to a greater or lesser extent.

Fourthly, and most importantly from the evolutionary perspective, it would not be reasonable to assume that each of these different modules constitutes its own distinct pattern. A large majority of firms within one cluster will neither rely, for example, only on innovation collaboration, nor will the firms in another cluster only rely on the use of advanced manufacturing techniques. Instead, it can be assumed that each of these modules can be found in each cluster, but to a different extent or with a different weighting. For instance, one cluster may be mainly characterised by a high degree of external orientation (e.g. through collaboration or external knowledge-sourcing) although it may also make use of advanced manufacturing techniques, while another cluster shows a very high use of such techniques, but does not have a marked propensity to collaborate. **Thus it also assumed that all of these elements (i.e. the so far existing findings about non-R&D-performing firms) may be constitutive for a distinct cluster, but are not exclusively limited to this cluster.** As parts of distinct innovation patterns of non-R&D-performing firms, previously contradictory findings no longer appear problematic. The identified innovation patterns are therefore believed to provide quantitative support to already discussed, fragmented qualitative and quantitative findings.

As a consequence, an explorative analysis of recurrent structures or patterns within evolutionary processes and the heterogeneity caused by them will always suffer from a certain paradox. On the one hand it seeks to identify distinct regularities in order to reduce and explain empirical complexity while, on the other hand, the analysis would have to return to single case studies to ultimately account for heterogeneity. Hence, the innovation patterns of non-R&D-performing firms which will hopefully be identified by cluster analysis will never be more than representatives or "ideal types" (in the sense of MAX WEBER) which help to recognise and understand recurrent structures in these firms' innovation behaviour.

5 Empirical analysis of non-R&D-performing firms in the German manufacturing industry

This chapter includes both a descriptive empirical analysis of the innovation characteristics of non-R&D-performing firms in the German manufacturing industry as well as a cluster analysis to explore their heterogeneous innovation patterns. First of all, the chapter will introduce the underlying database of the *German Manufacturing Survey* 2009 and identify the subsample of non-R&D-performing firms which is used for the further analysis. Secondly, it discusses the selection of indicators and measures in line with the theory. Subsequently, the descriptive analysis picks up the thread of the existing analytical approaches in the empirical literature and provides a first systematic overview of the innovation behaviour of non-R&D-performing firms. By additionally contrasting these findings with the group of R&D-performing firms should be possible to identify whether they already constitute a non-R&D-specific pattern of innovation or not. Moreover, the descriptive analysis ensures that the identified innovation indicators are adequate and feasible for describing the innovation behaviour of these firms before the subsequent cluster analysis goes one step further and elaborates on whether and how these innovation characteristics constitute heterogeneous innovation patterns among non-R&D-performing firms.

5.1 Database

The analyses are based on the *German Manufacturing Survey* (GMS) 2009, which is conducted by the Fraunhofer-Institute for Systems and Innovation Research (ISI) and is the German part of the *European Manufacturing Survey* (EMS). The objective of this regular, questionnaire-based postal survey conducted in Germany is to systematically monitor manufacturing industries. The survey targets firms with 20 or more employees from all manufacturing sectors (NACE 15-37) and is addressed to their general management, plant or manufacturing manager. Based on the population of 46,000 firms in the German manufacturing industry, the GMS 2009 drew a randomized net sample of 15,580 firms which were asked to fill in the questionnaire. Of this net sample, 1,484 firms returned useable responses. This corresponds to a response rate of 10% (JAEGER/MALOCA 2009).

Regarding content, the GMS is not designed to analyse only the technological activities of firms. Correspondingly, the firms' competitiveness is not primarily assessed by the performance of their products or services, but also by the characteristics of their manufacturing and assembling processes. On the innovation input side, it does not restrict its focus to innovative firms and firms' R&D activities, but includes questions about the implementation of innovative manufacturing technologies, organisational innovation, innovation cooperation, sources of innovation knowledge, human resources, product development, and product-related services. These features make it possible to venture beyond formal R&D activities and to consider the role of other innovation resources. Additionally, because it covers a significant percentage of firms without any R&D activity, the *German Manufacturing Survey* offers an advantage over standard databases which typically have a very large number of firms with high R&D intensity and thus could yield biased results (SANTAMARÍA ET AL. 2009: 510; also: BAYONA ET AL. 2001; CASSIMAN/VEUGELERS 2002; TETHER 2002). The fact that the *German Manufacturing Survey* does not suffer from this sample bias makes it particularly suitable for examining the innovation behaviour of non-R&D-performing firms, comparing this with that of R&D-performing firms, and identifying heterogeneous patterns of non-R&D-performing firms.

On the innovation output side, the GMS provides sufficient information on the economic impact of the firms' innovation activities, e.g. sales' share of new products or new product-related services, but also process-related performance measures like time to market, manufacturing lead time, productivity or quality. Additionally, there are measures to access long-term economic outcomes like employment trends, development of sales or export intensity, and to capture the general structural characteristics of firms like size, sector affiliation, product complexity or manufactured batch size.

5.1.1 Comparing the GMS 2009 sample to the total population of the German manufacturing industry

Before starting the empirical analyses, I would like to check to what extent the structure of the GMS 2009 sample corresponds to the structure of the total population of the German manufacturing industry. This can help to clarify whether the empirical findings drawn from the sample can be generalised to the basic population or whether they are rather subject to structural bias and can only be interpreted for the sample itself.

Comparing the firm sample characteristics of the GMS 2009 with the total population of firms in the German manufacturing industry[223] reveals that the GMS sample 2009 represents a good cross-section of the German manufacturing industry. With regard to firm size (Figure 18), the sample shows that both smaller and larger firms are well represented. Firms with fewer than 100 employees account for almost two thirds (63%) of the sample. Despite this strong representation of smaller firms within the sample, it has to be stated that especially very small firms are still slightly underrepresented. However, this problem is not solely restrictied to the present GMS survey: almost all surveys of comparable scope and topics face the drawback that micro-firms are less willing and interested in participating in them.

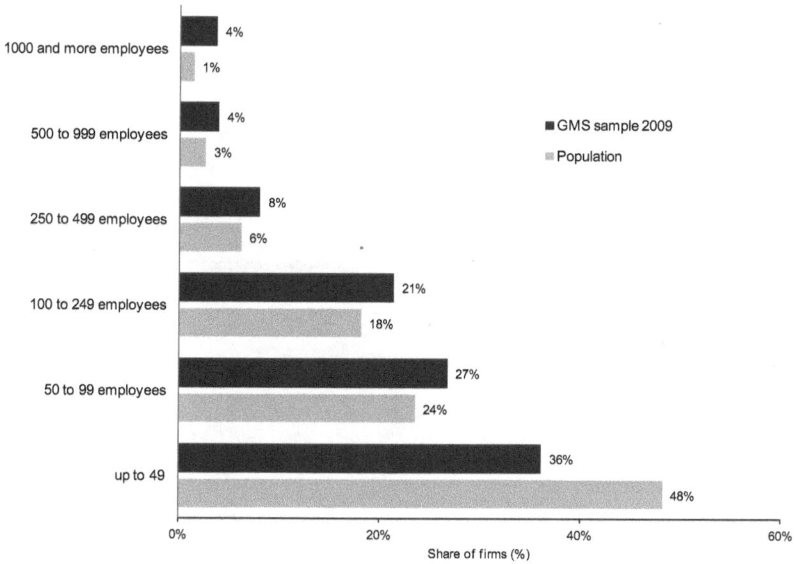

Figure 18: Firm size distribution of GMS 2009 sample compared to the total population (Source: Fraunhofer ISI)

223 Based on data of the German Statistical Office (2008), specialised publications ("Fachserie") No. 4, Vol. 4.1.2

The interpretation of the data should take into account that the empirical findings presented in the following sections may underestimate the effects for very small firms with fewer than 50 employees to some extent. If necessary, findings will be reported separately for different size categories.

Looking at the distribution over the German federal states (Figure 19), the GMS 2009 sample corresponds very well to the total population. The most obvious deviations can be found for Bavaria (slight overrepresentation) and North Rhine-Westphalia (slight underrepresentation). Thus, it can be stated that the empirical findings are unlikely to be biased by regional specificities.

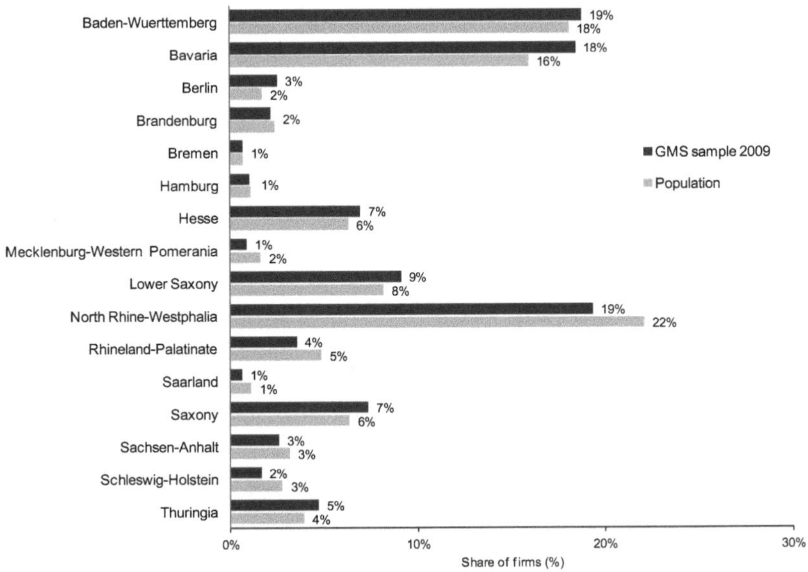

Figure 19: Federal distribution of GMS 2009 sample compared to the total population (Source: Fraunhofer ISI)

Last but not least, Figure 20 considers the sectoral distribution of the GMS 2009 sample. It is clear there is a wide congruence between the share of manufacturing industries in the sample and the total population. The most striking variations are in publishing, printing and reproduction of recorded media, whose share in the survey sample is about half as high as in the total population, and in food products and beverages whose share is 5 percentage points lower than the total population.

Of course, this is rather unsatisfactory, because both of these industries are non-R&D-intensive sectors with relatively low R&D intensity where non-R&D-performing firms may be affiliated. With regard to the empirical findings to be presented, this could mean that they only partially reflect the structures and characteristics of these two industries in the total population.

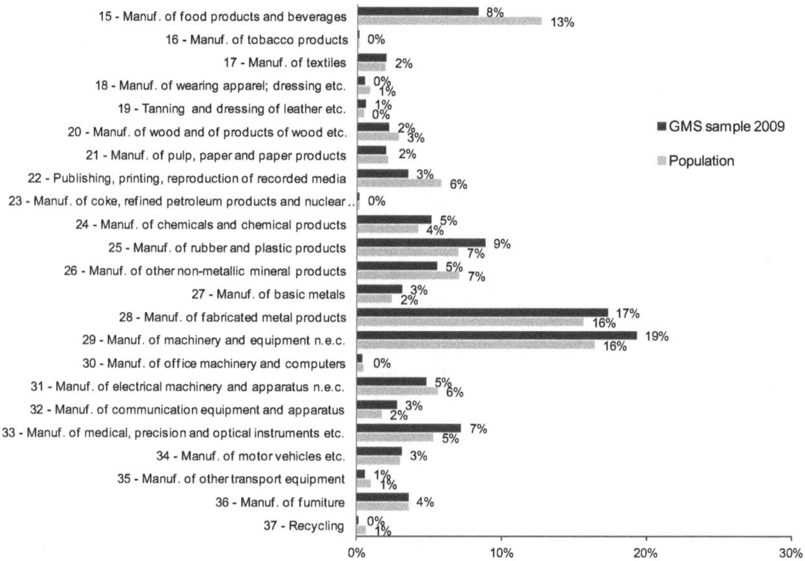

Figure 20: Sectoral distribution (NACE 2-digit) of GMS sample 2009 compared to the total population
(Source: Fraunhofer ISI)

Looking at the other industries, the general correspondence between the GMS 2009 sample and the total population suggests that the GMS sample provides a good cross-section of the German manufacturing industry. Regarding the few deviations in firm size and two industries, one possibility would be to weight the sample according to the relations in the total population. This was not done here for two reasons. First, as argued from the evolutionary perspective, innovation patterns are assumed to be widely independent of sectoral structure which means, that the slight over- or underrepresentation of one or two sectors will probably not influence the final results to a large extent. Second, sample weighting always runs the risk of overemphasizing outliers which could blur the results more than the slight variance between the sample and the population structure. Because the identified deviations are only slight and the theory does not require perfect sectoral congruence, it can be concluded that the data sample underlying the following empirical analyses is not significantly biased and does provide a reliable picture of the German manufacturing industry.

5.1.2 Identifying the subsample of non-R&D-performing firms

Because there is no basic population from which a subsample of non-R&D-performing firms may be drawn, these firms have to be manually selected using the indicator of total R&D expenditure given in the GMS 2009 (see Annex A.4.17). Based on the definition of non-R&D-performing firms provided in chapter 4.3.1, a subsample of *n=675 firms can be identified which have not spent any funds on internal or external R&D in the surveyed period,* which means that their total R&D expenditure amounts to 0% of sales (Table 2). As this subsample represents almost half of the GMS's 2009 total sample, there are sufficient valid cases of non-R&D-performing firms for the following analyses. Moreover, having noted that the GMS 2009 sample closely corresponds to the total population structure of German manufacturing industry, the large number of cases in the subsample of non-R&D-performers underlines the economic relevance of this group of firms.

Table 2: Subsample of non-R&D-performing firms
(Source: GMS 2009, own calculation)

	n	%	Valid %
Non-R&D-performing firms	675	45.5	46.0
R&D-performing firms	791	53.3	54.0
Total	1,466	98.8	100.0
Missing	18	1.2	
Total	1,484	100.0	

Based on this subsample, the following paragraphs first present descriptive statistical analyses of the innovation characteristics generally shown by non-R&D-performing firms and whether these characteristics are distinctive when compared to R&D-performing firms. Second, as described in the analytical framework, a cluster analysis will then be conducted on the innovation resource variables to identify heterogeneous patterns in the innovation behaviour of non-R&D-performing firms.

5.2 Selection of measures and indicators

In line with the requirements for the empirical analysis of non-R&D-performing firms' innovation patterns, the variables and indicators selected in this chapter have to consider the following aspects:

- Maximum consideration of possible heterogeneity in non-R&D-performing firms' innovation strategy.
- Maximum consideration of different firm-internal and -external resources for innovation.

As mentioned above, the *German Manufacturing Survey* 2009 (GMS 2009) provides a broad selection of measures and indicators to assess both a set of different innovation resources which meets the formulated requirements and a set of measures to analyse the innovation output and economic performance of non-R&D-performing firms. Of course, as in every secondary analysis, the selection of indicators is subjected to the constraints of the underlying dataset. On the one hand, this means that some measures and indicators fit the requirements better than others; while, on the other hand, not every desirable aspect is covered by the available indicators. The selection below thus represents a selection which – at least in my opinion – corresponds closest to the planned analysis. Moreover, the selected indicators are similar to the recommendations of the OSLO Manual (OECD 2005: 89ff.) or represent slight modifications to those mentioned by the manual as they focus on a more detailed level. The following sections briefly describe which measures and indicators have been selected from the GMS 2009. The corresponding extracts from the GMS 2009 questionnaire can be found in the annex (A.4).

Additional remarks on this set of indicators appear necessary as some of them have been adapted to the context of non-R&D-performing firms.

5.2.1 Innovation resources of non-R&D-performing firms

Based on the conceptual framework, the level of innovation resources is that dimension where the heterogeneity of non-R&D-performing firms' innovation patterns is expected to be observable as this corresponds to the evolutionary dimension of variation.

According to the integrated resource-based perspective and the basic definitions deduced from this in chapter 4.3, measures and indicators on the input side have to account for the various firm-internal and -external, material and immaterial, product- and process-related dimensions of non-R&D-performing firms' innovation resources. The following indicators have been chosen.

Product development

To start with the core dimension of manufacturing firms, their ability to successfully develop their own products is the main prerequisite for product innovation. Therefore, an indicator is included which asks whether the firm has successfully developed a new product within the past three years. This indicator is also frequently used as an output indicator for product innovation. But the mere development of a new product says little about its real economic effect (e.g. the new product's share of sales). Moreover, because the intensity of R&D, which is one of the most important input indicators for product innovation (OECD 2005: 91), has to be dropped in this context, the indicator of product development provides information about whether some of the non-R&D-performing firms are – despite the lack of any formal R&D – successfully developing new products. Likewise, this has been suggested for instance by KIRNER ET AL. (2009b) and others. As TIDD and BESSANT (2009: 6) point out, product development is an important capability by which firms are able to respond to constant changes in their environments (e.g. through shifts in what people believe, expect, want and earn) and to the introduction of new products by competitors which represent a major threat to their market positions.

Since the mere ability to successfully develop a product provides little information about the processes or routines behind product development, another indicator is included to measure how this takes place within the firm (e.g. according to customers' specifications, standardised programme). This measures the potential diversity of product development as well as the flexibility and openness of a firm's product development routines.

Furthermore, to assess the technological level of these products, a third indicator is included which captures whether the products contain "high-tech" elements such as micromechanical, -electrical, or -optical parts, high performance material, genetically-engineered or biotechnological elements. Of course, this is generally expected to be the case in R&D-performing firms. According to the relational view of the firm, such high-tech resources could also be accessed outside the firm, so there may be non-R&D-performing firms handling such high-tech elements (HIRSCH-KREINSEN 2008B; ARUNDEL ET AL. 2008).

Product-related services

While the ability to develop new products represents the material dimension of product innovation, product-related services (PRS) denote for the immaterial dimension. PRS imply that the original product portfolio of manufacturing firms is increasingly being supplemented by accompanying service offers (e.g. from traditional forms like installation and set-up, technical documentation, maintenance and repair to advanced service offers like leasing, renting, financing or service-based business models) so that firms can stand out from the competition as well strengthen their customer-connectivity (MATHIEU 2001a: 39f.; 2001b: 452; BAINES ET AL. 2009: 547f.). The literature distinguishes firms that merely add service offers to their existing product portfolio, where physical products remaining core business, firms with a distinctive portfolio of PRS, where services are an important competitive factor, and firms, characterised as manufacturing service-providers, where services are the core business and their products are configured in terms of customer-specific problem solutions (BECKENBAUER 2006: 27ff.). Depending on these characteristics, the service orientation of firms can be expected to vary between firms.

While the development of new products usually suggest formal R&D, developing and offering of PRS does not necessarily take place in an R&D context, but may be initiated in the areas of customer services, marketing, or distribution. Thus, it could be assumed, that PRS are an attractive opportunity particularly for non-R&D-performing firms to increase their competitive advantage. However, as PRS are often developed in close connection with a corresponding new product, it can also be expected that firms with no R&D are less active in this field of innovation (e.g. KIRNER ET AL. 2009a: 457).

To capture these aspects, two indicators have been selected for the field of PRS innovation. Similar to product development, the first indicator measures if the firm has developed or offered new PRS within the past three years. The second indicator covers the intensity of service orientation by measuring which share of sales has been directly and indirectly invoiced with services (i.e. *not* in terms of their amount – this would be an output measure). Thereby, it is assumed that the higher the share of directly invoiced services, the higher their strategic relevance and the more explicit the service orientation of the firm (GEBAUER 2004: 5ff.).

Use of advanced manufacturing techniques

Proceeding from the product- to the process-related dimension of innovation, the use of advanced manufacturing technology (AMT) represents a physical resource for process innovation. From the resource-based perspective, the use of AMT can be regarded either in terms of real investments or in terms of technological resources. Existing findings show that the use of AMT has turned out to be a crucial technological innovation resource for non-R&D-performing firms to obtain high levels of quality and cost efficiency in production (e.g. ARUNDEL ET AL. 2008; SANTAMARÍA ET AL. 2009; BARGE-GIL ET AL. 2008; TSAI/WANG 2009). However, it remains unclear so far, to which extent firms without R&D just adopt external existing, ready-to-use AMT, or whether they are able to develop or modify AMT according to their individual requirements.

To assess the use of AMT, the GMS 2009 has a highly detailed battery of variables. These not only cover whether new AMT has been introduced within the last three years, but also ask about concrete manufacturing techniques in the fields of automation and linkage, machining and production technologies, digital factory/IT cross-linkages, and energy efficiency. Additionally, the respondents were able to add further manufacturing technologies that are either also used by them or which were not covered by the given examples.[224]

Moreover, these AMT variables survey the year when the respective AMT was first introduced, the goals behind its introduction, and its intensity of use today. Respondents were asked to relate the extent of their actual utilisation (low, medium or high) to the maximum reasonable utilisation in their firm. Assessing the intensity of AMT use makes sense because it can be assumed that a broad use of different AMTs in a prototype state of application shapes the structure and performance of production and manufacturing to a lesser extent than an extensive, highly intensive use of automation systems or industrial robots, for instance.

From the resource-based perspective, it would also have been interesting to analyse the firms' financial expenditure on investments in machinery and equipment as well as whether they simply purchased ready-to-use AMT or de-

224 Each piece of additional information given by the respondents using this open category has been separately reviewed by the author and assigned to one of the fields of AMT. Those AMTs which could not be unambiguously assigned to one of the subcategories have been pooled in a category "other AMT".

veloped/modified it by themselves. Unfortunately, neither indicator is available in the underlying database of the GMS 2009.

Use of organisational and management concepts

In contrast to AMT, the use of organisational and management concepts (OMC) represents the immaterial counterpart of process innovation. According to resource-based theory, the design and character of intangible firm-internal processes and routines constitute are a central innovation resource which can explain the competitive advantage of firms.

Particularly in the case of non-R&D-performing firms, hard-to-imitate organisational routines and practices can make a major contribution to their competitive advantage because their development and implementation is not necessarily bound to any kind of R&D (KINKEL ET AL. 2004). This view has been supported by several studies highlighting the positive impact of OMC on the competitive advantage of non-R&D-performing firms (e.g. RAMMER ET AL. 2009; HEIDENREICH 2009; BARGE-GIL ET AL. 2008).

However, due to their intangible and often fuzzy nature, organisational routines and capabilities are difficult to measure empirically. Therefore, as mentioned above, this thesis decides to approximate the intangible process dimension using concrete organisational concepts and management tools. Similar to the dimension of AMT, the GMS 2009 offers a large, detailed item battery which asks about the use of 16 concrete organisational concepts and management tools within the fields of organisation of work, organisation of production, standardisation and knowledge management, working hours and payment schemes, and human resource management. Again, there is an additional open category to capture additional OMC which are not covered in the preselection.[225]

As ARMBRUSTER ET AL. (2008: 650ff.) point out, measuring organisational concepts using such a detailed list is strongly preferable to only asking whether any "organisational concept" has been implemented or not, or to their measurement coining ambiguous buzzwords like "teamwork".

225 Again: each additional piece of information given by the respondents using the open category has been carefully reviewed by the author and assigned to one of the subcategories. Those OMC that could not be unambiguously assigned to one of these subcategories have been pooled in a new category called "other OMC".

Intra-firm organisational processes or procedures are often assumed to represent a dichotomous split between having and not having them. But in analogy to individual skills, the introduction of new organisational skills – changing or adapting new organisational concepts – should to be a cyclic process of articulation and reinforcement of the new behaviour. This has to be repeated frequently over a certain period of time for it to take root in the enterprise and become effective (BESSANT ET AL. 2001). Thus, when empirically measuring the diffusion of organisational concepts, there is a methodological imperative to consider different stages or levels of intensity (ARMBRUSTER ET AL. 2008: 653f.). The actual extent of implementation of an organisational innovation mostly depends on the company's management as there are often no standards which define the exact steps and the "correct" way of its implementation. For this reason, the analysis presented below has also included an indicator for the OMC's intensity of use.

Last, but not least, there is one more point about the year of OMC introduction. While in the cases of new products, new product-related services or AMT, the indicator of introduction within the last three years usually follows the logic "the higher the recent activity, the higher the innovation activity based on this resource", its interpretation in the context of OMC is slightly different. It can be argued that the novelty of a new product or AMT may decline significantly over time as AMT more and more diffuse among industries or new products are imitated by competitors. But this is not necessarily true for organisational concepts. Thinking in evolutionary terms, the implementation of a certain organisational concept takes place as an adaptive response to environmental challenges and requirements. As long as these external frame conditions remain stable, the intended effects and benefits of the OMC will be maintained. This implies that organisational innovations - unlike product innovations or AMT - are not subject to an aging process per se (ARMBRUSTER ET AL. 2008: 651f.). Indeed, the implementation of organisational concepts usually implies significant changes to organisational structures and processes which is a time-consuming and cost-intensive process.

As a result, it is still of interest to measure the average year of the OMCs first implementation in the firm, but this has to be interpreted differently. If only those firms are taken into account which introduced a new OMC in the past three years, this would imply that all the other firms which have introduced a new OMC earlier, are considered to be less active in the field of OMC use. However, exactly the opposite may be correct. Assuming an equal usage ratio, those firms which have implemented OMC earlier are probably obtaining higher profits than "latecomer firms" which have only just introduced a new

OMC and have not yet obtained these profits because the new routines and behaviour have not yet become fully internalised.

Participation in innovation cooperation

As the relational-based view of the firm argues, innovation resources do not have to be necessarily located within the firm. Instead, inter-firm collaboration and alliances represent another possibility for firms to either access valuable innovation resources outside the firm or to create such valuable innovation resources by combining the complementary innovation resources of the firm and those of an external partner. As these valuable resources outside the firm can also encompass valuable knowledge for innovation, the variables of innovation cooperation and sources of innovation knowledge are familiar ones to some extent.

However, innovation cooperation requires active cooperation with external partners. This encompasses not only the reciprocal flow of knowledge but also requires social networking abilities and collaboration competences (LORENZONI/LIPPARINI 1999; DUSCHEK 1998; SIMONIN 1997) and is characterised by a much greater degree of mutual trust and interdependency of the cooperating partners. Innovation cooperation can either take place with other firms (e.g. customers, suppliers, competitors, service providers), or external R&D organisations (e.g. universities, R&D laboratories, non-profit R&D organisations). Thereby, innovation cooperation with external R&D organisations should not be confused with external R&D, as the quality of the reciprocal interaction in cooperations (exchange of knowledge) by definition goes far beyond mere R&D subcontracting (exchange of money for knowledge).

The variable for innovation cooperation that is incorporated in the GMS 2009 provides very detailed information about these issues. On the one hand, it measures whether the firms have participated in innovation collaboration, with which partner and how frequently. Additionally, it is possible to stratify each of these indicators across four fields of innovation (new, products, new AMT, new OMC, new PRS). Assuming that product innovation plays a less important role for non-R&D-performers, this feature makes it possible to observe the frequency of innovation collaboration in other fields of innovation too.

Sources of innovation knowledge

This dimension is also related to the external innovation resources of firms. But in contrast to innovation cooperation, it does not concern the inter-firm combination of resources based on the relational perspective. Instead, it asks which sources of innovation knowledge are exploited by the firm and is thus strongly rooted in the knowledge-based view of the firm. The relevant information for a firm's innovation activity can be located inside or outside the firm. Thus, several aspects have to be covered when looking at important knowledge sources for innovation as a valuable innovation resource for firms.

First, these resources reveal something about the nature of knowledge that is relevant for the firm's innovation activity. For instance, if the firm's own R&D department or external R&D organisations are the most important knowledge sources, one could argue that expert or science-based knowledge may plays a decisive role in the innovativeness of the firm. But firms without formal R&D may emphasise practical or experience-based forms of knowledge more (HEIDENREICH 2009; VON TUNZELMANN/ACHA 2005; HIRSCH-KREINSEN 2004, 2008a, 2008b; SUNDBO 1996). Hence, it would be interesting to see if such knowledge within non-R&D-performing firms generated by modes of "learning by doing, learning by using, learning by interacting, learning by producing and learning by searching" (LUNDVALL/JOHNSON 1994: 32) is linked to specific sources (e.g. other employees within the firm, customers, suppliers).

Second, they provide some information about the absorptive capacity (COHEN/LEVINTHAL 1989, 1990) of a firm. As described in the context of the dynamic capability of firms, absorptive capacity is defined as the firm's ability to recognise and evaluate external information and knowledge, and to implement and exploit this in such a way within the firm that it has a greater competitive advantage as a result. As this concept is originally linked to the R&D intensity of a firm, it might be assumed that non-R&D-performing firms lack absorptive capacity and thus rely on internal sources of knowledge to a greater extent. On the other hand, as they do not perform R&D, they lack one of the most important firm-internal sources for innovation knowledge and may be thus particularly dependent on external sources of information which requires a high level of absorptive capacity.

To measure these issues, the GMS 2009 offers a variable that asks for the most relevant information sources for triggering innovation activities within the firm. It is therefore possible to distinguish between firm-internal (e.g. own R&D, other employees) and firm-external knowledge sources (e.g. other firms like customers, suppliers, competitors, R&D organisations as well as conferences, trade fairs or other professional events). Similar to the indicator for innovation cooperation, it also allows the frequency of different knowledge sources to be stratified across the four fields of innovation (new, products, ne AMT, new OMC, new PRS). Of course, it should be assumed here for non-R&D-performing firms that the firm's own R&D does not play any role as a knowledge source.

Human resources

According to the integrated resource-based perspective, one of the central innovation resources of firms is their endowment with the necessary human resources. Especially within the knowledge-based view, the human resources of a firm are seen as the paramount carrier of both explicit and implicit knowledge which is gained either through formal education (i.e. expert knowledge) or through everyday practice and experience in the context of the working processes (i.e. practical knowledge). As previous results have shown, non-R&D-intensive firms are characterised by an above-average share of low- and unskilled personnel (KIRNER ET AL. 2009b) so they might be advised to draw on more experienced-based, practical forms of knowledge as an internal innovation resource (HEIDENREICH 2009; VON TUNZELMANN/ACHA 2005; HIRSCH-KREINSEN 2004, 2008b).

Moreover, manufacturing and production in non-R&D-intensive firms are often characterised as labour-intensive which means that such firms are assumed to follow more cost-oriented innovation strategies (HEIDENREICH 2009; HIRSCH-KREINSEN 2004). It would therefore be interesting to include a measure for the share of employees working in certain functional areas within the firm.

In the GMS 2009, two variables can be used to assess the field of human resources. The first measures the qualification structure of the firms' employees by asking about the share of highly skilled or graduate employees, technicians and masters, skilled tradesmen and skilled workers, low- and unskilled employees, and the share of apprentices and trainees. The second measures the distribution of employees across different functional areas (e.g. R&D, engineering construction and design, manufacturing and assembling, or customer services). Again, for the group of non-R&D-performing firms it is expected that the share of employees in R&D is zero.

Table 3: Indicators of the innovation resources of non-R&D-performing firms selected from the GMS 2009
(Source: own illustration, GMS 2009)

	Resource-based view	Relational-based view	Knowledge-based view
Origin of valuable resource	*Internal*	*External*	*Internal (external)*
Ownership/control of valuable resource	*Individual firm*	*Collaborating partners*	*Individuals (implicit and explicit knowledge)* *Firm (shared explicit and implicit knowledge)*
Selected indicators from the GMS 2009			
Own product development within last 3 years	X		
Type of product development	X		(X)
Products containing high-tech elements	X		(X)
New PRS within the last 3 years	X		
Share of PRS directly/indirectly invoiced	X		
Use/Intensity of AMT across different fields	X		
Use/Intensity of OMC across different fields	X		
Participation in innovation cooperation across different fields of innovation / different external partners		X	(X)
Source of innovation knowledge across different fields of innovation		(X)	X
Qualification structure of employees	(X)		X
Share of employees working in different functional areas within the firm	X		(X)

X = covered by corresponding indicator / (X) = partially covered by corresponding indicator

Table 3 summarises the set of indicators in the context of the integrated resource-based perspective that was elaborated to locate the potential innovation resources of non-R&D-performing firms. As can be seen, all three fields of the different resource-based perspectives are sufficiently covered by the selected indicators and variables, capturing the material and immaterial, internal and external, product- and process-related, implicit and explicit dimensions of innovation resources. Of course, because of the high degree of interrelation of the different resource-perspectives, some indicators account for more than one perspective simultaneously. If this is the case, this is indexed by an (X). Thus, this set of indicators provides a very dense network of measurements of non-R&D-performing firms' innovation resources.

5.2.2 Innovation output and economic performance of non-R&D-performing firms

Following the theoretical arguments as outlined in the conceptual framework, the variation in innovation resources only becomes an innovation when it causes economic effects. From the evolutionary perspective, the level of economic effects corresponds to the level of market selection, because market competition positively selects those variations which meet customer requirements and demands.

According to the fields of innovation defined in chapter 4.3.2, innovation output measures should encompass the fields of product innovation, technical process innovation, organisational innovation and product-related service innovation. Besides these direct effects of variations in innovation resources on innovation output, the long-term economic outcome of non-R&D-performing firms should also be assessed.

Looking at the GMS 2009 there are various measures and indicators of both the innovation output dimensions and the more general economic performance of firms. In the following, I want to briefly describe which indicators and measures have been chosen for the present analysis.

Innovation output

First, the output dimension of new products distinguishes between sales of products new to the firm and products new to the market. The latter represents a higher degree of novelty as the new product has not already been offered by other firms in the market (OECD 2005: 109f.). This differentiation allows the firms' innovation performance to be more accurately captured in terms of product novelty.

Secondly, in analogy to the share of sales with new products, the innovation output of PRS is measured by the share of sales of new service offers introduced within the past three years.

Thirdly, while the shares of sales of new products and of new product-related services represent direct monetary measures of the success of product and service innovations, the innovation output of technical and non-technical process innovation cannot be directly measured by a specific share of turnover. According to WHEELWRIGHT (1984: 81), WHEELWRIGHT and CLARK (1992: 4f.), or WILDEMANN (1997: 67), the three most important "competitive imperatives" for firms are speed, efficiency and quality. These performance measures can therefore be regarded as indicators of process innovation output (KIRNER ET AL. 2009a: 448). To describe the identified innovation patterns of non-R&D-performing firms, the GMS 2009 provides seven measures of process innovation output: the average time to market of new product developments, the average manufacturing lead time to manufacture a product, and the average delivery time indicators for speed; the firms' labour productivity (value-added per employee) and total factor productivity (which accounts for the intensity of labour *and* capital) for efficiency; and the adherence to confirmed delivery dates and the average percentage of products that have to be scrapped or reworked to indicate the quality level of internal processes.

Economic outcome

This dimension is used to describe the long-term economic performance or outcome which can be correlated with the use of certain innovation resources. As already described in chapter 2.6.5.2, the outcome dimensions mainly refer to the aspects of employment growth, development of sales, and export intensity.

As measures of this general economic outcome of firms, the GMS 2009 includes indicators for the monetary outcome (sales growth, return on sales), the employment outcome (employment growth) as well as the share of products exported in the previous year.

5.2.3 Structural characteristics of non-R&D-performing firms

In the following analyses, the measures of firms' structural characteristics are supposed to fulfil two main tasks. First they should serve as some kind of control variables. Theoretically, it has often been argued that heterogeneous innovation patterns of non-R&D-performing firms are expected to be widely independent of structural aspects, particularly of those of sector affiliation. Hence, it should be possible to prove whether this basic assumption about the firm level heterogeneity of non-R&D-performers is verified or not by analysing structural variables. Second, they serve the purpose of interpreting and validating the identified innovation patterns.

The GMS 2009 includes several variables and measures for the structural characteristics of firms. These comprise the usual ones of firm size and sector affiliation, but also survey the firm's position in the industrial value chain, its ownership structure, manufacture-related characteristics of product complexity and batch size, real net output ratio, year of establishment, or its most important competitive factor. These provide information about the firm's market environment in which it is active.

5.3 Descriptive analysis of non-R&D-performing firms' innovation characteristics in the German manufacturing industry

The descriptive analysis investigates the structural characteristics, innovation resources, innovation output and the general economic outcome of non-R&D-performing firms. To better interpret the findings, the analyses will contrast non-R&D-performing firms with R&D-performing firms.[226] However, it should be explicitly mentioned once again that it is *not* the prevalent aim of this chapter in particular, or of this thesis in general, to examine whether non-R&D-performers perform better or worse in terms of innovation than firms with R&D activities. Instead, the focus is to provide a better understanding of the characteristics and the importance of different innovation resources for non-R&D-performing firms as constitutive elements of their heterogeneous innovation patterns.

5.3.1 Structural characteristics of non-R&D-performing firms

This descriptive section starts by analysing the subsample of non-R&D-performing firms according to their underlying structural characteristics. It should then be possible to assess whether the theoretical assumption about non-R&D-performing firms' heterogeneity in terms of sector affiliation has to be abandoned or not at this relatively early stage. Moreover, the description of structural characteristics should help in assessing the findings about innovation resources and innovation output. Where it is assumed that firm size may

226 To check for statistical significance of mean value differences between R&D- and non-R&D-performing firms, findings are tested using Pearson's chi2 for nominal variables, non-parametric Mann-Whitney for ordinal variables and T-test for metric variables.

affect the respective variable, the analysis also differentiates between three firm size categories: very small firms (less than 50 employees), small and medium-sized firms (50-249 employees), and larger firms (more than 250 employees).

Table 4 below shows the non-R&D-performing firms' distribution according to sector affiliation and firm size. The descriptive analysis reveals that an industry classification based on R&D intensity (e.g. that of LEGLER/FRIETSCH 2007) only partially reflects the actual R&D intensity of the firms, because non-R&D-performing firms can be found in all three categories of sectoral R&D intensity.

Unsurprisingly, low-tech sectors comprise the biggest share of non-R&D-performing firms (58%). But it is remarkable that 27% of the firms in medium-tech sectors, and nearly 17% in high-tech sectors are firms without any R&D expenditures. This intra-sectoral heterogeneity continues in the distribution of non-R&D-performing firms within the manufacturing industry. While, as expected, typical low-tech sectors (e.g. food and beverages, textiles and leather, or paper industry) are dominated by non-R&D-performing firms (70%-80%), it is surprising that even high-tech (e.g. medical, precision and optical instruments) and R&D-intensive sectors (e.g. machinery and equipment) comprise a significant share of non-R&D-performers (17%-28%). Thus, non-R&D-performing firms are obviously managing to survive even in "high-tech" environments. These findings strongly support the theoretical assumption of the sectoral heterogeneity of non-R&D-performing firms and underline that the sector level may not be the appropriate level to start from when analysing firms' innovation behaviour.

With regard to firm size, non-R&D-performing firms are most likely to be found in the category of small and medium sized firms. Especially the category of very small firms with fewer than 50 employees is predominantly (62%) composed of non-R&D-performing firms. Furthermore, 43% of all firms within the group of larger SMEs (50-249 employees), are also non-R&D-performers, whereas only about 16% of firms with 250 and more employees did not perform any R&D.

Table 4: Sector affiliation and firm size of R&D- and non-R&D-performing firms (Source: GMS 2009; own calculations)

	R&D-performing firms (>0 % R&D expenditure)	Non-R&D-performing firms (0 % R&D expenditure)	Total
Sectoral affiliation (LEGLER/FRIETSCH 2007)***	*Share of firms (%)*		
Low-tech sectors (<2.5% R&D expenditure)	41.9	58.1	100.0 (n=948)
Medium-tech sectors (2.5 - 7% R&D expenditure)	73.0	27.0	100.0 (n=363)
High-tech sectors (> 7% R&D expenditure)	83.2	16.8	100.0 (n=155)
	(n=791)	(n=675)	(n=1466)
Sectoral affiliation within manufacturing industry (NACE 2-digit)***	*Share of firms (%)*		
15 - Manuf. of food products and beverages	27.0	73.0	100.0 (n=122
16 - Manuf. of tobacco products	100.0	0.0	100.0 (n=1)
17 - Manuf. of textiles	48.3	51.7	100.0 (n=29)
18 - Manuf. of wearing apparel; dressing etc.	42.9	57.1	100.0 (n=7)
19 - Tanning and dressing of leather etc.	50.0	50.0	100.0 (n=8)
20 - Manuf. of wood and of products of wood etc.	6.3	93.8	100.0 (n=32)
21 - Manuf. of pulp, paper and paper products	42.9	57.1	100.0 (n=28)
22 - Publishing, printing, reproduction of recorded media	8.2	91.8	100.0 (n=49)
23 - Manuf. of coke, refined petroleum products and nuclear fuel	100.0	0.0	100.0 (n=2)
24 - Manuf. of chemicals and chemical products	76.4	23.6	100.0 (n=72)
25 - Manuf. of rubber and plastic products	49.2	50.8	100.0 (n=128)
26 - Manuf. of other non-metallic mineral products	53.8	46.3	100.0 (n=80)
27 - Manuf. of basic metals	53.3	46.7	100.0 (n=45)
28 - Manuf. of fabricated metal products	36.9	63.1	100.0 (n=255)
29 - Manuf. of machinery and equipment n.e.c.	72.2	27.8	100.0 (n=281)
30 - Manuf. of office machinery and computers	100.0	0.0	100.0 (n=5)
31 - Manuf. of electrical machinery and apparatus n.e.c.	68.6	31.4	100.0 (n=70)
32 - Manuf. of communication equipment and apparatus	87.2	12.8	100.0 (n=39)
33 - Manuf. of medical, precision and optical instruments etc.	83.0	17.0	100.0 (n=106)
34 - Manuf. of motor vehicles etc.	70.5	29.5	100.0 (n=44)
35 - Manuf. of other transport equipment	62.5	37.5	100.0 (n=8)
36 - Manuf. of furniture, jewellery, sports equipment	43.4	56.6	100.0 (n=53)
37 - Recycling	0.0	100.0	100.0 (n=2)
	(n=791)	(n=675)	(n=1466)
Firm size***	*Share of firms (%)*		
up to 49 employees	37.9	62.1	100.0 (n=533)
50-249 employees	56.6	43.4	100.0 (n=707)
250 and more employees	83.6	16.4	100.0 (n=226)
	(n=791)	(n=675)	(n=1466)

Affiliation to low-, medium-, high-tech sectors calculated on basis of NACE 3-digit level
Sector affiliation within manufacturing industry on NACE 2-digit level (WZ 2003 Rev 1.1)
Significance levels: *** = p<0.000 / ** = p<0.005 / * = p<0.01 / + = p<0.1

These findings underline the argument that non-R&D-performing are often SMEs which refrain from R&D activity due to their restricted financial and personal resources. As R&D projects often require a certain minimum size, barriers to R&D include high fixed costs, high financial commitments, or high risk to firm survival in the case of R&D failure. These barriers account for the large numbers of non-R&D-performers among SMEs (RAMMER ET AL. 2009: 37). Nevertheless, despite the fact that the groups of non-R&D-performers and

SMEs do indeed show high congruency, the findings also suggest that not every non-R&D performer belongs to the group of SMEs since almost every fifth larger firm with 250 and more employees did not undertake any R&D in the surveyed period.

Regarding product complexity (Table 5), non-R&D-performing firms tend to manufacture products with low (28%) or medium complexity (51%), while about 21% of them turn out complex products. This contrasts significantly with R&D-performing firms, of which only 13% manufacture simple products and more than 40% produce complex products. Thus, R&D intensity is clearly reflected in the general product complexity. This also holds for different firm size categories. However, the varying product complexity is not reflected in the real net output ratio. Both, R&D- and non-R&D-performing firms show similar values, which means that both have comparable average input costs.

Looking at the average batch or lot size, one may assume that lower product complexity is aligned with larger batch sizes in manufacturing. But surprisingly, this is not the case. Only about one third (30%) of all non-R&D-performers in the sample reports large lot sizes. Instead, the majority (49%) produces in medium batch size, while about 22% even use single unit production. It is even more surprising that there is almost no difference between non-R&D-performing firms and R&D-performers. On the contrary, the share of firms with single unit production is even slightly higher for non-R&D-performers (30% vs. 24%), whereas, at the same time, the share of firms with large lot size manufacturing is only 1% higher for non-R&D-performers. When using firm size to further differentiate between the two groups, the most noticeable difference is in the group of very small non-R&D-performing firms with fewer than 50 employees, of which about 14% use large batch size production compared to 8% of R&D-intensive firms, and which therefore match the expected picture.

Table 5: Product complexity, batch size, vertical range of manufacture and year of establishment of R&D- and non-R&D-performing firms (Source: GMS 2009, own calculations)

	Non-R&D-performing firms (0 % R&D expenditure)			
	20-49 employees	50-249 employees	250 and more employees	Total
Product complexity* **		*Share of firms (%)*		
Simple products	31.2 ***	27.3 ***	13.5	**28.4**
Products with medium complexity	50.8	50.5	54.1	**50.9**
Complex products	18.0	22.2	32.4	**20.7**
				100.0 (n=641)
Manufactured batch or lot size +				
Single unit production	33.4	27.1 +	21.6	**29.9**
Small or medium batch/lot	52.3	50.5	48.6	**51.3**
Large batch/lot	14.3	22.3	29.7	**18.9**
				100.0 (n=636)
Real net output ratio		*(Total sales/input cost)/total sales*		
	0.583	0.573	0.479	**0.574** (n=513)
Year of establishment		*Share of firms (%)*		
founded before 1961	42.6 ***	46.9	59.5	**45.5**
founded between 1961-1970	6.0	8.1	8.1	**7.1**
founded between 1971-1980	11.5	6.2	5.4	**8.7**
founded between 1981-1990	16.3	13.7	5.4	**14.5**
founded between 1991-2000	21.1	20.5	18.9	**20.7**
founded after 2000	2.4	4.6	2.7	**3.4**
				100.0 (n=673)

	R&D-performing firms (>0 % R&D expenditure)			
	20-49 employees	50-249 employees	250 and more employees	Total
Product complexity		*Share of firms (%)*		
Simple products	12.2	14.5	9.8	**12.8**
Products with medium complexity	44.7	50.6	39.7	**46.5**
Complex products	43.1	34.9	50.5	**40.7**
				100.0 (n=774)
Manufactured batch or lot size				
Single unit production	33.5	22.1	17.5	**23.9**
Small or medium batch/lot	58.4	61.4	51.9	**58.4**
Large batch/lot	8.1	16.5	30.6	**17.7**
				100.0 (n=769)
Real net output ratio		*(Total sales/input cost)/total sales*		
	0.607	0.568	0.542	**0.572** (n=572)
Year of establishment		*Share of firms (%)*		
founded before 1961	28.2	40.3	67.2	**43.5**
founded between 1961-1970	8.4	8.8	8.5	**8.6**
founded between 1971-1980	9.4	12.3	6.9	**10.2**
founded between 1981-1990	20.3	15.8	7.4	**14.9**
founded between 1991-2000	26.7	18.5	10.1	**18.6**
founded after 2000	6.9	4.5	0.0	**4.0**
				100.0 (n=788)

Significance levels: *** = p<0.000 / ** = p<0.005 / * = p<0.01 / + = p<0.1

Recalling the perspective of growth theory, there was the assumption that non-R&D-performing firms are less competitive and more likely to fail market competition than firms with (high) R&D activities. The average year of establishment is a good indicator for how long non-R&D-performing firms are able to successfully survive market competition. The results are very clear: almost half the non-R&D-performers in the subsample have managed to survive more than 50 years. Of course, only panel data would be able to provide accurate information about whether these firms have really never performed any R&D over their whole lifetime. But considering the development of non-R&D sectors over time (see Annex A.3.2), it can be assumed that – even if they have performed R&D in the past – they probably did not have had any R&D expenditure at high-tech level. Accepting this point means that firms can be sustainably competitive without or with only occasional R&D activities.

Moreover, the data also show that the phenomenon of non-R&D activity is observable at even relatively young firms founded in the last ten to twenty years, which probably have less firmly established technological path dependencies. Non-R&D-performing firms should therefore not be considered a "phase-out" model which disappears over time.

There is one more finding which seems hard to interpret. Almost 43% of non-R&D-performing firms with up to 49 employees were founded before 1961. Viewed positively, this could mean that they have not needed to grow as this size allows them to survive the competition due to its various advantages (e.g. flexibility, speed). Seen negatively, however, this could also mean that these firms have shrunk over time due to their inferior competitiveness. Again, panel data would be needed to examine this problem more closely. The subsequent descriptive analysis of the innovation and economic performance of this group may be able to provide further insights.

Looking at the most important competitive factor of non-R&D-performing firms (see Table 6), growth theory assumes that firms without R&D rely primarily on cost competition, as they systematically lack the opportunity to shift their production to higher levels of productivity using R&D and technological knowledge. However, the findings presented below reject this assumption. It is true that about 23% of firms without R&D report that product price is their unique selling proposition, a figure which is about 10% higher than that of R&D-performing firms, but half of all non-R&D-performers also state that product quality is the most important competitive factor used to distinguish themselves from their competitors. Other relevant factors mentioned by non-R&D-performers are customisation of products (13%) and adherence to delivery times (8%).

Table 6: Competitive factors, position in value chain and ownership structure of R&D- and non-R&D-performing firms
(Source: GMS 2009, own calculations)

	R&D-performing firms (>0 % R&D expenditure)	Non-R&D-performing firms (0 % R&D expenditure)
Most important competitive factor* **	*Share of firms (%)*	
Product price	13.8	22.6
Product quality	43.3	49.7
Innovative products	18.8	4.6
Customisation to customers' demands	19.1	13.2
Adherence to delivery times / short delivery times	3.3	7.9
Product-related Service	1.9	1.9
	100.0	100.0
	(n=701)	(n=569)
Position in value chain	*Share of firms (%)*	
Producer of finished goods	67.5	65.7
Toll manufacturer	5.6	15.0
	(n=782)	(n=661)
If producer of finished goods...[1]		
...for customers***	26.1	41.7
...for industrial business**	78.0	68.4
	(n=528)	(n=434)
If not producer of finished goods...[2]		
...system supplier**	33.5	21.1
...parts/components supplier*	68.1	55.9
	(n=254)	(n=227)
Ownership structure* **	*Share of firms (%)*	
Private person / family	70.1	86.2
Financial investor (i.e. venture capital)	4.9	1.5
Other firm (not financial investor)	16.6	8.0
Foundation	1.7	0.0
Other owners	2.4	2.4
No major owner	4.4	1.8
	100.0	100.0
	(n=778)	(n=661)

Significance levels: *** = p<0.000 / ** = p<0.005 / * = p<0.01 / + = p<0.1
1, 2 = percentages above 100% due to the fact that question allowed for multiple answers

Comparing both groups, the development of innovative products is, as expected, clearly linked to R&D-performing firms (19%). Furthermore, product customisation seems to play an even bigger role for R&D-performing firms. This appears reasonable, as it can be argued that they develop and adapt their rather complex products (e.g. machinery, manufacturing equipment) in close cooperation with their customers. In the context of the existing findings about non-R&D-intensive firms, this result may be taken to indicate the prevailing process specialisation of these firms.

The indication of process specialisation builds on the assumption that non-R&D-performing firms are more frequently suppliers than producers of end-products. This assumption also fits in with the relatively lower product complexity of goods manufactured by non-R&D-performers presented above. However, the table reveals that this is not the case. About 66% of non-R&D-performing firms in the sample reported they were producers of finished products compared to about 68% of R&D-performers. Thus, on the one hand, non-R&D-performing firms are not dominantly suppliers for other (R&D-intensive) firms. On the other hand, this means that non-R&D-performing firms are also able to successfully compete on end-product markets and are not concentrated in protected supplier-niches in which competition is comparably weak. Firms without R&D more frequently (42%) produce goods for the consumer market ("B2C") than R&D-performers (26%), while the latter have a higher share of end-products for industrial business (78% vs. 68% - "B2B"). This seems plausible when considering firms in the field of mechanical engineering which often serve as "suppliers of productivity" by supplying other industries with highly efficient manufacturing equipment (KINKEL/SOM 2010: 7).

Additionally, in line with their higher product complexity, R&D-performers play the role of system suppliers more often (34%) than non-R&D-performers (21%). In contrast, the share of toll manufacturers who serve as a kind of "extended workbench" for their customers, executing rather simple manufacturing tasks (e.g. welding, coating or grinding), is much higher for non-R&D-performing firms (15%) than for their counterparts with R&D activities (6%). Such toll manufacturers often do not develop their own products, are highly dependent on their customers and correspond most closely to the image many people have in mind when thinking about firms without R&D.

Two main assumptions can be made regarding the ownership structure of non-R&D-performing firms. Because R&D often connotes high risk (RAMMER ET AL. 2009: 36), non-R&D-performing firms may be more attractive to financial investors. On the one hand, it would thus make sense for other, larger firms or companies to buy these non-R&D-performing firms in order to secure the supply of specialised, intermediate products. On the other hand, to assess additional technological knowledge, it can be assumed that R&D-performing firms are more attractive for financial investors and investments of other firms. Moreover, financial investment in R&D-intensive firms may yield higher profit margins if the R&D is successful. Looking at the data, the second assumption seems the more appropriate one. The big majority (86%) of non-R&D-performing firms is owned by a private person or family, in most cases the founding entrepreneur. This result is linked to the fact, that many of these firms without R&D are SMEs. The only slightly significant category is that some non-

R&D-performers (8%) are owned by another firm which is not a financial investor (e.g. subsidiary of a bigger firm). Again, there are only minor differences compared to R&D-performers. The most striking difference is the relatively high share of R&D-performers that are owned by another firm (17%) or financial investor (5%).

To briefly summarise, the analysis of certain structural characteristics of non-R&D-performing firms in the German manufacturing industry clearly supports the basic assumption of inter-firm heterogeneity. It could be shown that firms without R&D can be found in all sectors of the manufacturing industry and that there are significant shares of such firms even in industries with high or very high R&D intensity. Moreover, most of the considered variables did not reveal any "typical" structural pattern which indicates anything like a distinct, homogenous pattern of non-R&D-performers compared to R&D-performers. The most significant differences were found with regard to firm size distribution, product complexity, and innovative products as the most important competitive factor.

5.3.2 Innovation resources

The following sections describe the main findings concerning the use and importance of different firm-internal and firm-external innovation resources by non-R&D-performing firms.

Product development

As already stated, the ability to develop new products is a crucial innovation resource for product innovation in manufacturing firms. Becasue this ability is often linked to the existence of formal R&D, it is sometimes assumed that non-R&D-performing firms lack this ability and are not able to successfully develop their own products. However, empirical evidence reveals that a relevant share of firms without formal R&D is nevertheless able to develop new products

Table 7 shows that more than one third (35%) of the non-R&D-performing firms in our sample have successfully developed new products themselves. However, this also means that almost two thirds have not developed a new product within the past three years. Thus, they are either occasional product developers or non-product developers. This is prbably especially true for toll manufacturers. The discrepancy is very obvious when compared to R&D-performing firms, of which almost 77% developed their own products within the surveyed period. This shows that the propensity to undertake own product developments is significantly higher for firms with R&D activities.

Further subdividing the group of firms with their own product development shows that a vast majority of firms without R&D (55%) develop their products according to customers' specifications. At first sight, this may seem to be linked to their status as suppliers. But as the analysis has shown, non-R&D-performers are not more likely to be suppliers than R&D-performers. Thus, this high level of product customisation could be interpreted in terms of the knowledge and innovation stimulus that flows from the customers to the firms. As a consequence, these product developments are probably very specific and may have limited demand and market potential. Moreover, about 15% of the firms without R&D stated that they never develop their own products. This is a good match to the figures for toll manufacturers identified above. Regarding the group of firms with R&D, product customisation also plays the most important role even if its share is significantly lower (46%). In this group, the development of standard products with and without the possibility of ex post customisation has higher shares (38% and 15%, respectively) than in the group of non-R&D-performers (19% vs. 11%).

Table 7: Product development in R&D- and non-R&D-performing firms (Source: GMS 2009, own calculations)

	R&D-performing firms (>0 % R&D expenditure)	Non-R&D-performing firms (0 % R&D expenditure)
Share (%) of firms with own product development	76.8	35.1 ***
	(n=789)	(n=670)
Share (%) of firms that develop their products...		
...according to customers' specifications	46.0	54.8 *
...as a standardised basic programme into which customers' specific requirements are implemented	37.7	19.0 *
...for a standard programme from which the customers can select	14.7	10.8 *
...does not exist in the firm	1.7	15.4 *
	100.0	100.0
	(n=770)	(n=637)
High-Tech Components		
Share of firms (%) whose products contain high-tech elements[1]	32.4	7.3 ***
	(n=785)	(n=668)

Significance levels: *** = p<0.000 / ** = p<0.005 / * = p<0.01 / + = p<0.1
1 = e.g. micromechanical, microelectronic or microoptical parts, high performance material, gene- or biotechnological elements

When asked whether their products also contain high-tech elements (e.g. micromechanical, -electronic, -optical parts, high performance materials, genetically-engineered or biotechnological elements), 32% of R&D-performing firms respronded affirmatively in contrast to about 7% of non-R&D-performing firms. As such high-tech elements are probably not developed within non-R&D-performing firms, but are more likely to be purchased from external partners, this implies that this 7%-share of non-R&D-performers obviously has the ability to handle and deal with these high-tech parts in such a way that they fit into their existing product specifications. Hence, non-R&D does not necessarily mean the absence of high-tech.#

Product-related services

To continue with the immaterial dimension of product development, offering product-related services (PRS) is another strategy non-R&D-performing firms can enhance their competitive advantage since PRS are not necessarily linked to R&D activities.

With regard to recent innovation activities in the field of PRS, Table 8 shows that only a small minority (12%) of non-R&D-performers has offered a new or significantly improved PRS within the last three years. Because products with low complexity (e.g. one piece metal parts) do not leave much room for PRS, the variable has been additionally partitioned according to product complexity. The results clearly show that the higher the complexity of the manufactured products, the higher the share of firms that offered new PRS within the last three years. In the case of non-R&D-performing firms, this means that between 10% (simple products), 12% (medium complexity) and 17% of all firms in the sample have introduced new PRS. Hence, almost one fifth of firms without formal R&D have already recognised the competitive potentials of the resource PRS.

Looking more closely at those firms with new PRS, it becomes apparent that the strategic relevance of PRS is relatively high. The strategic relevance of PRS can be captured by analysing those services that have been directly invoiced. The higher the share of directly invoiced PRS, the more advanced is the transformation from a mere "producer of goods" to the new type of "producing service provider" (DEMUS/SPATH 2000: 70; GEBAUER 2004: 5ff.).[227] The average share of PRS which has been directly invoiced (8.3%) – in the sense of an autonomous, distinctive immaterial product – is notably higher than the share of indirectly invoiced PRS (5.4%). This implies that non-R&D-performing firms seem to assign higher strategic relevance to intentionally supplementing their product portfolio with PRS; particularly those non-R&D-performers with complex products.

Table 8: Offer of product-related services by R&D- and non-R&D-performing firms (Source: GMS 2009, own calculations)

	Non-R&D-performing firms (0 % R&D expenditure)				
	Simple products	Products with medium complexity	Complex products	Total	
New product-related services (PRS)					
Share of firms (%) that offered new PRS	9.7	12.2 *	17.1 +	**12.5** ***	(n=615)
Average share (%) of turnover of services directly invoiced	-	-	-	**5.4**	(n=346)
Average share (%) of turnover of services indirectly invoiced	-	-	-	**8.3**	(n=326)
	R&D-performing firms (>0 % R&D expenditure)				
	Simple products	Products with medium complexity	Complex products	Total	
New product-related services (PRS)					
Share of firms (%) that offered new PRS	15.2	20.2	27.4	**22.5**	(n=759)
Average share (%) of turnover of services directly invoiced	-	-	-	**6.0**	(n=599)
Average share (%) of turnover of services indirectly invoiced	-	-	-	**7.8**	(n=583)

Significance levels: *** = p<0.000 / ** = p<0.005 / * = p<0.01 / + = p<0.1
PRS = product-related services / (-) = empty cells due to insufficient number of cases

227 Both DEMUS/SPATH (2000) and GEBAUER (2004) distinguish the following three stages:

- *Producers:* firms that focus on the manufacturing of goods. Product-related services are understood as an „add-on" which is offered casually.

- *Service-offering producer:* if product characteristics and prices alone can no longer differentiate market competitors, offering PRS can become a distinguishing competitive factor, which justifies higher product prices. Such PRS are directly invoiced.

- *Producing service provider:* PRS represent the core business of these firms. The product is only used for the purpose of customer-connectivity. Market differentiation is achieved by individually customising each PRS. These hybrid bundles of product-service-solutions are sold as services and are highly embedded in the value creation processes of the customer.

Interestingly, the average share of directly invoiced PRS is even slightly higher than that for R&D-performers (7.8%), who, in turn, have a higher average share of indirectly invoiced PRS (6%). However, these differences are not statistically significant and can only be interpreted as general tendencies. Moreover, as expected due to the higher average product complexity of R&D-performing firms, the share of these firms that offered new PRS within the last three years is considerably and statistically higher (23%). This finding also supports those of KIRNER ET AL. (2009a), who argued that R&D-performers are more active in the use of PRS as new PRS are often developed parallel to new material products.

Use of advanced manufacturing techniques

Existing empirical findings show that the use of advanced manufacturing techniques (AMT) is a crucial innovation resource for non-R&D-intensive firms (e.g. ARUNDEL ET AL. 2008; SANTAMARÍA ET AL. 2009; BARGE-GIL ET AL. 2008; TSAI/WANG 2009). Since the use of AMT may be dependent on firm size due to the financial resources required to invest in AMT, the analysis also distinguishes between three firm size categories.

Table 9 shows that more than the half of all the surveyed non-R&D-performing firms use at least one AMT, which is similar to the corresponding share of R&D-performing firms (56%). However, when looking at the time of implementation, only 29% of non-R&Ds have introduced an AMT within the last three years in contrast to 42% of firms with R&D. This means that recent innovation activity in the field of AMT is significantly higher for R&D-performers than for non-R&D-performers, which could also imply that the AMT used by R&D-intensive firms are more state-of-the-art than those in non-R&D-performers.[228] Interestingly, firm size does not correlate positively with the use of AMT but a higher share of larger firms with more resources has introduced AMT since 2006.

Breaking down the use of AMT to different fields of application shows that non-R&D-performing firms use AMT primarily in the fields of automation and link-

228 Comparing the diffusion of specific AMT in R&D-intensive and non-R&D-intensive firms in the German manufacturing industry, KIRNER ET AL. (2009b) show that AMT-diffusion in non-R&D-intensive firms lags behind R&D-intensive firms by an average of three to five years. This suggests that non-R&D-intensive firms tend to put off implementing AMT until they are available at a mature status and moderate prices.

age (40%), digital factory and IT cross-linkage (42%) as well as energy efficiency (46%). Especially the use of automation and linkage is higher in large firms (67%) compared to medium (42%) and small firms (35%).

There are only a few significant differences when comparing the use of AMT in non-R&D- with R&D-performing firms. Non-R&D-performing firms use AMT much more rarely in the field of machining and production technology (25%) compared to R&D-performers (42%). One the one hand, this seems surprising, because it may be assumed that these firms' often-quoted process specialisation is linked to high levels of AMT use in production. On the other hand, however, when looking at the finding that a high share of employees work in manufacturing and assembling (KIRNER ET AL. 2009b), the lower rate of AMT in this field seems to make sense. AMT usage in the field of automation and linkage is higher (67%) for larger non-R&D-performing firms with more than 250 employees than for the same group of firms with formal R&D (46%), which may indicate greater cost orientation for the AMT use of these firms.

However, when regarding the prior goals of AMT introduction (see Annex A.5.1), it is apparemt that in both groups the majority of AMT were introduced to reduce manufacturing costs (37% for non-R&D firms, 42% for R&D-performers), followed by increasing quality (26% non-R&D, 27% R&D), flexibility (16% non-R&D, 19% R&D), and promoting product innovation (8% non-R&D, 15% R&D). While the order of AMT goals is the same, two differences between the groups should be kept in mind. First, the general assumption that non-R&D-performing firms compete more on a cost basis than R&D-intensive firms (PAVITT 1984: 356) could not be verified with regard to AMT use. Secondly, as expected, promoting product innovation is more frequently mentioned by R&D-performing firms as a goal of AMT use.

The results regarding non-R&D-performing firms with a high intensity use of AMT correspond to the previous findings about AMT use. They have particular relevance in the field of automation and linkage (28%), followed by the energy efficiency (about 19%). As can be seen, these shares increase considerably with firm size. This can perhaps be explained by the slightly higher share of non-R&D-performing firms involved in cost-centred competition. Although, in total, the share of non-R&D-performers with high intensity usage is not large (14%), the findings reveal that these fields are also of high relevance for the group of larger firms (38%), while only about 9%-12% of firms without R&D report high intensity use with regard to machining and production techniques. Obviously, it is only reasonable for a few non-R&D-performing firms to implement such techniques at a high level of intensity.

Table 9: Use of advanced manufacturing techniques by R&D- and non-R&D-performing firms
(Source: GMS 2009, own calculations)

	20-49 employees	50-249 employees	250 and more employees	Total	
Non-R&D-performing firms (0 % R&D expenditure)					
Use of advanced manufacturing technologies (AMT)					
Share of firms (%) that use of at least one AMT	52.5	57.0	47.1	**54.3**	(n=466)
Share of firms (%) that introduced AMT within past three years	22.7 +	33.2 **	43.2	**28.6** ***	(n=675)
Share of firms (%) that use AMT in the field of...					
...automation and linkage	35.4	42.4	66.7 +	**40.3** +	(n=663)
...machining and production technologies	24.0 **	24.7 ***	30.6 +	**24.7** ***	(n=649)
...digital factory and IT cross-linkage	36.3	48.9	30.8 +	**42.1**	(n=280)
...energy efficiency	44.9	45.9	55.6	**46.2**	(n=366)
...other AMT	-	-	-	**8.8**	(n=624)
Share of firms (%) with high intensity use in the field of...					
...automation and linkage	19.3	34.9	48.6	**28.0** +	(n=675)
...machining and production technologies	8.5 +	12.1 +	10.8 +	**10.2** ***	(n=675)
...digital factory and IT cross-linkage	9.1	16.9 +	37.8	**14.2** ***	(n=675)
...energy efficiency	14.8	20.5	35.1	**18.5** **	(n=675)
R&D-performing firms (>0 % R&D expenditure)					
	20-49 employees	50-249 employees	250 and more employees	Total	
Use of advanced manufacturing technologies (AMT)					
Share of firms (%) that use of at least one AMT	51.9	55.8	58.3	**55.6**	(n=682)
Share of firms (%) that introduced AMT within past three years	31.7	44.0	48.1	**41.8**	(n=791)
Share of firms (%) that use AMT in the field of...					
...automation and linkage	39.6	48.4	45.7	**45.5**	(n=784)
...machining and production technologies	36.1	39.9	51.1	**41.6**	(n=778)
...digital factory and IT cross-linkage	40.2	41.2	54.7	**45.4**	(n=491)
...energy efficiency	42.6	46.0	52.4	**47.2**	(n=513)
...other AMT	-	-	-	**11.2**	(n=708)
Share of firms (%) with high intensity use in the field of...					
...automation and linkage	20.3	31.8	47.6	**32.6**	(n=791)
...machining and production technologies	14.9	18.0	22.8	**18.3**	(n=791)
...digital factory and IT cross-linkage	10.4	23.5	41.3	**24.4**	(n=791)
...energy efficiency	18.3	25.3	32.3	**25.2**	(n=791)

Significance levels: *** = p<0.000 / ** = p<0.005 / * = p<0.01 / + = p<0.1
AMT = advanced manufacturing techniques
(-) =empty cells due to insufficient number of cases

Comparing both groups shows, that there are significantly higher shares of R&D-performing firms with an intensive use pf AMT in all fields. Only in the field of automation and linkage are both groups very similar to each other. Hence, non-R&D-performing firms in general do not seem to over-compensate their missing R&D by an excessive use of AMT. Both the number of AMT and their intensity of use are similar to or even lower than R&D-performers. Instead, against the backdrop of the comparable share of AMT users in both groups, firms without R&D seem to selectively choose the intensity of AMT use according to their specific needs, for example, automation and linkage in the context of a labour-intensive manufacturing structure. However, since intensity was assessed subjectively using the estimations of the interviewees, these differences may also result from structural differences in cognition.

In summary, it can be stated that non-R&D-performing firms have access to and make broad use of advanced manufacturing technologies. However, as supposed by some authors (e.g. VON TUNZELMANN/ACHA 2005; ARUNDEL ET AL. 2008; HEIDENREICH 2009), the data did not support the view that non-R&D-performing firms over-compensate their missing R&D by an above-average use of AMT. Compared to R&D-performing firms, there are no big differences in AMT use, with the exception of production technologies (in terms of use and intensity) and digital factory/IT cross-linkage (with regard to intensity). Moreover, the share of non-R&D firms that introduced AMT within the last three years is considerably lower than that of R&D-performers – despite a similar rate of use – which could mean that these firms tend to be AMT-followers rather than first-movers.

Use of organisational concepts and management tools

Intangible firm-internal processes and routines are a key element of resource-based theory used to explain the competitive advantage of firms. Several previous studies have highlighted the positive impact of the use of different organisational and management concepts (OMC) on the competitive advantage of non-R&D-performing firms (RAMMER ET AL. 2009; HEIDENREICH 2009; BARGE-GIL ET AL. 2008). As the feasibility of certain OMCs (e.g. especially in the fields of work and production organisation) heavily depends on firm size in some cases, the results are additionally differentiated by three firm size categories.

Table 10 shows that 61% of non-R&D-performing firms stated that they either use at least one of the cited OMCs or added an OMC in the open category. In addition, this overall share is roughly equal for all firm sizes. This means that even a high share of very small firms with fewer than 50 employees uses such OMC (58%). There is no significant difference here compared to R&D-performers.

However, only around 34% of the non-R&D-performing firms implemented an OMC within the last three years. The other non-R&D firms in the sample used OMC to enhance their competitive advantage at an earlier point in time. The sahre of OMC introducers within the last three years is higher (51%) for R&D-performing firms. With reference to the suggestion that OMC are not subject to the same technological aging processes as new products or new AMT, it can be assumed that those firms who have implemented OMC earlier obtain higher profits than "latecomer firms" who only recently introduced a new OMC and where the new routines and behaviours have not yet become fully internalised (ARMBRUSTER ET AL. 2008).

Differentiating between firm size categories reveals that the share of firms introducing OMC within the past three years increases with increasing size from 28% (up to 49 employees) through 39% (50-249 employees) up to 43% (250 and more employees). Hence, larger non-R&D firms have obviously been more active in recently introducing OMC than smaller ones. This basic correlation between firm size and year of OMC introduction also holds for R&D-performers, although their shares are significantly higher in each case.[229]

Looking closer at different fields of OMC, the findings suggest that non-R&D-performers primarily deploy OMC with regard to work organisation (48%) and standardisation and knowledge management (46%), followed by human resource management and leadership (38%) as well as working hours and payment schemes (37%), while the field of organisation of production is an outlier with about 24%. This is very similar to the findings for AMT use, where the field of production technology has also played a less important role for non-R&D-performers. Nevertheless, this finding is also slightly surprising, because, as showed by KIRNER ET AL. (2009b), the manufacturing and production in non-R&D-performing firms tend to be more labour-intensive and it could be expected that these firms' process expertise derives from their well-organised production. But on the other hand, non-R&D-performing firms producing rather simple products may not have the same options for production organisation as firms with more complex products.

In general, the propensity to use OMC increases with firm size in all fields. Despite the fact that the overall relationship between the different fields of OMC use is very similar to that of R&D-performing firms, the share of non-R&D firms

---

229 However, it should be taken into account that these findings cannot capture whether newer version of esxisting OMC may have been "updated" in the firms within the last three years. For instance, the concept of team work which has been deployed for many years within a firm could have been qualitatively modified during the past three years, for instance by empowering the teams with autonomous decision-making. But the available indicators in the dataset do not differentiate between different qualitative stages of OMC. Thus, it is impossible to analyse whether non-R&D-performing firms may be more focussed on modifications of their already existing OMC than R&D-performing firms, which could also explain the prevailing results.

using OMC in these fields is significantly lower. Only when differentiating by firm size are there types of OMC whose use is similar between R&D- and non-R&D-performing firms (e.g. organisation of work, standardisation and knowledge management).

When taking a closer look at the prior strategic goal (see Annex A.5.2) targeted with the implementation of OMC, the picture is very similar to that of AMT. The majority of non-R&D-performing firms in the sample state that the prior goal was reducing costs (39%), followed by increasing quality (32%), and increasing flexibility (29%), while promoting new product development only plays a subordinate role (4%). But again, these results do not differ significantly between the two groups of firms. Most interestingly, the largest share of R&D-performers (40%) also has stated that reducing costs was the main intention behind OMC implementation. Again, this finding does not uphold the assumption that non-R&D-performers are more strongly driven by cost oriented competition.

Because of the higher share of non-R&D-performers which implemented new OMC before 2006, it may be assumed that these also have a higher share of firms with high intensity use. But the opposite is true. In all the considered fields of OMC, the share of high intensity users is significantly larger for firms with formal R&D. This also holds for almost all firm size categories. One possible explanation might be the fact that R&D-performing firms have a larger share of highly qualified personnel which allows more advanced forms of work organisation, for instance the delegation of responsibilities and controlling tasks to lower hierarchical levels. Regarding non-R&D-performers, the most intensive users can be found in the field of flexible working hours and payment schemes (36%), standardisation and knowledge management (28%), and organisation of work (27%). Especially the first can be explained by these firms' higher labour intensity. Compared with the share of mere users, it seems that those firms who decided to implement flexible working hours and payment schemes do so with a higher intensity. In contrast, the high intensity use of production-oriented organisation concepts occurs in only a few of these firms (11%). Once again, intensity was assessed subjectively based on respondents' estimations so these differences could be due to differences in perception.

Table 10: Use of organisational concepts and management tools in R&D- and non- R&D-performing firms
(Source: GMS 2009, own calculations)

	Non-R&D-performing firms (0 % R&D expenditure)				
	20-49 employees	50-249 employees	250 and more employees	Total	
Use of innovative organisational concepts and management tools (OMC)					
Share of firms (%) that use at least one OMC	57.8	64.6 +	59.5	**61.0**	(n=667)
Share of firms (%) that introduced OMC within past three years	27.2 **	39.1 **	43.2 +	**33.5 *****	(n=675)
Share of firms (%) that use OMC in the field of...					
...organisation of work	44.8	49.2 *	62.2	**47.7 *****	(n=666)
...organisation of production	19.8 +	27.6 +	35.1	**24.2 *****	(n=658)
...standardisation, knowledge management	37.2 ***	52.5	62.2	**45.5 *****	(n=659)
...working hours, payment schemes	30.1	43.3	33.3 +	**36.5 +**	(n=490)
...human resource management, leadership	33.1 +	42.9	43.2	**38.1 *****	(n=666)
...other OMC	-	-	-	**5.1 *****	(n=651)
Share of firms (%) with high intensity OMC use in the field of...					
...organisation of work	22.7 +	29.3	40.5	**26.7 *****	(n=675)
...organisation of production	7.9	13.0 **	24.3	**11.1 *****	(n=675)
...standardisation and knowledge management	20.5 *	31.3 ***	59.5	**27.6 *****	(n=675)
...working hours, payment schemes	30.8	38.1	56.8	**35.6 ****	(n=675)
...human resource management, leadership	7.3	16.9	27.0	**12.7 *****	(n=675)
	R&D-performing firms (>0 % R&D expenditure)				
	20-49 employees	50-249 employees	250 and more employees	Total	
tools (OMC)					
Share of firms (%) that use at least one OMC	59.7	57.0	62.2	**58.9**	(n=789)
Share of firms (%) that introduced OMC within past three years	39.6	52.0	62.4	**51.3**	(n=791)
Share of firms (%) that use OMC in the field of...					
...organisation of work	46.5	59.9	68.8	**58.6**	(n=785)
...organisation of production	28.3	36.7	43.2	**36.1**	(n=775)
...standardisation, knowledge management	57.1	53.9	56.7	**55.4**	(n=782)
...working hours, payment schemes	38.0	37.8	52.3	**41.7**	(n=645)
...human resource management, leadership	43.7	48.9	52.2	**48.3**	(n=784)
...other OMC	-	-	-	**11.8**	(n=735)
Share of firms (%) with high intensity OMC use in the field of...					
...organisation of work	32.7	35.0	42.3	**36.2**	(n=791)
...organisation of production	11.4	21.3	34.9	**22.0**	(n=791)
...standardisation and knowledge management	31.2	44.5	55.0	**43.6**	(n=791)
...working hours, payment schemes	31.7	42.3	59.8	**43.7**	(n=791)
...human resource management, leadership	11.4	20.3	36.0	**21.7**	(n=791)

Significance levels: *** = p<0.000 / ** = p<0.005 / * = p<0.01 / + = p<0.1
OMC = organisational and management concepts
(-) = empty cells due to insufficient number of cases

To summarise, the findings concerning the use of OMC by non-R&D-performing firms are ambiguous. On the one hand, a high share of such firms uses OMC to enhance competitive advantage and even introduced them more than three years ago. OMC are widely used in all fields except the organisation of production, and especially in the organisation of work. In this sense, firms without R&D may be interpreted as first movers considering the use of OMC.

On the other hand, they do not overcompensate their lack of R&D with the intensive use of OMC. Although the number of firms in each field with high intensive use varies, the overall intensity of OMC use is considerably below that of R&D-performers.

Innovation cooperation

Because non-R&D-performing firms lack the internal innovation resource of formal R&D, it has frequently been argued that these firms compensate this by cooperating more intensively with external partners in innovation projects (e.g. BENDER ET AL. 2005, BENDER 2006, HIRSCH-KREINSEN 2004a, 2008; SANTAMARÍA ET AL. 2009). However, there are opposing findings based on empirical data showing that firms without R&D participate in innovation cooperation to a much lesser extent than R&D-intensive ones (e.g. ARUNDEL ET AL. 2008; TSAI/WANG 2009; KIRNER ET AL. 2009b). As the possibility to participate in inter-firm cooperation requires a certain level of personnel resources the following findings again differentiate between three categories of firm size.

The top half of Table 11 shows that more than half of all non-R&D-performing firms (53%) have participated in innovation cooperation with external partners. As assumed, the propensity to cooperate depends strongly on firm size with 47% of small firms, 58% of medium sized firms and about 76% of larger firms collaborating in such innovation projects.

The average number of innovation cooperations reveals that the majority of firms without R&D (23%) are active in three to five innovation cooperations with external partners[230], followed by 13% with two and 12% of one innovation cooperation. Only a a very small share of non-R&D-performing firms (4%) participates in six or more cooperations. Interestingly, this relation is valid for all firm size categories, while the absolute frequency of cooperation declines with smaller firm size. This means that, if a small or large firm without R&D decides to use innovation cooperation as an innovation resource, most of them participate in three to five collaborations.

230 This average number of cooperations was compiled using a count variable counting all "yes" answers across the different dimensions observed. As a result, it really represents distinct innovation collaborations, which means that only the first collaboration with one partner in one field of innovation is counted. Thus, this figure does *not* encompass the case of more than one collaboration with the same partner in one field of innovation (e.g. three collaborations with the same customer in the field of product innovation).

Table 11: Participation in innovation cooperation by R&D- and non-R&D-performing firms (Source: GMS 2009, own calculations)

	Non-R&D-performing firms (0 % R&D expenditure)				
	20-49 employees	50-249 employees	250 and more employees	Total	
Share of firms (%) participating in innovation cooperatio	46.7 ***	57.3 ***	75.7	**53.2** ***	(n=660)
Share of firms (%) that participate in...					
...in one innovation cooperation	10.9 ***	12.7 ***	13.5 **	**11.9** ***	
...in two innovation cooperations	12.4	14.3	13.5	**13.3**	
...in 3-5 innovation cooperations	18.7	24.8	43.2	**22.8**	
...in 6 and more innovation cooperations	3.3	4.6	5.4	**4.0**	(n=675)
	R&D-performing firms (>0 % R&D expenditure)				
	20-49 employees	50-249 employees	250 and more employees	Total	
Share of firms (%) participating in innovation cooperatio	74.0	76.9	86.7	**78.5**	(n=786)
Share of firms (%) that participate in...					
...in one innovation cooperation	12.4	6.5	5.3	**7.7**	
...in two innovation cooperations	20.3	17.5	13.8	**17.3**	
...in 3-5 innovation cooperations	30.7	42.0	40.2	**38.7**	
...in 6 and more innovation cooperations	9.9	10.5	27.0	**14.3**	(n=791)

Significance levels: *** = p<0.000 / ** = p<0.005 / * = p<0.01 / + = p<0.1

Comparing both groups, R&D-performing firms show a higher frequency of innovation cooperation with external partners (79%). Most surprisingly, this relatively high share is similar for all three categories of firm size (74% for small firms, 77% for medium-sized firms, and 87% of larger firms). Hence, R&D-performing firms are much more likely to participate in innovation cooperation than non-R&D-performers regardless of their size. Regarding only the group of non-R&D-performing firms, there is a much stronger relation between firm size and the propensity to participate in innovation collaboration. This means that non-R&D performers are generally more reluctant to cooperate, which supports the findings of ARUNDEL ET AL. (2008), TSAI/WANG (2009), and KIRNER ET AL. (2009b).

Besides R&D performers' higher general propensity to participate in innovation cooperation, the average number of innovation collaborations is actually similar to that of non-R&D-performers. Again, the majority (both in total and within size categories) undertakes between three and five cooperations (31-42%). This means that the main difference between R&D- and non-R&D-performing firms lies in the decision to participate in innovation collaboration at all.

To provide further insights, Table 12 shows with which partners firms cooperate in different innovation fields. For the more technologyl-driven fields of new products and advanced manufacturing technology (AMT), suppliers and customers (the latter particularly for new products) are by far the most important cooperation partners for non-R&D-performing firms. Nevertheless, 13-14% of them stated that they also cooperated with external R&D organisations to access specialised knowledge. A more differentiated picture results for the immaterial innovation dimension of organisational and management concepts (OMC) and product-related services (PRS). While service providers (e.g. business consultants) are an important partner for most of the firms (57%), for OMC the main cooperation partners for new PRS are customers (54%) followed by external service providers (38%). It is not clear whether service provider in this case means that they actually provide the service or whether they are consultants supporting the non-R&D-performing firms in implementing new PRS.

Table 12: Innovation cooperation of non-R&D-performing firms with different external partners across different fields of innovation
(Source: GMS 2009, own calculations)

	Non-R&D-performing firms (0 % R&D expenditure)			
	New products	New AMT	New OMC	New PRS
Share of firms (%) participating in innovation cooperation with...				
...customers	61.1	21.0	22.1 +	53.7 +
...suppliers	48.9	66.1	19.7	23.1
...competitors	12.2	16.7 ***	11.5 +	11.1 +
...service providers	13.3	15.5 *	56.6 +	38.0
...external R&D organisations	13.0 ***	13.7 ***	13.1 ***	3.7 ***
	(n=270)	(n=233)	(n=122)	(n=108)
	R&D-performing firms (>0 % R&D expenditure)			
	New products	New AMT	New OMC	New PRS
Share of firms (%) participating in innovation cooperation with...				
...customers	64.8	17.7	13.0	65.8
...suppliers	47.0	64.1	14.6	26.1
...competitors	9.8	4.8	4.6	5.4
...service providers	16.3	24.3	66.1	32.4
...external R&D organisations	53.5	42.3	30.1	16.2
	(n=559)	(n=440)	(n=239)	(n=222)

Significance levels: *** = p<0.000 / ** = p<0.005 / * = p<0.01 / + = p<0.1

Percentages exceeding 100% because of multiple answers

AMT = advanced manufacturing techniques / OMC = organisational and management concepts / PRS = product-related services

Basically, this picture also holds for R&D-performing firms as well, but there are some significant differences between the two groups. Obviously, the most striking and significant difference is the relevance of external R&D organisations as cooperation partners. For instance, in the field of product development, 54% of R&D-performing firms participate in such collaborations. In the case of AMT, 43% cooperate with external R&D organisations, and even in the fields of OMC and PRS, 30% and 16%, respectively, collaborate with such partners.

The second difference is not that obvious but nonetheless very interesting. The comparison shows that non-R&D-performing firms collaborate significantly more frequently in collaborations with their competitors, especially in the fields of AMT, OMC, and even PRS. Without the necessary further analysis, which cannot provided in this context, this finding is confusing and it is hard to interpret because cooperation always implies a win-win situation for both partners. One explanation could be that both partners cooperate to develop the basic solution to a problem before each of them then, individually adapts this solution to its own firm-specific situation. As this only makes sense as long as no fundamental technological assets or core businesses are involved, it could be assumed that such innovation cooperation with competitors tends to take place in peripheral business areas. Other possible explanations could be that, in their role as suppliers, non-R&D-performers are either more frequently forced by their main customers to collaborate with their competitors, or that they try to increase their bargaining power over customers by collaborating in certain fields. This is a very surprising peculiarity of non-R&D-performing firms which definitely requires further research.

While some authors (PAVITT 1984; HEIDENREICH 2009) have suggested that most innovations of non-R&D-performing firms stems from suppliers of equipment and materials, this could not be supported by the present findings. There are no significant differences between firms with and without R&D concerning innovation cooperation with suppliers in any field of innovation. Instead, the results show that non-R&D-performing firms – depending on the field of innovation - make use of the whole range of potential partners for innovation cooperation, even external R&D organisations to a minor extent. All in all, this supports the assumption of complementarity not only between internal and external R&D (see chapter 2.6.6) but also between the existence of formal R&D and cooperation with external R&D organisations. Hence, based on these findings, the absorptive capacity of external R&D knowledge through innovation cooperation strongly correlates with internal R&D capacity. But this is not to say that firms without formal R&D may not benefit from innovation cooperation with external R&D (about 13% obviously do). To conclude, non-R&D-performing firms do not limit their cooperation activities to specific partners.

Innovation knowledge

As previously mentioned, innovation in non-R&D-performing firms is thought to be driven more by customer-related, practice-oriented types of knowledge (HEIDENREICH 2009; VON TUNZELMANN/ACHA 2005; HIRSCH-KREINSEN 2004, 2008A, 2008B; SUNDBO 1996). Moreover, as ARUNDEL ET AL. (2008) showed, due to their lack of formal R&D, non-R&D innovators are likely to draw ideas from alternative sources within the firm like production engineers or design staff. Thus, while it can be expected that R&D-performing firms rely primarily on formal R&D to source knowledge relevant for innovation, non-R&D-performers are assumed to mainly draw on firm-external, non-R&D-based knowledge.

Table 13 contains the findings about the use of different sources of innovation knowledge by R&D- and non-R&D-performing firms. Right at the start, the results reveal a very curious finding: Despite the fact that non-R&D-performers were defined and sampled based on their zero-expenditure on formal R&D, significant shares of firms stated that their "own R&D department" is a relevant source of innovation knowledge. For instance, in the field of product innovation, 20% named internal R&D as relevant source of innovation knowledge, followed by about 13% in the field of AMT, and even 14% in the field of OMC.

On the one hand, it could be assumed that these firms are occasional R&D-performers, which did not undertake any R&D in the survey period, but did perform formal R&D at an earlier time. If this is the case then, R&D is definitely an important source of innovation knowledge. On the other hand, it could be assumed that these firms have some internal "area" which is regarded as functionally equivalent to an R&D department. This could mean that some employees within the firm are responsible for, or more deeply involved in new development (e.g. some key individuals in production and management). This underlines the methodological difficulties of the R&D construct with regard to how it is understood by firms in general and non-R&D-performers in particular.

Regardless of these possible misunderstandings, these firms still have to be dealt with in the cluster analysis. As the indicator is based on interviewees' subjective estimations about relevant sources of innovation knowledge in their subjective estimation, it is not implausible to assume that they may have blurred the functional role of some employees with that of an institutional R&D department. Moreover, if the firms really are occasional R&D-performers, the statement about R&D as an important knowledge source is fully valid. Because this indicator cannot be used to determine whether these firms really have some kind of institutionalised, formal R&D or not, I choose *not* to exclude them from the cluster analysis, and am aware that they might constitute a pattern consisting of occasional R&D-performers.

Table 13 also shows that the relevance of different knowledge source varies according to different fields of innovation. With regard to the development of new products and PRS about 74% and 48%, respectively, of the non-R&D-performing firms report that customers are a relevant source for innovation knowledge. This is closely connected to the previous finding that customisation plays a crucial role in product development. Obviously, the high-level customisation is not restricted only to physical products, but also applies to PRS.

Interestingly, conferences and trade fairs are considered to be the most relevant source of innovation knowledge in the fields of AMT (52%) and OMC (42%). This could mean that non-R&D-performing firms tend to be less embedded in specific knowledge arenas and expert networks, which makes it harder for them to access relevant knowledge for AMT and OMC. Hence, conferences, trade fairs and other events offer them the possibility to access new innovation knowledge at relatively low-costs. The fact that this knowledge source is also very relevant for a large share of firms concerning the fields of new products (40%) and new PRS (26%) confirms this interpretation, as well as the relatively low overall relevance of external R&D organisations as knowledge sources. The latter are the most relevant sources for innovation projects concerning the introduction of new OMC (13%) and AMT (11%). Last but not least, competitors are again seen as relevant knowledge sources mainly in the field of new products (34%). This may be an indicator for follower-strategies which aim to imitate innovations developed by competitors.

Surprisingly, other employees within the firms do not play any outstanding role as a knowledge source for innovation. Depending on the field of innovation, 6% to 18% of non-R&D-performers indicate that they have relevant innovation knowledge. This relativises the suggestion sometimes made that practice-oriented, tacit or implicit elements of employees' knowledge play a dominant role for non-R&D innovation (ARUNDEL ET AL. 2008; BENDER 2006; HIRSCH-KREINSEN 2008b).

Regarding the role of suppliers as a relevant knowledge source for non-R&D-performers, the GMS 2009 data show no signs of "supplier domination" as stated by PAVITT (1984) or HEIDENREICH (2009). Suppliers were only mentioned in the field of AMT by a relatively large share of firms (46%). This observation even holds when compared to relevant knowledge sources for R&D-performing firms. A significant difference between non-R&D-performing firms and R&D-performers can only be observed in the innovation field of new products, where non-R&D-performers rely more on supplier knowledge (25%) than R&D-performers (13%).

Table 13: R&D- and non-R&D-performing firms' sources of innovation knowledge by different knowledge sources across different fields of innovation (Source: GMS 2009, own calculations)

Sources of innovation knowledge	New products	New AMT	New OMC	New PRS
Non-R&D-performing firms (0 % R&D expenditure)				
Own R&D department	20.1 ***	12.7 ***	14.1 +	5.4 ***
Other employees	5.9	9.2 **	17.3 **	6.5 *
Customers	73.6	14.1	13.6	48.0 *
Suppliers	24.4 ***	45.8	12.6	13.7
Competitors	34.3	22.9 ***	15.2	20.3
External research organisations	8.2 ***	10.5 ***	13.1 ***	2.5 +
Conferences, trade fairs, other professional events	40.0 **	51.8 +	41.1 *	26.1 ***
	(n=632)	(n=629)	(n=626)	(n=629)
R&D-performing firms (>0 % R&D expenditure)				
Own R&D department	72.6	40.7	17.4	14.1
Other employees	5.7	15.1	24.3	10.6
Customers	77.3	15.1	14.5	56.3
Suppliers	12.7	42.7	11.4	10.8
Competitors	32.5	15.2	12.4	21.8
External research organisations	22.3	27.8	20.0	4.3
Conferences, trade fairs, other professional events	31.8	46.6	33.8	17.9
	(n=785)	(n=783)	(n=775)	(n=771)

Significance levels: *** = p<0.000 / ** = p<0.005 / * = p<0.01 / + = p<0.1

Percentages exceeding 100% because of multiple answers

AMT = advanced manufacturing techniques / OMC = organisational and management concepts / PRS = product-related services

The most striking differences between R&D- and non-R&D-performers show up - as expected – in the role of external R&D organisations and conferences / trade fairs. But it is interesting to see that the difference between the two groups regarding external R&D is not as marked as in the case of innovation cooperation. This could mean, that for R&D-performers, R&D knowledge is more likely to be generated internally, while cooperation with external R&D organisations is needed to realise concrete innovation projects. As a result it may be assumed that the innovation level (the degree of novelty or radicalness) is higher in R&D-performing firms. Besides this, the share of non-R&D-

performers that mentioned external R&D as a relevant knowledge source is almost equal to the share of these firms which cooperate with external R&D organisations.

Another result worth mentioning is that a considerable share of R&D-performers also mentioned competitors as a relevant source of knowledge. Their share (12%-33%) here does not differ significantly from firms without R&D (15%-34%), except in the field of AMT, where competitors play a significantly more important role for non-R&D-performers. Hence, innovation knowledge sourcing from competitors in terms of follower-strategies or imitation does not seem to be constitutive for only non-R&D-performing firms.

To summarise, it is obvious that the mere frequency of mentioned knowledge sources says little about their qualitative contribution to a firm's competitive advantage. Nonetheless, this does provide some insights into the general perceived relevance of different sources of innovation knowledge. It also reveals which source of innovation knowledge is perceived as the most important one. This analysis shows that external R&D knowledge only plays a subordinate role for non-R&D-performing firms. Instead, their absorptive capacity is mainly focused on other firms (customers, suppliers, competitors) and professional events (conferences, trade fairs) to access relevant innovation knowledge. Surprisingly, employees within the firm do not constitute a major knowledge source, a finding which contradicts the existing empirical results to some extent.

Most importantly with regard to heterogeneity, it has been shown that non-R&D-performers use almost the whole range of possible knowledge sources and are certainly not restricted to innovation knowledge provided by individual groups like suppliers.

Human resources

The final innovation resource considered in the descriptive analysis is a firm's workforce. Especially within the knowledge-based view, the human resources of a firm are seen as the paramount carrier of knowledge and competences gained either through formal education or through practical experience in the working context. As the results so far have shown, non-R&D-intensive firms are characterised by an above-average share of low- and unskilled personnel (KIRNER ET AL. 2009b) so they may be advised to draw on experienced-based, practical forms of knowledge as an internal innovation resource (HEIDENREICH 2009; VON TUNZELMANN/ACHA 2005; HIRSCH-KREINSEN 2004, 2008b). As it can be assumed that these different types of knowledge are weighted by educational background and functional area within the firm, Table 14 gives an overview of the distribution of different educational levels and of employees' distribution across different functional areas within the firm.

Table 14: Human resources of non-R&D-performing firms
(Source: GMS 2009, own calculations)

	R&D-performing firms (>0 % R&D expenditure)	Non-R&D-performing firms (0 % R&D expenditure)
Average share (%) of...		
...graduate / high skilled employees	15.0	6.9 ***
...technicians, masters	10.7	9.9
...skilled tradesmen, skilled workers	44.4	42.1 +
...low- and unskilled employees	22.2	31.8 ***
...apprentices, trainees	7.7	9.3 +
	100.0	100.0
	(n=791)	(n=675)
Average share (%) of employees working in...		
...R&D	7.5	1.9 ***
...Engineering, construction, design	7.8	5.3 ***
...manufacturing and assembling	56.0	66.6 ***
...customer service	6.7	8.1 +
...others	21.9	18.1 ***
	100.0	100.0
	(n=791)	(n=675)

Significance levels: *** = p<0.000 / ** = p<0.005 / * = p<0.01 / + = p<0.1

At first sight, it is clear that the majority of employees (42%) in non-R&D-performing firms fall into the group of skilled tradesmen and skilled workers, similar to firms with R&D (44%). Most interestingly, the group of low- and unskilled employees is next in size with an average share of almost 32%. Particularly when compared to this group's share in R&D-performing firms (22%), the high share of low- and unskilled employees does seem to be a dominant characteristic of many non-R&D-performing firms. This supports the assumption that these firms' low or medium complexity products are manufactured in simple tasks which are likely to be performed by this group of employees.

Supplementing the previous findings on the macro level, the share of graduate or highly skilled employees is also significantly lower in non-R&D-performing firms (7%) than in their R&D-performing counterparts (15%). This complies with the fact that most of these highly skilled employees are deployed as engineers or natural scientists in industrial R&D departments. Surprisingly, despite the fact that their share of highly skilled employees is relatively low, firms without R&D report that about 10% of their employees are masters or technicians, which is only marginally lower than in R&D-performing firms (11%). This may be an indicator for a higher relevance of practical, experience-based knowledge rather than explicit knowledge. Finally, non-R&D-performers also have a slightly higher share of apprentices and trainees (9%) than firms with R&D (8%). However, this difference seems too small to be interpreted as a higher relevance of experience-based knowledge in non-R&D firms.

Regarding the functional areas to which the firms' human resources are assigned shows that by far the highest share of employees in non-R&D-performing firms is in manufacturing and assemby (67%). This indicates the higher labour intensity of these firms' production processes. Compared to the significantly lower share in the case of R&D-performers (56%), this seems to be another characteristic of non-R&D-performing firms and matches the findings of HEIDENREICH (2009) and HIRSCH-KREINSEN (2004), who argue that manufacturing and production in non-R&D firms tend to be labour-intensive. As a consequence, these firms are supposed to follow more cost-oriented innovation strategies.

Entering both human resource variables into a contingency table (see Annex A.5.3), provides evidence that those firms with the highest share of low- or unskilled employees also report the highest share of employees working in manufacturing and assembling. Hence, it can be argued that most of these low- or unskilled employees are indeed deployed in simple, labour-intensive tasks in the production. It also shows that those firms with the highest share of graduate employees report the highest share of employees working in engineering, construction, and R&D.

But again, there is also a quite unexpected finding. Despite the fact that the non-R&D-performing firms in the subsample were defined by them having no R&D expenditure at all, some of these firms obviously have people working in R&D. Admittedly, this share is very low (2%) but it is still difficult to interprete and is problematic for the subsequent cluster analysis. Unlike the similar issue concerning knowledge sources, this finding really challenges the non-R&D definition. First, the explicit allocation of employees to R&D is a more serious indicator of formal R&D, which differs from subjective, diffused or past-oriented statements about internal R&D as a knowledge source due to its concrete monetary dimension. The number of employees working in R&D is also taken as an indicator for R&D intensity in the Frascati Manual (OECD 2002: 20). For this reason, I decide to exclude those firms that reported employees working in

R&D from the cluster analysis to ensure that only "real" non-R&D-performers (according to the OECD definition) are included.

Another finding concerning human resources was that the prominent role of engineering and construction proposed by some authors (e.g. EVANGELISTA ET AL. 2002; NASCIA/PERANI 2002; HIRSCH-KREINSEN 2004; BENDER 2004) for instance in the context of "architectural innovation" (BENDER 2004: 13), did not show up as expected. While non-R&D-performing firms report an average share of about 5% of employees working in this area, firms with R&D have an average share of 8%. Moreover, this difference is statistically significant. Of course, considering that R&D does not play an important role as a source of product or process novelty for non-R&D-performing firms, the share of employees in construction and engineering appears rather high in comparison.

Last, but not least, it is also interesting to see that firms without R&D report a significantly higher average share of employees in customer services (8%) than firms with R&D (7%). This fits quite well with the findings presented above about the high importance of product customisation, customers as a relevant knowledge source, and the more direct service orientation which has been recognised in the share of directly invoiced PRS.

In summary, both the qualification structure and the shares of employees working in the different functional areas (excluding R&D) do not differ fundamentally between firms with and those without R&D. However, there are some differences regarding the respective number of employees, like the higher share of graduates or highly skilled workers in R&D-performing firms and the higher share of low- and unskilled employees in non-R&D-performers. The latter correlates closely with the higher share of employees in manufacturing and assembly in these firms.

5.3.3 Innovation output and economic outcome

Having extensively considered the findings about the innovation resources' characteristics of non-R&D-performing firms, this chapter now proceeds to the output and outcome level of innovation. Thus, the empirical findings presented on the following pages show which innovation output and economic outcome are obtained by these firms. At this point, I want to point out once again, that it is not the purpose of this analysis to determine whether non-R&D-performing firms perform better or not, but to depict whether they have distinct output and performance characteristics when compared to firms with R&D. Due to economies of scale and scope, innovation output and economic performance measures are always subject to structural characteristics, particularly to firm size. Hence, all the findings differentiate between three firm size categories and product complexity where necessary (e.g. in the case of process-related measures like time to market). Moreover, many of the following measures and indicators have different scales which are not comparable with each other. Therefore, additional bar charts within the tables have been omitted.

Innovation output

Beginning with the dimensions of material and immaterial product innovations, Table 15 summarises the innovation output of R&D- and non-R&D-performing firms in terms of economic success in the fields of product and product-related service innovations.

As about 35% of non-R&D-performing firms have developed new products within the last three years, the average share of sales obtained with these new products is about 5%. However, it is clear that this share strongly depends on firm size. While in the case of small firms without R&D, the share with new products accounts for only 4.7% of sales, larger firms with 250 and more employees obtain about 8% of sales with their new products. As product innovations encompass both products that are only new to the firms and products that are new to market, it makes sense to explicitly examine the subcategory of market innovations as these require particular innovation effort. Despite the fact that the share of sales is distinctively lower at 3%, it is still surprising that firms without any formal R&D have developed such market novelties in the surveyed period. This share even rises to 4% in the case of large non-R&D-performing firms.

These shares are significantly higher for R&D-performing firms, however, which clearly underlines the major relevance of R&D as a driver of product and especially market innovation. Their average share of sales with product innovations amounts to about 13% and is thus more than double that for non-R&D-performers. It is especially astonishing that this share is only slightly lower for small firms with formal R&D (14%) than for larger R&D-performers (15%). This means that the impact of firm size on product innovation output is significantly weaker than was the case in the group of non-R&D-performers. As expected, the average share of sales of market novelties is also significantly higher (6%), with particularly small R&D-performing firms reaching a significantly higher share. Hence, R&D-performing firms are able to obtain much higher economic profits from product development, an observation which complies with the existing state of research.

Table 15: R&D- and non-R&D-performing firms' innovation output in the fields of new products and new product-related services
(Source: GMS 2009, own calculations)

	Non-R&D-performing firms (0 % R&D expenditure)				
	20-49 employees	50-249 employees	250 and more employees	Total	
Turnover with product innovation					
Average share of sales with new products [%]	4.7 ***	5.1 ***	8.1 +	**5.1** ***	(n=652)
Turnover with market innovation					
Average share of sales with market innovation [%]	2.5 ***	3.4 +	3.8	**3.0** ***	(n=230)
Turnover with new product-related services					
Average share (%) of sales with new product-related services	7.5	6.0	2.7	**6.4**	(n=76)
	R&D-performing firms (>0 % R&D expenditure)				
	20-49 employees	50-249 employees	250 and more employees	Total	
Turnover with product innovation					
Average share of sales with new products [%]	14.3	11.7	15.3	**13.2**	(n=758)
Turnover with market innovation					
Average share of sales with market innovation [%]	6.4	5.1	6.4	**5.7**	(n=589)
Turnover with new product-related services					
Average share (%) of sales with new product-related services	4.8	5.5	4.1	**5.0**	(n=175)

Significance levels (chi^2, T-test, Mann-Whitney): *** = p<0.000 / ** = p<0.005 / * = p<0.01 / + = p<0.1

There is a different picture when looking at the average share of sales of new product-related services (PRS). Recalling the innovation resource part of the analysis, non-R&D-performing firms turned out to be slightly more service-oriented than their R&D counterparts. This is reflected by the result that their average share with PRS is higher (6.4%) than that of firms with R&D (5%). This holds across all firm size categories although, especially small and medium sized non-R&D-performing firms seem able to profit from new PRS. These findings are even more impressive considering non-R&D-performers have less complex products which do not offer as many options for PRS as more complex ones. The control for product complexity (see Annex A.5.4) affirms these findings. Disregarding product complexity, non-R&D-performing firms obtain higher innovation output with PRS than R&D-performers. However, none of the differences between R&D- and non-R&D-performers are statistically significant. It can be definitely stated, however, that non-R&D-performing firms generally obtain higher innovation output with new PRS than with new products.

Proceeding to the process dimension of innovation output, non-R&D-performing firms widely reported using both technical (AMT) and non-technical (OMC) process resources to a similar extent as R&D-performers. But their process expertise is not fully revealed until the analysis of the corresponding output dimensions (Table 16).

Table 16: R&D- and non-R&D-performing firms' innovation output in the field of tech- nical and non-technical process innovation
(Source: GMS 2009, own calculations)

		Non-R&D-performing firms (0 % R&D expenditure)			
	20-49 employees	50-249 employees	250 and more employees	Total	
Time to market					
Average number of months	10.7 ***	12.4 ***	17.0 +	**12.1** ***	(n=219)
Manufacturing lead time					
Average number of working days à 8h	16.6 ***	30.6	19.3 **	**23.1** **	(n=609)
Delivery time					
Average number of calendar days	24.2 ***	34.1 *	35.1 +	**29.3** ***	(n=641)
Adherence to delivery dates					
Percentage of orders delivered on time	91.0	92.0	94.4	**91.6** +	(n=641)
Rework and scrap rate					
Percentage of (semi-)products that have to be scrapped or reworked due to quality problems	2.9	2.6	4.1	**2.8**	(n=633)
Labour productivity					
(turnover-input)/number of employees [tsd Euros]	81.11 +	97.37 +	83.95	**88.87** ***	(n=675)
Total factor productivity					
(turnover-input)/(amortisation+personnel costs)	1.81	1.93	1.77	**1.86** +	(n=463)
		R&D-performing firms (>0 % R&D expenditure)			
	20-49 employees	50-249 employees	250 and more employees	Total	
Time to market					
Average number of months	16.3	18.1	25.8	**19.6**	(n=595
Manufacturing lead time					
Average number of working days à 8h	37.1	33.0	43.1	**36.4**	(n=761)
Delivery time					
Average number of calendar days	51.2	47.9	60.7	**51.7**	(n=761)
Adherence to delivery dates					
Percentage of orders delivered on time	88.8	90.5	91.2	**90.3**	(n=761)
Rework and scrap rate					
Percentage of (semi-)products that have to be scrapped or reworked due to quality problems	3.6	2.9	3.0	**3.1**	(n=742)
Labour productivity					
(turnover-input)/number of employees [tsd Euros]	94.42	112.68	121.77	**109.93**	(n=791)
Total factor productivity					
(turnover-input)/(amortisation+personnel costs)	1.89	2.03	1.95	**1.98**	(n=507)

Significance levels (chi^2, T-test, Mann-Whitney): *** = p<0.000 / ** = p<0.005 / * = p<0.01 / + = p<0.1

Beginning with the output measures concerning speed, non-R&D-performing firms show superior results which are statistically significant compared to firms with R&D. For instance, the time needed to develop a market-ready new product (i.e. time to market) is on average more than 7 months shorter in non-R&D-performing firms. This holds even when controlled for different firm sizes and different complexity of products (see Annex A.5.4). The same is valid for the average manufacturing lead time needed to produce a good from the order to delivery. Here, non-R&D-performing firms show a manufacturing lead time that is on average 13 days shorter than that of firms with R&D. Again, this trend holds across all firm size categories.

Even in the case of complex products, the manufacturing lead time of firms without formal R&D is 5 days shorter on average then R&D-performing firms, but this is not statistical significant. Looking at the third measure of speed, the average delivery time, again, non-R&D-performing firms are faster than R&D-performers with significantly shorter delivery times. When considering variations of product complexity (see Annex A.5.4), these differences diminish for simple products. However, with regard to highly complex products, there is a significant difference in the average delivery time of about 30 calendar days in favour of non-R&D-performers. This means that these firms display their process expertise in terms of speed, particularly in the case of products with high complexity.

Continuing with the adherence to delivery dates as a measure for process quality, non-R&D-performing firms also show a slight advantage, but only at a low level of significance. When differentiating between firm size and levels of product complexity, the superior output of non-R&D-performers is maintained but becomes statistically insignificant. This may be due to the very right-tailed distribution of this measure and its low variance. The same is true for the second measure of quality, the average percentage of products and semi-products which have to be scrapped or reworked because of quality problems. Again, across almost all size categories and all levels of product complexity (see Annex A.5.4), firms without R&D report lower scrap rates than firms with R&D. Only in the size category of 250 and more employees is the scrap rate lower for R&D-performers than for non-R&D performers. But as this measure is characterised by a left-tailed value distribution, all these differences lack statistical significance.

Concluding with the innovation output of efficiency, the findings here are somewhat surprising in the light of the empirical evidence provided by some authors (e.g. KIRNER ET AL. 2009a) that firms with low R&D intensity achieve similar production efficiencies as more R&D-intensive firms.[231] With regard to production efficiency the question about whether or not a firm performs R&D

231 However, their analysis was based on the 2.5 % threshold of R&D expenditure.

seems to be crucial. Looking at the measure of labour productivity, the non-R&D firms' process excellence does not show up in terms of higher efficiency. The R&D-performing firms in the sample obtain a significant higher average production output per employee (110,000 euros) than firms without formal R&D (89,000 euros). However, this measure may be biased towards firm size (as the production factor of labour is calculated using the number of employees in the divider) and thus neglects the intensity of capital investment. The total factor productivity (TFP) as a second measure of efficiency, accounts for both labour and capital intensity. The TFP results show that the differences in efficiency between R&D- and non-R&D-performers diminish which underlines the role of capital investment for non-R&D-performers. Nevertheless, their total average TFP is still lower than that of R&D-performers, but on a lower level of statistical significance. Moreover, differentiating between size classes shows that particularly small and medium-sized firms without R&D have a TFP which is almost equal to their R&D-performing counterparts.

Economic performance

To finish the descriptive analysis of the innovation characteristics of non-R&D-performing firms, Table 17 shows the results concerning employment growth, sales growth, return on sales, and export intensity. As described in chapter 2.6.5.2, all of these performance measures are usually positively correlated with firms' R&D intensity.

These positive correlations are also supported by my results. The most striking and significant difference is that the average share of exported products in R&D-performing firms is double that of non-R&D-performers (40% compared to 20%). Thus, at this general level, the higher export relevance of R&D-based products is confirmed once again. Positive correlations between the existence of formal R&D and the economic performance of firms can also be found with regard to the monetary measures of total sales growth and return on sales. In both cases, firms with formal R&D perform better. On average, they report sales growth of 11% and more than half of them show a return on sales of between 5 and 10% or more. In contrast, the non-R&D-performers obtain 9% sales growth on averagee and more than 50% report 5% or less with regard to the return on sales. Only non-R&D-performers with 250 or more employees show slightly higher sales growth than their R&D counterparts, but not at a statistically significant level.

Table 17: R&D- and non-R&D-performing firms' economic performance (Source: GMS 2009, own calculations)

	Non-R&D-performing firms (0 % R&D expenditure)				
	20-49 employees	50-249 employees	250 and more employees	Total	
Employment growth					
Change in number of employees from 2006-2008 (% per year)	3.3	5.6	6.5	**4.6**	(n=627)
Growth of total sales					
Change in total sales between 2006-2008 [% per year]	7.9 +	9.8	10.4	**8.9** +	(n=609)
Return on Sales (before tax)					
negative	-	-	-	**9.9**	
0-2 %	-	-	-	**16.5**	
>2 - 5 %	-	-	-	**31.1**	
>5 - 10 %	-	-	-	**29.0**	
>10 %	-	-	-	**13.5**	
				100.0 ***	(n=563)
Export intensity					
Average share of products (%) that have been exported in 2008	15.0 ***	23.3 ***	33.1 ***	**19.7** ***	(n=614)
	R&D-performing firms (>0 % R&D expenditure)				
	20-49 employees	50-249 employees	250 and more employees	**Total**	
Employment growth					
Change in number of employees from 2006-2008 (% per year)	4.0	5.1	4.7	**4.7**	(n=741)
Growth of total sales					
Change in total sales between 2006-2008 (% per year)	11.6	11.4	9.9	**11.1**	(n=725)
Return on Sales (before tax)					
negative	-	-	-	**8.1**	
0-2 %	-	-	-	**12.6**	
>2 - 5 %	-	-	-	**26.9**	
>5 - 10 %	-	-	-	**30.4**	
>10 %	-	-	-	**22.0**	
				100.0	(n=665)
Export intensity					
Average share of products (%) that have been exported in 2008	31.6	38.6	54.3	**40.4**	(n=739)

Significance levels (chi^2, T-test, Mann-Whitney): *** = p<0.000 / ** = p<0.005 / * = p<0.01 / + = p<0.1
(-) = empty cells due to insufficient number of cases

However, there are no significant differences between R&D- and non-R&D-performers performers in terms of employment growth: both groups have increased their employment rate by between 3 and 7% depending on firm size. This could be explained by the fact that non-R&D-performing firms tend to have more low- and unskilled employees. As a result, they can probably increase the number of employees more easily and flexibly than R&D-intensive firms which have greater difficulties in recruiting comparable numbers of highly skilled personnel. This result matches previous findings which suggested that the positive employment effects which are usually linked to higher R&D intensity have not yet been shown (HIRSCH-KREINSEN 2004: 5; HAHN 2009: 306; SANDVEN ET AL. 2005: 36f.; BENDER 2006: 2).

Despite the comparatively lower long-term performance measures of non-R&D-performers, these measures also indicate heterogeneity. For instance, about 14% of them are located in the highest category of return on sales. Moreover, large firms without R&D report more than 10% growth in sales, and 20% in exported productsm which shows that a considerable share of non-R&D-performers is able to manufacture products that are successful in direct international market competition.

It has to mentioned, however, that these long-term measures of firms' general economic performance are subject to a broad set of other influencing variables apart from innovation and are also highly dependent on general economic dynamics. Therefore, these measures are only able to give a general impression of the economic outcome that may be correlated to formal R&D or to no formal R&D.

5.3.4 Summary

The research question addressed by the descriptive analysis was "which resources (as not formal R&D) contribute to enhancing and maintaining non-R&D-performing firms' competitive advantage in German manufacturing industry".

The findings presented above regarding the use of different innovation resources by non-R&D-performing firms revealed that these firms do not display a homogeneous pattern in compensating for their lack of formal R&D. One expection, for instance, was that they compensate their lack of R&D knowledge by excessive use of advanced manufacturing techniques, organisational concepts, or above-average use of alternative, not R&D related knowledge sources. But none of the findings on the general level suggests the empirical existence of such homogeneous compensation strategies common to all non-R&D-performers in the sense that certain innovation resources are only or mainly used by them compared to R&D-performers. Hence, "typical" innovation strategies of non-R&D-performing firms are not recognisable on the basis of these descriptive results. This again stresses the aspect of inter-firm heterogeneity.

Considering the different kinds of innovation resources, it can be seen that all of these innovation resources beyond formal R&D are also used by non-R&D-performing firms to various extents. Of course, the descriptive analysis did identify some general tendencies. First, the development of new products seems to play a less important role. Second, the products manufactured by non-R&D-performers are mainly characterised by low and medium product complexity. Third, the share of low- and unskilled employees tends to be higher and corresponds to the higher share of employees working in manufacturing and assembly. Last, but not least, non-R&D-performers have a lower overall propensity to participate in innovation cooperation.

On the other hand, there are also relevant shares of non-R&D-performers which are highly service-oriented, develop new products or even market novelties, manufacture products of high complexity, have survived market competition for many decades, or which collaborate with external R&D organisations. Furthermore, the absence of formal R&D should not be equated to the absence of technology or technological knowledgee, not least because of the numbers of non-R&D-performers using advanced manufacturing techniques or whose products contain high-tech elements. Moreover, the data was not able to verify some previous findings such as the significantly stronger competition based on product prices or the supplier domination of non-R&D-performers.

Instead, the superior performance of R&D-performers with regard to monetary measures of innovation output (e.g. sales with product or market innovations), economic performance (sales growth, return on sales), and in particular export intensity has been empirically verified. In contrast, non-R&D-performers do better in terms of sales of product-related services and especially in terms of the speed and quality of their process performance. But as in the case of innovation resources, there are also firms without formal R&D that obviously manage to obtain sales of market novelties, have high export rates and achieve high sales growth, and thus exhibit a performance level that is quite comparable to that of R&D-performers. Thus, when revisiting the question "which innovation output and economic performance are achieved by non-R&D-performing firms based on their innovation resources", the answer is ambiguous. On the one hand, non-R&D-firms generally show relatively poor innovation output in terms of product or market innovations. On the other hand, these firms have superior process performance. Therefore, their use of innovation resources beyond formal R&D does enable non-R&D-performers to maintain competitive advantage and be successful on the market.

To conclude, in the context of the overall research question of this thesis, the results of the descriptive analysis conducted above turn out to be very fruitful for two reasons:

- **Firstly, they strongly support the concept of inter-firm heterogeneity of non-R&D-performing firms because no dominant, "typical" innovation pattern has been revealed:** The results clearly show that non-R&D-performing firms are by no means a homogenous group. The descriptive analysis of their structural characteristics shows that relevant shares of such firms are represented in all sectors of the German manufacturing industry. Besides their considerable share in so-called medium-tech industries (e.g. machinery and equipment), they can even be found in very R&D-intensive sectors like medical, precision and optical instruments. Hence, it does not seem adequate to start from the sectoral level of "low-tech" industries when analysising non-R&D-performing firms' innovation strategies as this would narrow down the possible scope of their innovation behaviour a priori.

- **Secondly, with regard to the innovation resources deployed, the results indicate how different innovation patterns of non-R&D-performing firms may be structured:** On the one hand, it is apparent that all these innovation resources can serve as distinct elements of non-R&D-performers' innovation patterns. On the other hand, their varying degrees of importance indicate which elements will probably turn out to be common across innovation patterns and which may function as distinctive elements of innovation patterns. For example, some characteristics which have turned out to be relatively stable on a general level like the high share of low- and unskilled employees, or the high customer orientation are probably important elements across different innovation patterns, while those resources which are only used by relatively few non-R&D-performers (e.g. product development, innovation cooperation with external R&D, intensive use of advanced manufacturing techniques) can be expected to vary greatly between innovation patterns.

Hence, the following cluster analysis explores how the different elements of non-R&D-performing firms' innovation resources can be grouped into distinctive innovation patterns.

5.4 Cluster analysis of innovation patterns of non-R&D-performing firms in the German manufacturing industry

Subsequent to the descriptive analysis, this chapter takes up the innovation resources used by non-R&D-performing firms and connects them to the concept of heterogeneous innovation patterns developed on the basis of evolutionary theory in the third chapter. It aim is to identify the innovation patterns of non-R&D-performing firms and address the main question of this thesis:

> *To what extent are heterogeneous innovation patterns of non-R&D-performing firms in the German manufacturing industry reflected in the dimension of their use of different innovation resources? With*

which innovation output and economic performance can these heterogeneous innovation patterns be characterised?

As described in chapter 4.4, the firms are not preselected by sector affiliation, innovation output or by firm size in order to try and capture the broadest level of heterogeneity. The firms are clustered solely according to their characteristics regarding the innovation input dimensions described in chapter 5.2.1. After the cluster analysis, each of the explored clusters is described, interpreted and checked for feasibility by an enlarged set of both structural and innovation performance indicators as identified in chapters 5.2.2 and 5.2.3.

After discussing several methodological aspects of cluster analysis, the following paragraphs describe which innovation patterns of non-R&D-performing firms – based on their innovation resources – have been identified by cluster analysis and how they can be interpreted and described based on their structural characteristics, innovation output and economic performance.

5.4.1 Variables for cluster analysis

Following the resource-based theoretical perspective, the innovation activities of non-R&D-performing firms rely on a broad set of innovation resources such as the implementation of technical and non-technical (i.e. organisational) process innovations, the development of product-related services, external and internal sourcing of information and knowledge relevant for innovation, collaboration in innovation projects, as well as the management of human resources (OECD 2005).

The different innovation resource variables closely correspond to those described in the context of operationalisation (see 5.2.1). Several have been separated into dummy variables where necessary (e.g. categorical variables) to harmonise scale levels among them. Moreover, some of the variables were modified to make them both feasible for cluster analysis and to preserve their valuable information.

Firstly, to handle the use and intensity of use of advanced manufacturing techniques (AMT) a mean value-centred index (TECHPROC) was compiled for the corresponding indicator battery. It covers both, a) the use (yes/no) of certain manufacturing technologies in the fields of automation and linkage, machining and production technologies, digital factory/IT cross linkage, and energy efficiency, as well as b) the intensity of use measured by a three-step scale (low/medium/high) based on the respondents' estimated extent of actual utilisation compared to the most reasonable potential utilisation in their firm. Moreover, as mentioned above, the respondents were able to add other manufacturing technologies which are either used additionally by them, or which were not covered by the given answer categories. Following the argumentation that a higher intensity of AMT use leads to stronger economic effects, the index shows higher values for higher intensities of use, which means that a high intensity use in one field may result in a higher index value than low intensity use in two or three fields. After summing up the intensity values across the fields of AMT described, the overall value is related to the average index value of total manufacturing industry. The value range is between -2 and +2. Thus, the value of this index indicates whether a firm is more (positive values) or less (negative value) innovative with regard to its intensity of AMT use compared to the average of manufacturing industry (0).

Secondly, exactly the same procedure was applied to make the use of different organisational and management concepts (OMC) suitable for cluster analysis using one variable. The non-technical process dimension is captured by a mean value-centred index (NONTECPROC) about the use and intensity of use of organisational concepts in the areas of human resource management, work organisation, organisation of production, and intensity of innovation cooperation. Again, the respondents were able to add further organisational practices which they use or which were not covered by the given categories. As a result, this index also provides information, about whether a firm is more (positive values) or less (negative values) innovative with regard to its intensity of OMC use compared to the average of manufacturing industry. Similar to the index for the intensity of AMT use, this index also covers a value range from -2 to +2 with 0 representing the average of manufacturing industry.

In consequence, these two indices give much more detailed information in the context of cluster analysis about the use of AMT and OMC than a simple dichotomous modification of the original variables (use "yes/no") would be able to provide.

Thirdly, as the mere participation in innovation cooperation and its intensity is already included in the index of non-technical process innovation, the cooperation variable used in the cluster analysis focuses on the number (i.e. the broadness) of cooperation areas across the different fields of innovation.

Fourthly, as the variable representing the important sources for innovation knowledge originally allowed multiple replies, more than 90% of the non-R&D-performing respondents indicated that other firms (e.g. suppliers, customers or competitors) are a relevant source of innovation knowledge. As a result, these categories have only limited, differentiating power and were excluded from cluster analysis. Instead, the variable concerning the sources of innovation knowledge used in the cluster analysis distinguishes whether an important source of innovation knowledge is located inside (e.g. own employees) or outside the firm in terms of external R&D organisations (e.g. universities, R&D laboratories, non-profit R&D). Table 18 provides an overview of the variables of non-R&D-performing firms' innovation resources that were finally included in the cluster analysis. Testing has shown that none of these variables are intercorrelated with each other.[232]

It can be seen in the table, that all dimensions of non-R&D-performing firms' innovation resources that turned out to be relevant in the descriptive analysis are included in cluster analysis.

Table 18: Indicators of innovation input used in the cluster analysis (Source: GMS 2009; own calculations)

Indicator of innovation input	Mean	Std.Dev.	Unit	n
Product devlopment				
Development of new products [*ProdDev*]	0.36	0.48	%*100	435
Customisation of products [*ProdDev_Cus*]	0.57	0.50	%*100	435
High-tech components [*HighTech*]	0.08	0.27	%*100	435
Product-related services				
New product-related services [*NewPRS*]	0.13	0.34	%*100	435
Use of advanced manufacturing techniques				
Technical process innovation [*TechProc*]	0.03	0.77	Metric	435
Use of organisational and management concepts				
Organisational innovation [*NonTechProc*]	0.02	0.75	Metric	435
Innovation cooperation				
Innovation cooperation [*InnoCoop*]	1.19	1.29	#	435
Sources of innovation knowledge				
Innovation knowledge (firm internal) [*Know_Int*]	0.50	0.50	%*100	435
Innovation knowledge (external research) [*Know_ExtR&D*]	0.26	0.44	%*100	435
Human resource management				
Personnel in Manufacturing and Assembling [*M&A_Employ*]	65.92	20.88	%	435
Highly skilled employees [*HS_employ*]	7.06	7.80	%	435
Low- or unskilled employees [*LS_employ*]	29.88	27.68	%	435

232 For this reason, a preliminary reduction of the variable set by factor analysis (as proposed for example by HOLLENSTEIN 2002; LEIPONEN/DREJER 2007; SRHOLEC/VERSPAGEN 2008) does not seem appropriate in this case.

5.4.2 Firm sample for cluster analysis

As defined in the previous chapter, non-R&D-performing firms are those which do not undertake any R&D activities in terms of intramural or extramural (contracted) R&D. In the GMS 2009 data set, n=675 of 1,484 firms (=45.5%) comply with this definition and can thus be identified as non-R&D-performing firms. However, because of variables' missing values across the cluster variables of innovation resources, the subsample of non-R&D-performing firms available for cluster analysis is reduced from n=675 to n=435. Moreover, as the descriptive analysis revealed, some non-R&D-performing firms report that a certain share of their employees is working in R&D. Although this share is very low (2%) it nevertheless poses some difficulties to the intended cluster analysis because it challenges the definition of non-R&D-performing firms. Having employees in R&D is a serious indicator for formal R&D because of its concrete monetary dimension. This is the reason why the number of employees working in R&D is also mentioned as an indicator for R&D intensity in the Frascati Manual (OECD 2002: 20). Thus, it was decided to exclude those firms that reported employees working in R&D from the cluster analysis to ensure that only "real" non-R&D-performers (according to the OECD definition) are included.

To make sure that this sample reduction is not accompanied by structural bias, the following table (Table 19) compares the original sample of 675 non-R&D-performing firms with the reduced sample of 435 for the cluster analysis.

The comparison shows that the reduced sample of 435 non-R&D-performing firms for the cluster analysis does not differ from the original sample of 675 firms to a large extent. Hence, the sample reduction due to missing values does not result in any significant structural bias in terms of sector affiliation or firm size. This is especially important as it is thereby assured that the sample used for cluster analysis is characterised by the same level of inter-sectoral heterogeneity as the originally identified sample for the descriptive analysis and that it thus meets the requirements for empirical analysis derived from the theory.

Table 19: Comparison between the subsamples of non-R&D-performing firms for the descriptive and the cluster analysis (Source: GMS 2009, own calculations)

Sectoral affiliation within manufacturing industry (NACE 2-digit)	Sample of non-R&D-performing firms for descriptive analysis n=675		Sample of non-R&D-performing firms for cluster analysis n=435	
	%	n	%	n
15 - Manuf. of food products and beverages	13.2	89	12.9	56
17 - Manuf. of textiles	2.2	15	2.1	9
18 - Manuf. of wearing apparel; dressing etc.	0.6	4	0.5	2
19 - Tanning and dressing of leather etc.	0.6	4	0.2	1
20 - Manuf. of wood and of products of wood etc.	4.4	30	4.1	18
21 - Manuf. of pulp, paper and paper products	2.4	16	2.5	11
22 - Publishing, printing, reproduction of recorded media	6.7	45	7.1	31
24 - Manuf. of chemicals and chemical products	2.5	17	2.5	11
25 - Manuf. of rubber and plastic products	9.6	65	10.3	45
26 - Manuf. of other non-metallic mineral products	5.5	37	5.5	24
27 - Manuf. of basic metals	3.1	21	3.9	17
28 - Manuf. of fabricated metal products	23.9	161	21.8	95
29 - Manuf. of machinery and equipment n.e.c.	11.6	78	11.5	50
31 - Manuf. of electrical machinery and apparatus n.e.c.	3.3	22	3.0	13
32 - Manuf. of communication equipment and apparatus	0.7	5	0.9	4
33 - Manuf. of medical, precision and optical instruments etc.	2.7	18	3.2	14
34 - Manuf. of motor vehicles etc.	1.9	13	2.5	11
35 - Manuf. of other transport equipment	0.4	3	0.5	2
36 - Manuf. of furniture, jewellery, sports equipment	4.4	30	4.6	20
37 - Recycling	0.3	2	0.2	1
	100.0	675	100.0	435
Firm size	%	n	%	n
Up to 49 employees	49.0	331	47.1	205
50-249 employees	45.5	307	47.1	205
250 and more employees	5.5	37	5.8	25
	100.0	675	100.0	435

5.4.3 Variables for cluster interpretation and verification

Briefly, all the variables that were previously identified in GMS 2009 to either describe innovation output or long-term economic performance will also be utilized to interpret, characterise and verify the identified innovation patterns of non-R&D-performing firms (see Table 20).

After the clusters of innovation patterns among non-R&D-performing firms have been identified, they will be further checked for plausibility and described by some basic structural variables to sharpen the underlying characteristics. The first check is to establish whether the identified clusters are predominantly determined by single sectors, respectively to which degree they correspond to sectors of manufacturing industry. It will also be determined to what degree they correspond to the sectoral classification of low-, medium-, and high-tech sectors. Secondly, the clusters will be characterised according to the average size of the firms constituting the innovation pattern. The clusters will then be described according to the predominat product complexity, manufactured batch or lot size, and the firms' position in the value chain. Last but not least, the firms were asked to indicate their most important competitive factor from a list of six competitive factors (price, quality, innovativeness, customisation, adherence to delivery times/short delivery times, and service).

Table 20: Indicators for innovation output and economic performance (Source: GMS 2009, own illustration)

Indicators of innovation output	Unit
New products	
Average share of sales with products new to the firm	%
Average share of sales with products new to the market	%
New product-related services	
Average share of sales with new product-related services	%
Technical and non-technical process innovation	
Average time to market of product developments	months
Average manufacturing lead time	Number of working days à 8h
Average delivery time	Number of calendar days
Average share of order delivered on time	%
Average scrap and rework rate	%
Labour productivity ((Turnover-input)/number of employees)	K€/employee
Total factor productivity - (Turnover-input)/(amortisation+personnel costs)	-
Indicators of overall economic performance	
Employment growth (change in number of employees from 2006-2008)	% per year
Sales growth (change in total sales between 2006-2008)	% per year
Export intensity (average share of firms' products that are exported)	%

5.4.4 Methodology and proceeding of cluster analysis

Within the guiding conceptual framework of empirical analysis (see chapter 4.3), two main requirements were derived to empirically analyse of non-R&D-performing firms' innovation patterns:

- maximum consideration of heterogeneity in non-R&D-performing firms' innovation strategy
- maximum consideration of different firm-internal and firm-external resources for innovation

It is precisely the great variety of non-R&D-performing firms' behaviour which motivates the present search for regularities and systematic empirical patterns by creating classifications. As defined before, the term "patterns" in an evolutionary context particularly means the occurrence of recurrent good, viable practice in the ongoing process of evolution and environmental adaptation. Thus exploring innovation patterns on the one hand allows taking account of inter-firm heterogeneity, while on the other hand, simultaneously forcing to be selective with regard to relevant dimensions in which heterogeneity may occur.

As this selection was provided by the integrated resource-based perspective, the analysis takes a relatively large number of variables into account in order to derive robust empirical evidence. Hence, this thesis applies explorative cluster analysis which is described as "the art of finding groups in data" so that the degree of natural association is high among members within the same group and low between members of different groups (BACHER 2002: 2ff.; KOZELKA 1982: 6). Cluster analysis thereby links the dimensions of homogeneity and heterogeneity by exploring groups or patterns which are characterised by a high level of homogeneity within them and a simultaneously high level of heterogeneity between them.

Generally, cluster analysis represents an explorative methodology to identify more or less plausible, useful, or insightful groups of statistical units. In contrast to discriminant analysis or logistic regression models, it does not allow for any verification based on interference statistics. Additionally, cluster analysis also not does incorporate any forecast model which may predict the cluster affiliation of statistical units that were not part of the analysis sample. Strictly speaking, the results of cluster analysis are only valid for the subjects of analysis. However, of one presumes that the sample is to some extent representative of its underlying population, it can be assumed that the findings are also able to provide a picture of the population.

To meet the above requirements, the cluster analysis will proceed in two steps by combining hierarchical and non-hierarchical techniques of cluster analysis (MILLIGAN/SOKOL 1980; PUNJ/STEWART 1983). Starting from the broadest level of heterogeneity, a first hierarchical, agglomerative cluster analysis assigns each observation to a single cluster before they are successively grouped into bigger clusters (BACHER 2002: 297).

Before conducting cluster analysis, two main decisions have to be made. The first is to decide which proximity measure should be used to calculate the relative distances between the objects' characteristics, while the second decision is about the cluster algorithm that should be used to group the objects according to their similarity or dissimilarity. I used squared Euclidean distance as the proximity measure (5-1). It is one of the most frequently used distance measures and has the advantage of separating outliers particularly well because squaring puts more weight on larger differences while smaller differences are less weighted (BACKHAUS ET AL. 2006: 504, 527; BACHER 2002: 222):

(5-1)

$$QEUKLID_{g.g*} = \sum_{i} (X_{g,i} - Y_{g*,i})^2$$

As the formula shows, the squared Euclidean distance defines each difference between the characteristics of a pair of objects (g, g*), squares them and adds up all these squared distances. As the emphasis lies on the difference between the characteristics of the two objects, the Euclidean distance is a measure of dissimilarity.[233] Strictly speaking, Euclidean distance is only applicable to metric variables because it is based on the calculation of arithmetic mean values. However, the selected variables for cluster analysis also include ordinal and nominal variables. To include them in cluster analysis, they were transformed into dichotomous dummy variables (0/1) which allows them to be treated as metric variables (BACHER 2002: 186, 297).

To generate the initial solutions hierarchical analysis was carried out using the Ward's method, which is based on squared Euclidian distances and generally provides good results compared to other clustering methods

233 For a detailed overview of different proximity measures, see BACKHAUS ET AL. (2006: 493ff.).

(MILLIGAN/COOPER 1987: 336ff.). Homogeneous groups are built so as to minimise the distance in scores of firms within a single cluster where the group centroid is equally affected by all objects, and to maximise the distance in scores between companies from the various clusters. Thereby, Ward's method tends to distinguish comparable groups of approximately similar size (Backhaus et al. 2006: 528) and generally provides good results compared to other clustering methods (MILLIGAN/COOPER 1987: 336).

The main problem with cluster analysis is to decide on the number of clusters, to balance the need to represent the data appropriately and, at the same time, to keep the results manageable. To ensure such manageable, plausible interpretations and for future applications of the taxonomy, between two and up to twelve groups were considered initially, thereby anticipating a clear cutting point to define the number of clusters. Although most of similar analyses limited their results to two to six groups (e.g. DE JONG/MARSILI 2006; EVANGELISTA 2000; PENEDER 2002), it was decided to leave more room for possible cluster solution to account for maximum heterogeneity.

However, the visual inspection of Ward's dendrogram (Figure 21) shows no clear solution. Instead, a two, four, five or six cluster solution could be considered possible. Hence, in order to statistically identify the appropriate number of clusters, the two stopping rules of CALINSKI and HARABASZ (1974) (pseudo-F index) and the DUDA/HART (1973) (Je(2)/Je(1) index) were computed for each cluster solution at each level of the hierarchy in cluster analysis. For both rules, larger values indicate more distinct clustering. Presented with the DUDA-HART Je(2)/Je(1) values are pseudo-T-squared values. Smaller pseudo-T-squared values indicate more distinct clustering. While CALINSKI/HARABASZ'S pseudo-F index rather indicated a two cluster solution, DUDA-HART index favoured a five cluster solution (see Annex A.6).

As a solution with two clusters did not make much sense against the background of theoretical expectations about inter-firm heterogeneity, I decided to focus on the five cluster solution that was indicated by the DUDA-HART index. But in order to also account for the neighbouring solutions, the results with four, five and six clusters produced by Ward were taken as starting points for the second step, non-hierarchical cluster analysis using k-means algorithm and squared Euclidean distance as the measure of dissimilarity. Thus, for each number of groups (k), between four and six, a k-means, non-hierarchical cluster analysis was performed, in which the firms are iteratively classified based on their distance to some initial starting points of dimension k. While some k-means methods use randomly selected starting points, I employed the centroids of the initial hierarchical solutions for this purpose. This approach has generally been demonstrated as superior in terms of cluster recovery and is therefore recommended (MILLIGAN/ SOKOL, 1980; PUNJ/STEWART, 1983).

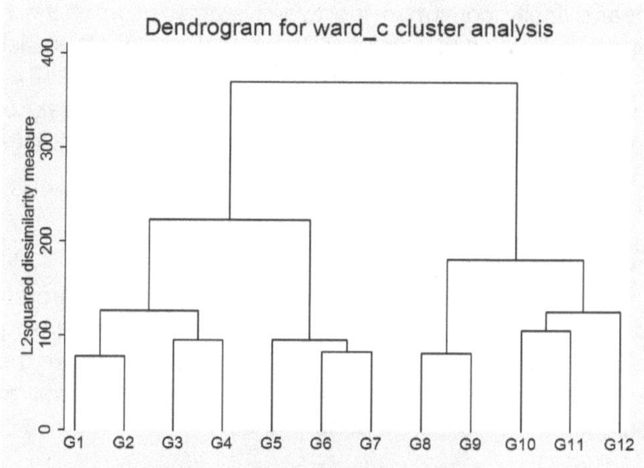

Note: G1-G12 = maximum number of initially considered cluster solutions
L2squared dissimilarity measure = squared Euclidean distance

Figure 21 Dendrogram for Ward's hierarchical cluster analysis with Euclidean distance measure
(Source: GMS 2009, own calculation)

Further, the advantage of the non-hierarchical cluster algorithm is that firms are iteratively classified on the basis of their distance to cluster centroids which are constantly recalculated after reassignments took place. Therefore, and in contrast to hierarchical clustering, this method permits that once classified firms may change their group membership through the cluster formation process until they fit best into one of the clusters (DE JONG/MARSILI 2006: 222; LEIPONEN/DREJER 2007: 1227).

As the basic structure of cluster types was very similar for all three solutions with four, five and 6 clusters provided by non-hierarchical k-means, three criteria were taken into account to determine the final cluster solution[234]: the statistical properties in terms of the relationship between intra-cluster and inter-

234 The cluster solutions that could not be reported for reasons of space are available from the author on request.

cluster variance, the plausibility and interpretability in terms of distinct innovation patterns of the cluster identified, and a sufficient number of firms per cluster. While the first and third criteria were both met by all solutions, the five-cluster solution was preferred for reasons of plausibility and interpretability. Hence, k-means finally confirmed the five-cluster solution of the hierarchical cluster analysis and provided a solution that is satisfactory in statistical terms. As it was shown by Kruskal-Wallis tests (KRUSKAL/WALLIS 1952)[235], each of the cluster variables is constitutive for and differentiates at least one cluster. More importantly, this five-cluster solution can be reasonably interpreted as innovation patterns of non-R&D-performing firms.

5.4.5 Empirical findings

The following paragraphs and subchapters will describe, interpret and verify the identified innovation patterns of non-R&D-performing firms, proceeding from the description of the identified results of cluster analysis to their interpretation and verification on the basis of structural variables as well as variables of innovation output and economic performance.

5.4.5.1 Identification of non-R&D-performing firms' innovation patterns by innovation resource variables

The cluster analysis revealed five distinguishable innovation patterns of non-R&D-performing firms (Table 21). The question, whether they can indeed be interpreted as distinctive innovation patterns of non-R&D-performers and thus confirm the theoretically expected heterogeneity will be elaborated in the following paragraphs.

Basically, it can be seen that non-R&D-performing firms can indeed be described by a broad set of innovation input variables in terms of a Schumpeterian understanding of innovation that goes beyond formal R&D. All of the non-R&D-performing firms in the subsample were successfully assigned to one of these innovation patterns. This corrresponds to the evolutionary perspective that all firms existing at the point of analysis can be regarded as more or less innovative, as they obiously managed to survive (i.e. be positively selected) in

---

235 As the variance of the variables is not homoscedastic, it is necessary to deploy a non-parametric test for significance. The applied Kruskal-Wallis test is similar to the non-parametric Mann-Whitney test, but allows more than two groups to be compared.

market competition by making use of innovation resources to adapt themselves to external challenges and requirements.

Table 21: Result of cluster analysis – five innovation patterns of non-R&D-performing firms in German manufacturing industry (Source: GMS 2009, own analysis)

Cluster variables	Cluster I		Cluster II		Cluster III		Cluster IV					
	1		2		3		4		5			
	Mean	Std.Dev	Mean	Std.Dev	Mean	Std.Dev	Mean	Std.Dev	Mean	Std.Dev	Total	p-value
ProdDev	**0.94**	0.24	0.26	0.44	0.29	0.46	0.05	0.22	0.10	0.31	0.36	.001 [1]
ProdDev_Cus	0.58	0.50	**0.92**	0.27	0.01	0.09	0.71	0.45	0.86	0.35	0.57	.001
NewPRS	**0.29**	0.46	0.13	0.34	0.04	0.19	0.06	0.25	0.12	0.33	0.13	.001
TechProc	0.25	0.78	**0.52**	0.74	-0.14	0.69	-0.44	0.59	-0.13	0.64	0.04	.001
Non-TechProc	**0.54**	0.56	0.35	0.62	-0.19	0.70	-0.83	0.50	0.17	0.37	0.02	.001
HighTech	**0.20**	0.40	0.05	0.23	0.05	0.23	0.01	0.11	0.03	0.18	0.08	.001
InnoCoop	**1.81**	1.25	1.51	1.38	1.13	1.18	0.06	0.47	1.26	1.22	1.19	.001
Know_Int	**0.71**	0.46	0.54	0.50	0.60	0.49	0.16	0.37	0.34	0.48	0.50	.001
Know_ExtR&D	**0.41**	0.49	0.33	0.47	0.16	0.37	0.16	0.37	0.19	0.40	0.26	.001
M&A_Employ	50.07	19.99	70.55	15.29	61.52	21.62	75.04	14.49	**81.03**	15.92	65.90	.001
HS_Employ	**13.21**	9.99	5.56	3.15	5.63	5.89	7.35	8.95	1.66	2.54	7.06	.001
LS_Employ	17.09	20.97	34.55	27.57	33.06	23.10	27.78	31.73	**40.24**	32.51	29.90	.001
n	96		93		111		77		58		435	

Cluster algorithm = non-hierarchical k-means analysis based on k=5 cluster centroids identified by previous hierarchical Ward analysis
Measure of dissimilarity = Euclidean distance
Scales vary between 0 - 1 except for INNOCOOP (1-4), TECHPROC and NON-TECHPROC (open metric scale from -2 to +2) and M&A_, HS-, and LS_EMPLOY (0-100).
1 = F-test, all other variables are not homoscedastic and thus non-parametric Kruskal-Walis-Test was used / Bold = highest mean/percentage for each variable.

Moreover, as theoretically expected, the different innovation resources vary considerably in their degree of severity across the innovation patterns, as none of the innovation patterns is exclusively built upon one distinct innovation resource. Thus, the findings so far correspond to the previously raised guiding theses. While clusters one and two are characterised by a relatively broad range of innovation resource usage, clusters three to five are mainly represented by high shares in fewer resource variables.

In detail, the cluster analysis identified the following five distinct innovation patterns of non-R&D-performing firms in German manufacturing firms:

Innovation pattern #1 – Knowledge-intensive product developers

This cluster most strongly corresponds to an innovation pattern which one would not expect when analysing the innovation behaviour of non-R&D-

performing firms because of its similarity to R&D-intensive, high-tech patterns of innovation. It is predominantly characterised by a high share of product developers (almost 94% of the firms). These new products are frequently accompanied by product-related services which represent the strong service orientation of the firms. Moreover, an above-average share of firms (20%) state that their products consist of high-tech components, which suggests the assumption that these firms mainly produce ambitious, complex products which can be well combined with product-related services. To develop new products, these firms strongly rely on multiple sources of knowledge. On the one hand, most of these firms state that they rely on internal sources of innovation knowledge (about 71%), whereby the share of highly skilled personnel of on average 13% plays a crucial role. On the other hand, about 41% of the firms indicate that external R&D organisations such as universities or R&D laboratories are another important source of innovation knowledge. This reliance on multiple knowledge sources of external R&D knowledge, as well as on internal knowledge embodied in employees, underlines the high knowledge intensity of the innovation pattern. Regarding the process dimension of innovation, this innovation pattern is characterised by a far above-average level of non-technical process innovation, which means that these firms show a broad and intensive use of organisational and management concepts of work organisation and production organisation, as well as a high intensity of innovation cooperation. This likelihood of innovation cooperation is also reflected in the above-average number of innovation fields in which the firms collaborate, whereby it can be assumed that these collaborations mainly take place to develop new products. All in all, this cluster represents an innovation pattern of high innovation intensity regarding product, service, and organisational innovation and is strongly based on internal and external expert knowledge. Interestingly, with 96 non-R&D-performing firms belonging to this innovation pattern, it represents the second largest group.

Innovation pattern #2 – Customer-driven, technical process specialists:

Firstly, this pattern is mainly characterised by the strong customer orientation of its firms. Almost all of them (92%) indicate that they develop their products according to customer needs or specifications. Secondly, they are highly innovative regarding the deployment of advanced manufacturing techniques as their intensity of use is considerably above the average in manufacturing industry (0.52). This may be interpreted in terms of manufacturing highly customer-specific products whose specifications are realised through the use of advanced manufacturing equipment and machinery. For this purpose, they have a major share of employees working in manufacturing and assembly of whom a considerable number is low-skilled or unskilled. Most interestingly, despite the comparably high share of low-skilled personnel (35%), an above-average share of firms (54%) state that their own employees are an important source of innovation, which leads to the assumption that their underlying knowledge-base also consists of experience-based knowledge which is generated through the individual customisation of manufacturing processes to realise highly specific products. But, additionally, the firms of this innovation pattern also report a comparably above-average share of firms (33%) which source innovation knowledge from external R&D organisations. Moreover, these firms are also likely to cooperate in innovation projects with external partners, where it can be reasonably assumed that these innovation collaborations probably deal with technical process innovation in most cases. To summarise, this innovation pattern shows high innovation intensity regarding customised, technical process innovation and is probably based on a mixture of internal, experienced-based knowledge and external R&D-related expert knowledge which is sourced on demand. 93 non-R&D-performing firms were assigned to this innovation pattern by cluster analysis.

Innovation pattern #3 – Occasional business-to-customer (B2C) product developers:

At first glance, this innovation pattern seems to be rather diffuse because, compared to the innovation patterns one and two, there is no obvious manifestation of a distinct use of innovation resources. But if one takes a closer look, especially at the interplay of the variables inside this cluster, it is possible to derive at least some assumptions of what may be going on. On the one hand, there are a considerable number of product developers (29%) in this cluster, which indicates at least occasional new product development. On the other hand, more than 60% of the firms report that knowledge relevant for innovation comes mainly from inside (i.e. own employees). This is all the more astonishing as their average share of highly skilled employees is below-average. But most interestingly, (almost) none of these firms develop their products according to customer specifications, and product-related service only seldom - if ever - plays an important role. This suggests that this innovation pattern is mainly about business-to-consumer relationships. Against this background, as well as considering the fact that these firms can obviously not be captured fully by the remaining variables of innovation input, I would presume that other dimensions of innovation which were not explicitly addressed in this analysis (e.g. creativity, design or marketing) may be of greater importance within this cluster. As far as possible, I will try to substantiate the presumption made here in the following description according to structural and innovation output indicators. Such additional information would be of particular interest, especially as this innovation pattern represents the largest group of non-R&D-performing firms with 111 firm assigned to it.

Innovation pattern #4 – Low-innovative, labour-intensive manufacturers:

This cluster is mainly characterised by an above-average share of firms (71%) that customise their products according to external specifications. But in contrast to the innovation patterns one to three, they almost never develop new products on their own. Moreover, their lack of product development is compensated neither in terms of product-related service innovation, nor in terms of technical or non-technical process innovation. Unsurprisingly, there is also no prevalent source of innovation knowledge or propensity to collaborate which can be interpreted in a meaningful way. Most strikingly, these firms report a comparably high share of employees (75%) working in manufacturing and assembly, which indicates a high labour intensity of production as well as a considerable share of highly skilled personnel. The latter does not really fit into this picture. One possible interpretation of this share of graduate employees may lie in their function as boundary spanners (KIRNER 2005) who are located at the interfaces to customers and "translate" customer needs into feasible information for the firm. Another possibility is that these firms already have a broad offer of product-related services which may be reflected in the fact that they show the lowest activity in product-related services within the last three years and which requires highly skilled employees in service. Another possible interpretation could be that this cluster is significantly composed of toll manufacturers in the sense of an "extended workbench" to which their customers source out labour-intensive manufacturing tasks. All in all, this cluster of non-R&D-performing firms comes very close to the image which is usually associated with the term of "missing R&D activities" or "low-tech", so far. But is has to be mentioned that the interpretation as "low-innovative" solely applies to the underlying set of innovation input variables deployed in this analysis. This means that these firms may turn out to be highly innovative, if a broader set of other variables is considered.

Innovation pattern #5 – Volume-flexible, specialised suppliers:

The last innovation pattern which was revealed in this analysis shows some accordance with the fourth cluster, as most of the innovation input variables do not show outstanding values compared to the first three clusters, and the basic distribution of cluster loadings is very similar. Thus one could label this cluster again as “low-innovative”, but a closer look at and especially a direct comparison with the fourth cluster show that such a labelling would underestimate the innovative ability of this cluster. First, the share of firms which customise their products is almost as high as in the second cluster (86%) which indicates very high customer orientation. Secondly, the firms inside this cluster show - compared to clusters three, four and the total mean - an above-average value of non-technical process innovation. Thirdly, compared to all clusters, these firms deploy the highest share of employees in manufacturing and assembly (81%), and have by far the highest share of low-skilled and unskilled employees (40%). Thus, their high production intensity is accompanied by a high labour intensity of primarily low-skilled and unskilled personnel, which indicates that the manufacturing and assembly processes are probably characterised by rather simple tasks. Moreover, this high share of low-skilled and unskilled employees may allow these firms to react flexibly to changing order situations and occupancy rates. This assumption is underpinned by the value on organisational innovation that may encompass the use of flexible working-time accounts or the use of homework. Moreover, this flexibility may also be realised by employing contract or temporary workers. Compared to the fourth cluster, this fifth cluster as a whole shows a higher level of innovativeness which manifests itself, for instance, in the higher share of product developers (10% vs. 5%), the higher share of product-related service providers (12% vs. 7%), or the higher relevance of internal knowledge sources for innovation; probably in terms of mainly experience-based knowledge, as the share of highly skilled employees is the lowest among all observed patterns of innovation. Considering all these observations in total, this cluster corresponds well to the idea of a volume-flexible, customer-specialised supplier based on experience-based knowledge and medium overall innovation intensity.

The innovation patterns one to three identified up to this point clearly represent separate patterns of non-R&D-based innovation because they basically differ in their strategic orientation. This is indicated by Roman numerals I, II and III. In contrast, the remaining clusters four and five show considerable similarities in their basic strategic orientation. Both of them are production- or labour-intensive manufacturers that are highly oriented towards their customers' individual needs and specifications. Therefore, Roman numeral IV groups them into a strategic subset of non-R&D-performing, production-intensive manufacturers. But as the description according to the innovation input measures have shown, the distinction into two different patterns regarding innovation activity can be reasonably justified. Thus, I will continue to treat them as separate innovation patterns of non-R&D-performing firms in the following consideration of structural and innovation output dimensions.

5.4.5.2 General structural characteristics of non-R&D-performing firms' innovation patterns

After having described the innovation patterns of non-R&D-performing firms explored by cluster analysis above, I will further elaborate on these patterns by shedding some light on their structural characteristics.

Recalling the evolutionary arguments as to why even within the common selection criteria of industrial sectors considerable inter-firm heterogeneity of innovation behaviour can be expected, it would be interesting to see whether single sectors are dominated by a certain type of innovation pattern or not. If this were the case, the theoretical suggestion of evolutionary theory needs to be withdrawn. Moreover, this would imply that including sector variables would have led to similar results, as the use of certain innovation resources would be highly sector-specific.

Firstly, the results (Table 22) show that none of the manufacturing industries are dominated by a single innovation pattern. All identified innovation patterns of non-R&D-performing firms can be found in relevant shares in all sectors. This inter-sectoral heterogeneity of non-R&D-performing firms' innovation patterns in German manufacturing industry is underlined by the fact that they can be even found in typical R&D-intensive sectors (e.g. medical, precision, and optical precision instruments). Furthermore, this reveals the strategic adaptability of non-R&D-performing firms as they are able to exist even in niches of such highly R&D-intensive sectors. This is also true if one thinks the other way round: none of these innovation patterns is solely assigned to individual manufacturing industries (see Annex A.7.1). Nevertheless, the innovation patterns are obviously not equally distributed among the sectors, as some innovation patterns are more likely to be found in certain sectors which will be pointed out by the following considerations.

This intra-sectoral heterogeneity is additionally underpinned by mapping the sectoral classification of R&D intensity provided by LEGLER and FRIETSCH (2007) across the identified innovation patterns. This clearly shows that such a classification is not able to capture the heterogeneity of non-R&D-performing firms' innovation heterogeneity in its entirety.

Table 22: Distribution of German manufacturing sectors across non-R&D-performing firms' innovation patterns
(Source: GMS 2009, own calculations)

Sectoral affiliation	Cluster I 1	II 2	III 3	IV 4	5	Total	
Manufacturing industries (NACE 2-digit)* **							
15/16 - Manuf. of food products, beverages and tobacco	21.4	16.1	**48.2**	10.7	3.6	100.0	(n=56)
17-19 - Manuf. of textiles, textile products or leather	**25.0**	8.3	**25.0**	16.7	**25.0**	100.0	(n=12)
21/22 - Manuf. of paper products, publishing or printing	14.3	**33.3**	19.1	23.8	9.5	100.0	(n=42)
24 - Chemical industries	27.3	18.2	**36.4**	9.1	9.1	100.0	(n=11)
25 - Manuf. of rubber and plastic products	22.2	**35.6**	15.6	8.9	17.8	100.0	(n=45)
27/28 - Manuf. of basic metals and fabricated metal products	12.5	20.5	24.1	**28.6**	14.3	100.0	(n=112)
29 - Manuf. of machinery und equipment	**46.0**	14.0	8.0	10.0	22.0	100.0	(n=50)
30-32 - Manuf. of electrical and optical equipment (excl. NACE 33)	**29.4**	17.7	11.8	17.7	23.5	100.0	(n=17)
33 - Manuf. of medical, precision, and optical instruments etc.	**57.1**	7.1	21.4	14.3	0.0	100.0	(n=14)
34/35 - Manuf. of transport equipment	0.0	**38.5**	23.1	23.1	15.4	100.0	(n=13)
Other sectors (20/23/26/36/37)	19.1	19.1	**36.5**	14.3	11.1	100.0	(n=63)
Sectoral classification (LEGLER/FRIETSCH 2007)*							
Low-Tech sectors (average share of BERD < 2.5%)	18.1	22.1	**27.9**	18.4	13.5	100.0	(n=348)
Medium-Tech sectors (average share of BERD 2.5 - 7%)	**35.3**	17.7	14.7	16.2	16.2	100.0	(n=68)
High-Tech sectors (average share of BERD > 7%)	**47.4**	21.1	21.1	10.5	0.0	100.0	(n=19)
n	96	93	111	77	58	435	

All values are percentages of non-R&D-performing firms / Bold = highest mean/percentage for each variable
Significance levels: *** = p < 0.001 / ** = p < 0.005 / * = p < 0.01
Key: Cluster 1: Knowledge-intensive product developers / Cluster 2: Customer-driven, technical process specialists / Cluster 3: Occasional B2C product developers / Cluster 4: Low-innovative, labour-intensive manufacturers / Cluster 5: Volume-flexible specialised suppliers

Secondly, Table 23 shows that the large majority of non-R&D-performing firms belong to the group of small and medium-sized enterprises (93%). In contrast to public opinion, their most relevant competitive factor is *not* price or cost, but quality (50%). In addition to price (which was mentioned by 21%), the customisation of products is the most crucial competitive dimension for about 14% of the firms. Thereby, the non-R&D-performers are far from merely producing simple products (28%). The majority (51%) indicated that they manufacture products of medium complexity, such as multi-part products (e.g. pumps). Moreover, it is remarkable that 22% of the non-performers manufacture products with high complexity (e.g. machinery) which is usually connected with R&D-intensive or even high-tech manufacturers. This finding is also reflected in the average manufacturing batch size of the non-R&D-performers. In total, only 18% of them manufacture their products in large batch sizes. Most of them (55%) use medium lot sizes for the manufacturing of their products, while almost a third (27%) manufactures in single unit production. Finally, the share of suppliers (49%) and end producers (52%) among the non-R&D-performers is very balanced, indicating that non-R&D-performers are equally able to survive in direct end product competition compared to R&D-performers.

In order to stress the unique characteristics of each innovation pattern, I will focus on the main differences and idiosyncrasies of each cluster in comparison to the others. But as all data is contained in the tables below, every reader is able to individually access additional information.

To start with the *knowledge-intensive product developers* of cluster one, it can be seen that they account for the highest share of non-R&D-performing firms in the R&D-intensive sectors of manufacturers of machinery and equipment (46%), manufacturing of electrical and optical equipment (29%), and manufacturers of medical, precision, and optical instruments (57%). If we remember their very high level of innovativeness and intensity of knowledge sourcing, it seems as if they have perfectly adapted to the high-tech environment in which they are operating in. But it is also interesting to see that one quarter of the firms within the (low-tech) sector of textile products also correspond to this innovation pattern. This may be reasonable if one thinks of producers of high-tech textiles which comprise, for example, micro-electrical elements (e.g. pulse or thermo sensors), nano-particles (e.g. functional textiles) or ultra-powerful types of fibres. Another considerable part of this innovation pattern can be identified within the sector of chemical industry (27%) that is also usually characterised as R&D-intensive. Although most of these firms are small and medium-sized, these knowledge-intensive product developers show the highest share (10%) of large firms with 250 or more employees. Compared to the other clusters, this cluster is composed of the highest share of firms that mentioned innovativeness (9%) and service (4%) as their most important competitive factor. This corresponds closely to the fact that this innovation pattern has the highest share of firms producing mainly products of high complexity in single unit production, which makes sense against the backdrop of their sectoral distribution described above. To summarise, the structural characteristics fit almost perfectly in this pattern of innovation which represents the "high-tech" type (supplier as well as end producer) of non-R&D-performing firms. According to these findings, one would expect above-average innovation performance, for instance, in terms of share of sales with new products and product-related services.

Table 23: Structural characteristics of non-R&D-performing firms' innovation patterns (Source: GMS 2009, own calculations)

Structural characteristics	Cluster I 1	II 2	III 3	IV 4	5	Total
Firm size**						
Up to 49 employees	42.7	30.1	55	**62.3**	46.6	47.1
50 -249 employees	46.9	**63.4**	40.5	33.8	51.7	47.1
250 and more employees	**10.4**	6.5	4.5	3.9	1.7	5.8
	100	100	100	100	100	(n=435)
Most important competitive factor						
Product price	14.4	22.1	21.7	19.7	**27.5**	20.5
Product quality	48,9	40,3	52,6	54,1	**54,9**	49,7
Innovative products	**8,9**	5,2	5,2	1,6	0,0	4,8
Customisation to customers' demands	13,3	**22,1**	8,3	14,8	9,8	13,6
Adherence to delivery times	10,0	9,1	**10,3**	8,2	5,9	9,0
Product-related Service	**4,4**	1,3	2,1	1,6	2,0	2,4
	100,0	100,0	100,0	100,0	100,0	(n=376)
Product complexity***						
Simple products	14,7	26,1	**40,4**	30,3	22,8	27,5
Medium complexity	52,6	53,3	44,0	42,1	**70,2**	51,1
High complexity	**32,6**	20,7	15,6	27,6	7,0	21,5
	100,0	100,0	100,0	100,0	100,0	(n=429)
Batch / lot size**						
Single unit production	33,7	24,2	19,1	**34,3**	28,6	27,3
Small / medium batch / lot	56,8	47,3	56,4	**58,9**	53,6	54,6
Large batch / lot	9,5	**28,6**	24,6	6,9	17,9	18,1
	100,0	100,0	100,0	100,0	100,0	(n=425)
Position in value chain**						
End-producer	55,2	43,0	**66,4**	48,1	35,1	51,5
Supplier	44,8	57,0	33,6	52,0	**64,9**	48,5
	100,0	100,0	100,0	100,0	100,0	(n=433)

All values are percentages of non-R&D-performing firms / Bold = highest mean/percentage for each variable
Significance levels: *** = $p < 0.001$ / ** = $p < 0.005$ / * = $p < 0.01$
Key: Cluster 1: Knowledge-intensive product developers / Cluster 2: Customer-driven, technical process specialists / Cluster 3: Occasional B2C product developers / Cluster 4: Low-innovative, labour-intensive manufacturers / Cluster 5: Volume-flexible specialised suppliers

The second innovation pattern of *customer-driven, technical process specialists* can be predominantly found in the sectors of paper products, publishing or printing (33%), rubber and plastic products (36%), as well as transport equipment (39%). This appears to be reasonable if one considers their high intensity of technical manufacturing process innovation and customisation of products; sectors in which the mastery of manufacturing technology and adherence to customer specifications (e.g. exact specifications of plastics, rubber or automotive parts) is central for competitiveness. In comparison with the others, this cluster reports customisation most frequently as the most important competitive factor (22%). The high level of technical process specialisation enables them to produce their customised products in an above-average number of cases in large batch or lot sizes (29%). Finally, there is a small overhang of suppliers in this cluster (57%) which fits well into the picture described. As they were labelled as technical process specialists, they could be expected to exhibit superior innovation output in terms of outstanding product quality or manufacturing speed.

The third innovation pattern was a bit of a challenge regarding how it could be interpreted. Based on the occasional development of new products, the high share of highly skilled employees, and the unimportance of product customisation, I named this cluster *occasional B2C product developers.* Now one might be curious if this interpretation holds against its structural characteristics. And indeed, considering the sectoral structure, this cluster is most frequently to be found in the consumer-oriented industries of food, beverages, and tobacco (48%), other sectors (37% - like the wood industry (NACE 20), the manufacturing of glass, ceramics, or natural stone (NACE 26) as well as the manufacturing of furniture, jewellery, musical instruments, sports equipment and toys (NACE 36). As these sectors not do represent the heart of manufacturing industry, they may indeed be inadequately captured by an innovation survey which mainly focuses on the core of manufacturing industry and which does not directly account for creativity, art-work, design and marketing as innovation dimensions which are probably of high relevance for this innovation pattern. Regarding the rest of the structural variables, the non-R&D-performers in this cluster comparably report the adherence to (short) delivery times as an important competitive factor (10%), while product customisation (as expected) plays the least important role (8%). Compared to the other clusters, this pattern shows the highest share of firms producing simple products (40%). Last but not least, two thirds of this cluster are end producers; a result which again supports the assumption of a B2C market. All in all, this cluster is certainly one of the most appealing findings of the analysis and definitely requires further research work. This also applies to the high share of this innovation pattern in chemical industry (36%) which can hardly be interpreted without closer scrutiny. Against this backdrop, it is difficult to expect a certain innovation performance.

In contrast, the fourth cluster of *low-innovative, labour-intensive manufacturers* can be found within all manufacturing industries to a similar extent. The highest shares can be identified in the sectors of basic metals and fabricated metal products (29%), paper products, printing and publishing (24%) and manufacturers of transport equipment (23%). This seems reasonable, as the first two sectors are typical non-R&D-intensive, "low-tech" sectors. Within the sector of transport equipment it can be assumed that they play the role of toll manufacturers (i.e. contracted manufacturers) to which automotive enterprises have outsourced their labour-intensive or specialised working tasks. This would also explain the high share of firms manufacturing by single unit production (34%), while the majority of firms in this cluster deploy small or medium batch sizes (59%). The variable of product complexity remains diffuse; however, this cluster shows relatively high shares of firms manufacturing either simple products (30%) or highly complex products (28%). This indicates a considerable amount of residual heterogeneity within the innovation pattern. Most of these firms do not have more than 50 employees (62%) which can explain their weak use of advanced manufacturing technology or organisational concepts. The latter especially are highly dependent on the firm size, as many of these cannot reasonably be implemented in such micro-firms (e.g. some forms of work or production organisation need a critical mass of employees to be rationally implemented) (KIRNER ET AL. 2009b: 43). To summarise, these firms seems to have remarkable (tacit) manufacturing expertise according to customer specifications. However, the structural description of this cluster supports the first impression of low innovativeness and it can be reasonably assumed that this underperformance regarding innovation will also be evident in the innovation output dimensions.

Finally, the fifth cluster was interpreted as *volume-flexible, specialised suppliers*. As argued above, this pattern is similar to the fourth cluster in its basic strategic orientation. Nevertheless, it can be stated that this cluster represents an "advanced" type of the fourth cluster, as throughout it shows higher values among the various innovation input dimensions. This previous assumption is underpinned by the result that this cluster is mainly located within the sectors of textiles, textile products or leather (25%), manufacturing and equipment (22%), and electrical and optical equipment (24%). Thus, these non-R&D-performing firms are also active in sectors of considerable R&D intensity which is one possible explanation for their stronger orientation towards innovation. The assumption that mainly suppliers are involved in this cluster is clearly confirmed by the average share of suppliers (65%) for whom price is the comparably most frequent competitive factor (28%). This also corresponds to the high share of low- or unskilled employees. With this picture in mind, superior innovation output would be expected in the fields of manufacturing speed, quality, or adherence to delivery times.

5.4.5.3 Innovation output of non-R&D-performing firms' innovation patterns

This section focuses on the innovation patterns identified in the previous cluster analysis by shedding some light on their output dimension of innovation success and economic performance to see which economic effects correspond to each pattern (Table 24), and thereby to round off the interpretation of each innovation pattern. Due to the issue of missing values, the number of cases varies across the output variables. Therefore, the number of cases is reported separately for each indicator.

Table 24: Innovation output of non-R&D-performing firms' innovation patterns (Source: GMS 2009, own calculations)

Innovation output	Cluster I 1	II 2	III 3	IV 4	5	Total	
Product and service innovation							
Average share of sales with products new to the firm (%)	**16.0**	3.4	4.8	0.3	1.3	5.5 ***	(n=424)
Average share of sales with products new to the market (%)	**3.4**	0.5	0.8	0.0	0.0	1.1 ***	(n=434)
Average share of sales with new product-related services (%)	1.2	**1.3**	0.4	1	0.6	0.87	(n=434)
Technical and non-technical process innovation							
Average time to market of product developments (months)	13.0	14.0	**8.8**	13.0	15.0	12.0	(n=147)
Average manufacturing lead time (working days)	29.0	22.0	**11.0**	43.0	16.0	24.0 ***	(n=404)
Average delivery time (calendar days)	37.0	33.0	**18.0**	34.0	25.0	29.0 ***	(n=421)
Average share of orders delivered on time (%)	92.0	**95.0**	91.0	92.0	91.0	92.0	(n=428)
Average. scrap and rework rate (%)	2.8	**1.9**	2.4	3.5	2.5	2.6	(n=420)
Average labour productivity (K€/employee)	**96.9**	84.1	96.6	76.9	74.6	84.4 +	(n=341)
Average total factor productivity	1.83	1.81	**1.95**	1.86	1.78	1.85	(n=309)

Bold = highest/best mean for each variable
Significance levels: *** = p < 0.001 / ** = p < 0.005 / * = p < 0.01
Key: Cluster 1: Knowledge-intensive product developers / Cluster 2: Customer-driven, technical process specialists / Cluster 3: Occasional B2C product developers / Cluster 4: Low-innovative, labour-intensive manufacturers / Cluster 5: Volume-flexible specialised suppliers

As expected, the first cluster of *knowledge-intensive product developers* shows the highest innovation performance regarding share of sales with products new to the firm (16%) and products new to the market (3%).[236] Thus,

236 This average share of sales of product innovation in this cluster (16%) is even higher than the average share of sales of product innovation that was reported by the R&D-performing firms in the previous descriptive analysis (13%).

these firms are even able to successfully develop market novelties and implement them as first movers in the market; a performance characteristic which one would not intuitively expect when considering the innovation performance of non-R&D-intensive firms. As this cluster also shows the second highest share of sales with product-related services introduced in the last three years, it can be concluded that these firms simultaneously develop such service offers to match their new products. Nevertheless, these market novelties represent only a fraction of new product developments. KIRNER ET AL. (2009a, 2009b) showed that the average share of sales with market novelties or product-related services is much higher for firms with R&D activities, respectively higher R&D intensity.

But there is another performance dimension which appears quite remarkable in this cluster: against the background, that the firms within this pattern show the highest average product complexity compared to all clusters, they achieve a faster average time to market (13 months) than other clusters which develop less complex products (cluster 2 and 5). Thus, the high intensity of non-technological, organisational process innovation obviously enables these firms to manage their product innovation processes in terms of high efficiency and optimal sourcing of internal and external knowledge. Moreover, this speed advantage is also illustrated by the considerably shorter manufacturing lead time (averagely 29 working days) compared to cluster four (on average 43 working days) which is characterised by a similar product complexity. Finally, the high level of efficiency results in superior labour productivity (97 K€ per employee) which far above total average (84 K€). This is all the more impressive as these firms have a higher average number of employees than the firms in clusters three with similar efficiency in terms of labour productivity. However, their higher personnel cost, caused by their high share of highly skilled employees, is probably reflected in their total factor productivity (1.83) which lies slightly below the average value of all clusters (1.85). In sum, the high intensity of innovation input of this innovation pattern is clearly reflected in superior economic performance regarding product, market and service innovations and results in outstanding process speed and efficiency against the background of their product and firm characteristics.

In contrast, the *customer-driven, technical process specialists* in the second cluster show a different picture of innovation output. As expected, their high intensity regarding the use of cutting-edge manufacturing technology and equipment results in the superior share of orders delivered on time (95%) as well as the lowest scrap or rework rate (1.9%). Thus, these process specialists have a really good handle on their manufacturing processes and are capable of fulfilling their customers' highest requirements and expectations in terms of reliability and quality. Additionally, their high customer orientation is also shown by the higher economic relevance of product-related services in terms of share of sales with new product-related services. This cluster has the highest share compared to all other innovation patterns (1.3%) although cluster one has twice as many firms that offered new services within the last three years. This may be due to the fact, that these process specialists do not only perform manufacturing tasks, but also provide engineering services like simulation, planning and design of manufacturing processes. In short, this cluster perfectly fits into the picture of non-R&D-performing firms that gained manufacturing mastery as they are able to engineer and perform highly specific and demanding tasks for their customers which the customers cannot realise by themselves. Hence, innovation output in terms of efficiency is not so much in the foreground as the average level of productivity measures show.

The third innovation pattern named *occasional B2C product developers* instead shows high innovation output in terms of speed. Compared to the other clusters, these firms have the shortest time to market of new product developments (9 months), the shortest average manufacturing lead time (11 working days), and shortest average delivery time (18 calendar days), which matches the relatively high importance of the competitive factor "adherence to (short) delivery times" reported above. However, these findings have to be put into perspective regarding the considerably lower product complexity in this innovation pattern. The performance characteristic is even more surprising because this cluster shows below-average use of technical or non-technical process innovation. As was been already assumed above, this may be caused by the fact, that this cluster has not been captured very well by the innovation input indicators deployed in the cluster analysis. Moreover, this innovation pattern shows higher innovation performance in terms of share of sales of new products (5%) compared to cluster two (3%), despite their similar share of firms reporting new product developments. But it still remains unclear which innovation input this product innovation performance is achieved by. One can only assume at this point that other issues related to innovation such as creativity, artistic design, or marketing play a crucial role within this cluster. Finally, this innovation pattern displays the highest level of efficiency in terms of total factor productivity (1.95) and the second highest value of labour productivity (97K€ per employee). Besides some other aspects, which cannot be defined in this context, this may also be due to their relatively smaller firm size and their lower level of production input or expenditures on cutting-edge manufacturing technology. But unfortunately the analysis of innovation output did not reveal the hoped for insights to clarify the interpretation of this non-R&D-performing innovation cluster.

Considering the fourth cluster of *low-innovative, labour-intensive manufacturers*, it was assumed that its comparatively low innovation orientation regarding the input dimension will be reflected in its innovation output characteristics. As can be seen, this expectation is indeed answered by the fact that this cluster of non-R&D-performing firms shows the lowest share of sales regarding product innovations (0.3%) and market innovations (0%), as well as the worst values for average manufacturing lead time (43 working days) and scrap and rework rate (4%). Likewise, most of the remaining performance measures are below or equal to the total average. The most striking positive effect can be seen in the share of sales with product-related services (1%) which is comparable to clusters one and two (1.2 / 1.3). Considering the fact that only about 7% of the firms within this cluster have introduced new product-related services since 2006, this supports our previous assumption that some of the firms must have been already active in this field before 2006, which would also explain the relatively high share of highly skilled workers that may be involved in such product-related services. But this assumption definitely needs some further research and thus - and particularly because service is not the most important competitive factor compared to the other clusters – does not appear solid enough to change the basic interpretation of their relatively low innovativeness into something like "service-oriented manufacturers".

Last, but not least, the fifth innovation pattern of *volume-flexible, specialised suppliers* was expected to show superior innovation output in the fields of manufacturing speed, quality, or adherence to delivery times. However, this assumption had to be withdrawn because there is no clear performance dimension which shows a specifically outstanding characteristic. But perhaps this is not even necessary because these firms successfully survive in their niche as flexible and specialised suppliers without any outstanding strength. Moreover, it should be kept in mind that this innovation pattern is very widespread among manufacturing sectors, and provides mainly products of medium complexity. Against this backdrop, it is remarkable that most of the speed indicators as well as the indicator for orders delivered on time (reliability of processes) are above the total average. For instance, they report the second-best average manufacturing lead time (16 working days). Only cluster three which is strongly characterised by simple products reported a shorter lead time (11 working days). The same applies to the average delivery time (25 calendar days in this cluster versus 18 calendar days in cluster three). Thus, regarding their structural background and leaving out cluster three which is somehow special, this cluster shows the best performance regarding the speed and reliability of their manufacturing processes. This performance characteristic may be connected to their above-average use of organisational or management concepts to optimise their working processes. Despite the fact that they even manage to develop new products in some cases, this is certainly not their core business, as the output indicators of product innovation and time to market show. Additionally, they are definitely not among the most efficient non-R&D-performing firms. Both labour and total factor productivity is considerably inferior compared to the other clusters which is probably due to their high production inputs (raw material, operating supplies). To summarise, the analysis of the innovation output of cluster five underpinned the basic expectations and confirms the original interpretation in terms of volume-flexible, specialised suppliers. Moreover, the consideration of both the input and output dimensions confirmed the assumption that this cluster is more oriented towards innovation than its low-innovative counterpart represented by cluster four.

In total, these results support that each innovation pattern results to some extent in a corresponding level of performance regarding innovation output. Moreover, it was shown that one output dimension can be triggered by different sources of innovation, which underlines the earlier argumentation that there is not necessarily a unidirectional relationship between innovation input and innovation output that could be assumed as given.

5.4.5.4 General economic performance of non-R&D-performing firms' innovation patterns

To complete the interpretation of the identified innovation patterns of non-R&D-performing firms, this paragraph briefly glances at the overall economic performance of the non-R&D-performing firms within the innovation patterns (Table 25).

As the collected data refer to the years 2006 to 2008, they are almost not biased by the global financial and economic crisis whose economic turnaround badly affected the German manufacturing sector in 2009. As the German Statistical Office reports in their economy indicators, the break-in of incoming orders in Germany did not take place until the middle of 2008, and turnover did not start to decline seriously until the last quarter of 2008 (see Annex A.7.2 and A.7.3). Against this background, it is not surprising that all performance indicators show positive trends, as the economic situation in German manufacturing industry was superb in the prior years.

Nevertheless, it is remarkable that all types of non-R&D-performing firms reported positive trends in employment and development of total turnover while one might have expected that certain groups of non-R&D-performing firms would show negative employment outcomes, as low-tech sectors tend to move their production into low-wage countries (e.g. LU 2007; HEIDENREICH 2009). Most strikingly, the two patterns of production- and labour-intensive suppliers (clusters four and five) exhibit the highest increase of employees. This is reasonable in so far as these firms have to adapt to increasing demands by increasing their dominant production factor: labour. On the other hand, the innovation pattern of *customer-driven, technical process specialists* which is most strongly based on manufacturing technology, reports the lowest increase in employment, as they have to adapt by higher capital investment.

Table 25: Overall economic performance of non-R&D-performing firms' innovation patterns (Source: GMS 2009, own calculations)

Economic performance variables	Cluster I	Cluster II	Cluster III	Cluster IV		Total	
	1	2	3	4	5		
Employment growth Change in number of employees from 2006-2008 (% per year)	4,6	3,9	3,4	**7,8**	6,3	4,9	(n=416)
Growth of total sales Change in total sales between 2006-2008 (% per year)	9,5	8,2	7,0	8,7	**11,2**	8,7	(n=402)
Export intensity Average share of products (%) that have been exportet in 2008	27,1	20,8	16,7	15,6	**27,5**	21,1 ***	(n=403)

Bold = highest/best mean for each variable
Significance levels: *** = p < 0.001 / ** = p < 0.005 / * = p < 0.01
Key: Cluster 1: Knowledge-intensive product developers / Cluster 2: Customer-driven, technical process specialists / Cluster 3: Occasional B2C product developers / Cluster 4: Low-innovative, labour-intensive manufacturers / Cluster 5: Volume-flexible specialised suppliers

Secondly, the fifth cluster of *volume-flexible, specialised suppliers* shows the most positive trend in total turnover. As described above, these firms can be frequently found in sectors like machinery and equipment (NACE 29) or electrical and optical equipment (NACE 30-32), which represent one of the core industries of German manufacturing industry. Thus, their considerable increase in total turnover in 2006 to 2008 may reflect their strong interlinkage to end producers within these sectors. Moreover, this type of non-R&D-performers represents the most widespread pattern with regard to sector affiliation, so that they were able to profit more from the general economic situation than firms which are focused on niche industrial or consumer markets (e.g. clusters two and three).

Finally, the identified innovation patterns of non-R&D-performing firms vary significantly in their average share of products which are exported abroad. It is interesting to see that - disregarding cluster five - the export ratio basically declines with the intensity of innovation activities. Cluster one which was characterised by the highest innovation intensity exports about 27% of its products, while cluster two reports about 21%, cluster three about 17%, and the low-innovative cluster four about 16%. This aligns with the argumentation of international trade theory (POSNER 1961; HIRSCH 1965; VERNON 1966), that knowledge- or innovation-intensive goods are more likely to be exported. But when considering cluster five, this line of reasoning is not applicable. But as this cluster is dominated by specialised suppliers for R&D-intensive sectors like machinery and equipment, their high export ratio may be caused by their interrelations to international customers within this sector. It would be interesting to investigate this hypothesis further by conducting input-output analysis on this group of non-R&D-performing firms.[237]

To sum up, the overall performance characteristics well fit the interpretation and description of the identified innovation patterns of non-R&D-performing firms. On one side, they show that even non-R&D-performers are capable of achieving good economic performance and may - in the case of production or labour-intensive patterns - be able to generate job growth, not only for highly skilled employees, but also for low-skilled and unskilled people. On the other side, they help to underpin some of our starting assumptions and interpretations based on the initial cluster analysis.

In addition, these results also show, on the one hand, that different patterns of innovation input may lead to partially similar economic performance (e.g. the export ratios of innovation patterns one and five). On the other hand, the performance dimensions are not necessarily interrelated with each other. For instance, employment growth is not automatically correlated with a proportional growth in turnover - see clusters one and four). Nevertheless, each pattern obviously represents a distinct strategic option for non-R&D-performing firms to survive in economic competition. However, when interpreting such overall economic performance measures one should not forget that these measures are highly dependent on other variables of firm behaviour, structure, or strategy besides the issue of innovation. Thus, they may only provide a rough approximation of the relationship between the identified innovation patterns of non-R&D-performers and their overall economic performance.

---

237 A very recent attempt at such an input-output analysis was been made by SOM ET AL. (2010).

5.4.5.5 Additional validation of non-R&D-performing firms' innovation patterns

In the course of the previous description and interpretation of the five explored innovation patterns of non-R&D-performing firms, some questions occurred, respectively remained unanswered. Therefore, this paragraph will provide additional information to enhance the interpretation and verification of the identified innovation patterns of non-R&D-performing firms.

Firstly, one of the most controversially discussed findings of existing empirical studies is the propensity of non-R&D-performing firms to participate in innovation cooperation. As for reasons of feasibility the issue of innovation cooperation was included in terms of the intensity and scope of different partners, it would be interesting to have some additional information about the respective cooperation partners that are linked with the different innovation patterns (Table 26).

Table 26: Participation in innovation cooperation with different collaboration partners across non-R&D-performing firms' innovation patterns (Source: GMS 2009, own calculation)

Additional descriptive variable	Cluster I 1	Cluster II 2	Cluster III 3	Cluster IV 4	Cluster 5	Total	
Average share of firms (%) taking part in innovation collaboration with...							
...customers	**0.47**	0.41	0.24	0.26	0.36	0.31 **	(n=435)
...suppliers	**0.55**	0.44	0.41	0.01	0.38	0.38 ***	(n=435)
...competitors	0.08	0.15	**0.21**	0.00	0.12	0.12	(n=435)
...service providers	**0.35**	0.25	0.19	0.00	0.16	0.20 **	(n=435)
...external R&D organisations	**0.27**	0.14	0.08	0.00	0.10	0.12 ***	(n=435)

Bold = highest/best mean for each variable / percentages exceeding 100% due to multiple answers
Significance levels: *** = $p < 0.001$ / ** = $p < 0.005$ / * = $p < 0.01$
Key: Cluster 1: Knowledge-intensive product developers / Cluster 2: Customer-driven, technical process specialists / Cluster 3: Occasional B2C product developers / Cluster 4: Low-innovative, labour-intensive manufacturers / Cluster 5: Volume-flexible specialised suppliers

This picture fits perfectly with the previous interpretation of the innovation patterns as the high knowledge intensity of cluster one (*knowledge-intensive product developers)* is impressively supported. It significantly shows the highest shares of firms that cooperate with customers, suppliers, service providers or external R&D organisations. Especially the latter corresponds to the high share of firms within this cluster that cited external R&D organisations as relevant sources for innovation knowledge. Moreover, it can also be recognised that the use of innovation cooperation to access external innovation resources is almost not used by the firms belonging to the cluster of *low-innovative, labour-intensive manufacturers*. If they collaborate at all, they do so solely with customers. Although this perfectly agrees with the high importance of product customisation that was reported earlier, this also indicates the high customer dependency of these firms which may turn out to be a major challenge for their future competitiveness.

It can be seen that a more differentiated analysis based on the identified innovation patterns brings more light into the darkness of the existing ambiguous findings of other empirical studies about the propensity of non-R&D-performing firms to participate in innovation cooperation. Thus, the formerly contradictory findings about the cooperation behaviour of non-R&D-performing firms no longer appear inconsistent, as they represent different innovation patterns. There are firms without formal R&D (clusters one and two) that indeed have a relatively high propensity to participate in innovation cooperations with a broad range of different external partners; even with external R&D organisations. Likewise, there are non-R&D-performers that are most likely to cooperate with their suppliers (which corresponds somewhat to the suggestions of PAVITT (1984) and HEIDENREICH (2009)), firms with a very low propensity to participate in innovation cooperation (cluster four), as well as firms that are most likely to collaborate with their customers and suppliers to equal extent (cluster five).

A second question that occurred during the cluster description was whether the volume flexibility of innovation pattern number five (*volume-flexible, specialised suppliers)* may really be rooted in their intensive use of flexible working hours and employment of temporary workers.

Table 27 presents the corresponding findings. At a first glance, there is little support for this assumption. With regard to the employment of temporary workers, the first innovation pattern shows the highest share of firms employing temporary staff (64%), while the fourth cluster reports the highest average share of these temporary workers in total personnel (8%). But both variables should be combined to assess the actual intensity of use of temporary workforce. As a result, the pattern of volume-flexible, specialised suppliers shows the highest intensity, with 63% of these firms employing temporary workers who on average account for about 7% of total employees. Thus, the assumption of the use of temporary workers as one enabler for their volume flexibility appears reasonable. Furthermore, this fifth cluster depicts the second highest share of firms (together with the second pattern) that deploy collective arrangements for flexible working hours. Especially in contrast to the *low-innovative, labour-intensive manufacturers* located in cluster four, this underlines the distinct, advanced characteristic of this cluster, although they share a common basic strategic orientation. As can be seen, the interpretation of cluster five as "volume-flexible" is thus clearly justified by these additional findings.

Table 27: Employment of temporary workers and use of flexible working hours across non-R&D-performing firms' innovation patterns
(Source: GMS 2009, own calculations)

	Cluster						
	I	II	III		IV		
Additional descriptive variables	1	2	3	4	5	Total	
Employment of temporary workers							
Share of firms (%) employing temporary workers	**0.64**	0.58	0.50	0.54	0.63	0.57	(n=418)
Average share of temporary workers on total personnel	0.05	0.06	0.07	**0.08**	0.07	0.06	(n=414)
Share of firms (%) that use OMC in the field of...							
...collective arrangements for flexible working hours	**0.70**	0.67	0.54	0.44	0.67	0.60 **	(n=434)

Bold = highest/best mean for each variable
Significance levels: *** = p < 0.001 / ** = p < 0.005 / * = p < 0.01
Key: Cluster 1: Knowledge-intensive product developers / Cluster 2: Customer-driven, technical process specialists / Cluster 3: Occasional B2C product developers / Cluster 4: Low-innovative, labour-intensive manufacturers / Cluster 5: Volume-flexible specialised suppliers

The third major question that remained open so far is about the service orientation of the fourth innovation pattern. The previous analysis showed, that this group of *low-innovative, labour-intensive manufacturers* is characterised by a relatively high share of highly skilled employees which does not really fit into the picture. In the context of the high relevance of product customisation that was reported by this cluster, it was thus assumed that these employees act as "boundary spanners" (KIRNER 2005) located at the interfaces between customer and firm to "translate" and communicate their needs and demands into feasible, compatible information which can be processed by their firm. To shed some more light on the service orientation of these firms, the following table (Table 28) contains some data about the average share of product-related services that was directly invoiced (as an indicator for the distinct service orientation), as well as the average share of employees working in customer services.

Table 28: Service orientation across non-R&D-performing firms' innovation patterns (Source: GMS 2009, own calculations)

	Cluster					Total	
	I	II	III	IV			
Additional descriptive variables	1	2	3	4	5		
Service orientation							
Average share of sales with PRS directly invoiced	6.0	5.6	3.9	5.0	**6.2**	5.4	(n=243)
Average share of employees (%) working in customer service	10.9	4.9	**13.1**	4.0	3.5	8.0 ***	(n=435)

Bold = highest/best mean for each variable
Significance levels: *** = p < 0.001 / ** = p < 0.005 / * = p < 0.01
Key: Cluster 1: Knowledge-intensive product developers / Cluster 2: Customer-driven, technical process specialists / Cluster 3: Occasional B2C product developers / Cluster 4: Low-innovative, labour-intensive manufacturers / Cluster 5: Volume-flexible specialised suppliers

The result is clear: this cluster neither shows the highest share of product-related services being directly invoiced nor an above average share of employees working in customer services. But this is not to say that these firms are not at all service-oriented. Actually, they show a remarkable share of services directly invoiced in relation to the number of employees working in services. However, this result is not strong enough to reasonably assume that the highly skilled employees in these firms are inevitably linked to customer services. Moreover, there is no empirical evidence based on the available data that a higher share of highly skilled employees is correlated with a higher share of personnel in customer services (see Annex A.5.3). Hence, the question about the role of highly skilled personnel within this fourth cluster has to remain unanswered.

Nevertheless, as a side effect, the analysis reveals an interesting picture about the service orientation of the fifth pattern. As can be seen, the fifth cluster of *volume-flexible, specialised suppliers* shows the strongest intentional service orientation of all innovation patterns, with more than 6% of their service sales being directly invoiced and having the lowest average share of employees working in this field (3.5%). This means that these suppliers are also specialised in terms of offering their customers not only highly customised products, but also accompanying service bundles to maintain and enhance their competitiveness. On the contrary, the third cluster of *occasional B2C product developers* reports the lowest share of directly invoiced product-related services (4%). But at the same time, these firms have by far the highest share of employees working in customer services (13%). Both numbers thereby strongly underline the interpretation that these firms are active in consumer markets, because this result meets the assumption that these firms have a dense network of agents and salesmen working in field service. Last but not least, the very high service orientation in both variables showed by the first cluster (*knowledge-intensive product developers)* on the one side, closely corresponds to their service innovativeness stated by cluster analysis, and their rather complex products, on the other side, which are linked to these services so that they can offer their customers innovative product-service bundles as hybrid problem solutions. Although the second pattern of *customer-driven, technical process specialists* reported the highest share of sales of product-related service above, they do not show the strongest strategic service orientation as their particular innovation strategy is focused on technical expertise in manufacturing and production.

Last but not least, two more questions arose during cluster interpretation which should be answered by the following additional analysis. The first question was about the assumption that the firms in innovation pattern four (*low-innovative, labour-intensive manufacturers)* have a higher propensity to be toll or contracted manufacturers for their customers. Secondly, it is still not ultimately clear, if the B2C interpretation used for the third cluster is really appropriate. Indeed, there is a lot of evidence that this may be true, but as the *German Manufacturing Survey* 2009 offers a variable for more detailed description, it would be a pity not to consider it in this context (Table 29).

With regard to the first question, the findings impressively confirm the higher consumer orientation of cluster four compared to the other clusters. The share of firms which are end producers of finished goods for consumers is significantly almost twice as high (43%) as in the remaining innovation patterns (9%-22%).

Likewise, the second assumption that cluster four (*low-innovative, labour-intensive manufacturers)* is more likely to be composed of toll manufacturers can be affirmed too, as this cluster depicts the highest share of toll manufacturers (23%) which serve as an extended workbench to perform labour- or cost-intensive manufacturing tasks that were outsourced by their customers. However, as can be seen by the considerable share of toll manufacturers in cluster five (*volume-flexible, specialised suppliers)*, the role of toll manufacturers should not necessarily be linked to a low innovativeness in general. This again provides support for the interpretation that clusters four and five are based on a similar strategic foundation, with cluster five representing a more innovative, advanced variant of cluster four.

Table 29: Positioning in industrial value chain across non-R&D-performing firms' innovation patterns
(Source: GMS 2009, own calculations)

	Cluster						
	I	II	III		IV		
Additional descriptive variables	1	2	3	4	5	Total	
Share of firms (%) that are producers of finished products...							
...for consumers	0.22	0.18	**0.43**	0.18	0.09	0.24 ***	(n=433)
...for other firms	**0.60**	0.45	0.43	0.51	0.42	0.48 +	(n=433)
Share of firms (%) that are...							
...systems supplier	**0.19**	0.17	0.05	0.09	0.16	0.13 +	(n=433)
...parts / components supplier	0.26	**0.37**	0.20	0.31	0.35	0.29 +	(n=433)
...toll manufacturers	0.07	0.10	0.15	**0.23**	0.19	0.14 +	(n=433)
Year of foundation							
Average year of firm's establishment	1960	1948	1952	1957	1969	1956 +	(n=420)

Bold = highest/best mean for each variable / percentages exceeding 100% due to multiple answers
Significance levels: *** = $p < 0.001$ / ** = $p < 0.005$ / * = $p < 0.01$
Key: Cluster 1: Knowledge-intensive product developers / Cluster 2: Customer-driven, technical process specialists / Cluster 3: Occasional B2C product developers / Cluster 4: Low-innovative, labour-intensive manufacturers / Cluster 5: Volume-flexible specialised suppliers

Looking at the remaining clusters, their previous interpretation and description is also confirmed. For instance, cluster one (*knowledge-intensive product developers)* in comparison shows the highest share of firms acting as end producers for other firms (business-to-business – "B2B") or system suppliers for their customers which again fits perfectly with their rather complex products, whereas cluster two (*customer-driven, technical process specialists)* reasonably reports the highest share of part and component supplying firms.

To sum up, the additional analyses presented above clarified two important points. First, they strongly supported the basic interpretation of the innovation patterns by means of the innovation resource variables deployed in cluster analysis. Second, as the case of innovation cooperation attested again, consideration of different non-R&D-performing firms' innovation patterns provides additional explanatory value and insight into their innovation behaviour.

5.5 Summary – five innovation patterns of non-R&D-performing firms

The findings presented in the previous paragraphs conclusively show that the theoretical assumption of the heterogeneity of non-R&D-performing firms' innovation behaviour has been substantiated by cluster analysis. Thus, it can be stated that the firm level approach which was chosen on the basis of the evolutionary argumentation, was suitable for analysing the innovation patterns of non-R&D-performing firms. To theoretically capture this heterogeneity in terms of the possible innovation sources of non-R&D-intensive firms, it was decided to take an extended resource-based perspective. This perspective is a) highly compatible with the firm level evolutionary framework, b) can include resources that are located outside the firm, and c) thus broadens the range of possible innovation resources for non-R&D-performing firms beyond mere R&D. The set of possible innovation resources of non-R&D-performing firms derived from the integrated resource-based perspective thereby proved a useful analytical framework for analysing firms' innovation resources beyond formal R&D.

The research design followed these theoretical requirements in its underlying holistic, Schumpeterian definition of firm level innovation as well as its selection of cluster analysis as the method to explore innovation patterns of non-R&D-performing firms. Simultaneously, the analysis allowed for maximal heterogeneity in the subset of the 435 non-R&D-performers as the firms were neither selected nor grouped in advance according to sector affiliation or other structural or output/performance dimensions. Otherwise, firm heterogeneity would have been constrained a priori. Hence, cluster analysis proved to be an adequate methodological instrument to handle both maximum freedom in inter-firm heterogeneity and the broad set of innovation input variables. The two-step, hierarchical and non-hierarchical cluster analysis based on the *German Manufacturing Survey* 2009 revealed five distinct innovation patterns of non-R&D-performing firms in German manufacturing industry, which are summarised in Table 30:

- Knowledge-intensive product developers
- Customer-driven, technical process specialists
- Occasional B2C product developers
- Low-innovative, labour-intensive manufacturers
- Volume-flexible, specialised suppliers

With regard to the underlying evolutionary framework, these innovation patterns cover the whole range of non-R&D-performing firms' variations in their innovation application. On the one hand, there is a recognisable pattern of *knowledge-intensive product developers:* highly innovative firms characterised by a "science-based mode of innovation" (MOWERY/ROSENBERG 1998), which features, for instance, the importance of expert knowledge, a high share of highly skilled employees, and the production of high-tech products, all features more usually associated with R&D-performing firms or R&D-intensive sectors (e.g. JACOBSON/MUSYCK 2008). On the other hand, there is also a pattern of *low-innovative, labour-intensive manufacturers,* whose utilisation intensity of innovation resources is mostly below average and who run the risk of losing their competitive advantage in the future because of their predominant focus on customers as their sole source of external innovation knowledge.

In between these two extremes, there is a group of non-R&D-performers which profits from the high expert use of advanced manufacturing technology that is mainly incorporated in the practical and experienced-based knowledge of their employees. Based on this technical mastery, they are able to provide their customers with tailor-made process solutions that give them an edge in market competition. In addition, these *customer-driven, technical process specialists* show a certain degree of openness to other sources of innovation knowledge by occasionally cooperating not only with customers but also with suppliers and if necessary even external R&D organisations. A less technically-oriented type of innovation pattern is that of *volume-flexible specialised suppliers*, which are characterised by a very high ability to customise their products. But in contrast to the technical expertise characterising the previous innovation pattern, these firms specialise more in organising their production in terms of superior speed and volume flexibility. An impressive example for the heterogeneity of non-R&D-performing firms is given by the third innovation pattern of *occasional B2C product developers*. This pattern distinguishes itself from the other patterns by the greater activity in consumer markets like food, beverages, musical instruments, furniture, sports equipment or jewellery. As a consequence, it is hard to describe them with the innovation input variables used for cluster analysis, because other aspects like creativity, marketing or package design are also very important. This is indicated by their high level of reliance on firm-internal sources of innovation knowledge, on the one side, and the smaller relevance of technical resources or product customisation, on the other hand.

Table 30: Overview of the innovation patterns of non-R&D-performing firms identified by cluster analysis of innovation input variables (Source: own illustration)

Innovation patterns of non-R&D-performing firms				
I	II		III	IV
1 *Knowledge-intensive product developers*	2 *Customer-driven, technical process specialists*	3 *Occasional B2C product developers*	4 *Low-innovative, labour-intensive manufacturers*	5 *Volume-flexible, specialised suppliers*
Highest ability to develop new products	Highest level of customer orientation / customisation of products	Considerable ability to develop new products	Considerable level of customer orientation / customisation of products	High level of customer orientation / customisation of products
Strongest orientation towards product-related services	High use of innovative manufacturing equipment and technologies	No customisation of products	Almost no development of new products	Considerable use of innovative organisational concepts and management tools
Manufacture of advanced products containing high-tech elements	High share of employees working in manufacturing and assembling	Product-related services only of very little importance	Product-related services only of very little importance	Highest share of employees working in manufacturing and assembly
Multiple sources of internal and external sources of innovation knowledge	Considerable share of low and unskilled employees	Own employees as the most important source of innovation knowledge	Weak use of technical and non-technical process innovation; no prevalent source of innovation knowledge	Highest share of low and unskilled employees
Above-average share of highly skilled employees	Own employees as the most important source of innovation knowledge	Above-average share of low and unskilled employees	Low propensity to participate in innovation cooperation	Weak ability to develop new products
Highest level of participation in innovation cooperation with external partners	Occasional knowledge sourcing from external R&D organisations	Hypothesis: high relevance of creativity, design, and marketing	Very high share of employees working in manufacturing and assembly	Considerable orientation towards product-related services
Highest use of innovative organisational concepts and management tools	Participation in innovation cooperation with external partners		Considerable share of highly skilled employees	Above-average propensity to participate in innovation cooperation with external partners

The identified innovation patterns lend quantitative support to the already discussed qualitative and quantitative findings as their elements can be identified in the existing empirical literature. Non-technological, organisational process innovation (e.g. RAMMER ET AL. 2009) is of equal importance for certain innovation patterns as the use of cutting-edge manufacturing technology (e.g. BENDER 2006; ARUNDEL ET AL. 2008; SANTAMARÍA ET AL. 2009), sourcing of expert knowledge in terms of highly skilled employees or external R&D knowledge (e.g. RAMMER ET AL. 2009; SANTAMARÍA ET AL. 2009), as well as internal, probably more experience-based or customer-related stocks of knowledge (e.g. HEIDENREICH 2009; VON TUNZELMANN/ACHA 2005; HIRSCH-KREINSEN 2008). Regarding the issue of innovation cooperation, which has sometimes led to contradictory results, the identified patterns show that some firms make heavy use of innovation cooperation, while others do not or not with the same intensity. Thus, such previous contradictory findings no longer appear problematic as they can be integrated as reasonable elements of the different, heterogeneous innovation patterns of non-R&D-performing firms. Moreover, the identified innovation patterns can also be linked to the typology of non-R&D-intensive firms' innovation strategy presented by HIRSCH-KREINSEN (2007). His type of "problem solver" can be found in the patterns of "knowledge-intensive product developers" and "volume-flexible, specialised suppliers", the type of "process specialist" obviously corresponds to the second pattern of "customer-driven, technical process specialists" and the "producer of standardised goods" is probably close to the fourth pattern of "low-innovative, labour-intensive manufacturers".

At the level of market selection, it has been shown that each of these innovation patterns results in a distinct economic output. For instance, while the first pattern of *knowledge-intensive product developers* shows a higher output of product and market innovations, the second pattern of *customer-driven, technical process specialists* has superior innovation output in terms of process quality. Thereby, from the evolutionary perspective, it is very interesting to see that a similar innovation output (e.g. process efficiency, time to market) can be obtained to a certain degree based on different innovation resources. Thus, as assumed by evolutionary theory, there is no single, one-best-way for all firms to achieve a competitive advantage. To conclude, by accounting for the inter-firm heterogeneity of non-R&D-performing firms, the present analysis provides a much sharper picture of these firms' innovation strategies and their innovativeness than studies which only consider the top layer of the non-R&D-performers' agglomerate.

The conceptual framework which laid out the main cornerstones for the empirical analysis (see chapter 4) has raised guiding theses (*not:* hypotheses!) to validate whether the identified innovation patterns of non-R&D-performing firms' correspond to the basic underlying theory. These guiding theses were:

- The innovation resources of non-R&D-performers identified by the integrated, resource-based perspective serve as elements or modules from which their innovation patterns are composed.
- All of these elements can be identified in the distinct innovation patterns of non-R&D-performing firms to a greater or lesser extent.
- All of these elements may admittedly be constitutive for a distinct cluster, but are not exclusively limited to this cluster.

Without going into the details again here, it can be stated that all three theses have been confirmed by the presented results; or better: as the results of the cluster analysis correspond to the previously formulated theses derived from the conceptual framework, they can be verified. The prime focus on innovation resources which was developed based on the integrated, resource-based perspective turned out to be a fruitful basis from which to explore and describe the distinct innovation patterns of non-R&D-performing firms. It opened up dimensions of innovation input beyond the limited perspectives of growth theory, formal R&D, or market-based approaches of strategic management which focus only on the economic effects of innovation. Furthermore, all of the innovation patterns derived on the basis of innovation input variables have been reviewed and their consistency supported by a broad set of structural variables, variables of innovation output, and measures of general economic performance. This explicitly addresses the criticism frequently made of resource-based approaches that they neglect the value of innovation resources which depends on their contribution to firm survival in market competition.

With regard to the second and the third theses, the findings show that each innovation pattern is composed of a distinct mix of innovation resources. On the one hand, each of the considered innovation resources is – at least to a certain extent – found in each innovation pattern and is constitutive for at least one of them. But, on the other hand, none of the innovation patterns can be described solely by the exclusive occurrence of only one innovation resource. All the patterns are composed of differently weighted innovation inputs and are all characterised by a certain degree of "residual heterogeneity" within them.

As a result – and this should be explicitly mentioned – the explored and described innovation patterns of non-R&D-performing firms are not more – but also not less – than representatives or "ideal types" which can help to identify recurrent structures in these firms' innovation behaviour and provide additional insights to the central question formulated by ARUNDEL ET AL. (2008) and others: "How do firms that do not perform R&D innovate?" Taking these innovation patterns into acount offers the chance to paint a much more detailed and differentiated picture than the mere analysis of non-R&D-performers as an erroneously homogeneous group produces (e.g. as the case of innovation cooperation documents). In doing so, it provides additional explanatory insights as well as a starting point for a more detailed analysis of the innovation behaviour of these firms (see 6.3).

Of course, this analysis also suffers from limitations. It is certainly necessary to enlarge both the number of non-R&D-performing firms and the number of innovation input indicators. For instance, unfortunately, no monetary indicator was available to assess the investments in machinery and equipment of non-R&D-performing firms, which could have helped to support the relevance of this innovation dimension. Likewise, it would be desirable to include measures of the most relevant type of knowledge within each innovation pattern (e.g. how competitively-relevant knowledge is protected against imitation), or indicators which differentiate a broader range of external knowledge sources or co-operation partners like suppliers, customers or competitors in order to fine-tune the findings obtained so far. Last, but not least, it has to be clarified whether the first innovation pattern of *knowledge-intensive product innovators* is really a distinct pattern of non-R&D based innovation, or whether it represents a group of occasional R&D-performers, which have been erroneously odentified as non-R&D-performers because they did not happen to have any R&D expenditures in the surveyed time period. This would mean that they are still profiting from past R&D activity. Regrettably, it was not possible to control for this issue using the indicators available in the datasetr and to distinguish between occasional and habitual non-R&D-performers.

6 Consolidation, interpretation and outlook

Firstly, this part of the thesis summarises the conclusions, results and findings gained in the previous chapters through theoretical discussion and empirical analysis. Secondly, the major results and contributions of this thesis should be interpreted from the perspectives of academic innovation research, innovation management, and innovation policy, the three basic motivations mentioned at the beginning of the thesis. Finally, the thesis closes with a brief outlook of further fields requiring research.

6.1 Summary

During the last decades, research and development (R&D) activities have become substantial pillars of innovation research in the quest to explain the innovation ability and competiveness of sectors and enterprises. Today, the majority of scholarly research and policy documents on innovation still focuses almost entirely on R&D, ignoring other methods that firms use to innovate. As a consequence, firms that do not perform formal R&D were largely neglected by academic research and the policy community because R&D-focused approaches often overlook the fact that a major part of firms' innovation does not necessarily originate from institutionalised R&D activities.

For this reason, this thesis explicitly focused on the group of non-R&D-performing manufacturing firms as an evolutionary subset of enterprises which systematically lack the resource of R&D expenditure ***(chapter 1)***. Based on a comprehensive overview of existing empirical literature, the starting point of this thesis was the observation that the few empirical studies dealing with innovation strategies of non-R&D-performing manufacturing firms provide a fragmented, incomplete, and sometimes even inconsistent picture. While there is some consensus about the particular relevance of practical, experience-based knowledge, their strong process orientation, or their relatively high labour intensity, there is disagreement, for instance, about whether non-R&D-performers are driven rather by technical or non-technical process innovation, or if they have a higher or lower propensity to participate in innovation cooperations. This is due to the fact, for instance, that these studies use different definitions of what should be considered non-R&D-intensive or "low-tech" firms. But most importantly, the vast majority of existing studies assume that non-R&D-intensive firms represent a more or less homogeneous group of firms either because they are affiliated to a certain category of "low-tech" industries or they all lack the same innovation resource of R&D.

The central question of this thesis is whether non-R&D-performing manufacturing firms demonstrate characteric, heterogeneous innovation patterns or not. In order to address this research question theoretically and empirically, the research question was divided into a structural aspect dealing with the question if and why inter-firm heterogeneity of non-R&D-performing firms can be expected, and a substantive aspect addressing which innovation dimensions this heterogeneity could manifest itself in.

To provide a basic understanding of the theoretical roots and relevance of the prevailing R&D focus in innovation research, ***chapter 2*** elaborated on economic growth theory as well as the firm level effects of R&D activities on innovativeness and economic performance. It was shown that R&D still represents one of the most important determinants to explain why some firms persistently gain higher competitiveness and better economic performance than others. Recent OECD data proved, however, that despite their frequently inferior performance compared to R&D-intensive sectors[238], there is no justification to assume a declining economic relevance of non-R&D-intensive sectors in Germany by absolute numbers. On the contrary, the data indicates that their economic importance may further increase during the next years.[239] But by discussing the underlying linear model of innovation as well as dealing with the definition and measurement of R&D, the chapter builds a conceptual bridge between macro-economic growth theory and firm level R&D. Thus it provides the starting point for the developing a theoretical framework to explore the innovation behaviour of firms without formal R&D which was the subgoal of *chapter 3*.

Based on evolutionary innovation theory, ***chapter 3*** first provided theoretical arguments as to why inter-firm heterogeneity regarding their innovation strategies can be still expected, even within low-tech industries, respectively the evolutionary subset of non-R&D-intensive firms. Unlike other contributions, the arguments do not primarily refer to ex post manifestations of heterogeneity, like firm-specific behaviour as a result of evolutionary processes. Instead, the arguments were developed based on the interplay between the evolutionary mechanisms of variation and selection to explain why inter-firm heterogeneity is a structural characteristic of economic evolution. Three major reasons were

---

238 So far, no statistical data particularly dealing with non-R&D-performing firms exists. Hence, this analysis has to take non-R&D-intensive sectors as a proxy to approach their economic relevance.

239 See, for example: Gehrke/Legler (2010), Som et al. (2010).

identified why heterogeneity created by variation is not necessarily wiped out by market selection. Firstly, external selection criteria as a part of the firms' environment may also be highly variable. Secondly, selection works on the tangible effects and consequences of variation and not on the nature of the variations. Thirdly, in analogy to biological evolution, evolutionary game theory argues that only a mixed strategy composed of heterogeneous sub-strategies is an evolutionary stable strategy within a given population. Thus, with regard to the structural aspect there are good reasons according to evolutionary theory why even within similar framework conditions (in the sense of similar selection criteria) heterogeneity in firms' innovation behaviour can be expected.

In order to capture possible dimensions of the innovation behaviour of non-R&D-performing firms in which this heterogeneity may become visible in a second step, chapter 3 employed an extended resource-based perspective which was highly compatible with the previous firm-level evolutionary framework. It includes both resources that are located inside and outside the firm, and thus broadens the range of possible resources for non-R&D-performing firms beyond mere internal R&D. The most important result was that, despite (or because of) the fact that they lack the specific resource of R&D, non-R&D-performing firms can basically draw on many other types of innovation input to enhance and maintain their innovativeness and competitive advantage: technical, non-technical, physical, immaterial, firm-internal, firm-external innovation resources as well as different types of knowledge. Thus it can be assumed in principle that the heterogeneous innovation patterns of non-R&D-performing firms will be revealed via these different resources of innovation and competitive advantage.

Chapter 4 merged these explorative findings from the theoretical discussion into a conceptual framework for empirical analysis. The theoretical approaches of evolutionary economic theory and an integrated resource-based perspective which accounts for internal and external resources of innovation as well as for the particular role of knowledge, has thereby provided a fruitful conceptual framework to explain why inter-firm heterogeneity can be expected and in which innovation resources this heterogeneity may be observed. Following from theory, the chapter developed basic definitions concerning the terms of "non-R&D-performing firms", "innovation", and "innovation patterns". In contrast to many of the previous studies in this field, it suggested a strict definition of non-R&D-performing firms by only including firms which have no intra- or extramural R&D expenditures. Furthermore, it developed a broad, "SCHUMPETERIAN" understanding of innovation which encompasses not only the product and process dimension of firm level innovation, but also a technical and non-technical dimension, and is thus able to grasp the innovation sources of firms beyond the scope of formal R&D. Most importantly, by strictly referring to the theoretical requirements, the chapter set out the methodological design for empirical analysis. To account for maximal heterogeneity in the innovation behaviour of non-R&D-performers it was decided to identify their innovation patterns at the evolutionary level of variation by solely focusing on their innovation resources. This means that they were not preselected according to their sector affiliation or to other structural or performance dimensions. Instead, it was decided to use structural variables as well as measurements of their innovation output and economic performance only to interpret and verify the identified innovation patterns.

The basic objective of this study was to identify and explore non-R&D-performing firms' innovation patterns by empirical data. This was done in ***chapter 5*** by drawing on recent data from the *German Manufacturing Survey* 2009 conducted by the Fraunhofer ISI which provides comprehensive information on the innovation behaviour of firms in the German manufacturing industry. As a first step, a descriptive analysis provided a systematic overview of the characteristics of non-R&D-performing firms' sources of innovation, their innovation output as well economic performance, and supplemented the existing qualitative case studies with quantitative data. On the one hand, it emerged that the dimensions of innovation resources derived from resource-based theory can serve as a fruitful basis to describe non-R&D-performing firms' innovation behaviour and can thus be reasonably used for cluster analysis. In addition, the assumed of inter-firm heterogeneity was also strongly supported, as non-R&D-performing firms were identified in all sectors of the manufacturing industry; even in sectors of higher R&D intensity and high-tech industries. Hence the sectoral classification of R&D intensity does not provide a consistent picture of the actual distribution of non-R&D-performing firms in German manufacturing industry. In consequence, no predominant, single type of innovation behaviour could be identified by referring to descriptive analysis.

Therefore, a hierarchical cluster analysis using Ward's method on the innovation resource variables went one step further and identified five distinct innovation patterns which were confirmed and further shaped by non-hierarchical k-means clustering. As the core result of this thesis, five innovation patterns of non-R&D-performing firms in the German manufacturing industry were successfully identified:

- knowledge-intensive product developers
- customer-driven, technical process specialists
- occasional B2C product developers

- low-innovative, labour-intensive manufacturers
- volume-flexible, specialised suppliers

Thus, despite their common framework condition of missing formal R&D, evolutionary selection still left considerable inter-firm heterogeneity among these firms which strongly corresponds to the previous evolutionary argumentation. It emerged that these innovation patterns cover the whole range of firms' innovation intensity. On the one hand, there is a group of knowledge-intensive, highly innovative firms showing a "science-based mode of innovation", for instance by the importance of expert knowledge, a high share of highly skilled employees, and the production of high-tech products, which are usually more likely to be found in R&D-performing firms or R&D-intensive sectors. On the other hand, there is a group of firms that closely corresponds to the ideal type of a low-innovative, "low-tech" firm. Last but not least, each of these innovation patterns could be reasonably interpreted, verified and supported by reference to measures of innovation output, economic performance, and structural characteristics. Once more, these results reveal that there are many other innovation sources for firms' innovativeness besides mere R&D to gain competitive advantage and economic success. The heterogeneity within the use of these innovation inputs shows that -even in the case of non-R&D-performing firms - there is no single one-best-way to achieve competitive advantage in terms of innovation.

The identified innovation patterns provide novel quantitative support to the already discussed qualitative and quantitative findings as they closely correspond to the existing empirical literature. Non-technological, organisational process innovation (e.g. RAMMER ET AL. 2009) is of equal importance for certain innovation patterns as the use of cutting-edge manufacturing technology (e.g. BENDER 2006; ARUNDEL ET AL. 2008; SANTAMARÍA ET AL. 2009), sourcing of expert knowledge in terms of highly skilled employees or external R&D knowledge (E.G. RAMMER ET AL. 2009; SANTAMARÍA ET AL. 2009), as well as internal, experience-based or customer-related stocks of knowledge (e.g. HEIDENREICH 2009; VON TUNZELMANN/ACHA 2005; HIRSCH-KREINSEN 2008). Regarding the issue of innovation cooperation which sometimes led to contradictory results, the identified patterns show that some firms heavily rely on innovation cooperation, while others do not, or not with similar intensity. Thus, the previous contradictory findings no longer appear problematic, as they can be integrated as reasonable elements of the different, heterogeneous innovation patterns of the non-R&D-performing firms explored in this thesis.

To sum up, the thesis provides comprehensive and novel theoretical and empirical insights into the innovation behaviour of non-R&D-performing firms in general, and German manufacturing industry in particular. Starting from macro-economic growth theory the thesis forged a bridge to firm level R&D activities and their role for firms' innovativeness and competitiveness. Taking this as a starting point, a theoretical framework taking an evolutionary, extended resource-based perspective was developed to elaborate on the heterogeneity of non-R&D-performing firms' innovation patterns, as well as to set out the guiding principles and methodological issues for exploratory empirical analysis. These "guidelines", especially the need to account for maximum inter-firm heterogeneity, were met by the methodological design of firm level, hierarchical and non-hierarchical cluster analysis. Additionally, the prior descriptive analysis provides a systematic overview of the innovation behaviour of non-R&D-performing firms in German manufacturing industry. As a final result, the identification of heterogeneous innovation patterns of non-R&D-performing firms represents the first attempt in empirical innovation research to quantitatively explore the different types of innovation strategies they employ and thereby close the research gap that was defined at the beginning.

6.2 Interpretation of results

Following the motivations outlined in the introduction, this thesis aimed to enhance the understanding of non-R&D-performing firms' innovation behaviour from the three perspectives of academic innovation research, innovation management, and innovation policy. The following paragraphs will point out some main findings for each of these perspectives which can be derived from this thesis.

6.2.1 ...from the perspective of innovation research

Firms that do not invest in formal R&D activities pose a challenge to mainstream innovation theory as they question its predominant belief that R&D intensity is the element constituting the innovativeness and competitiveness of manufacturing firms. Like most of the previously existing studies on non-R&D-intensive or non-R&D-performing firms, this thesis provided empirical evidence of the fact that firms are indeed able to achieve innovativeness and economic success by using innovation resources besides formal R&D. Hence, the findings strongly support the need to continue the shift towards a more holistic and systemic understanding of firms' innovation processes which FREEMAN and SOETE (2009: 587) have named the "21st century innovation paradigm".

According to this paradigm, firms are embedded in socio-economic systems of innovation. Innovation should no longer be assumed to take place isolated *only* within firms or – even less – in sealed-off R&D departments. Instead, firms innovate in close collaboration and interdependence with other firms and organisations, institutions and other social entities and thereby make use of their surrounding, external innovation resources through collaborations and strategic alliances. Looking inside the firm, firms' innovation resources may not only encompass R&D knowledge, but also competences in engineering, design, production and distribution activities, as well as the acquisition of knowledge embodied in new technologies and capital investment. Consequently, and this is most prominently shown by the innovation patterns of *knowledge-intensive product developers, customer-driven, technical process specialists*, and the group of *volume-flexible, specialised suppliers* the absence of R&D does by no means imply the absence of knowledge.

However, it is important to stress that this should *not* be so interpreted that R&D no longer plays an important role for firm level innovativeness. R&D represents a major source of explicit knowledge because it strongly relies on the systematic and intentional deployment of scientific methods which can be documented and intersubjectively transferred between individuals. Moreover, R&D may also fit the conditions of valuable resources in the sense of resource-based theory, because the process, routines and competencies necessary for R&D activities themselves can prove difficult for other firms to imitate. Thus, the firm-specific nature of the R&D process impedes imitation: when the R&D process has an important tacit or intangible element, it is difficult for others outside the firm to replicate the process (HELFAT 1994). Moreover, R&D is not subject to the same kind of economic constraints as those characterising the regular production of goods and services, the ability to perform exploratory activities that would otherwise not be possible in everyday business life is a key factor supporting rapid knowledge advances in firms. But it can be stated that, based on a resource-based perspective, firm-level R&D is just one - and in many cases not even the most important - resource within a broad variety of other sources of firms' competitive advantage.

In this context, non-R&D-performing firms are ideal objects of innovation research in the sense of a "21st century innovation paradigm", because such alternative resources of firm level innovation can be analysed without being blurred by the "noise" of R&D activities, because they automatically control for the circumstance that the positive effects on innovation output are not caused by parallel R&D. However, as the identified innovation patterns of non-R&D-performing firms impressively underline, researchers should not expect a homogeneous population when dealing with non-R&D-performers either because they are frequently located within non-R&D-intensive "low-tech" sectors, or because they commonly lack the same innovation resource of formal R&D. Nevertheless, despite the fact that many of the studies so far criticise and aim to overthrow the linear R&D paradigm, they are based on the implicit assumption that firms which systematically lack an identical resource (R&D) must display an identical innovation behaviour to replace the R&D resource. On the one hand, this appears to be an inversion of the originally criticised linear R&D paradigm of innovation. On the other hand, and this is even more serious, the resulting empirical findings based on such a homogeneous perspective could be partially inconsistent or contradict to each other at a first glance. But this thesis shows that such inconsistent and contradictory findings can be reasonably integrated into a more comprehensive picture which provides a much more differentiated view. It reveals individual characteristics of distinct types of non-R&D-performing innovation patterns not only in terms of innovation resources, but also with regard to their innovation output and economic success.

Against this background, another conclusion of this thesis is that innovation research must account for inter-firm heterogeneity, especially when deploying such a broad understanding of firms' innovation behaviour as it is set out by the 21st century innovation paradigm to adequately consider the multiplicity of heterogeneous sources of innovation and competitive advantage. While this insight has widely diffused within innovation theory, empirical research is still dominated by a narrow, R&D-focused perspective which is largely due to the lack of alternative indicators and measurements to formal R&D. As discussed in chapter 2.6.7, the R&D indicator creates substantial problems for measuring firms' innovativeness. On the one hand, it captures only the input side of innovation and thus reveals little about the innovation output that was achieved due to R&D. On the other hand, SMEs especially fail to report R&D because of the lack of a clear definition of R&D. Problems in determining what is and what is not R&D could be common among managers in firms that have neither a formal R&D department nor full-time researchers, and therefore do not publish specialised R&D investments in their key figures.

Hence, one of the major challenges for future innovation research is to develop and improve appropriate measures and indicators to assess sources of innovativeness besides formal R&D. One very recent and very promising approach, which is for instance used in the Community Innovation Survey (CIS), is to gradually replace the indicator "R&D expenditures" by the much broader indicator "innovation expenditures". This indicator includes all expenditures that were devoted to innovation projects by a firm within a surveyed period, including expenditures for acquiring machinery and equipment, engineering, product design, marketing, licences and patenting or job training of employees.

However, as discussed in chapter 4.5, in the case of rather latent, often non-technical innovation resources like "routines" or "capabilities", developing new innovation indicators remains difficult so far. This is because these constructs are difficult to measure, not only because of their diffuse nature, but also because of their impact on a firm's innovativeness and competiveness is blurred and hard to separate from other effects.

In general, based on the findings of this study, I would agree with the suggestion made by several authors (e.g. DAVID/FORAY 1995; BENDER 2004; HIRSCH-KREINSEN 2008b; FREEMAN/SOETE 2009; ARUNDEL 2007) that the innovativeness of firms within the 21st century should be seen less in terms of the ability to discover new technical solutions, but more in terms of the ability to systematically exploit the effects produced by new combinations, uses of components in their existing stock of knowledge, and – if no solution is available in existing knowledge – the ability to purposefully tap new sources of knowledge inside or outside the firm. By focussing on these abilities, the thesis showed that each of these may count as a constitutive element of firms' heterogeneous innovation patterns. Thus evolutionary economic theory and an integrated resource-based perspective - despite their many shortcomings which will have to be dealt with in the future - may serve as a promising starting point for further theories about these issues. This research can thus be considered as an exploratory step to gain more theoretical insights and empirical knowledge about the different, heterogeneous innovation sources of firms within a 21st century innovation paradigm.

6.2.2 ...from the perspective of innovation management

This short section deals with some managerial implications that could be derived from the thesis' findings, both for non-R&D-performing firms as well as firms with formal R&D in manufacturing industry.

To start with manufacturing firms in general, the results reveal some key aspects that help bridge the cognitive gap between R&D- and non-R&D-based innovation, by strongly supporting the view that non-R&D activities are crucial to understanding the innovation process of any firm. Against the backdrop of the overwhelming presence of R&D in past empirical innovation research and the manifold evidences of formal R&D as the driving force in manufacturing firms, many firms adopted this narrow, R&D-focussed view in their strategic innovation management. Consequently, as many firms and strategic decision-makers intuitively think about reducing the number of their employees to lower manufacturing costs, many firms consider formal R&D as the one-best-way to enhance their innovativeness. Conversely, the overemphasis that was placed on R&D as the major source of innovation may also have discouraged many firms from attempting to be innovative because they perceive R&D as a very costly and uncertain undertaking, demanding very specific and high investments.

Both options involve the risk of cognitive "lock-in" effects in the longer term. In the first case, such a cognitive lock-in situation means that the firm is stuck in the R&D-mode of innovation, concentrating all its innovation efforts on realising an uncertain and risky product development and overlooking other, less risky opportunities to realise the innovation project. In the second case, the cognitive lock-in effect is caused by the lack of innovation activities to renew and modify the firm-internal stock of knowledge. In both cases, the firms will lose their ability to quickly and flexibly adapt themselves to changing environments by losing the ability to – in evolutionary terms - generate "variation". As a result, their propensity to survive in market competition is probably lower in the long run. Regarding the empirical findings, an example of a non-R&D-performing firm which runs the risk of such a lock-in effect is most likely to be found in the fourth innovation pattern of *low-innovative, labour-intensive manufacturers* which reported neither a high importance of internal sources of knowledge nor a high propensity to collaborate in innovation cooperation.

However, the integrated resource-based perspective as well the empirical results of this thesis both showed that activities other than R&D which are maybe even closer to daily routines may also serve as sources of innovation. Thus, one conclusion can be that firms in general should widen their search space for opportunities for innovation, for instance, by exploiting the knowledge bases already existing in the heads of their employees, or by accessing external knowledge and resources through cooperation and strategic alliances. Moreover, as some of the identified innovation patterns show, innovativeness is not limited to the technical dimension. Instead, non-technical process and product innovation, like the use of organisational concepts and management tools, or the offer of product-related services can also help to maintain and increase a firm's innovativeness and competitive advantage. This became clear in the analysis of innovation output. Despite their lack of any formal R&D, the firms located in the innovation patterns two (*customer-driven, technical process specialists)* and three (*occasional B2C product developers)* are able to achieve superior performance in terms of process speed and quality. Moreover, the descriptive analysis of non-R&D-performing firms revealed that their process performance is significantly better across most performance indicators. This holds even if controlled for different categories of firm size and product complexity. The same counts for the first pattern of *knowledge-intensive product developers* with regard to their share of sales of product innovation.

There are far more opportunities for manufacturing firms to achieve innovativeness and competitive success than mere formal R&D. The results showed that none of these different innovation paths is necessarily superior to the others. Thus, managers who are primarily interested in fostering the innovation performance of their firms should be aware that undertaking R&D or not, is not the only relevant decision to be made, as many other actions and sources also yield favourable innovation results. To put it in a nutshell: the absence of R&D does not imply the absence of innovation or competitive success.

Taking a closer look at the group of non-R&D-performing firms, the results of this thesis can contribute in another respect. From a market-based perspective, the strategic options for non-R&D-performing firms are frequently reduced to the level of cost- or price-driven market competition, as the only option open to them. This view is also deeply rooted in growth theory. Accordingly, firms without formal R&D are said to be unable to obtain superior performance in regard to other competitive aspects like innovative products, quality or flexibility because of their lack of formal R&D activities. Non-R&D-performing firms and their products are thus assumed to be locked in in traditional markets. Within these markets, strong price competition with rather stagnating volumes forces them to repeatedly reduce costs/prices.

But this not necessarily true, as both the descriptive and the innovation patterns of non-R&D-performing firms have revealed. There is almost no difference in the relevance of costs as a competitive factor of R&D- and non-R&D-performing firms in the German manufacturing industry. And the comparison of non-R&D-performing firms' innovation patterns did not exhibit a distinct cost-driven type of non-R&D-based innovation. Hence, it can be stated for German manufacturing industry that some of the non-R&D-performing firms (at least of those who were included in the analysis sample) obviously managed to occupy attractive niche markets which allows them to play their trumps of process expertise, superior quality, speed, and flexibility, and thereby to successfully survive in – even international - market competition (e.g. most typically the patterns are *customer-driven, technical process specialists* or *volume-flexible, specialised suppliers)*.

For managers and strategic decision-makers in non-R&D-performing firms it may be interesting to see that a non-R&D based competitive strategy must by no means be limited to the "low road" (BRÖDNER/LATNIAK 2004: 360) of competitive strategy which is primarily aimed to lower costs by downsizing or outsourcing to reduce staff and wage costs. Instead, the empirical findings of this thesis show that non-R&D-performing firms are also able to take the "high road" (BRÖDNER/LATNIAK 2004: 361) of competitive strategy which is directed towards quality competition and the extension of markets and services. This strategy is primarily built on an extended use of existing competencies and human resources within a more efficient organisational structure. The paramount objective is to explore and develop new business areas rather than reduce costs, by simultaneously increasing firm-internal stocks of knowledge and competencies, and making use of reorganised work processes to obtain higher process performance in terms of quality, speed, and efficiency (also PORTER 1985: 164ff.). To summarise briefly, the absence of formal R&D in manufacturing firms should therefore not be equated with cost-driven competitive strategies.

6.2.3 ...from the perspective of innovation policy

As discussed in the second chapter of this thesis, scholars of the endogenous growth theory (ROMER 1986, 1990) explained economic growth through endogenous, intentional research and development (R&D) activities of firms as the source of technological progress. The relationship between technological progress and economic growth is thereby characterised as a linear, steady-state growth pattern which can be adjusted relatively easily by "turning the knobs of the R&D process" (VERSPAGEN 2005: 504). The continued popularity of this linear, R&D-based model of innovation in public debate and political discussion is mainly due to its very simplicity and rhetorical entity. “It is a thought figure that simplifies and affords administrators and agencies a sense of orientation when it comes to thinking about allocations of funding to R&D” (GODIN 2006a: 659f.). As technical progress and innovation as determinant source of economic growth start to define the science and economic policy agenda, the already available statistics on R&D were taken as a legitimate proxy to measure technological innovation, because they included the development of new products and technical processes. Over time the R&D-based model became entrenched in political and public discourses with the help of economic research, statistics and methodological rules. Until today, the linear R&D-based model of innovation has become a “social fact” (GODIN 2006a: 660).

For these reasons, the R&D-based model of innovation and economic growth is very appealing to policy makers.[240] From the perspectives of political science, a policy measure can derive its legitimacy basically from two sides. On one side, there is the dimension of input legitimation which means that a policy is believed to be normatively "right" as it corresponds to the ideology, values and norms of the people. For this purpose, the linear model offers a readily understandable and reasonable causality. Almost no one would seriously question the agumentation that new technical solutions to the major challenges of modern, developed societies have their origins in basic and industrial R&D activities. On the other side, there is the dimension of output legitimation. This means that a policy measure is legitimised as it causes exactly the intended, positive effects it was expected to cause. Hence, output legitimation derives from the actual effects and impacts of a certain policy. In this case, output legitimation of R&D is provided by the numerous empirical findings of innovation and economic research which suggest predictable and ex-ante calculable, positive outcomes of R&D. In contrast, if one considers the complex, systemic and interwoven picture created by the modern, systemic innovation approaches within the 21st century innovation paradigm with its problems in clearly separating cause-and-effect relationships, it is obvious why policy-makers prefer the linear, R&D-based model. This is all the more understandable because the positive relationship between R&D and innovation, respectively economic growth, is a fact that cannot be denied. Hence, R&D provides a good opportunity for policy-makers to justify their normative-political goals by using statistical and empirical research data.

But as some authors state, this preference for the simple, linear, R&D-based model of innovation has turned into something like a "high-tech obsession" (HIRSCH-KREINSEN 2008b) or, optionally, an "obsession for competitiveness" (GODIN 2004) in innovation policy during the last years. This development is reflected in the dominance of supply-side R&D support programmes in national and European innovation policy, for example, the EU Lisbon Agenda to transform the European community into the most competitive, knowledge-

240 Between 2004 and 2006, Anthony Arundel and his colleagues at UNU-MERIT interviewed 67 members of the policy community - 55 from 15 European countries and 12 from Canada, Japan, Australia and New Zealand - on their use of and need for innovation indicators. R&D indicators were the most widely used and were considered to be the most valuable. In contrast, only a minority of respondents referred to the use of indicators drawn from the CIS or similar innovation surveys in policy-making or evaluation (Arundel 2007: 50).

based economy in the world[241], the European Council's Barcelona initiative to counteract the EU's decline in competitiveness with a proposal to increase European R&D intensity to 3% of GDP by the year 2010, or the so-called "High-tech Strategie" of the German government to secure and enhance the future international competiveness of German industry. The more or less explicit common feature of these innovation policies is that they systematically aim to stimulate firms' R&D activities while overlooking that R&D is not the only source of innovation.

While the diagnosis of such a "high-tech obsession" (HIRSCH-KREINSEN 2008b; GODIN 2004) may be controversially discussed, the findings of this thesis provide empirical insight that a solely R&D-based innovation policy does not appear to encompass the complex and multi-layered reality of industrial innovation activities. In order not to stray too far beyond this context, the following major aspects can be summarised from the innovation policy perspective.

First, the division of industry into sectors of low, medium and high technology according to R&D intensity can only distinguish a more or less valid entity of firms with different levels of R&D in terms of statistical data or economic research. But it certainly does not reflect the actual empirical reality. In reality, there are no true "non-R&D-intensive sectors". Instead, what we observe is a significant share of non-R&D-intensive or non-R&D-performing firms which permeate to varying degrees into sectors with high or very high R&D. The contribution of non-R&D-performing firms to the German economy's international competitiveness thus takes place both directly and indirectly. Besides their direct contribution, in successfully exporting a relevant share of their products abroad, non-R&D-performers also indirectly contribute to the international competitiveness and economic growth by supplying R&D-intensive firms with high quality intermediate or end products. Therefore, the strong international competitiveness of German high-tech industry is partially also based on non-R&D-performing firms. Moreover, by acquiring advanced manufacturing technology, non-R&D-performing firms are frequently customers for high-tech products like automation systems, industrial robots or other advanced techniques and thereby again indirectly stimulate innovation activities in these "high-tech" firms.

---

241 However, the underlying theoretical reasoning as well the measurement strategy of the Lisbon Agenda are closely tied to R&D intensity (HAHN 2007: 3ff.; HIRSCH-KREINSEN 2008b: 18).

Second, as mentioned above, the absence of R&D does not imply the absence of knowledge, innovation, or competitive success. Non-R&D-performing firms use many ways to exploit technical and non-technical, product- and process-related, firm-internal and -external innovation resources and stocks of knowledge. All in all, they allow the firms to successfully survive in market competition which is not necessarily only driven by costs and product prices. Hence, an innovation policy which only aims to stimulate R&D activities tends to overlook the particular strengths of these firms by forcing them into an innovation strategy that does not fit their individual situation. It is thus a key task of innovation policy to adopt a broader understanding of innovation and the insight that innovative ability should no longer be equated with R&D activities alone. It should support activities and measures to raise the awareness for non-R&D-intensive firms and their specific needs and conditions according to their functionally differentiated, systemic interrelations within industrial value chains. One possibility could be to supplement R&D-focused policy programmes with policy instruments that better support innovation activities beyond formal R&D (e.g. to support for technology diffusion between firms or between firms and R&D organisations, support for generating of capabilities, e.g. in the fields of engineering, marketing, and design, or promoting the development of non-technical innovations like product-related services or organisational concepts and management tools).[242] Further, the descriptive analysis in this thesis showed that these fields of innovation are by no means specifically reserved for non-R&D-performers, as R&D-performing firms also make use of them, sometimes to an even greater extent. Hence, a broader range of innovation policy instruments would benefit both groups, and would maybe lead to increasingly, fruitful interrelationships.[243]

Third, it is also insufficient, to simply provide data on the number of firms that perform R&D or not. The present results revealed that the group of firms that innovate without performing R&D should also be disaggregated, as policy-makers should not expect a homogeneous group when they think of non-R&D-

242 A first attempt to shift European innovation policy from a solely R&D-based view to a broader understanding of innovation can be found in the recent EU 2020 strategy: "It is also clear that by looking at R&D and innovation together we would get a broader range of expenditure which would be more relevant for business operations and for productivity drivers. The Commission proposes to keep the 3% target while developing an indicator which would reflect R&D and innovation intensity" (EUROPEAN COMMISSION 2010: 8f.).

243 See also the policy recommendations in the studies of SOM ET AL. (2010) and RAMMER ET AL. (2011).

intensive or non-R&D-performing firms. Against the background of their heterogeneous innovation strategies, a "one size fits all" approach to stimulating innovation activities within these firms may thus be misguided. Instead, this type of detailed information could assist policy analysts to identify where weaknesses in the innovative ability of non-R&D-performers lie and produce policies to encourage firms to enhance their innovativeness. While, for instance, the first innovation pattern *knowledge-intensive product developers* may be best served by policy support that stimulates innovation collaboration with external R&D organisations (e.g. within collaborative projects), most firms in the third cluster named *occasional B2C product developers* would not be, as they are perhaps rather in need of support to enhance their marketing competences or design abilities.

Last but not least, the results showed that some of the innovation patterns of non-R&D-performing firms are characterised by a high labour intensity and a high share of low and unskilled employees. On the one hand, against the background of employment policy, this higher labour intensity may cause more positive effects than solely stimulating of R&D-performing firms. On the other hand, by employing a significantly higher share of low and unskilled employees, non-R&D-performing firms are one of the last remaining segments in German manufacturing industry which still provides attractive jobs for people with this level of qualification.

To conclude, if they rely only on a linear, R&D-focused model of innovation, policy-makers are liable to neglect a very important group of firms. But as industrial knowledge bases are increasingly distributed across highly interwoven innovation systems of multiple actors, solely focussing innovation policy on the innovation resource "R&D" means disregarding the alternative strategies of both R&D-performers *and* non-R&D-performers to obtain successful innovation in the 21st century, and thus overlooking valuable potentials to increase the economy's competitiveness and economic growth in the long run. Hence, the paradigm of a 21st-century innovation and technology policy should support the overall *innovation ability* instead of the mere *R&D intensity* of firms and sectors. But with regard to the necessary justification for such policy measures, it is up to innovation research to provide future policy-makers with more scientific insight into the non-R&D based patterns of innovation, as well as the corresponding measurement tools and indicators to identify which innovation policies are needed and to control ex post whether the intended effects were achieved.

6.3 Limitations and need for further research

Finally, I would like to shed some light on the limitations of this study and how they may serve as starting points for further research.

Firstly, despite the fact that this thesis produced five innovation patterns and tried to describe, interpret and verify them by some structural characteristics, innovation output and economic performance measures, these efforts can still only scratch the surface. This applies all the more as these patterns should not be understood as types of non-R&D-performing firms' innovation strategy which can be resampled one-on-one in empirical reality. Instead, these patterns point to some superordinate, dominantly recurring structures and regularities in the innovation behaviour of the aggregate of non-R&D-performing firms, which exactly corresponds to the paramount objective of this thesis. But as the results show, a certain residual heterogeneity remains within each of the innovation patterns, which is of course not surprising. Only case studies would be able to resolve the ultimate level of evolutionary inter-firm heterogeneity.

Secondly, the findings presented in this work can thus give some first indications of what is evolving in these innovation patterns and hence provide ideas for future, more detailed research on these clusters. For example, it would be interesting to see how the *customer-driven, technical process specialists* (pattern two) manage to recognise, adapt and implement their cutting-edge machinery in terms of their - obviously high - absorptive capacity. And, if so, how does the existence of this absorptive capacity fit into the original concept by COHEN and LEVINTHAL (1989, 1990) who argued that technological absorptive capacity is determined by the existence of in-house R&D? Moreover, with regard to their high-level product and process customisation another interesting question would be whether they purchase ready-to-use manufacturing machinery and equipment or if they autonomously modify the techniques available in the market for their own, specific needs and purposes.

To give another example, in the third cluster of *occasional B2C product developers,* more clarification is required concerning the assumed role of creativity, design or marketing as an additional source of innovation for this specific group of non-R&D-performing firms which frequently operate in consumer markets. In the cluster description, it was already argued that this group of firms remains exotic to some extent as they do not really fit into the existing picture of non-R&D-performing firms as suppliers or end producers of simple or medium-complex products in business-to-business markets. As a consequence, we still know very little about these firms' innovation strategies in consumer markets, like music instruments, furniture, fashion or jewellery. Last but not least, it still has to be clarified whether the first innovation pattern of *knowledge-intensive product innovators* really forms a distinct type of non-R&D based innovation. It could also be assumed that this pattern, at least partially comprises a group of occasional R&D-performers who were accidentally characterised as non-R&D-performers because they did not register any R&D expenditures in the surveyed time period. This could mean that they may still be profiting from past R&D activities. In consequence, the interpretation of this innovation pattern would have to be wholly reconsidered.

Thirdly, this pioneer attempt to address the issue of heterogeneous innovation patterns of non-R&D-performers should also be refined and further clarified by including a larger set of non-R&D-performing firms and more innovation input indicators in the cluster analysis. The findings are of course limited by the constraints of the underlying database of the *German Manufacturing Survey* 2009, despite its numerous advantages and high level of applicability. With regard to the last aspects, it was unfortunately not possible, for example, to distinguish between "habitually non-R&D-performers" which usually do not undertake formal R&D and "occasional non-R&D-performers" as a subset of firms which usually perform formal R&D. But as such an additional differentiation could basically change the findings as well as the interpretation of this first innovation pattern, it represents one of the most urgent needs for further research. Another limitation was the lack of an indicator of "innovation expenditure" (which was already described above in the academic interpretation). This indicator would have been practical as it provides a considerable amount of information about financial expenditures by firms in the broader context of innovation besides mere R&D (e.g. expenditures for licensing, patenting, employee training etc.). It would also have provided concrete information about the financial investments in machinery and equipment which would have supported the relevance of this innovation dimension, not only for the second innovation pattern of *customer-driven, technical process specialists*.

Likewise, it would be useful to include some measures which would provide better information about the nature of knowledge primarily relevant within each cluster, e.g. how relevant competitive knowledge is protected against imitation. Moreover, it would be interesting to subdivide the indices on technical and non-technical process innovation into their subcategories in order to gain more detailed information about the technical and non-technical process dimensions of innovation. But this was not possible in the present analysis because of the insufficient number of cases. Last, but not least, it would also be desirable for further research to be able to differentiate between a broader range of external knowledge sources or cooperation partners like suppliers, customers or competitors as well as their geographical distribution in order to fine-tune the findings in the context of the patterns' socio-economic embeddedness in regional, national or international innovation systems. Furthermore, the analytical integration of more qualitative indicators which enable direct capture of rather latent, resource-based variables (e.g. absorptive capacity, certain routines, capabilities) as well as indicators that are capable of accessing stocks of practical or experience-based knowledge would be highly recommended.

Another limitation of this study is that so far its findings are only valid for the German manufacturing industry. Hence, another important issue for further research would be to test whether equivalent innovation patterns can be also found in other European or international manufacturing industries, or whether they represent a phenomenon only specific to the German manufacturing industry. Of course, this is not to say that non-R&D-performing firms are assumed to constitute a homogenous group in other countries, but it is possible that their innovation patterns constitute themselves along different resource dimensions. But as the requirements of a database for such an international analysis are very high, in terms of coverage innovation resources and output besides formal R&D, and sufficient numbers of cases for cross-country comparison this would be easy to implement. The only international database which might begin to be to meet these requirements at present is the CIS survey of the European Commission.

Thirdly, and this is important to mention, the relationship between the identified innovation patterns based on the specific use of a set of innovation resources and their corresponding innovation output and economic performance are, with respect to the exploratory character of this analysis, only of correlative nature. Thus, a necessary next research step could be to take a closer look at the causalities between the innovation patterns and their innovation and economic performance by building and testing corresponding multivariate models. Possibly the same innovation resources will cause different innovation outputs and economic effects in different innovation patterns, or some innovation resources will only account for innovation output in the context of a specific innovation pattern. Thus, analysing the moderating effect of the non-R&D-performing firms' pattern affiliation on the causalities between innovation input and output may be very interesting in clarifying the impact of strategic orientation.

Fourthly, the identified innovation patterns of non-R&D-performing firms represent a more or less static snapshot in time, whereby, as discussed before, static should be understood in an equilibrium sense. Therefore, they do not tell us anything about the potential evolutionary dynamics between them. It would be very interesting to see, not only from an evolutionary perspective (e.g.

through dynamic modelling), but also from a managerial point of view, what possible dynamics might cause firms to change their pattern affiliation by modifying their use of innovation resources. I want to shortly clarify this point taking the example of the fourth pattern which was named *low-innovative, labour-intensive manufacturers*. Due to their very low use of the surveyed innovation resources, the firms in this pattern are very likely to risk losing their ability to adequately adapt to challenges and changes in their environment, and thus be negatively selected in market competition. But as they represent the same basic strategic orientation as pattern number five (*volume-flexible, specialised suppliers)* it would probably be possible for these firms to improve their innovativeness by just slightly modifying their search space for required innovation resources (e.g. by strengthening their collaborative orientation or more strongly exploiting existing firm-internal stocks of knowledge). Thus evolutionary paths of strategic development probably exist between the innovation patterns that could be intentionally pursued by strategic decisions-makers and managers within those firms. Instead, from an evolutionary perspective, it could be exciting to see how these patterns develop over periods of time, which of them are more likely to prosper, respectively, which of them are more likely to disappear or merge with other patterns to perhaps create a new innovation pattern of non-R&D-performers.

Despite these limitations and needs for further research, this thesis has provided novel empirical evidence based on quantitative data to prove that one should not expect a homogeneous group of firms when targeting so-called low-tech industries or non-R&D-performing firms. At best, this thesis helps to raise the awareness that innovation in the 21st century has many sources which are not necessarily bounded to formal R&D. Against the background of the global challenges which face us today, the obvious question is whether we can afford any longer to focus on only one of them.

List of references

A

Aaker, D.A. (1989): Managing Assets and Skills: The Key to Sustainable Competitive Advantage. In: California Management Review, Vol. 31 (2), pp. 91-106.

Abernathy, W.J., Utterback, J.M. (1978): Patterns of Industrial Innovation. In: Technology Review, Vol. 80, pp. 41-47.

Abramovsky, L., Kremp, E., López, A., Schmidt, T., Simpson, H. (2009): Understanding Co-operative Innovative Activity: evidence from four European Countries. In: Economics of Innovation and New Technology, Vol. 18 (3), pp. 243–265.

Acs, Z.J., Audretsch, D.B. (1987): Innovation, Market Structure and Firm Size. In: The Review of Economics and Statistics, Vol. 71, pp. 567-574.

Acs, Z.J., Audretsch, D.B. (1988): Innovation in Large and Small Firms: an empirical analysis. In: American Economic Review, Vol. 78 (4), pp. 678–690.

Acs, Z.J., Audretsch, D.B. (1991a): Innovation and Technological Change: An International Comparison. Ann Arbor, University of Michigan Press.

Acs, Z.J., Audretsch, D.B. (1991b): R&D Firm Size and Innovative Activity. In: Acs, J.Z., Audretsch, D.B. (eds.): Innovation and Technological Change: An International Comparison. Ann Arbor, University of Michigan Press.

Acs, Z.J., Audretsch, D.B. (1993): Small Firms and Entrepreneurship: An East–West Perspective. Cambridge (UK), Cambridge University Press.

Aghion, P., Bloom, N., Blundell, R., Griffith, R., Howitt, P. (2005): Competition and Innovation: An inverted-U Relationship. In: Quarterly Journal of Economics, Vol. 120, 701-728.

Aghion, P., Howitt, P. (1992): A Model of Growth Through Creative Destruction. In: Econometrica, Vol. 60, pp. 323-351.

Aghion, P., Howitt, P. (1998): Endogenous Growth Theory, Cambridge: MIT Press.

Aghion, P., Howitt, P. (2006): Joseph Schumpeter Lecture – Appropriate Growth Policy: A Unifying Framework. In: Journal of the European Economic Association, Vol. 4, pp. 269-314.

Ahuja, G. (2000): The Duality of Collaboration: Inducements and Opportunities in the Formation of Interfirm Linkages. In: Strategic Management Journal, Vol. 21, pp.317-343.

Al-Laham, A. (2004): Wettbewerbsvorteile aus Wissen? Was leistet der wissensbasierte Ansatz für die strategische Unternehmensführung? In: Die Unternehmung, Vol. 58 (6), pp. 405-433.

Alchian, A.A. (1950): Uncertainty, Evolution, and Economic Theory. In: Journal of Political Economy, Vol. 58, pp. 211-221.

Aldrich, H.E., Ruef, M. (2006): Organisations Evolving. 2nd Edition, London, Sage Publications.

Alfranca, O., Rama, R., von Tunzelmann, N. (2004): Innovation Spells in the Multinational Agri-Food Sector. In: Technovation, Vol. 24, pp. 599–614.

Amit, R., Shoemaker, P.J.H. (1993): Strategic Assets and Organisational Rent. In: Strategic Management Journal, Vol. 14, pp. 33-46.

Anderson, E.S. (2009): Schumpeter's Evolutionary Economics. A Theoretical, Historical and Statistical Analysis of the Engine of Capitalism. London, New York, Anthem Press.

ANDREWS, K. (1980): The Concept of Corporate Strategy. Homewood (ILL), Richard D. Irwin.

ARBUSSA, A., COENDERS, G. (2007): Innovation Activities, Use of Appropriation Instruments and Absorptive Capacity: Evidence from Spanish Firms. In: Research Policy, Vol. 36, pp. 1545-1558.

ARCHIBUGI, D. (1988A): In Search of a Useful Measure of Technological Innovation. In: Technological Forecasting and Social Change, Vol. 34, pp. 253-77.

ARCHIBUGI, D. (1988B): The Inter-Industry Distribution of Technological Capabilities. A case study in the application of the Italian patenting in the USA. In: Technovadon, Vol. 7 (3), pp. 259-274.

ARCHIBUGI, D. (2001): Pavitt's taxonomy sixteen years on: A review article. In: Economics of Innovation and New Technology, Vol. 10, pp. 415-425.

ARGYRIS, C., SCHÖN, D.A. (1978): Organisational Learning: A Theory of Action Perspective. Reading, Massachusetts, Addison-Wesley.

ARMBRUSTER, H., BIKFALVI, A., KINKEL, S., LAY, G. (2008): Organisational Innovation: The challenge of measuring non-technical innovation in large-scale surveys. In: Technovation, Vol. 28, pp. 644-657.

ARNOLD, J. M. HUSSINGER, K. (2005): Export Behavior and Firm Productivity in German Manufacturing. A Firm level Analysis. In: Review of World Economics, Vol. 141 (2), pp. 219-243.

ARNOLD, L. (1997): Wachstumstheorie. München, Franz Vahlen Verlag.

ARNOLD, L. (1998): Growth, Welfare, and Trade in an Integrated Model of Human-Capital Accumulation and Research. In: Journal of Macroeconomics, Vol. 20 (1), pp. 81-105.

ARORA, A., GAMBARDELLA, A. (1994): Evaluating Technological Information and Utilizing It. Scientific Knowledge, Technological Capability, and External Linkages in Biotechnology. In: Journal of Economic Behavior and Organisation, Vol. 24 (1), pp. 91–114.

ARROW, K.J. (1962A): Economic Welfare and the Allocation of Resources for Invention. In: Nelson, R.R. (ed.): The Rate and Direction of Incentive Activity: Economic and Social Factors. Princeton, Princeton University Press.

ARROW K.J. (1962B): The Economic Implications of Learning by Doing. In: Review of Economic Studies, Vol. 29, pp. 155-173.

ARUNDEL, A., (1997): Why Innovation Measurement Matters. In: Arundel, A., Garrelfs, R. (eds.): Innovation Measurement and Policies. Luxemburg, EU, EIMS 94/197.

ARUNDEL, A., HOLLANDERS, H. (2004): European Innovation Scoreboard - EXIS: An Exploratory Approach to Innovation Scoreboards. European Commission, Brussels.

ARUNDEL, A. (2007): Innovation Survey Indicators: What Impact on Innovation Policy? In: Science, Technology and Innovation Indicators in a Changing World: Responding to Policy Needs. Paris, OECD publishing, pp. 49-64.

ARUNDEL, A, BORDOY C, MOHNEN P, SMITH, K. (2007): Innovation Surveys and Policy: Lessons from the CIS. In: Nauwelaers, C. Wintjes, R. (eds.): Innovation Policy in Europe. Cheltenham, Edward Elgar, pp. 3-28.

ARUNDEL, A., BORDOY, C., KANERVA, M. (2008): Neglected Innovators: How Do Innovative Firms That Do Not Perform R&D Innovate?. Results of an analysis of the Innobarometer 2007 survey No. 215. INNO-Metrics Thematic Paper.

ASHEIM, B.T., ISAKSEN, A. (1997): Location, Agglomeration and Innovation: Towards Regional Innovation Systems in Norway. In: European Planning Studies, Vol. 5 (3), pp. 299–330.

AUNGER, R. (2002): The Electric Meme: A New Theory of How We Think. Free Press, New York.

AW, B. Y, ROBERTS, M. J., WINSTON, T. (2005): The Complementary Role of Exports and R&D Investments as Sources of Productivity Growth. NBER Working Paper #11774.

AW, B. Y., ROBERTS, M. J., WINSTON, T. (2007): Export market participation, investments in R&D and worker training, and the evolution of firm productivity. In: The World Economy, Vol. 30 (1), p. 83.

B

BACHER, J. (2002): Clusteranalyse. Eine anwendungsorientierte Einführung. 2nd Completed Edition, München, Wien, R. Oldenbourg Verlag.

BACKHAUS, K., ERICHSON, B., PLINKE, W., WEIBER, R. (2006): Multivariate Analysemethoden. Eine anwendungsorientierte Einführung. 11th fully revised and completed edition, Berlin, Heidelberg, New York, Springer-Verlag.

BAGOZZI, R.P. (1982): The Role of Measurement in Theory Construction and Hypothesis Testing: Toward a Holistic Model. In: Fornell, C. (ed.): A Second Generation of Multivariate Analysis, Vol. 2, New York, Praeger, pp. 5-23.

BAIN, J.S. (1956): Barriers to New Competition. Cambridge (MA), Harvard University Press.

BAIN, J.S. (1968): Industrial Organisation. 2nd Edition, New York, John Wiley & Sons.

BAINES, T.S., LIGHTFOOT, H.W., BENEDETTI, O., KAY, J.M. (2009): The Servitization of Manufacturing. A Review of Literature and Reflection on Future Challenges. In: Journal of Manufacturing Technology Management, Vol. 20 (5), pp. 547-567.

BALDWIN, J.R., HANEL,P. (2003): Innovation and Knowledge Creation in an Open Economy: Canadian Industry and International Implications. Cambridge (UK), Cambridge University Press.

BALTER, M. (2000): Genetics. Was Lamarck just a little bit right? In: Science, Vol. 288 (5463), p. 38.

BAMBERGER, I., WRONA, T. (1996a): Der Ressourcenansatz und seine Bedeutung für die strategische Unternehmensführung. In: Zeitschrift für betriebswirtschaftliche Forschung (ZfbF), Vol. 48 (2), pp. 130-153.

BAMBERGER, I., WRONA, T. (1996b): Der Ressourcenansatz im Rahmen des Strategischen Managements. In: Zeitschrift für betriebswirtschaftliche Forschung (ZfbF), Vol. 25 (8), 386-391.

BARGE-GIL, A., NIETO, M. J., SANTAMARÍA, L. (2008): Hidden Innovators: The role of non-R&D activities. Paper presented at the 25th Celebration Conference 2008 on Entrepreneurship and Innovation – Organisations, Institutions, Systems and Regions. Copenhagen, CBS, Denmark, June 17 - 20, 2008, available at: http://www2.druid.dk/conferences/viewpaper.php?id=3352&cf=29 (last accessed on 10th of June 2010).

BARGE-GIL. A. (2010): Cooperation-based Innovators and Peripheral Cooperators: An empirical analysis of their characteristics and behavior. In: Technovation, Vol. 30, pp. 195-206.

BARKEMA, H., SHENKAR, O., VERMEULEN, F., BELL, J. (1997): Working Abroad, Working With Others: how firms learn to operate international joint ventures. In: Academy of Management Journal, Vol. 40 (2), pp. 426–442.

BARNEY, J.B. (1986a): Strategic Factor Markets: Expectations, Luck, and Business Strategy. In: Management Science, Vol. 32 (10), pp. 1231-1241.

BARNEY, J.B. (1986b): Organisational Culture: Can It Be a Source of Sustained Competitive Advantage? In: Academy of Management Review, Vol. 11, pp. 656-665.

Barney, J. B. (1989): Asset Stocks and Sustained Competitive Advantage: A comment. In: Management Science, Vol. 35, pp. 1511-1513.

Barney, J.B. (1991a): Firm Resources and Sustained Competitive Advantage. In: Journal of Management, Vol. 17, pp. 99-120.

Barney, J.B. (1991b): The Resource Based View of Strategy: Origins, Implications, and Prospects. In: Journal of Management, Vol. 17, pp. 97-211.

Barney, J.B. (1997): Gaining and Sustaining Competitive Advantage. New York, Prentice Hall.

Barney, J.B. (2001): Is The Resource-Based "View" A Useful Perspective For Strategic Management Research? Yes. In: Academy of Management Review, Vol. 26 (1), pp. 41-56.

Barney, J. B., Wright, M., Ketchen, D. J. (2001): The resource-based view of the firm: Ten years after 1991. In: Journal of Management, Vol. 27, pp. 625-641.

Barney, J.B., Clark, D.N. (2007): Resource-Based Theory. Creating and Sustaining Competitive Advantage. Oxford, University Press.

Barro, R.J. (1990): Government Spending in a Simple Model of Endogenous Growth. In: Journal of Political Economy, Vol. 98 (5), pp. 103-125.

Baumol, W.J. (2002): The Free-Market Machine: Analyzing the Growth Miracle of Capitalism. Princeton, Princeton University Press.

Bayona, C., García-Marco, T., Huerta, E. (2001): Firms Motivations For Cooperative R&D: an empirical analysis of Spanish firms. In: Research Policy, Vol. 30, pp. 1289–1307.

Beck, T.C. (1998): Kosteneffiziente Netzwerkkooperationen: Optimierung komplexer Partnerschaften zwischen Unternehmen. Wiesbaden.

Becheikh, N., Landry, R., Nabil, A. (2006): Lessons from Innovation Empirical Studies in the Manufacturing Sector: A Systematic Review of the Literature from 1993-2003. In: Technovation, Vol. 26, pp. 644-664.

Beckenbauer, B. (2006): Internationales Management industrieller Dienstleistungen. Dissertation Nr. 3153, Bamberg, Difo-Druck GmbH.

Becker, A., Brauner, E., Duschek, S. (2006): Transaktives Wissen, Kompetenzen und Wettbewerbsvorteile: Der Akteur als strategischer Faktor. In: Schreyögg, G., Conrad, P. (eds.): Management von Kompetenz. Managementforschung, Vol. 16, Wiesbaden, Gabler Verlag, pp. 201-230.

Becker, M.C. (2001): The role of routines in organisations – an empirical and taxonomic investigation. PhD Thesis, Judge Institute of Management, University of Cambridge.

Becker, M. C. (2004): The Concept of Routines Twenty Years after Nelson and Winter (1982): A review of the Literature. In: Industrial and Corporate Change, Vol. 13 (4), pp. 643-677.

Becker, M.C. (2005): The Concept of Routines: Some Clarifications. In: Cambridge Journal of Economics, Vol. 29, pp. 249-262.

Becker, M. C., Zirpoli, F. (2006): Innovation Routines. Exploring the Role of Routines for Innovation. Paper presented at the International Schumpeter Society Conference, Sophia-Antipolis, 21-24 June 2006.

Becker, A., Brauner, E., Duschek, S. (2006): Transaktives Wissen, Kompetenzen und Wettbewerbsvorteile: Der Akteur als strategischer Faktor. In: Schreyögg, G., Conrad, P. (eds.): Management von Kompetenz. Wiesbaden, Gabler, pp. 201-230.

Becker, S. O., Egger, P. H. (2009): Endogenous product versus process innovation and a firm's propensity to export. In: Empirical Economics, DOI 10.1007/s00181-009-0322-6 (forthcoming).

Becker, W., Dietz, J. (2004): R&D Cooperation and Innovation Activities of Firms - Evidence for the German Manufacturing Industry. In: Research Policy, Vol. 33, pp. 209–223.

Becker, W., Peters, J. (2000): Technological Opportunities, Absorptive Capacities, and Innovation. In: Volkswirtschaftliche Diskussionsreihe der Universität Augsburg No. 195.

Behrens, G. (1993): Wissenschaftstheorie und Betriebswirtschaftslehre. In: Wittman, W. (ed.): Handwörterbuch der Betriebswirtschaft. 5. Auflage, Stuttgart, pp. 4763-4772.

Belderbos, R., Carree, M., Lokshin, B. (2004): Cooperative R&D and Firm Performance. In: Research Policy, Vol. 33, pp. 1477-1492.

Belitz, H., Clemens, M., Gornig, M., Schiersch, A., Schumcher, D. (2010): Wirtschaftsstrukturen, Produktivität und Außenhandel im internationalen Vergleich. Berlin, Expertenkomission für Forschung und Innovation (EFI).

Bender, G. (2004): Innovation in Low-Tech – Considerations based on a few case studies in eleven European Countries. Soziologisches Arbeitspapier Nr. 6/2004, Universität Dortmund.

Bender, G., Laestadius, S. (2005): Non-Science Based Innovativeness. On Capabilities Relevant to Generate Profitable Novelty. In: Journal of Mental Changes, Vol. 11 (1-2), pp. 123-170.

Bender, G., Jacobson, D., Robertson, P. L. (2005): Non-Research-Intensive Industries in the Knowledge Economy. Perspectives on Economic, Political and Social Integration. Special Issue I, Catholic University Lublin.

Bender, G. (2006): Peculiarities and Relevance of Non-Research-Intensive Industries in the Knowledge-Based Economy. Final Report of the Project "Policy and Innovation in Low-Tech – Knowledge Formation, Employment & Growth Contributions of the 'Old Economy' Industries in Europe – PILOT", Technische Universität Dortmund.

Bender, G., Laestadius, S. (2007): Innovationen ohne Wissenschaft und Forschung? Erfahrungen aus „Lowtech-Unternehmen". In: Abel, J., Hirsch-Kreinsen, H. (eds.): Lowtech-Unternehmen am Hightech-Standort. Berlin, edition sigma, pp. 193-228.

Bender, G. (2008): How to Grasp Innovativeness of Organisations: outline of a conceptual tool. In: Hirsch-Kreinsen, H., Jacobson, D. (eds.): Innovation in Low-Tech Firms and Industries. Cheltenham, Northampton, Edward Elgar, pp. 25-42.

Beneito, P. (2002): Technological Patterns Among Spanish Manufacturing Firms. In: Entrepreneurship & Regional Development, Vol. 14, pp. 89–115.

Beneito, P. (2006): The Innovative Performance of In-house and Contracted R&D in Terms of Patents and Utility Models. In: Research Policy, Vol. 35, pp. 502-517.

Bergstrom, C.T. Godfrey-Smith, P. (1998): On the Evolution of Behavioral Heterogeneity in Individuals and Populations. In: Biology and Philosophy, Vol. 13, pp. 205-231.

Berman, S.L., Down, J., Hill, C.W.L. (2002): Tacit Knowledge as a Source of Competitive Advantage in the National Basketball Association. In: Academy of Management Journal, Vol. 45, pp. 13-31.

Bertschek, I., Entorf, H. (1996): On Nonparametric Estimation of the Schumpetrian Link Between Innovation and Firm Size: evidence from Belgium, France and Germany. In: Empirical Economics, Vol. 21, pp. 401-426.

Bessant, J., Caffyn, S. (1997): High-Involvement Innovation Through Continuous Improvement. In: International Journal of Technology Management, Vol. 14 (1), pp. 7-28.

Bessant, J., Caffyn, S., Gallagher, M. (2001): An Evolutionary Model of Continuous Improvement Behaviour. In: Technovation, Vol. 21, 67-77.

Bhatt, G.D. (2001): Knowledge Management in Organisations: Examining the Interaction Between Technologies, Techniques, and People. In: Journal of Knowledge Management, Vol. 5 (1), pp. 68-75.

Birchall, D.W., Chanaron, J.J., Soderquist, K. (1996): Managing Innovation in SMEs: a comparison of companies in the UK, France and Portugal. In: International Journal of Technology Management, Vol. 12 (3), pp. 291–305.

Black, S., Lynch, L. (2000): What's Driving the New Economy: The benefits of workplace innovation. NBER Working Paper No. 7479.

Blackler, F. (1995): Knowledge, knowledge work and organisations: an overview and interpretation. In: Organisation Studies, Vol. 16 (60), pp. 1021-1046.

Blanchard, O.J. (1999): Macroeconomics. Second Edition, Upper Saddle River, New Jersey, Prentice Hall.

Bloch, C. (2005): R&D Investment and Internal Finance: The cash flow effect. In: Economics of Innovation and New Technologies, Vol. 14, pp. 213–223.

Bollmann, P. (1990): Technischer Fortschritt und wirtschaftlicher Wandel. Heidelberg, Physica-Verlag.

Boschma, R.A. (2005): Proximity and Innovation: a Critical Assessment. In: Regional Studies, Vol. 39, pp. 61–74.

Bougrain, F., Haudeville, B. (2002): Innovation, Collaboration and SMEs' Internal Research Capacities. In: Research Policy, Vol. 31, pp. 735–747.

Bound, J., Cummins, C., Griliches, Z., Hall, B., Jaffe, A. (1984): Who Does R&D and Who Patents? In: Griliches, Z. (ed.): R&D, Patents, and Productivity. Chicago (IL), University of Chicago Press, pp. 21-54.

Bowman C, Ambrosini V. (2000): Value Creation Versus Value Capture: Towards a Coherent Definition of Value in Strategy. In: British Journal of Management, Vol. 11 (1), pp. 1–15.

Bresnahan, T.F., Brynjolfsson, E., Hitt, L.M. (2002): Information Technolgy, Workplace Organisation and the Demand for Skilled Labor: Firm level evidence. In: Quarterly Journal of Economics, Vol. 177, pp. 339-376.

Bretschger, L. (2004): Wachstumstheorie. 2. durchgesehenen Auflage, München, Wien, Oldenbourg.

Brödner, P., Latniak, E. (2004): Recent Findings on Organisational Changes in German Capital Goods Producing Industry. In: Journal of Manufacturing Technology Management, Vol. 15 (4), pp. 360-368.

Bronwyn, H.H., Mairesse, J., Mohnen, P. (2010): Measuring the Returns to R&D. UNU-MERIT Working Paper No. 2010-006, available at: www.merit.unu.edu/publications/wp.php?year_id=2010 (last access on 8th of June 2010).

Brouwer, E., Kleinknecht, A. (1993): Technology and a firms export intensity: the need for adequate innovation measurement. In: Konjunkturpolitik, Vol. 39 (5), pp. 315–325.

Brouwer, E., Kleinknecht, A. (1996): Firm Size, Small Business Presence and Sales of Innovative Products: A micro-econometric analysis. In: Small Business Economics, Vol. 8, pp. 189–201.

Brouwer, E., Kleinknecht, A., Reijnen, J.O.N. (1993): Employment Growth and Innovation at the Firm Level: An empirical study. In: Journal of Evolutionary Economics, Vol. 3 (2), pp. 153-159.

Brown, J.S., Duguid, P. (1991): Organisational Learning and Communities-of-Practice: Toward a Unified View of Working, Learning and Innovation. In: Organisation Science, Vol. 2, pp. 40-57.

Brown, J. S., Duguid P. (2001): Knowledge and Organisation: A Social-Practice Perspective. In: Organisation Science, Vol. 12, pp. 198–213.

Brown, J.S., Eisenhardt, K.M. (1995): Product development: past research, present findings and future directions, Academy of Management Review, Vol. 20 (2), pp. 343-378.

Bruland, K., Mowery, D.C. (2005): Innovation Through Time. In: Fagerberg, J., Mowery, D.C., Nelson, R.R. (eds.): The Oxford Handbook of Innovation. New York, Oxford University Press, pp. 349-379.2.

Brumagim, A.L. (1994): A Hierarchy of Corporate Resources. In: Shrivastava, P., Huff, A.S., Dutton, J.E. (eds.): Advances in Strategic Management. Greenwich (CT), London, JAI Press, pp. 81-112.

Brunsson, N. (1985): The irrational Organisation. New York, John Wiley.

Burr, W. (2004): Innovationen in Organisationen. Stuttgart, Kohlhammer Verlag.

Burr, W., Stephan, M., Soppe, B., Weisheit, S. (2007): Patentmanagement. Strategischer Einsatz und ökonomische Bewertung von technologischen Schutzrechten. Stuttgart, Schäffer-Poeschel Verlag.

Burt, R. (2003): The Social Structure of Competition. In: Ostrom, E.; Ahn, T.K. (eds.): Foundations of Social Capital. Cheltenham, Edward Elgar, pp. 197-219.

C

Calinski, T., Harabasz, J. (1974): A Dendrite Method for Cluster Analysis. In: Communications in Statistics, Vol. 3, pp. 1-27.

Campbell, D.T. (1965): Variation, Selection and Retention in Sociocultural Evolution. In: Barringer, H.R., Blanksten, G.I., Mack, R.W. (eds.): Social Change in Developing Areas: A Reinterpretation of Evolutionary Theory. Schenkman, Cambridge (MA), pp.19-49.

Campbell, D.T. (1994): How individual and face-to-face group selection undermine firm selection in organisational evolution. In: Baum, J.A., Singh, J.V. (eds.): Evolutionary dynamics of organisations, Oxford, Oxford University Press, pp. 23-38.

Campbell, A., Goold, M., Alexander, M. (1995): Corporate Strategy: The Quest for Parenting Advantage. In: Harvard Business Review, Vol. March-April, pp. 120-132.

Cantner, U., Hanusch, H., Pyka, A. (1996): Routinized Innovations - Dynamic Capabilities in a Simultation Study. In: Volkswirtschaftliche Diskussionsreihe, Beitrag No. 148, Institut für Volkswirtschaftslehre der Universität Augsburg.

Cantner, U., Hanusch, H. (1998): Industrie-Evolution. In: Schweitzer U., G. Silverberg (eds.): Selbstorganisation. Jahrb. Bd. 9, Evolution und Selbstorganisation in der Ökonomie, Berlin, Duncker&Humblot, pp. 265-293.

Caputo, A., Palumbo, M. (2005): Manufacturing Re-Insourcing in the Textile Industry: A case study. In: Industrial Management and Data Systems, Vol. 105 (2), pp. 193–207.

Carayannopoulos, S., Auster, E.R. (2010): External Knowledge Sourcing in Biotechnology through Acquisition Versus Alliance. In: Research Policy, Vol. 39, pp. 254-267.

Cassiman, B., Veugelers, R. (2002): R&D Cooperation and Spillovers: some empirical evidence from Belgium. In: American Economic Review, Vol. 92 (4), pp.1169– 1184.

Cassiman, B., Golovko, E. (2007): Innovation and the Export-productivity Link. CEPR Discussion Papers 6411.

CASSIMAN, B., MARTINEZ-ROS, E. (2007): Product Innovation and Exports. IESE working paper, mimeo.

CASTANIAS, R.P., HELFAT, C.E. (1991): Managerial Resources and Rents. In: Journal of Management, Vol. 17 (1), pp. 155-171.

CASTELLACCI, F. (2008a): Innovation and the Competitiveness of Industries: Comparing the Mainstream and the Evolutionary Approaches. In: Technological Forecasting & Social Change, Vol. 75, pp. 984-1006.

CASTELLACCI, F. (2008b): Technological paradigms, regimes and trajectories: manufacturing and service industries in a new taxonomy of sectoral patterns of innovation. In: Research Policy, Vol. 39, pp. 1139–1158.

CESARATTO, S., MANGANO, S. (1993): Technological profiles and economic performance in the Italian manufacturing sector. In: Economics of Innovation and New Technology, Vol. 2, pp. 237–256.

CHAMBERLIN, T., DOUTRIAUX, J. (2010): Sourcing Knowledge and Innovation in a Low-Technology Industry. In: Industry and Innovation, Vol. 17 (5), pp. 487-510.

CHANDLER, A.D. (1962): Strategy and Structure: Chapters in the History of the American Industrial Enterprise. Cambridge (MA), MIT Press.

CHANDLER, A.D. (1984): Comparative Business History. In: Coleman, D.C., Mathias, P. (eds.): Enterprise and History. Essays in Honor of Charles Wilson. Cambridge, Cambridge University Press, pp. 3-26.

CHEN, C.J., HUANG, J.W. (2009): Strategic Human Resources Practices and Innovation Performance – The Mediating Role of Knowledge Management Capacity. In: Journal of Business Research, Vol. 62, pp. 104-114.

CHENNELLS, L., VAN REENEN, J. (2002): The Effects of Technical Change on Skills, Wages and Employment: A survey of the microeconomic evidence. In: Greenan, N., L'Horty, Y., Mairesse, J. (eds.): Productivity, Inequality and the Digital Economy. Cambridge, London, MIT Press, pp. 172-223.

CHESBROUGH, H.W. (2003): Open Innovation: The New Imperative for Creating and Profiting from Technology. Cambridge (MA), Harvard Business School Publishing.

CHESBROUGH, H.W. (2007): Why Companies Should Have Open Business Models. In: Sloan Management Review, Vol. 48 (2), pp. 22–28.

CHIAROMONTE, F., DOSI, G. (1993): Heterogeneity, Competition, and Macroeconomic Dynamics. In: Structural Change and Economic Dynamics, Vol. 4, pp. 39-63.

CHILD, J. (1972): Organisational Structure, Environment and Performance – The Role of Strategic Choice. In: Sociology, Vol. 6 (1), pp. 1-22.

CHMIELEWICZ, K. (1993): Forschungskonzeptionen der Wirtschaftswissenschaft. 3. Auflage, Stuttgart.

CHRISTENSEN, J. F. (2002): Corporate strategy and the management of innovation and technology. In: Industrial and Corporate Change, Vol. 11 (2), pp. 263-288.

CHRISTENSEN, J.F., OLESEN, M.H., KJÆR, J.S. (2005): The Industrial Dynamics of Open Innovation – evidence from the transformation of consumer electronics. In: Research Policy, Vol. 34 (10), pp. 1533–1549.

CHRISTENSEN, J.L. (2010): Joint Competition: The World Dominance of Danish Fur Production. Paper to be presented at the International Schumpeter Society Conference 2010 on Innovation, Organisation, Sustainability and Crisis. Aalborg, June 21-24, 2010. (available on request by the author).

Christiaans, T. (2004a): Neoklassische Wachstumstheorie. Darstellung, Kritik, Erweiterung. Norderstedt, Books on Demand.

Christiaans, T. (2004b): Types of Balanced Growth. In: Economic Letters, Vol. 82 (2), pp. 253-258.

Chudnovsky, D., López, A., Pupato, G. (2006): Innovation and Productivity in Developing Countries: A Study of Argentine Manufacturing Firms' Behaviour (1992-2001). In: Research Policy, Vol. 35, pp.266-288.

Cincera, M., Kempen, L., van Pottelsberghe, B., Veugelers, R., Villegas, C. (2003): Productivity Growth, R&D and the Role of International Collaborative Agreements: Some evidence from Belgium manufacturing companies. In: Brussels Economic Review, Vol. 46 (3), pp.107-140.

Coase, R.H. (1960): The Problem of Social Cost. In: Journal of Law and Economics, Vol. 1, pp. 3-44.

Coase, R.H. (1974): The Lighthouse in Economics. In: Journal of Law and Economics, Vol. 17 (22), pp. 357-376.

Cobb, C.W., Douglas, P.H. (1928): A Theory of Production. In: American Economic Review, Vol. 18, (1), pp. 139-165.

Cockburn, I.M., Henderson, R.M., Stern, S. (2000): Untangling the Origins of Competitive Advantage. In: Strategic Management Journal, Vol. 21, (10/11), pp. 1123-1145.

Cohen, B. (1993): Analogy, Homology, and Metaphor in Interactions between the Natural Sciences and the Social Sciences, especially Economics. In: History of the Political Economy, Vol. 25, Annual Supplement: Non-Natural Social Science (ed. by N. Marchi), pp. 7-44.

Cohen, W.M., Richard C.L., Mowery, D. (1987): Firm Size and R&D intensity: A Re-Examination. In: Journal of Industrial Economics, Vol. 35, pp. 543-563.

Cohen, W.M., Levinthal, D.A. (1989): Innovation and Learning: the Two Faces of R&D. In: Economic Journal, Vol. 99, pp. 569–596.

Cohen, W.M., Levinthal, D.A. (1990): Absorptive Capacity: A New Perspective On Learning and Innovation. In: Administrative Science Quarterly, Vol. 35, pp. 128-152.

Cohen, W.M., Klepper, S. (1996a): A Reprise of Size and R&D. In: Economic Journal, Vol. 106, pp. 925-951.

Cohen, W.M., Klepper, S. (1996b): Firm size and the Nature of Innovation within Industries: The case of process and product R&D. In: The Review of Economics and Statistics, Vol. 78, pp. 232–243.

Cohen, W.M., Goto, A., Nagata, A., Nelson, R.R., Walsh, J. (2002): R&D Spillovers, Patents and the Incentives to Innovate in Japan and the Unites States. In: Research Policy, Vol. 31, pp. 1349–1367.

Collis, D.J. (1991): A Resource-Based Abalysis of Global Competition: The Case of the Bearings Industry. In: Strategic management Journal, Vol. 12 (special issue), pp. 49-68.

Collis, D.J. (1994): Research Note: How Valuable are Organisational Capabilities? In: Strategic Management Journal, Vol. 15, pp. 143-152.

Collis, D.J. (1996): Organisational Capability as a Source of Profit. In: Moingeon, B., Edmondson, A. (eds.): Organisational Learning and Competitive Advantage. London, Sage, pp. 139-163.

Collis, D.J., Montgomery, C.A. (1995): Competing on Resources. In: Harvard Business Review, Vol. 73, pp. 118-128.

Collis, D.J., Montgomery, C.A. (1998): Corporate Strategy. A Resource-Based Approach. Boston (MA), Irwin/McGraw-Hill.

COLOMBO, M.G., GRILLI, L., PIVA, E. (2006): In Search of Complementary Assets: The Determinants of Alliance Formation of High-tech Start-ups. In: Research Policy, Vol. 35 (8), pp. 1166–1199.

CONNOR, K.R. (1991): A Historical Comparison of Resource-Based Theory and Five Schools of Thought Within Industrial Organisation Economics: Do We Have a new Theory of the Firm? In: Journal of Management, Vol. 17 (1), pp. 121-154.

CONNOR, K.R., PRAHALAD, C.K. (1996): A Resource-Based Theory of the Firm. Knowledge Versus Opportunism. In: Organisation Science, Vol. 7, pp. 477-501.

CONTRACTOR, F. J., LORANGE, P. (2002): The Growth of Alliances in the Knowledge-Based Economy. In: International Business Review, Vol. 11 (4), pp. 485-502.

COOK, S.D.N., BROWN, J.S. (1999): Bridging Epistemologies: The Generative Dance Between Organisational knowledge and Organisational Knowing. In: Organisation Science, Vol. 10 (4), pp. 381-400.

CORBETT, L.M. (2008): Manufacturing strategy, the business environment, and operations performance in small low-tech firms. In: International Journal of Production Research, Vol. 46 (20), pp. 5491-5513.

CORNES, R., SANDLER, T. (1986): The Theory of Externalities. Public Goods, and Club Goods. Cambridge, Cambridge University Press.

CORNWALL, J. (1976): Diffusion, Convergence and Kaldor's Law. In: Economic Journal, Vol. 85, pp. 307-314.

CORNWALL, J. (1977): Modern Capitalism. Its Growth and Transformation.

COWAN, R., FORAY, D. (1997): The Economics of Codification and the Diffusion of Knowledge. In: Industrial and Corporate Change, Vol. 6 (3), pp. 595-622.

COWAN, R., VAN DE PAAL, G. (2000): Innovation Policy in the Knowledge-Based Economy. Brussels, European Commission, DG Enterprise.

COX, H., FRENZ, M., PREVEZER, M. (2002): Patterns of Innovation in UK industry: Exploring the CIS Data to Contrast High and Low Technology Industries. In: The Journal of Interdisciplinary Economics, Vol. 13, pp. 267-304.

COZZI, G., SPINESI, L. (2006): How Much Horizontal Innovation is Consistent With Vertical Integration? In: Research in Economics, Vol. 60 (1), pp. 47-53.

CROSBY, P. B. (1979): Quality is Free: The Art of Making Quality Certain. New York, McGraw-Hill.

CROSBY, P.B. (1984): Quality Without Tears: The Art of Hassle-Free Management. New York, McGraw-Hill.

CUERVO-CAZURRA, A., UN, C.A. (2010): Why Some Firms Never Invest In Formal R&D. In: Strategic Management Journal, Vol. 31 (7), pp. 759-779.

CYERT, R.M., MARCH, J.G (1992): A Behavioral Theory of the Firm. (2nd Edition, Malden (MA), Oxford, Carlton: Blackwell Publishing.

CZARNITZKI, D. (2006): Research and Development in Small and Medium-sized Enterprises: The role of financial constraints and public funding. In: Scottish Journal of Political Economy, Vol. 53, pp. 257–335.

CZARNITZKI, D., WASTYN, A. (2009): Does Professional Knowledge Management Improve Innovation Performance at the Firm Level? ZEW Discussion Paper No. 09-067, Centre for European Economic Research, available at: ftp://ftp.zew.de/pub/zew-docs/dp/dp09067.pdf (last access on 07|04|2010).

Cziko, G. (1995): Without Miracles. Universal Selection Theory and the Second Darwinian Revolution. Cambridge (MA), London.

D

D'Aveni, R.A. (1995): Hyperwettbewerb: Strategien für die neue Dynamik der Märkte. Frankfurt a. Main, New York, Campus Verlag.

Damanpour, F. (1992): Organisational Size and Innovation. In: Organisation Studies, Vol. 13 (3), pp. 375-402.

Damanpour, F. (1996): Organisational Complexity and Innovation: Developing and Testing Multiple Contingency Models. In: Management Science, Vol. 42 (5), pp. 693–716.

Damanpour, F., Evan, W.M. (1984): Organisational innovation and performance: the problem of "Organisational Lag". In: Administrative Science Quarterly, Vol. 29, pp. 392-409.

Damanpour, F., Wischnevsky, J.D. (2006): Research On Innovation in Organisations: Distinguishing Innovation-generating From Innovation-Adopting Organisations. In: Journal of Engineering and Technology Management, Vol. 23, pp. 269-291.

Darroch J. (2005): Knowledge Management, Innovation and Firm Performance. In: Journal of Knowledge Management, Vol. 9 (3), pp. 101-115.

Das, T.K., Teng, B.S. (2000): A Resource-based Theory of Strategic Alliances. In: Journal of Management, Vol. 26 (1), pp.31–60.

Das, T.K., Teng, B.S. (2002): Alliance Constellations: A Social Exchange Perspective. In: Academy of Management Review, Vol. 27 (3), pp. 445-456.

Das, T.K., Teng, B.S. (2003): Partner Analysis and Alliance Performance. In: Scandinavian Journal of Management, Vol. 19, pp. 279-308.

Dasgupta, P., Stiglitz, J. (1980): Uncertainty, Industrial Structure and the Speed of R&D. In: Bell Journal of Economics, Vol. 11, pp. 1-28.

Dasgupta, P., Maskin, E.S. (1987): The Simple Economics of Research Portfolios. In: Economic Journal, Vol. 97, pp. 581-595.

David, P. (1996): Science Reorganized? Postmodern Visions of Research and the Curse of Success. MERIT Research Memoranda, No. 2-96-002, available at: http://www.merit.unu.edu/publications/rmpdf/1996/rm1996-002.pdf (last access on 13th of June 2010).

David, P., Foray, D. (1995): Accessing and Expanding the Science and Technology Knowledge Base. In: STI Review, Vol. 16, pp. 16-38.

Dawkins, R. (1976): The Selfish Gene. Oxford University Press, Oxford.

Dawkins, R. (1983): Universal Darwinism. In: Bendall, D.S. (ed.): Evolution from Molecules to Man. Cambridge University Press, Cambridge (MA), pp. 403-425.

Debackere, K., Clarysse, B., Rappa, M.A. (1996): Dismantling the Ivory Tower: The Influence of Networks on Innovative Output in Emerging Technologies. In: Technological Forecasting and Social Change, Vol. 53, pp. 139-154.

DeCarolis, D.M., Deeds, D.L. (1999): The Impact of Stocks and Flows of Organisational Knowledge on Firm Performance: an Empirical Investigation of the Biotechnology Industry. In: Strategic Management Journal, Vol. 20, pp. 953-968.

De Geus, A. (1988): Planning as Learning. In: Harvard Business Review, Vol. 66 (2), pp. 70-74.

De Jong, J.P.J., Marsili, O. (2006): The fruit flies of innovations: a taxonomy of innovative small firms. In: Research Policy, Vol. 35 (2), pp. 213–229.

De Jong, J.P.J., Vermeulen, P.A.M. (2006): Determinants of Product Innovation in Small Firms. In: International Small Business Journal, Vol. 24 (6), pp. 587–609.

De Toni, A., Tonchia, S. (2003): Strategic Planning and Firms' Competencies - Traditional Approaches and New Perspectives. In: International Journal of Operations & Production Management, Vol. 23, (9), pp. 947-976.

De Wit, B., Meyer, R. (2004): Strategy, Process, Content, Context – An International Perspective. 3rd Edition, London, Thomson Business Press.

Deming, W. E. (1982): Quality, Productivity, and Competitive Position. Massachusetts Institute of Technology, Center for Advanced Engineering Study. Cambridge (MA), West Publishing.

Deming, W. E. (1986): Out of the Crisis. Cambridge (MA), MIT Press.

Demsetz, H. (1973): Industry Structure, Market Rivalry, and Public Policy. In: Journal of Law and Economics, Vol. 16, pp. 1-9.

Demsetz, H. (1988): The Theory of the Firm Revisited. In: Journal of Law, Economics, and Organisation, Vol. 4 (1), pp. 141-161.

Demsetz, H. (1991): The Theory of the Firm Revisited. In: Williamson, O., Winter, S.G. (eds.): The Nature of the Firm - Origin, Evolution and Development. Oxford, Blackwell, pp. 159-178.

Denz, H. (1982): Analyse latenter Strukturen. München,Francke.

Depew, D., Weber, B. (1995): Darwinian Evolving. Cambridge (MA), London, MIT Press.

Depew, D., Weber, B. (1996): Natural Selection and Self-Organisation. In: Biology and Philosophy, Vol. 11, pp. 33-65.

Dierickx, I., Cool, K. (1989): Asset Stock Accumulation and Sustainability of Competitive Advantage. In: Management Science, Vol. 25, pp. 1504-1511.

Dierkes, M., Antal, A.B., Child, J., Nonaka, I. (2001): Handbook of Organisational Learning and Knowledge. Oxford, Oxford University Press.

Dijk, van B., Den Hertog, R., Menkveld, B., Thurik, R. (1997): Some New Evidence on the Determinants of Large and Small-Firm Innovation. In: Small Business Economics, Vol. 9, pp. 335–343.

Disselkamp, M. (2005): Innovationsmanagement. Instrumente und Methoden zur Umsetzung im Unternehmen, Wiesbaden, Gabler.

Dixit, A.K., Stiglitz, J.E. (1977): Monopolistic Competition and Optimum Product Diversity. In: American Economic Review, Vol. 67, pp. 297-308.

Djellal, F., Francoz., D., Gallouj, C., Gallouj, F., Jacquin, Y. (2003): Revising the Definition of Research and Development in the Light of the Specificities of Services. In: Science and Public Policy, Vol. 30 (6), pp. 415-429.

Dodgson, M., Rothwell, M. (1994): The Handbook of Industrial Innovation. Aldershot, Edward Elgar Publishing.

Dodgson, M. (2000): The Management of Technological Innovation. Oxford, Oxford University Press.

Dodgson, M., Gann, D., Salter, A. (2005): Think, Play, Do. Technology, Innovation, and Organisation. Oxford.

DODGSON, M., GANN, D., SALTER, A. (2008): The Management of Technological Innovation. Completely revised and updated edition, Oxford, Oxford University Press.

DOMAR, E.D. (1946): Capital Expansion, Rate of Growth and Employment. In: Econometrica, Vol. 14, pp. 137-147.

DOMAR, E.D. (1957): Essays in the Theory of Economic Growth. Oxford: Oxford University Press.

DOMS, M., DUNNE, T., ROBERTS, M.J. (1995): The Role of Technology Use in the Survival and Growth of Manufacturing Plants. In: International Journal of Industrial Organisation, Vol. 13 (4), pp. 523-542.

DOSI, G. (1982): Technological Paradigms and Technological Trajectories: A Suggested Interpretation of the Determinants and Directions of Technical Change. In: Research Policy, Vol. 11, pp. 147-162.

DOSI, G. (1988a): Sources, Procedures and Microeconomic Effects of Innovation. In: Journal of Economic Literature, Vol. 25, pp. 1120–1171.

DOSI, G. (1988b): The Nature of Innovation Processes. In: Dosi, G., Freeman, C., Nelson, R.R., Silverberg, G., Soete, L. (eds.): Technical Change and Economic Theory. London, New York, Pinter, pp. 221–238.

DOSI, G., TEECE, D.J., CHYTRY, J. (1998): Technology, Organisation, and Competitiveness. Perspectives on Industriaö and Corporate Change. Oxford, Oxford University Press.

DOSI, G., PAVITT, K., SOETE, L. (1990): The Economics of Technical Change and International Trade. Brighton, Wheatsheaf.

DOSI, G., MARENGO, L. (1994): Some elements of an Evolutionary Theory of Organisational Competencies. In: England, R.W. (ed.): Evolutionary Concepts in Contemporary Economics. University of Michigan Press, Ann Arbour.

DOSI, G., ORSENIGO, L. (1994): Macrodynamics and Microfoundations: an Evolutionary Perspective. In: Granstrand, O. (ed.): Economics of Technology. Amsterdam, Elsevier North Holland, pp. 91-123.

DOSI, G., NELSON, R.R., WINTER, S.G. (2000): The Nature and Dynamics of Organisational Capabilities. Oxford, Oxford University Press.

DREHER, C., LAY, G. (1992): Diversity On Firm Level. Different Modes of Usage of NC/CNC-Machine Tools in German Mechanical Engineering. In: Revue d'Économie Industrielle, Vol. 59 (1), pp. 86-98.

DREHER, C. (1997): Technologiepolitik und Technikdiffusion. Auswahl und Einsatz von Förderinstrumenten am Beispiel der Fertigungstechnik. Baden-Baden, Nomos Verlagsgesellschaft.

DREHER, C., EGGERS, T., KINKEL, S., MALOCA, S. (2005a): Gesamtwirtschaftlicher Innovationswettbewerb und betriebliche Innovationsfähigkeit. In: Bullinger, H. J. (ed.): Fokus Innovation. Kräfte bündeln, Prozesse beschleunigen. Stuttgart, Hanser, pp. 1-28.

DREHER, C., KIRNER, E., MALOCA, S. (2005b): Möglichkeiten zur Innovationsbeschleunigung aus Sicht der Unternehmen. In: Bullinger, H. J. (ed.): Fokus Innovation. Kräfte bündeln, Prozesse beschleunigen. Stuttgart, Hanser, pp. 137-158.

DUDA, R.O., HART, P.E. (1973): Pattern Classification and Scene Analysis. New York, Wiley.

DUSCHEK, S. (1998): Kooperative Kernkompetenzen - Zum Management einzigartiger Netzwerkresourcen. In: Zeitschrift Führung und Organisation, Vol. 68 (4), pp. 230-236.

DUSCHEK, S. (2002): Innovation in Netzwerken: Renten - Relationen - Regeln. Wiesbaden, Gabler.

Duschek, S., Sydow, J. (2002): Ressourcenorientierte Ansätze des Strategischen Managements: Zwei Perspektiven auf Unternehmungskooperation. In: Wirtschaftswissenschaftliches Studium, Vol. 8, pp. 426-431.

Dyer, J.H. (1996): Does Governance Matter? Keiretsu Alliances and Asset Specificity as Sources of Japanese Competitive Advantage. In: Organisation Science, Vol. 7, pp. 649-666.

Dyer, J.H. (1997): Effective Interfirm Collaboration: How Firms Minimize Transaction Costs and Maximize Transaction Value. In: Strategic Management Journal, Vol. 18, pp. 553-556.

Dyer, J.H., Singh, H. (1998): The Relational View: Cooperative Strategy and Sources of Interorganisational Competitive Advantage. In: Academy of Management Review, Vol. 23 (4), pp. 660-679.

Dyer, J.H., Nobeoka, K. (2000): Creating and Managing a High-Performance Knowledge Sharing Network: The Toyota Case. In: Strategic Management Journal, Vol. 21, pp. 345-367.

Dyer, J.H., Hatch, N.W. (2006): Relation-Specific Capabilities and Barriers to Knowledge-Transfers: Creating Advantage Through Network Relationships. In: Strategic Management Journal, Vol. 27, pp. 701-719.

Dyer, J.H., Singh, H., Kale, P. (2008): Splitting the Pie: Rent Distribution in Alliances and Networks. In: Managerial and Decision Economics, Vol. 29, pp. 137–148.

E

Easterly, W., Rebelo, S. (1993): Fiscal Policy and Economic Growth: An Empirical Investigation. In: Journal of Monetary Economics, Vol. 32 (3), pp. 417-458.

Edquist, C. (1997): Systems of Innovation: Technologies, Institutions and Organisations. London, Pinter.

Edquist, C., Texier, F. (1998): Innovations, Systems and European Integration (ISE). Linköping, Linköping University.

Eggers, T., Engelbrecht, A. (2005): Kooperation – Gründe und Typologisierung. In: Wiendahl, H.P., Dreher, C., Engelbrecht, A. (eds.): Erfolgreich kooperieren. Best-Practice-Beispiele ausgezeichneter Zusammenarbeit. Heidelberg, Physica-Verlag, pp. 1-12.

Eisele, J. (1995): Erfolgsfaktoren des Joint-Venture-Management. Wiesbaden.

Eisenhardt, K.M., Galunic, D.C. (2000): Coevolving: At Last, a Way to Make Synergies Work. In: Harvard Business Review, Vol. 87 (1), pp. 91-101.

Eisenhardt, K. M., Martin, J. A. (2000): Dynamic Capabilities. What are they? In: Strategic Management Journal, Vol. 21, pp. 1105-1121.

Eisenhardt, K.M., Santos, F.M. (2002): Knowledge-Based View: A New Theory of Strategy? In: Pettigrew, A., Thomas, H., Whittington, R. (eds.): Handbook of Strategy & Management. London, Sage, pp. 139-164.

Endres, E., Wehner, T. (1995): Störungen zwischenbetrieblicher Kooperation – Eine Fallstudie zum Grenzstellenmanagement in der Automobilindustrie. In: Schreyögg, G., Sydow, J. (eds.): Managementforschung, Vol. 5, Berlin, New York, Walter de Gruyter, pp. 1-45.

Enkel, E., Gassmann, O., Chesbrough, H. (2009): Open R&D and Open Innovation: exploring the phenomenon. In: R&D Management, Vol. 39 (4), pp. 311-316.

Entorf, H., Pohlmeier, W. (1990): Employment, Innovation and Export Activity: Evidence from firm level data. In: Florens, J.P., Ivaldi, M, Laffont, J.J., Laisney, F. (eds.): Microeconometrics: Surveys and Applications. Oxford, Basil Blackwell, pp. 394-415.

Escribano, A., Fosfuri, A., Tribó, J.A. (2009): Managing External Knowledge Flows: The moderating role of absorptive capacity. In: Research Policy, Vol. 38, pp. 96-105.

Essletzbichler, J., Rigby, D.L. (2005): Technological evolution as creative destruction of process heterogeneity: evidence from US plant-level data. In: Economic Systems Research, Vol. 17. pp. 25-45.

Ethier, W.J. (1979): Internationally Decreasing Costs and World Trade. In: Journal of International Economics, Vol. 9, pp. 1-24.

European Commission (2010): Europe 2020 - A strategy for smart, sustainable and inclusive growth. European Commission, Brussels, available at: http://ec.europa.eu/eu2020/pdf/COMPLET%20EN%20BARROSO%20%20%20007%20-%20Europe%202020%20-%20EN%20version.pdf (last accessed on 17th of August 2010).

Evangelista, R., Sandven, T., Sirilli, G., Smith, K. (1998): Measuring innovation in European Industry. In: International Journal of the Economics of Business, Vol. 5 (3), pp. 311–333.

Evangelista, R. (1999): Knowledge and Investment: The Sources of Innovation in Industry. Cheltenham (UK), Edward Elgar.

Evangelista, R. (2000): Sectoral Patterns of Technological Change in Services. In: Economics of Innovation and New Technology, Vol. 9, 183-221.

Evangelista, R., Iammarino, S., Mastrostefano, V., Silvani, A. (2002): Looking for regional systems of innovation: Evidence from the Italian innovation Survey. In: Regional Studies, Vol. 36 (2), pp. 173-186.

F

Faems, D., van Looy, B., Debackere, K. (2005): The Role of Inter-Organisational Collaboration within Innovation Strategies: towards a portfolio approach. In: Journal of Product Innovation Management, Vol. 22, pp.238-251.

Fagerberg, J. (1987): A Technology Gap Approach to Why Growth Rates Differ. In: Research Policy, Vol. 16, pp. 87-99.

Fagerberg, J. (1988): Why Growth Rates Differ. In: Dosi, G., Freeman, C., Nelson, R., Silverberg, G., Soete, L. (eds.): Technical Change and Economic Theory. London, Pinter Publishers, pp. 432-457.

Fagerberg, J. (1991): Innovation, Catching Up and Growth. In: OECD, Technology and Productivity: The Challenge for Economic Policy. OECD, Paris.

Fagerberg, J. (1994): Technology and International Differences in Growth Rates. In: Journal of Economic Literature, Vol. 32, pp. 1147-1175.

Fagerberg, J. (2003): Schumpeter and the revival of evolutionary economics: an appraisal of the literature. In: Journal of Evolutionary Economics, Vol. 13, pp. 125-159.

Fagerberg, J. (2005): Innovation: A Guide to the Literature. In: Fagerberg, J., Mowery, D., Nelson, R. R., (eds.): The Oxford Handbook of Innovation, Oxford University Press, pp. 1-27.

Fagerberg, J., Godinho, M.M. (2005): Innovation and Catching-Up. In: Fagerberg, J., Mowery, D.C., Nelson, R.R. (eds.): The Oxford Handbook of Innovation. New York, Oxford University Press, pp. 514-542.

Fagerberg, J., Srholec, M., Verspagen, B. (2009): Innovation and Economic Development. UNU-Merit, Working Paper #2009-032, Maastricht, United Nations University, (available at: http://www.merit.unu.edu/publications/wppdf/2009/wp2009-032.pdf, last accessed on 8th of August 2010).

Falk, M. (2009): High-tech Exports and Economic Growth in Industrialized Countries. In: Applied Economic Letters, Vol. 16 (10), pp. 1025-1028.

Federation of British Industries (1947): Scientific and Technical Research in British Industry. London, FBI.

Federation of British Industries (1961): Industrial Research in Manufacturing Industry. London, FBI.

Feigenbaum, A.V. (1983): Quality Control. 3rd Edition, New York, McGraw-Hill.

Felin, T., Hesterly, W.S. (2007): The Knowledge-based View, Nested Heterogeneity, and the New Value Creation: Philosohpical Considerations on the Locus of Knowledge. In: Academy of Management Review, Vol. 32 (1), pp. 195-218.

Filatotchev, I., Piesse, J. (2009): R&D, Internationalization and Growth of Newly Listed Firms: European Evidence. In: Journal of International Business Studies, Vol. 40 (8), pp. 1260-1276.

Filippaios, F., Papanastassiou, M., Pearce, R., Rama, R. (2009): New forms of organisation and R&D internationalisation among the world's 100 largest food and beverages multinationals. In; Research Policy, Vol. 38, pp. 1032-1043.

Financial Times (2008): The Nuts and Bolts of Innovation. Financial Times, 5th of March 2008.

Fisher, F.M., Temin, P. (1973): Returns to Scale in Research and Development: what does the Schumpeterian hypothesis imply? In: Journal of Political Economy, Vol. 81, pp. 56–70.

Flemming, J., Götz, G. (1993): Externalitäten, Nichtkonvexitäten und endogener technischer Fortschritt: Ein Einblick in die Funktionsweise der „neuen" Wachstumstheorie. In: Jahrbuch für Sozialwissenschaft, Vol. 44, Göttingen, Vandenhoeck & Ruprecht, pp. 203-215.

Flor, M.L., Oltra, M.J. (2004): Identification of Innovating Firms Through Technological Innovation Indicators: An Application to the Spanish Ceramic Tile Industry. In: Research Policy, Vol. 33, pp. 323-336.

Foray, D. (1998): The Economics of Knowledge Openness: Emergence, Persistence and Change of Conventions in the Knowledge Systems. In: Lazaric, N., Lorenz, E. (eds.): Trust in Economic Learning, London, Edward Elgar, pp. 162-189.

Foray, D. (2006): The Economics of Knowledge. Cambridge (MA), London, The MIT Press.

Fosfuri, A., Tribo, J.A. (2008): Exploring the Antecedents of Potential Absorptive Capacity and its Impact on Innovation Performance. In: Omega, Vol. 36, pp. 173-187.

Foss, N.J., Knudsen, C., Montgomery, C.A. (1995): An Exploration of Common Ground: Integrating Evolutionary and Strategic Theories of the Firm. In: Montgomery, C.A. (ed.): Resource-Based and Evolutionary Theories of the Firm: Towards a Synthesis. Boston, Dordrecht, London, Kluwer Academic Publishers, pp. 1-18.

Foss, N.J. (1996): Knowledge-based Approaches to the Theory of the Firm: Some Critical Comments. In: Organisation Science, Vol. 7 (5), pp. 470-476.

Foss, N.J. (1997a): The Resource-based Perspective: An Assessment and Diagnosis of Problems. DRUID Working Paper No. 97-1, Danish Research Unit for Industrial Dynamics.

Foss, N.J. (1997b): Equilibrium vs. Evolution in the Resource-based Perspective: The Conflicting Legacies of Demsetz and Penrose. DRUID Working Paper No. 97-10, Danish Research Unit for Industrial Dynamics.

Foss, N.J. (1997c): Resources and Strategy: Problems, Open Issues, and Ways Ahead. In: Foss, N.J. (ed.): Resources, Firms, and Strategies. Oxford, Oxford University Press, pp. 345-365.

Foss, N.J. (1998): The New Growth Theory: Some Intellectual Growth Accounting. In: Journal of Economic Methodology,Vol. 5 (2), pp. 223-246.

Foss, N.J. (2002): Edith Penrose: Economics and Strategic Management. In: Pitelis, C. (ed.): The Growth of the Firm; The Legacy of Edith Penrose. Oxford, Oxford University Press, pp. 147-164.

Foss, N.J., Knudsen, T. (2003): The Resource-Based Tangle: Towards a Sustainable Explanation of Competitive Advantage. In: Managerial and Decision Economics, Vol. 24, pp. 291-307.

Foss, K., Foss, N.J. (2004): The Next Step in the Evolution of the RBV: Integration with Transaction Cost Economics. In: Management Revue, Vol. 15 (1), pp. 107-121.

Foster, J. (2000): Competitive Selection, self-organisation and Joseph A. Schumpeter. In: Journal of Evolutionary Economics, Vol. 10, pp. 311-328.

Freddi, D. (2009): The integration of old and new technological paradigms in low- and medium-tech sectors: The Case of Mechatronics. In: Research Policy, Vol. 38, pp. 548-558.

Freel, M.S. (2000a): Do Small Innovating Firms Outperform Noninnovators? In: Small Business Economics, Vol. 14, pp. 195–210.

Freel, M.S. (2000b): Strategy and Structure in Innovative Manufacturing SMEs: The case of an English region. In: Small Business Economics, Vol. 15, pp. 27–45.

Freeman, C, Fuller, J. K., Young, A. J. (1963): The Plastics Industry: A comparative study of research and innovation. In: National Institute Economic Review, Vol. 26, pp. 22-62.

Freeman, C. (1974): The Economics of Industrial Innovation. Harmondsworth, England, Penguin.

Freeman, C. (1982): The Economics of Industrial Innovation. 2nd Edition, London.

Freeman, C. (1987a): Output Measurement in Science and Technology, Amsterdam, North-Holland.

Freeman, C. (1987b): Technology Policy and Economic Performance: Lessons from Japan, London, Frances Pinter.

Freeman, C. (1988): Introduction. In: Dosi, G., Freeman, C., Nelson, R.R., Silverberg, G., Soete, L. (eds.): Technical Change and Economic Theory. London, New York, pp. 1-8.

Freeman, C. (1990): Schumpeter's Business Cycles' Revisited. In: Heertje, A., Perlmann, M. (eds.): Evolving Technology and Market Structure. Studies in Schumpeterian Economics. Ann Arbor, The University of Michigan Press, pp. 17-38.

Freeman, C. (1994a): The Economic of Technical Change. In: Cambridge Journal of Economics, Vol. 18 (5), pp. 463-514.

Freeman, C. (1994b): Innovation and Growth. In: Dodgson, M., Rothwell, R. (eds.): The Handbook of Industrial Innovation. Gower House et al., Edward Elgar, pp. 78-93.

Freeman, C. (2003): A Schumpeterian Renaissance? In: SPRU Electronic Working Paper Series, No. 102, Falmer, Brighton, University of Sussex, available at: http://www.sussex.ac.uk/spru/documents/sewp102 (last accessed on 12th of august 2010).

Freeman, C., Soete, L. (1997): The Economics of Industrial Innovation. 3rd Edition, Cambridge (MA), MIT Press.

FREEMAN, C., SOETE, L. (2009): Developing Science, Technology and Innovation Indicators: What We Can Learn from the Past. In: Research Policy, Vol. 38, 583-589.

FREILING, J. (2001): Resource-Based View und ökonomische Theorie – Grundlagen und Positionierung des Ressourcenansatzes. Wiesbaden, Gabler.

FREILING, J. (2004): A Competence-Based Theory of the Firm. In: Croucher, R., Kabst, R., Kellerman, R., Matiaske, W. (eds.): Management Revue. The International Review of Management Studies. Special issue: Beyond Resource Based View (edited by Manfred Moldaschl), Vol. 14 (1), pp. 27-52.

FREILING, J., GERSCH, M., GOEKE, C. (2006): Eine "Competence-based Theory of the Firm" als marktprozesstheoretischer Ansatz. In: Schreyögg, G., Conrad, P. (eds.): Managementforschung 16 (2006). Wiesbaden, Gabler, pp. 37-82.

FRENKEL, M., HEMMER, H.R. (1999): Grundlagen der Wachstumstheorie. Vahlen, München.

FRIED, A. (2007): Was erklärt die Resource-based View of the Firm? In: Moldaschl, M. (ed.): Immaterielle Ressourcen. Nachhaltigkeit von Unternehmensführung und Arbeit I, 2nd Edition, Mering, Hampp Verlag, pp. 185-218.

FRIEDMAN, M. (1953): Essays in Positive Economics, Chicago, University of Chicago Press.

G

GAITANIDES, M. (2007): Prozessorganisation: Entwicklung, Ansätze und Programme des Managements von Geschäftsprozessen. 2. Edition, München, Vahlen

GALBRAITH, J. K. (1952): American Capitalism. The concept of countervailing power. Boston, Houghton Mifflin.

GALENDE, J., SUAREZ, I. (1999): A resource-based analysis of the factors determining a firm's R&D activities. In: Research Policy, Vol. 28, pp. 891–905.

GALLIE, E.P., ROUX, P. (2008): Forms and Determinants of R&D Collaborations. New Evidence from French Data. DRUID Working Paper No. 08-15, available at: www3.druid.dk/wp/20080015.pdf (last access on 4th of June 2010).

GALUNIC, D.C., RODAN, S. (1998): Resource Recombinations in the Firm: Knowledge Structures and the Potential for Schumpeterian Innovation. In: Strategic Management Journal, Vol. 19, pp. 1193-1201.

GALUNIC, D.C., EISENHARDT, K.M. (2001): Architectural Innovation and Modular Corporate Forms. In: Academy of Management Journal, Vol. 44 (6), pp. 1229-1249.

GAVETTI, G. (2005): Cognition and Hierarchy: Rethinking the Microfoundations of Capabilities' Development. In: Organisation Science, Vol. 16, pp. 599-617.

GEBAUER, H. (2004): Die Transformation vom Produzenten zum produzierenden Dienstleister. Dissertation Nr. 2873, Bamberg, Difo-Druck GmbH.

GEHRKE, B., KRAWCZYK, O., LEGLER, H. (2007): Forschungs- und wissensintensive Wirtschaftszweige in Deutschland: Außenhandel, Spezialisierung, Beschäftigung und Qualifikationserfordernisse. Aktualisierung und Überarbeitung unter Berücksichtigung der NIW/ISI-Listen 2006. In: Bundesministerium für Bildung und Forschung (BMBF) (ed.): Studien zum deutschen Innovationssystem, Nr. 17-2007, Berlin.

GEHRKE, B., LEGLER, H. (2010): Forschungs- und Wissensintensive Wirtschaftszweige – Außenhandel, Spezialisierung, Produktion, Beschäftigung und Qualifikationserfordernisse in Deutschland, Studien zum deutschen Innovationssystem Nr. 04-2010, Hannover.

GEISENDORF, S. (2004): Der ökonomische Evolutionsbegriff: Selbstorganisation und Selektion. In: Lehmann-Waffenschmidt, M., Ebner, A., Forndahl, D. (eds): Institutioneller Wandel, Marktprozesse und dynamische Wirtschaftspolitik. Perspektiven der Evolutorischen Ökonomik. Marburg, Metropolis, pp.79-96.

GEORGESCU-ROEGEN, N. (1971): The Entropy Law and the Economic Process. Cambridge (MA), Harvard University Press.

GEROSKI, P. (1995): Markets for Technology: Knowledge, Innovation, and Appropriability. In: Stoneman, P. (ed.): Handbook of the Economics of Innovation and Technological Change. Oxford, Blackwell, pp. 89-131.

GERSCH, M. (1998): Vernetze Geschäftsbeziehungen: die Nutzung von EDI als Instrument des Geschäftsbeziehungsmanagement. Wiesbaden, Gabler.

GERSCH, M., FREILING, M., GOEKE, C. (2005): Grundlagen einer "Competence-based Theory of the Firm". Die Chance zur Schließung einer "Realisierungslücke" innerhalb der Marktprozesstheorie. Arbeitsbericht Nr. 100, Institut für Unternehmensführung, Ruhr-Universität Bochum.

GERYBADZE, A. (1982): Innovation, Wettbewerb und Evolution. Tübingen, JCB Mohr.

GIDDENS, A. (1976): New Rules of Sociological Method: A Positive Criticism of Interpretative Sociologies. London, Hutchinson.

GIDDENS, A., (1984): The Constitution of Society: Outline of the Theory of Structuration. Cambridge, London, Polity Press.

GIRMA, S., GREENAWAY, D., KNELLER. R. (2004): Entry to export markets and productivity. A microeconometric analysis of matched firms. In: Review of International Economics, Vol. 12, pp. 855–866.

GIRMA, S., GÖRG, H., HANLEY, A. (2008): R&D and Exporting: A Comparison of British and Irish Firms. In: Review of World Economics, Vol. 144 (4), pp. 750-773.

GLASERSFELD, E.V. (1987): Wissen, Sprache und Wirklichkeit. Arbeiten zum Konstruktivismus. Braunschweig, Wiesbaden, Vieweg.

GODFREY-SMITH, P. (2000): The Replicator in Retrospect. In: Biology and Philosophy, Vol. 15, pp. 403-423.

GODIN, B. (2001a): Defining R&D: Is Research Always Systematic? Project on the History and Sociology of S&T Statistics. Paper No. 7, Montreal, Canada, available at www.csiic.ca (last access on 1st of June 2010).

GODIN, B. (2001b): The Emergence of Science and Technology Indicators: Why Did Governments Supplements Statistics With Indicators? Project on the History and Sociology of S&T Statistics, Paper No. 8, Montreal, Canada, available at: www.csiic.ca (last access on 1st of June 2010).

GODIN, B. (2001c): Tradition and Innovation: The Historical Contingency of R&D Statistical Classifications. Project on the History and Sociology of S&T Statistics. Working Paper No. 11, Montreal, Canada.

GODIN, B. (2004): The Obsession for Competitiveness and its Impact on Statistics: the Construction of High-Technology Indicators. Working Paper No. 25, Project on the History and Sociology of S&T Statistics, Montréal, Canadian Science and Innovation Indicators Consortium (CSIIC).

GODIN, B. (2006a): The linear model of innovation: The historical construction of an analytical framework. Project on the history and sociology of S&T statistics. In: Science, Technology, and Human Values, Vol. 31 (6), pp. 639-667.

GODIN, B. (2006b): Research and Development: How the 'D' got into R&D. In: Science and Public Policy, Vol. 33 (1), pp. 59-76.

GOMULKA, S. (1971): Inventive Activity, Diffusion and Stages of Economic Growth. Skrifter fra Aarhus Universitets Okonomiske institute, Vol. 24, Aarhuis.

GORDON, R.A., HOWELL, J.E. (1959): Higher Education for Business. New York, Columbia University Press.

GOSHAL, S., HAHN, M., MORAN, P. (2002): Management Competence, Firm Growth and Economic Progress. In: Pitelis, C. (ed.): The Growth of the Firm; The Legacy of Edith Penrose. Oxford University Press, pp. 279-309.

GOTO, A., SUZUKI, K. (1989): R&D Capital, Rate of Return on R&D Investment and Spillover of R&D in Japanese Manufacturing Industries. In: The Review of Economics and Statistics, Vol. 71, pp. 555–564.

GOULD, S. (2002): The Structure of Evolutionary Theory. Cambridge, MA, London.

GOURLAY, S. (2006): Conceptualizing Knowledge Creation: A Criticism of Nonaka's Theory. In: Journal of Management Studies, Vol. 43 (7), pp. 1415-1436.

GRANT, R.M. (1991): The Resource-Based Theory of Competitive Advantage: Implications for Strategy Formulation. In: California Management Review, Vol. 33 (3), pp. 114-135.

GRANT, R.M. (1996a): Towards a Knowledge-Based Theory of the Firm. In: Strategic Management Journal, Vol. 17 (Special Issue), pp. 109-122.

GRANT, R.M. (1996b): Prospering in Dynamically-Competitive Environments: Organisational Capability as Knowledge Integration. In: Organisation Science, Vol. 7, pp. 375-387.

GRANT, R.M. (2002): The Knowledge-Based View of the Firm. In: Choo, C.W., Bontis, N. (eds.): The Strategic Management of Intellectual Capital and Organisational Knowledge: A Collection of Readings. Oxford, Oxford University Press, pp. 133-148.

GRANT, K.A. (2007): Tacit Knowledge Revisited – We Can Still Learn from Polanyi. In: The Electronic Journal of Knowledge Management, Vol. 5 (2), pp. 173 - 180, available online at www.ejkm.com (last access on 29th of March 2010).

GRAVES, S. B., LANGOWITZ, N. S. (1996): R&D Productivity: a Global Multi-Industry Comparison. In: Technological Forecasting and Social Change, Vol. 53, 125-137.

GREENAN, N., GUELLEC, D. (2000): Technological Innovation and Employment Reallocation. In: Labour, Vol. 14 (4), pp. 547-590.

GREENAWAY, D. KNELLER, R. (2007): Firm heterogeneity, exporting and foreign direct investment: A survey. In: The Economic Journal, Vol. 117, pp. F134-F161.

GREENLEE, P., CASSIMAN, B. (1999): Product Market Objectives and the Formation of Research Joint Ventures. In: Managerial and Decision Economics, Vol. 20 (3), pp.115–130.

GRIEBEN, W.H. (2001): Paul Michael Romer. Die neue Wachstumstheorie und die Aufgaben moderner Wirtschaftspolitik. Mimeo.

GRILLICHES, Z. (1957): Hybrid Corn: An Exploration in the Economics of Technological Change. In: Econometrica, Vol. 25, pp. 501-522.

GRILICHES Z. (1958): Research Cost and Social Return: Hybrid Corn and Related Innovations. In: Journal of Political Economy, Vol. 66 (5), pp. 419-31.

GRILICHES Z. (1964): Research Expenditures, Education and the Aggregate Agricultural Production Function. In: American Economic Review, Vol. 54 (6), pp. 961-74.

GRILICHES Z. (1979): Issues in Assessing the Contribution of R&D to Productivity Growth. In: Bell Journal of Economics, Vol. 10 (1), pp. 92-116.

Griliches Z. (1986): Productivity, R&D and Basic Research at the Firm Level in the 1970s. In: American Economic Review, Vol. 76 (1), pp. 143-154.

Griliches, Z. (1984): R&D, Patents and Productivity, Chicago, Chicago University Press.

Griliches Z. (1988): Productivity Puzzles and R&D: Another Nonexplanation, In: Journal of Economic Perspectives, Vol. 2 (4), pp. 9-21.

Griliches, Z. (1990): Patent Statistics as Economic Indicators: A survey. In: Journal of Economic Literature, Vol. 28 (4), pp. 1661-1707.

Griliches, Z. (1991): The Search for R&D Spillovers. NBER Working Paper Series, Working Paper No. 3768, Cambridge (MA), National Bureau of Economic Research.

Griliches, Z. (1995): R&D and Productivity: Econometric results and measurement issues. In: Stoneman, P. (ed.): Handbook of the Economics of Innovation and Technological Change. Oxford (UK), Cambridge (MA), Blackwell, pp. 53-89.

Griliches Z., Mairesse, J. (1984): Productivity and R&D at the Firm Level. In: Griliches, Z. (ed.): R&D, Patents and Productivity, Chicago: University of Chicago Press, pp. 339-374.

Grimpke, C., Sofka, W. (2009): Search Patterns and Absorptive Capacity: Low- and High-Technology Sectors in European Countries. In: Research Policy, Vol. 38, pp. 495-506.

Grochla, E. (1978): Einführung in die Organisationstheorie. Stuttgart, Poeschel.

Grossekettler, H. (1985): Wettbewerbstheorie. In: Borchert, M., Grossekettler, H. (eds.): Preis- und Wettbewerbstheorie. Marktprozesse als analytisches Problem und ordnungspolitische Gestaltungsaufgabe. Stuttgart, Verlag W. Kohlhammer, pp. 113-335.

Grossman, G.M., Helpman, E. (1990): Comparative Advantage and Long-Run Growth. In: American Economic Review, Vol. 80 (4), pp. 796-815.

Grossman, G.M., Helpman, E. (1991a): Innovation and Growth in the Global Economy. Cambridge, MIT-Press.

Grossman, G.M., Helpman, E. (1991b): Quality Ladders in the Theory of Growth. In: Review of Economic Studies, Vol. LVIII, pp. 43-61.

Grossman, G.M., Helpman, E. (1994): Endogenous Innovation in the Theory of Growth. In: Journal of Economic Perspectives, Vol. 8, pp. 23-44.

Groth, C., Schou, P. (2002): Can Non-Renewable Resources Alleviate the Knife-Edge-Character of Endogenous Growth? In: Oxford Economic Papers, Vol. 54 (3), pp. 386-411.

Grunfeld, L.A. (2004): The Multiple Faces of R&D: Absorptive capacity effects and Different sources of spillovers. Norwegian Institute of International Affairs.

Grupp, H. (1992): Dynamics of Science-Based Innovation, Berlin, Springer-Verlag.

Grupp, H. (1997): Messung und Erklärung des technischen Wandels. Grundzüge einer empirischen Innovationsökonomik. Berlin, Heidelberg, New York, Springer.

Grupp, H. (2008): Critical comments on the 'moral economy of technology indicators'. In: Hirsch-Kreinsen, H., Jacobson, D. (eds.): Innovation in Low-Tech Firms and Industries. Cheltenham (UK), Northampton (MA), Edwaed Elgar, pp. 85-90.

Gulati, R. (1998): Alliances and Networks. In: Strategic Management Journal, Vol. 19 (4), pp. 293-317.

Gulati R, Nohria N, Zaheer A. (2000): Strategic networks. In: Strategic Management Journal, Vol. 21, pp. 203–215.

Gupta, A.K., Govindarajan, V. (2000): Knowledge Flow Within Multinational Corporations. In: Strategic Management Journal, Vol. 21, pp. 473-496.

H

Haas, M., Hansen, M.T. (2005): When Using Knowledge Can Hurt Performance: The Value of Organisational Capabilities in a Management Consulting Company. In: Strategic Management Journal, Vol. 26 (1), pp. 1-24.

Ha, J., Howitt, P. (2007): Accounting for Trends in Productivity and R&D: A Schumpetrian Criticism of Semi-Endogenous Growth Theory. In: Journal of Money, Credit and Banking, Vol. 39 (4), pp. 733-774.

Habakkuk, H.J. (1962): American and British Technology in the Nineteenth Century. Cambrige, Cambridge University Press.

Hahn, K. (2007): Der Lissabon-Prozess: Das Innovationskonzept und die Auswirkungen auf die Politikgestaltung. Soziologisches Arbeitspapier Nr. 20/2007, Technische Universität Dortmund.

Hahn, K. (2009): Der Lissabon-Prozess: Warum eine Hightech-Strategie zur Innovationsförderung nicht ausreicht. In: WSI Mitteilungen, Vol. 6/2009, pp. 302-308.

Hall B.H. (1993): Industrial Research during the 1980s: Did the Rate of Return Fall? In: Brookings Paper on Economic Activity, Microeconomics, 1993 (2), pp. 289-330.

Hall, B.H. (2002): The Financing of Research and Development. In: Oxford Review of Economic Policy, Vol. 18, pp. 35–51.

Hall, B.H., Mairesse, J. (1995): Exploring the Relationship Between R&D and Productivity in French Manufacturing Firms. In: Journal of Econometrics, Vol. 65, pp.263–293.

Hall, L.A., Bagchi-Sen, S. (2002): A Study of R&D, Innovation and Business Performance in the Canadian Biotechnology Industry. In: Technovation, Vol. 22, pp. 231-244.

Hall, B.H., Lotti, F., Mairesse, J. (2009): Innovation and Productivity in SMEs: Empirical Evidence for Italy. in: Small Business Economics, Vol. 33, pp. 13–33.

Hall, B.H., Mairesse, J., Mohnen, P. (2010): Measuring the Returns to R&D. UNU-MERIT Working Papers, No. 2010-006, available at: www.merit.unu.edu/publications/wppdf/2010/wp2010-006.pdf (last access on 2nd of June 2010).

Hamel, G. (1991): Competition for Competence and Interpartner Learning within International Strategic Alliances. In: Strategic Management Journal, Vol. 12, Special Issue, Summer, pp. 83-104.

Hamel, G., Prahalad, C.K. (1997): Wettlauf um die Zukunft. Wie Sie mit bahnbrechenden Strategien die Kontrolle über Ihre Branche gewinnen und die Märkte von morgen schaffen. 2nd Edition, Wien, Wirtschaftsverlag Carl Ueberreuter.

Hammann, P., Freiling, J. (2000): Die Ressourcen- und Kompetenzperspektive des Strategischen Managements. Wiesbaden, Gabler.

Hannan, M.T., Freeman, J. (1984): Structural inertia and organisational change. In: American Sociology Review, Vol. 49, pp. 149-164.

Hansen, G.S., Wernerfelt, B. (1989): Determinants of Firm Performance: The Relative Importance of Economics and Organisational Factors. In: Strategic Management Journal, Vol. 10, pp. 399-411.

Hansen, M., Hoskisson, B., Barney, J. (1997): Trustworthiness: A Cooperative Resource. Working Paper, Brigham Young University, Provo, UT.

Hansen, P. A., Serin, G. (1997): Will Low Technology Products Disappear? The Hidden Innovation Process in Low Technology Industries. In: Technology Forecasting and Social Change, Vol. 55 (2), pp. 179-191.

Hansen, M.T., Nohria N., Tierney T. (1999): What's Your Strategy for Managing Knowledge? In: Harvard Business Review, Vol. 77 (2), pp. 106-116.

Hanusch, H., Cantner, U. (1993): Neuere Ansätze in der Innovationstheorie und der Theorie des Technischen Wandels – Konsequenzen für eine Industrie- und Technologiepolitik. In: Meyer-Krahmer, F. (ed.): Innovations- und Technologiepolitik. Forschungsansätze und politische Konsequenzen. Schriftenreihe des Fraunhofer-Instituts für Systemtechnik und Innovationsforschung (ISI), Heidelberg, Physica-Verlag, pp. 11-46.

Harhoff, D. (1998): R&D and Productivity in German Manufacturing Firms. In: Economic of Innovation and New Technology, Vol. 6, pp. 29-49.

Harris, R., Li, Q.C. (2009): Exporting, R&D, and Absorptive Capacity in UK Establishments. In: Oxford Economic Papers - New Series, Vol. 61 (1), pp. 74-103.

Harrod. R.F. (1954): Towards a Dynamic Economics. London.

Hauknes, J., Knell, M. (2009): Embodied knowledge and sectoral linkages: An input-output approach to the interaction of high- and low-tech industries. In: Research Policy, Vol. 38, pp. 459-469.

Heckman, J.J. (1976): A Life-Cycle Model of Earnings, Learning, and Consumption. In: Journal of Political Economy, Vol. 84 (4), pp. 11-44.

Heckscher, E.F. (1919): The Effect of Foreign Trade on the Distribution of Income. Originally in: Ekonomisk Tidskrift, Vol. 21 (2), pp. 1-32, reprinted in: AEA (1950): Readings in the Theory of International Trade. Philadelphia, Blakiston Co, p. 272-300.

Heidenreich, M. (2009): Innovation patterns and location of European low- and medium-technology industries. In: Research Policy, Vol. 38, pp. 483-494.

Helfat, C.E. (1994): Firm-specificity in Corporate Applied R&D. In: Organisation Science, Vol. 5 (2), pp. 173-184.

Helfat, C. E., Peteraf, M. A. (2003): The Dynamic Resource-Based View. Capability Lifecycles. In: Strategic Management Journal, Vol. 24, pp. 997-1010.

Helfat, C.E., Peteraf, M.A. (2009): Understanding Dynamic Capabilities: Progress Along a Developmental Path. In: Strategic Organisation, Vol. 7, pp. 91-102.

Helfat, C.E., Finkelstein, S., Mitchell, W., Peteraf, M.A., Singh, H., Teece, D.J., Winter, S.G. (2007): Dynamic Capabilities. Understanding Change in Organisations. Oxford: Blackwell.

Helleloid, D.A., Simonin, B.L. (1994): Organisation Learning and a Firm's Core Competence. In: Hamel, G., Heene, A. (eds.): Competence-based Competition. Chichester, New York, John Wiley & Sons, pp. 213-239.

Helpman, E. (1994): Technical Progress and Economic Calculus. In: Pasinetti, L.L., Solow, M. (eds.): Economic Growth and the Structure of Long-Term Development. London, Macmillan Press, pp. 369-372.

Henderson, R.M., Clark, K.B. (1990): Architectural Innovation: The Reconfiguration of Existing Product Technologies and the Failure of Established Firms. In: Administrative Science Quartely, Vol. 35 (1), pp. 9-30.

Herrman-Pillath, C. (2002): Grundriss der Evolutionsökonomik. München, Wilhelm Fink Verlag.

Hessels, S.J.A. (2007): Innovation and international involvement of Dutch SMEs. In: International Journal of Entrepreneurship and Small Business, Vol. 4 (3), pp. 234–255.

Hill, C.W.L., Jones, G.R. (1992): Strategic Management Theory: An Integrated Approach. Boston, Houghton, Mifflin.

Hipp, C. (2000): Information Flows and Knowledge Creation in Knowledge-Intensive Business Services: Scheme for a Conceptualization. In: Metcalfe, J.S., Miles, I. (eds.): Innovation Systems in the Service Economy, Kluwer Academic Publishers, Boston, Dordrecht, London, pp. 149-167.

Hipp, C., Grupp, H. (2005): Innovation in the service sector: The demand for service-specific innovation measurement concepts and typologies. In: Research Policy, Vol. 34 (4), pp. 517-535.

Hippel, E., von (1988): The Sources of Innovation. New York, Oxford University Press.

Hirsch, S. (1965): The United States Electronics Industry in International Trade. In: National Institute Economic Review, November, pp. 39-60.

Hirsch, S., Bijaoui, I. (1985): R&D intensity and Export Performance: A micro view. In: Weltwirtschaftliches Archiv, Vol. 121 (2), pp. 238-251.

Hirsch-Kreinsen, H. (2004): "Low-Technology" – Ein innovationspolitisch vergessener Sektor. In: Hirsch-Kreinsen, H., Weyer, J. (eds.): Soziologisches Arbeitspapier Nr. 2/2004, Technische Universität Dortmund.

Hirsch-Kreinsen, H. (2007): "Lowtech". Innovationsmuster und Entwicklungschancen. In: Abel, J., Hirsch-Kreinsen, H. (eds.): Lowtech-Unternehmen am Hightech-Standort. Berlin, edition sigma, pp. 253-280.

Hirsch-Kreinsen, H. (2008a): "Low-Tech" Innovations. In: Industry and Innovation, Vol. 15 (1), pp.19-43.

Hirsch-Kreinsen, H. (2008b): Innovationspolitik: Die Hightech-Obsession. In: Hirsch-Kreinsen, H., Weyer, J. (eds.): Soziologisches Arbeitspapier Nr. 22/2008, Technische Universität Dortmund.

Hirsch-Kreinsen, H., Jacobson, D., Laestadius, S., Smith, K. (2003): Low-Tech Industries and the Knowledge Economy. State of the Art and Research Challenges. EU 5th Framework project "Pilot: Policy and Innovation in Low-tech".

Hirsch-Kreinsen, H., Jacobson, D., Laestadius, S., Smith, K. (2005): Low-Tech Industries in the Knowledge Economy. The Analytical Isssues. In: Hirsch-Kreinsen, H., Jacobson, D., Laestadius, S. (eds.): Low-tech Innovation in the Knowledge Economy. Frankfurt a. M., Peter Lang, pp. 11-31.

Hirsch-Kreinsen, H., Jacobson, D., Robertson, P.L. (2006): 'Low-Tech' Industries: Innovativeness and Development Perspectives – A Summary of a European Research Project. In: Prometheus, Vol. 24 (1), pp. 3-21.

Hitt, M.A., Ireland, R.D. (1985): Corporate Distinctive Competence , Strategy, Industry and Performance. In: Strategic Management Journal, Vol. 6 (3), pp. 273-293.

Hitt, M.A., Irleland, R.D. (1986): Relationships among Corporate Level Distinctive Competencies, Diversification Strategy, Corporate Strategy and Performance. In: Journal of Management Studies, Vol. 23, pp. 401-416.

Hitt, M.A., Ireland, R.D., Hoskisson, R.E. (1997): Strategic Management: Competitiveness and Globalization. 2nd Edition, New York, West.

Hodgson, G.M. (1993): Economics and evolution. Bringing life back into economics. Cambridge UK, Polity Press.

Hodgson, G.M. (1995): The Evolution of Evolutionary Economics. In: Scottish Journal of Political Economy, Vol. 42, pp. 469-488.

Hodgson, G.M. (1996): The Challenge of Evolutionary Economics. In: Journal of Institutional and Theoretical Economics, Vol. 152, pp. 697-706.

Hodgson, G.M. (1997): The Evolutionary and non-Darwinian economics of Joseph Schumpeter. In: Journal of Evolutionary Economics, Vol. 7, pp. 131-145.

Hodgson, G.M. (2002): Darwinism in economics: from analogy to ontology. In: Journal of Evolutionary Economics, Vol. 12, pp. 259-281.

Hodgson, G.M. (2004a): The Evolution of Institutional Economics: Agency, Structure and Darwinism in American Institutionalism. London, New York, Routledge.

Hodgson, G.M. (2004b): Darwinian Economic Evolution and the Concept of the Routine. In: Lehmann-Waffenschmidt, M., Ebner, A., Forndahl, D. (eds.): Institutioneller Wandel, Marktprozesse und dynamische Wirtschaftspolitik. Metropolis-Verlag, Marburg, pp. 53-78.

Hodgson, G.M., Knudsen, T. (2006): Why we need a generalized Darwinism, and why generalized Darwinism is not enough. In: Journal of Economic Behaviour and Organisation, Vol. 61, pp. 1-19.

Hollanders, H. (1965): The Sources of Increased Efficiency: A Study of the Du Pont Rayon Plants. Cambridge (MA), MIT Press.

Hollenstein, H. (2002): Innovation Modes in the Swiss Service Sector: a Cluster Analysis Based on Firm level Data. In: Research Policy, Vol. 32 (5), pp. 715–885.

Hollenstein, H. (2003): Innovation modes in the Swiss service sector: a cluster analysis based on firm level data. In: Research Policy, Vol. 32, pp. 845–863.

Holm, K. (1970): Gültigkeit von Skalen und Indizes. In: Kölner Zeitschrift für Soziologie und Sozialpsychologie, Vol. 22, pp. 693-714.

Holmstrom, B. (1989): Agency Costs and Innovation. In: Journal of Economic Behavior and Organisation, Vol. 12, pp. 305-327.

Hoskisson, R., Hitt, M., Wan, W., Yiu, D. (1999): Theory and Research in Strategic Management: Swings of A Pendulum. In: Journal of Management, Vol. 25 (3), pp. 417-456.

Hounshell, D.A. (2000): The Medium is the Message, or How Context Matters: The RAND Corporation Builds an Economics of Innovation, 1946-1962. In Hughes, T.P., Hughes, A. (eds.): Systems, Experts, and Computers. The Systems Approach in Management and Engineering. Cambridge (Mass.), MIT Press, pp. 255-310.

Hrebiniak, L.G., Snow, C.C. (1982): Top-Management Agreement and Organisational Performance. In: Human Relations, Vol. 35 (12), pp. 1139-1157.

Huang, Y.A., Chung, H.J., Lin, C. (2009): R&D Sourcing Strategies: Determinants and consequences. In: Technovation, Vol. 29, pp. 155-169.

Huang, C., Arundel, A., Hollanders, H. (2010): How Firms Innovate: R&D, Non-R&D, and Technology Adoption. UNU-Merit Working Paper #2010-027. Maastricht, United Nations University, available at: http://www.merit.unu.edu/publications/wppdf/2010/wp2010-027.pdf (last accessed on 23th of July 2010).

Huber, G.P. (1991): Organisational Learning: the Contributing Processes and Literatures. In: Organisation Science, Vol. 2, pp. 88-115.

Hübner, H. (2002): Integratives Innovationsmanagement: Nachhaltigkeit als Herausforderung für ganzheitliche Erneuerungsprozesse. Berlin, Erich Schmidt Verlag.

Hull, D.L. (1988): Science as a Process: An Evolutionary Account of the Social and Conceptual Development of Science. University of Chicago Press, Chicago.

Humphreys, J. (1996): Lamarck and the general theory of evolution. In: Journal of Biological Education, Vol. 30 (4), pp. 295-303.

Hunt, S.D. (2000): A General Theory of Competition: Resources, Competences, Productivity, Economic Growth. London, Thousand Oaks.

I

Iizuka, M. (2009): "Low-Tech" industry: a new path for development? The case of the salmon farming industry in Chile. In: Malerba, F., Mani, S. (eds.): Sectoral Systems of Innovation and Production in Developing Countries. Actors, Structure and Evolution. Cheltenham, Northampton, Edward Elgar, pp. 232-258.

Inada, K.I. (1963): On a Two-Sector Model of Economic Growth: Comments and a Generalization". In: Review of Economic Studies, Vol. 30, pp. 119-127.

Imai, M. (1986): Kaizen: The Key to Japan's Competitive Success. New York, Random House.

Inkpen, A. (2000): Learning Through Joint Ventures: a Framework of Knowledge Acquisition. In: Journal of Management Studies, Vol. 37 (7), pp. 1019–1043.

Inkpen, A.C., Currall, S.C. (2004): The Coevolution of Trust, Control, and Learning in Joint Ventures. In: Organisation Science, Vol. 15 (5), pp. 586-599.

Ishikawa, K. (1985): What is Total Quality Control? The Japanese Way. Englewood Cliffs (NJ), Prentice-Hall.

Itami, H. (1987): Mobilizing Invisible Assets. Cambridge (MA), Harvard University Press.

Iwai, K. (2000): A Contribution to the Evolutionary Theory of Innovation, Imitation and Growth. In: Journal of Economic Behavior & Organisation, Vol. 43, pp. 167-168.

J

Jacobson, D., Musyck, B. (2008): LMT innovations in a high-tech environment: human factor 'tools' for the airline industry. In: Hirsch-Kreinsen, H., Jacobson, D. (eds.): Innovation in Low-Tech Firms and Industries. Cheltenham (UK), Northampton (MA), Edward Elgar, pp. 118-139.

Jaeger, A., Maloca, S. (2009): Dokumentation der Umfrage Modernisierung der Produktion 2009 des Fraunhofer-Instituts für System - und Innovationsforschung. Karlsruhe. Available on request at: www.isi.fraunhofer.de/isi-en/i/projekte/erhebung_pi2.php.

Janz, N., Lööf, H., Peters, B. (2003): Firm Level Innovation and Productivity - is there a common story across countries? Mannheim: ZEW-Working Paper dp0326.

Jensen, M., Johnson, B., Lorenz, E., Lundvall, B.A. (2007): Forms of knowledge and modes of innovation. In: Research Policy, Vol. 36, pp. 680-693.

Jirjahn, U., Kraft, K. (2006): Do Spillovers Stimulate Incremental or Drastic Product Innovations? Hypotheses and evidence from German establishment data. ZEW Discussion Paper No. 06-023.

Jones, C.I. (1995a): Time Series Tests of Endogenous Growth Models. In: The Quarterly Journal of Economics, Vol. 110, pp. 495-525.

Jones, C.I. (1995b): R&D-based Models of Economic Growth. In: The Journal of Political Economy, Vol. 103, pp. 759-784.

Juran, J. M. (1988): Juran on Planning for Quality. New York, The Free Press.

Juran, J.M., Gyrna, F. M. (1993): Quality Planning for Analysis: From Product Development Through Use. New York, McGraw-Hill.

K

Kaldor, N. (1957): A model of Economic Growth. In: Economic Journal, Vol. 67, pp. 591-624.

Kaldor, N., Mirrlees, J. (1962): A New Model of Economic Growth. In: Review of Economic Studies, Vol. 29 (3), pp. 174-192.

Kale, P., Dyer, J.H., Singh, H. (2002): Alliance Capability, Stock Market Response, and Long Term Alliance Success: The role of alliance function. In: Strategic Management Journal, Vol. 23 (8), pp. 747–767.

Kamien, M.I., Schwartz, N.L. (1982): Market Structure and Innovation. Cambridge (Mass.).

Kantzenbach, E. (1967): Die Funktionsfähigkeit des Wettbewerbs. 2. Auflage, Göttingen.

Katsoulacos, Y.S. (1986): The Employment Effect of Technical Change. University of Nebraska Press.

Katsoulakos,Y., Ulph, D. (1998): Endogenous Spillovers and the Performance of Research Joint Ventures. In: Journal of Industrial Economics, Vol. 46, pp.333–354.

Kay, J. (1996): Oh Professor Porter, Whatever Did You Do? In: Financial Times, No. 10, May, 17.

Keizer, J.A., Dijkstra, L., Halman, J.I.M. (2002): Explaining Innovative Efforts of SMEs. An exploratory survey among SMEs in the mechanical and electrical engineering sector in the Netherlands. In: Technovation, Vol. 22, pp. 1–13.

Kemp, R.G.M., Folkeringa, M., De Jong, J.P.J., Wubben, E.F.M. (2003): Innovation and Firm Performance: Scientific Analysis of Entrepreneurship and SMEs (SCALES). Research Report H200207: EIM, Zoetermeer.

Keynes, J.M. (1923): A Tract on Monetary Reform. London, Macmillan.

Keynes, J.M. (1936): The General Theory of Employment, Interest and Money. London, Macmillan

Keynes, J.M. (1937): The General Theory of Employment. In: The Quartely Journal of Economics, Vol. 51 (2), pp. 209-223.

Keynes, J.M. (1939): Relative Movements of Real Wages and Output. In: Economic Journal, Vol. 49 (193), pp. 34-51.

Khandwalla, P. N. (1977): The design of organisations. New York, Harcourt Brace Jovanivich.

Kim, H.E., Pennings, J.M. (2009): Innovation and Strategic Renewal in Mature Markets: A Study of the Tennis Racket Industry. In: Organisation Science, Vol. 20 (2), pp. 368-383.

Kim, L., Nelson, R.R. (2000): Technology, Learning and Innovation: Experiences of Newly Industrialized Economies. Cambridge, Cambridge University Press.

Kimura, M. (1983): The Neutral Theory of Molecular Evolution. Cambridge University Press, Cambridge.

King, R.G., Rebelo, S. (1987): Business Cycles with Endogenous Growth. Working Paper, Rochester, New York, University of Rochester.

KING, A.W., ZEITHAML, C.P. (2001): Competencies and Firm Performance: Examining the Causal Ambiguity Paradox. In: Strategic Management Journal, Vol. 22 (1), pp. 75-99.

KINKEL, S., LAY, G., WENGEL, J. (2004): Innovation: Mehr als Forschung und Entwicklung. Mitteilungen aus der Produktionsinnovationserhebung, Nr. 33, Fraunhofer ISI, Karlsruhe.

KINKEL, S., SOM, O. (2010): Internal and External R&D Collaboration as Drivers oft he Product Innovativeness oft he German Mechanical Engineering Industry. In: International Journal of Product Development, Vol. 12 (1), pp. 6-20.

KIRNER, E. (2005): Arbeit in neuen Unternehmensstrukturen. Eine empirische Untersuchung zu Arbeit in Kooperationen produzierender Unternehmen unter besonderer Berücksichtigung von "Boundary Spanners". Stuttgart, Fraunhofer Verlag (IRB).

KIRNER, E., SOM, O., DREHER, C., WIESENMAIER, V. (2006): Innovation in KMU. Der ganzheitliche Innovationsansatz und die Bedeutung von Innovationsroutinen für den Innovationsprozess. Karlsruhe, Fraunhofer Institut für System- und Innovationsforschung.

KIRNER, E., KINKEL, S., JÄGER, A. (2007): Innovationspfade von Low-, Medium- und High-Tech-Unternehmen in der deutschen Industrie. In: Abel, J., Hirsch-Kreinsen, H. (eds.): Lowtech, Unternehmen am Hightech-Standort. Berlin, Sigma, pp.165-192.

KIRNER, E., LAY, G., KINKEL, S. (2008): The relevance of services for high-, medium- and low-tech firms – an empirical analysis in German industry. In: Hirsch-Kreinsen, H., Jacobson, D. (eds.): Innovation in Low-Tech Firms and Industries. Cheltenham, Northampton, Edward Elgar, pp. 175-198.

KIRNER, E., KINKEL, S., JAEGER, A. (2009a): Innovation Paths and the Innovation Performance of Low-Technology Firms – An Empirical Analysis of German Industry. In: Research Policy, Vol. 38, pp. 447-458.

KIRNER, E., SOM, O., JÄGER, A. (2009b): Vernetzungsmuster und Innovationsverhalten von nicht forschungsintensiven Betrieben. Empirische Ergebnisse aus der deutschen Industrie. Stuttgart, Karlsruhe, Fraunhofer-Verlag.

KITCH, E. (1977): The Nature and Function of the Patent System. In: Journal of Law and Economics, Vol. 20, pp. 265-290.

KLEIN, B., CRAWFORD, R.G., ALCHIAN, A.A. (1978): Vertical Integration, Appropriable Rents and the Competitive Contracting Process. In: Journal of Law and Economics, Vol. 21, pp. 297-328.

KLEINKNECHT, A. (1987): Measuring R&D in Small Firms: How Much Are We Missing? In: The Journal of Industrial Economic, Vol. 36 (2), pp. 253-256.

KLEINKNECHT, A. (1989): Firm Size and Innovation. Observations in Dutch manufacturing Industries. In: Small Business Economics, Vol. 1, pp. 215–222.

KLEINKNECHT, A. (1996): Determinants of Innovation, Macmillan Press, London.

KLEINKNECHT, A., VAN MONTFORT, K., BROUWER, E. (2002): The Non-Trivial Choice Between Innovation Indicators. In: Economics of Innovation and New Technology, Vol. 11 (2), pp. 109-121.

KLEPPER, S. (1997): Industry Life Cycles. In: Industrial and Corporate Change, Vol. 6 (1), pp. 145-182.

KLETTE T.J., GRILICHES, Z. (1998): Empirical Patterns of Firm Growth and R&D Investment: A quality ladder model interpretation. In: Economic Journal, Vol. 110 (463), pp. 363-387.

KLETTE, T.J., JOHANSEN, F. (1998): Accumulation of R&D Capital and Dynamic Firm Performance: A Not-So-Fixed Effect Model. In: Annales de Economie et de Statistique, Vol. 49/50, pp. 389–419.

KLETTE, T.J., KORTUM, S. (2004): Innovating Firms and Aggregate Innovation. In: Journal of Political Economy, Vol. 112 (5), pp. 986-1018.

Kline, S. J., Rosenberg, N. (1986): An Overview of Innovation. In: Landau, R., Rosenberg, N., (eds.): The Positive Sum Strategy. Harnessing Technology for Economic Growth. Washington D.C., National Academy Press, pp. 275-305.

Klodt, H. (1990): Technologietransfer und internationale Wettbewerbsfähigkeit. In: Außenwirtschaft, Vol. 45 (1), pp. 57-79.

Knaese, B. (1996): Kernkompetenzen im strategischen Management von Banken. Wiesbaden, Gabler.

Knudsen, T. (2001): Nesting Lamarckism with Darwinian Explanations: Necessity in Economics and Possibility in Biology? In: Laurent, J., Nightingale, J. (eds.): Darwinism and Evolutionary Economics. Cheltenham, Northampton (MA), pp. 121-159.

Knudsen, T. (2002): Economic Selection Theory. In: Journal of Evolutionary Economics, Vol. 12, pp. 443-470.

Köhler, H.D. (2008): Profit and Innovation Strategies in Low-Tech Firms. In: Estudios De Economía Aplicada, Vol. 26 (3), pp. 73-88.

König, H., Buscher, H.S., Licht, G. (1995): Employment, Investment and Innovation at the Firm Level. In: OECD (ed.): The OECD Jobs Study: Evidence and explanations. Paris, OECD, pp. 57-80.

Kogut, B. (1987): Country Patterns in International Competition: Appropriability and Oligopolistic Agreement. In: Hood, N., Vahlne, J. (eds.): Strategies in Global Competition. London, Croom-Helm, pp. 315-340.

Kogut, B. (1988): Joint Ventures: theoretical and empirical perspectives. In: Strategic Management Journal, Vol. 9, pp.319–332.

Kogut, B. (1989): The Stability of Joint Ventures: reciprocity and competitive rivalry. In: The Journal of Industrial Economics, Vol. 38 (2), pp. 183–198.

Kogut, B., Zander, U. (1992): Knowledge of the Firm, Combinative Capabilities, and the Replication of Technology. In: Organisation Science, Vol. 3, pp. 383-397.

Kogut, B., Zander, U. (1993): Knowledge of the Firm and the Evolutionary Theory of the Multinational Corporation. In: Journal of International Business Studies, Vol. 4, pp. 625–645.

Kogut, B., Zander, U. (1996): What Firms Do? Coordination, Identity, and Learning. In: Organisation Science, Vol. 7, pp. 502-518.

Koka, B.R.; Prescott, J. (2001): Strategic Alliances as Social Capital. In: Strategic Management Journal, Vol. 22, pp. 795-816.

Kooths, S. (2004): Wachstum durch Wissenschaft? In: Dettling, D., Prechtl, Ch. (eds.): Weißbuch Bildung – Für ein dynamisches Deutschland. Wiesbaden, Verlag für Sozialwissenschaften, pp.31-41.

Koschatzky, K., Bross, U., Stanovnik, P. (2001): Development and Innovation Potential in the Slovene Manufacturing Industry: Analysis of an Industrial Innovation Survey. In: Technovation, Vol. 21, pp. 311–324.

Koza, M.P., Lewin, A.Y. (2000): The Co-Evaluation of Strategic Alliances. In: Organisational Science, Vol. 9, pp. 255-265.

Kozelka, R.M. (1982): How to Work Through a Clustering Problem. In: Hudson, H.C. (ed.): Classifying Social Data. San-Francisco, Washington, Londen, Jossey-Bass Publishers, pp. 1-12.

Kretschmann, J. (1990): Die Diffusion des Kritischen Rationalismus in der Betriebswirtschaftslehre. Stuttgart, Peschel.

KROMREY, H. (1998): Empirische Sozialforschung. 8. Auflage, Opladen, Leske+Budrich.

KRÜGER, W., HOMP, C. (1997): Kernkompetenzen. In: Wirtschaftswissenschaftliches Studium (WiSt), Vol. 27 (10), pp. 529-531.

KRUGMAN, P.R. (1990): The Age of Diminished Expectations. Cambridge, MIT Press.

KRUSKAL, W.H., WALLIS, W.A. (1952): Use of Ranks in One-Criterion Variance Analysis. In: Journal of the American Statistical Association, Vol. 47, pp. 583-621.

KUBICEK, H. (1977): Heuristische Bezugsrahmen und heuristisch angelegte Forschungsdesigns als Elemente einer Konstruktionsstrategie empirischer Forschung. In: Köhler, R. (ed.): Empirische und handlungstheoretische Forschungskonzeption in der Betriebswirtschaftslehre. Stuttgart, Poeschel, pp. 3-36.

KUMAR, N., SIDDHARTHAN, N.S. (1994): Technology, firm size and export behaviour in developing countries: the case of Indian enterprises. In: Journal of Development Studies, Vol. 31 (2), pp. 289–309.

KUMAR, S., RUSSELL, R.R. (2002): Technological change, technological catch-up, and capital deepening: relative contributions to growth and convergence. In: American Economic Review, Vol. 92, pp. 527–548.

KURUKAWA, S. (1997): Make-or-Buy Decisions in R&D: Small Technology Based Firms in the United States and Japan. In: Transactions on Engineering Management, Vol. 44 (2), 124-134.

KUZNETS, S. (1959): Economic Change: Selected essays in business cycles, national income and economic growth. London, Heineman.

KUZNETS, S. (1973): Modern Economic Growth: Findings and Reflections. In: American Economic Review, Vol. 63, pp. 247-258.

L

LACHENMAIER, S. (2007): Effects Of Innovation on Firm Performance. ifo Beiträge zur Wirtschaftsforschung. München, ifo Institut für Wirtschaftsforschung.

LAESTADIUS, S. (2007): Empirisches Wissen in einem Lowtech-Unternehmen. In: Abel, J., Hirsch-Kreinsen, H. (eds.): Lowtech-Unternehmen am Hightech-Standort. Berlin, edition sigma, pp. 57-68.

LAM, A. (2005): Organisational innovation. In: Fagerberg, J., Mowery, D.C., Nelson, R.R. (eds.): The Oxford Handbook of Innovation. Oxford University Press, Oxford, pp. 115-147.

LANDES, D. (1970): The Unbound Prometheus. London, Cambridge University Press.

LANDRY, R., AMARA, N., LAMARI, M. (2002): Does Social Capital Determine Innovation? To What Extent? In: Technological Forecasting and Social Change, Vol. 69, pp. 681–701.

LANGLOIS, R.N., ROBERTSON, P.L. (1995): Firms, Markets, and Economic Change: A Dynamic Theory of Business Institutions. London, Routledge.

LANGLOIS, R.N., ROBERTSON, P.L. (1996): Stop Crying over Split Knowledge: A Critical Look at the Theory of Spillovers and Technical Change. Economic Working Papers, Vol. 1996-06, University of Connecticut.

LAVIE, D. (2006): The Competitive Advantage of Interconnected Firms: an Extension of the Resource-based View. In: Academy of Management Review, Vol. 31 (3), pp. 638-65.

LAWRENCE, P. R., LORSCH, J. W. (1967): Organisation and Environment. Managing, Differentiation, and Integration. Boston: Harvard Business School Press.

LEARNED, E.P., CHRISTENSEN, C.R., ANDREW, K.R., GUTH, W. (1969): Business Policy. Homewood (ILL.), Irwin.

LEE, J. (1995): Small Firms' Innovation in Two Technological Settings. In: Research Policy, Vol. 24, pp. 391-401.

LEE, C. (2004): The Determinants of Innovation in the Malaysian Manufacturing Sector. In: ASEAN Economic Bulletin, Vol. 21 (3), pp. 319-329.

LEE, J.D., PARK, C. (2006): Research and Development Linkages in a National Innovation System: Factors affecting Success and Failure in Korea. In: Technovation, Vol. 26, pp. 1045-1054.

LEE, C.Y. (2002): Industry R&D intensity Distributions: Regularities and underlying determinants. In: Journal of Evolutionary Economics, Vol. 12, pp. 307–341.

LEFEBVRE, É., LEFEBVRE, L.A., BOURGAULT, M. (1998): R&D-Related Capabilities as Determinants of Export Performance. In: Small Business Economics, Vol. 10 (4), pp. 365-377.

LEGLER, H., FRIETSCH, R. (2007): Neuabgrenzung der Wissenswirtschaft – forschungsintensive Industrien und wissensintensive Dienstleistungen (NIW/ISI-Listen 2006) (= Studien zum deutschen Innovationssystem Nr. 22–2007). NIW, Fraunhofer ISI. Hannover, Karlsruhe.

LEI, D., HITT, M. A., BETTIS, R. (1996): Dynamic Core Competencies Through Meta-Learning and Strategy. In: Journal of Management, Vol. 22 (4), pp. 549-569.

LEIPONEN, A. (2005): Organisation of knowledge and innovation: The case of Finnish business services. In: Industry and Innovation, Vol. 12 (2), pp. 185-203.

LEIPONEN, A., DREJER, I. (2007): What exactly are technological regimes? Intra-Industry Heterogeneity in the Organisation of Innovation Activities. In: Research Policy, Vol. 36, pp. 1221–1238.

LEONARD-BARTON, D. (1992): Core Capabilities and Core Rigidities. A Paradox in Managing New Product Development. In: Strategic Management Journal, Vol. 13, pp. 111-126.

LEONARD-BARTON, D. (1995): Wellsprings of Knowledge: Building and Sustaining the Sources of Innovation. Boston (MA), Harvard Business School Press.

LEVIN R.C., REISS, P.C. (1984): Tests of a Schumpeterian Model of R&D and Market Structure. In: Griliches Z. (ed.): R&D, Patents, and Productivity. Chicago, London: University of Chicago Press.

LEVIN, R.C., KLEVORICK, A.K., NELSON, R.R., WINTER, S.G., GILBERT, R., GRILICHES, Z. (1987): Appropriating the Returns from Industrial Research and Development. In: Brookings Papers on Economic Activity, Vol. 3, Special Issue On Microeconomics, pp. 783-831.

LEVIN, R.C., REISS, P.C. (1988): Cost-Reducing and Demand-Creating R&D with Spillovers. In: The RAND Journal of Economics, Vol. 19 (4), pp. 538-556.

LEVINE, R., RENELT, D. (1992): A Sensitivity Analysis of Cross-Country Growth Regressions. In: American Economic Review, Vol. 82, pp. 942-963.

LEVITT, B., MARCH, J.G. (1988): Organisational Learning. In: Annual Review of Sociology, Vol. 14, pp. 319-340.

LHUILLERY, S., PFISTER, E. (2009): R&D Cooperation and Failures in Innovation Projects: Empirical Evidence from French CIS data. In: Research Policy, Vol. 38, pp. 45-57.

LI, C.W. (2000): Endogenous vs. Semi-Endogenous Growth in a Two-R&D-Sector Model. In: Economic Journal, Vol. 110 (462), pp. c109-c122.

LI, M., SIMERLY, R.L. (2002): Environmental Dynamism, Capital Structure and Innovation: an empirical test. In: International Journal of Organisational Analysis, Vol. 10 (2), pp. 156–171.

LICHTENBERG, F.R., SIEGEL, D. (1991): The Impact of R&D Investment on Productivity: New Evidence Using Linked R&D-LRD data. In: Economic Enquiry, Vol. 29, pp. 203–228.

LICHTENTHALER, U. (2009): The role of corporate technology strategy and patent portfolios in low-, medium- and high-technology firms. In: Research Policy, Vol. 38, pp. 559-569.

LIEBESKIND, J.P. (1996): Knowledge, Strategy and the Theory of the Firm. In: Strategic Management Journal, Vol. 17 (2), pp. 93-108.

LIPPMAN, S.A., RUMELT, R.P. (1982): Uncertain Imitability: An Analysis of Interfirm Differences in Efficiency Under Competition. In: The Bell Journal of Economics, Vol. 13, pp. 418-438.

LIPSETT, M.S., LIPSEY, R.G. (1995): Benchmarks, Yardsticks and New Places to Look for Industrial Innovation and Growth. In: Science and Public Policy, Vol. 22 (4), pp. 259-265.

LIST, F. (1841): Das nationale System der Politischen Ökonomie, Erster Band: der internationale Handel, die Handelspolitik und die deutsche Zollunion. Stuttgart, Tübingen.

LÖÖF, H., HESHMATI, A. (2006): On the Relationship Between Innovation and Performance: A sensitivity analysis. In: Economics of Innovation and New Technology, Vol. 15 (4-5), pp. 317-344.

LOOSE, A., SYDOW, J. (1994): Vertrauen in Ökonomie und Netzwerkbeziehungen – Strukturationstheoretische Betrachtungen. In: Sydow, J., Windeler, A. (eds.): Management interorganisationaler Beziehungen. Opladen, Westdeutscher Verlag, pp. 160-193.

LÓPEZ, R.A., (2005): Trade and growth: Reconciling the macroeconomic and microeconomic evidence. In: Journal of Economic Surveys, Vol. 19 (4), pp. 623.

LORENZONI, G., LIPPARINI, A. (1999): The Leveraging of Interfirm Relationships as a Distinctive Organisational Capability: A Longitudinal Study. In: Strategic Management Journal, Vol. 20 (4), pp. 317-338.

LOVE, J.H., ROPER, S. (1999): R&D, Technology Transfer and Networking Effects On Innovation Intensity. In: Review of Industrial Organisation, Vol. 15 (1), pp. 43–64.

LOVE, J.H., ROPER, S. (2001): Location and Network Effects on Innovation Success: evidence for UK, German and Irish manufacturing plants. In: Research Policy, Vol. 30 (4), pp. 643–661.

LU, C.H. (2007): Moving up or moving out? A Unified Theory of R&D, FDI, and Trade. In: Journal of International Economics, Vol. 71, pp. 324-343.

LUCAS, R.E. (1972): Expectations and the Neutrality of Money. In: Journal of Economic Theory, Vol. 4, pp. 103-124.

LUCAS, R. E. (1988): On the Mechanics of Economic Development. In: Journal of Monetary Economics, Vol. 22, pp. 3-42.

LUNDVALL, B.Å. (1985): Product Innovation and User-Producer Interaction. In: Industrial Development Research Series, Vol. 31, Aalborg, Aalborg University Press.

LUNDVALL, B. Å. (1988): Innovation as an Interactive Process – from User-Producer Interaction to the National System of Innovation. In: Dosi, G., Freeman, C., Nelson R., Silverberg, G., Soete, L. (eds.): Technical Change and Economic Theory. London, pp. 349-369.Lundvall, B.Å. (1992): National Systems of Innovation: Towards a Theory of Innovation and Interactive Learning. London, Pinter.

LUNDVALL, B.Å. (1992): National Systems of Innovation. Towards a Theory of Innovation and Interactive Learning. London, Pinter Publishers.

LUNDVALL, B.Â., JOHNSON, B. (1994): The Learning Economy. In: Journal of Industry Studies, Vol. 1 (2), pp. 23-42.

M

MACKEY, A. (2006): Dynamics in Executive Labor Markets: CEO Effects, Executive-firm Matching, and Rent Sharing. The Ohio State University: Unpublished doctoral dissertation, available at: http://etd.ohiolink.edu/send-pdf.cgi/Mackey%20Alison.pdf?osu1148305593 (last accessed on 20th of August 2010).

MACPHERSON, A. (1997): The Contribution of External Service Inputs to the Product Development Efforts of Small Manufacturing Firms. In: R&D Management, Vol. 27 (2), pp. 127–144.

MADSEN, J.B. (2008): Semi-Endogenous Growth Versus Schumpeterian Growth Models: Testing the Knowledge Production Function Using International Data. In: Journal of Economic Growth, Vol. 13 (1), pp. 1-26.

MAHONEY, J.T., PANDIAN, J.R. (1992): The Resource-Based View Within The Conversation of Strategic Management. In: Strategic Management Journal, Vol. 13, pp. 363-380.

MAHONEY, J.T. (1993): Strategic Management and Determinism: Sustaining the Conversation. In: Journal of Management Studies, Vol. 30 (1), pp. 173-191.

MAHONEY, J.T. (1995): The Management of Resources and the Resource of Management. In: Journal of Business Research, Vol. 33, pp. 91-101.

MAIRESSE J. (1990): Time-Series and Cross-National Estimates on Panel Data: Why Are They Different and Why Should They Be Equal? In: Hartog, J., Ridder, G., Theeuwes, J. (eds.): Panel Data and Labour Market Studies. Amsterdam, North-Holland, pp. 81-95.

MAIRESSE J. SASSENOU, M. (1991): R&D and Productivity: A Survey of Econometric Studies at the Firm Level. In: Science - Technology - Industry Review, OECD, Vol. 8, pp. 9-43.

MAIRESSE, J., HALL, B. (1996): Estimating the Productivity of Research and Development: An Exploration of GMM Methods Using Data on French and United States Manufacturing Firms. NBER Working Paper No. 5501.

MAJUMDAR, S.K. (1995): The Determinants of Investment in New Technology: an Examination of Alternative Hypothesis. In: Technological Forecasting and Social Change, Vol. 50, pp. 153-165.

MAKADOK, R. (2001): Towards a Synthesis of the Resource-Based and Dynamic-Capability Views of Rent Creation. In: Strategic Management Journal, Vol. 22, pp. 387-401.

MALERBA, F., ORSENIGO, L. (1995): Schumpeterian patterns of innovation. In: Cambridge Journal of Economics, Vol. 19, pp. 47–65.

MALERBA, F., ORSENIGO, L. (1996): Schumpeterian Patterns of Innovation are Technology specific. In: Research Policy, Vol. 25, pp. 451-478.

MALERBA, F., NELSON, R.R., ORSENIGO, L., WINTER, S., (1999): "History-Friendly" Models of Industry Evolution: The Computer Industry. In: Industrial and Corporate Change, Vol. 8, pp. 3-40.

MALTHUS, T.R. (1986): An Essay on the Principle of Population. Original 1798, reprint: Pickering, London.

MANSFIELD E. (1961): Technical Change and the Rate of Imitation. In: Econometrica, Vol. 29 (4), pp. 741-66.

MANSFIELD, E. (1962): Entry, Gibrat's Law, Innovation and Firm Growth. In: Journal of Financial Economics, Vol. 40, pp. 3-29.

MANSFIELD E. (1965): Rates of Return from Industrial R&D. In: American Economic Review, Vol. 55 (2), pp. 310-22.

MANSFIELD, E. (1968a): Industrial Research and Technological Innovation. New York, Norton.

MANSFIELD, E. (1968b): The Economics of Technological Change. New York, Norton.

MANSFIELD, E. (1984): R&D and Innovation: Some empirical findings. In: Griliches, Z. (ed.): R&D, Patents and Productivity, Chicago, Chicago University Press, pp. 127-154.

MARCH, J.G., SIMON, H.A. (1958): Organisations. New York, John Wiley & Sons.

MARCH, J.G. (1981): Footnotes to Organisational Change. In: Administrative Science Quarterly, Vol. 26 (2), pp. 563-577.

MARCH, J.G. (1991): Exploration and Exploitation in Organisational Learning. In; Organisation Science, Vol. 2 (1), pp. 71-87.

MARCH, J.G. (1994): A Primer on decision making: How decisions happen. New York, Free Press.

MARCH, J.G., SUTTON, R.I. (1997): Organisational Performance as a Dependent Variable. In: Organisation Science, Vol. 8 (6), pp. 698-706.

MARCH, J. (1999): The evolution of evolution. In: March, J. G. (ed.): The pursuit of organisational intelligence. Malden. MA, Blackwell, pp. 100-113.

MARGLIN, S.A. (1984): Growth, Distribution, and Prices. Cambridge (MA), Harvard University Press.

MARSILI, O. (2001): The Anatomy and Evolution of Industries. Technological Change and Industrial Dynamics. Cheltenham, Edward Elgar.

MARSILI, O., VERSPAGEN, B. (2002): Technology and the dynamics of industrial structures: an empirical mapping of Dutch manufacturing. In: Industrial and Corporate Change, Vol. 11, pp. 791-815.

MARSHALL, A. (1920): Principles of Economics. 8th Edition, London, Macmillan Press.

MARTIN, A. (1989): Die empirische Forschung in der Betriebswirtschaftslehre. Stuttgart, Schäffer-Poeschel.

MARTIN, S. (1994): Industrial Economics. Economic Analysis and Public Policy. Englewood Cliffs, NJ.

MARX, K. (1987): Das Kapital. Band I-II (Originals 1867, 1885 und 1894), reprint: Dietz, Berlin.

MASKELL, P. (1998): Low-Tech Competitive Advantages and the Role of Proximity: The Danish Wooden Furniture Industry. In: European Urban and Regional Studies, Vol. 5 (2), pp. 99-118.

MASSINI, S., LEWIN, A. Y., GREVE, H. R. (2005): Innovators and imitators: Organisational reference groups and adoption or organisational routines. In: Research Policy, Vol. 34, pp. 1550-1569.

MATHEWS, J.A. (2006): Catch-Up Strategies and the Latecomer Effect in Industrial Development. In: New Political Economy, Vol. 11 (3), pp. 313-335.

MATHIEU, V. (2001a): Product Services: From a Service Supporting the Product to a Service Supporting the Client. In: Journal of Business and Industrial Marketing, Vol. 16 (1), pp. 39-58.

MATHIEU, V. (2001b): Service Strategies Within the Manufacturing Sector: Benefits, Costs and Partnership. In: International Journal of Service Industry Management, Vol. 12 (5), pp. 451-475.

MAYR, E. (1972): Lamarck Revisited. In: Journal of the History of Biology, Vol. 5 (1), pp. 55-94.

MAYR, E. (1991): Eine neue Philosophie der Biologie. München, Zürich, Piper.

MEADE, J.E. (1961): A Neo-Classical Theory of Economic Growth. London, Allen & Unwin.

MEADOW CONSORTIUM (2010): The MEADOW Guidelines, Project funded within the 6th Framework Programme of the European Commission's DG Research, Grigny, France, http://www.meadow-project.eu/index.php?/Article-du-site/Guidelines.html.

MEDCOF, J.W. (2000): Dynamic Capabilities: A Neo-Contingency Theory? Working Paper No. 96, Management of Innovation and New Technology Research Centre, The Michael G. DeGroote School of Business, McMaster University, Hamilton (Canada).

MEISTER, C., VERSPAGEN, B. (2004): European Productivity Gaps: Is R&D the Solution? Oslo, University of Oslo Centre for Technology, Innovation and Culture.

MENDONÇA, S. (2009): Brave Old World: Accounting for 'High-Tech' Knowledge in 'Low-Tech' Industries. In: Research Policy, Vol. 38, pp. 470-482.

MESSMAN, K. (1977): A study of key concepts and norms for the international collection and presentation of science statistics. Paris, UNESCO.

METCALFE, J.S. (1994): Competition, Fisher's Principle and Increasing Returns in the Selection process. In: Journal of Evolutionary Economics, Vol. 4, pp. 327-346.

METCALFE, J.S. (1995): The Economic Foundations of Technology Policy: Equilibrium and Evolutionary Perspectives. In: Stoneman, P. (ed.): Handbook of the Economics of Innovation and Technical Change. Oxford, Cambridge (MA), pp. 409-512.

METCALFE, J. (1998): Evolutionary Economics and Creative Destruction. London, New York.

MICHIE, J. (1998): Introduction. The Internationalisation of the Innovation Process. In: International Journal of the Economics of Business, Vol. 5 (3), pp. 261–277.

MIKIC, S., WHITE, G.R.T., RAZAK, A.A. (2009): A Knowledge Review: Implications for Future Research and Practical Application. In: International Journal of Business and Management, Vol. 4 (1), pp. 26-31.

MILES, I. (2007): Research and development beyond manufacturing: the strange case of service R&D. In: R&D Management, Vol. 37, (3), pp. 249-268.

MILLER, A. G., MAIR, D. (1991): A modern guide to economic thought. Aldershot, Edgar Miller.

MILLIGAN, G.W., SOKOL, L.M. (1980): A Two-Stage Clustering Algorithm With Robust Recovery Characteristics. In: Educational and Psychological Measurement, Vol. 40, pp. 755–759.

MILLIGAN, G.W., COOPER, M.C. (1987): Methodology Review: Clustering Methods. In: Applied Psychological Measurement, Vol 11 (4), pp. 329–354.

MINTZBERG, H. (1994): The Decline and Fall of Strategic Planning. New York, Prentice-Hall.

MÖLLER, K. (2006): Unternehmensnetzwerke und Erfolg - Eine empirische Analyse von Einfluss- und Gestaltungsfaktoren. In: Zeitschrift für betriebswirtschaftliche Forschung (zfbf), Vol. 58, pp. 1051-1076.

MOLDASCHL, M., FISCHER, D. (2004): Beyond the management view. A Resource-Centered Socio-Economic Perspective. In: Croucher, R., Kabst, R., Kellerman, R., Matiaske, W. (eds): Management Revue. The international review of management studies. Special issue: Beyond resource based View (edited by Manfred Moldaschl), Vol. 14 (1), pp. 122-151.

MOLDASCHL, M. (2006): Innovationsfähigkeit, Zukunftsfähigkeit, Dynamic Capabilities. Moderne Fähigkeitsmystik und eine Alternative. In: Schreyögg, G., Conrad, P. (eds.): Management von Kompetenz. Managementforschung, Vol. 16, Wiesbaden, Gabler Verlag, pp. 1-36.

MONTGOMERY, C.A. (1995): Of Diamonds and Rust: A New Look at Resources. In: Montgomery, C.A. (ed.): Resource-Based and Evolutionary Theories of the Firm: Towards a Synthesis. Boston, Kluwer Academic Publishers, pp. 251-268.

Moran, P., Ghoshal, S. (1999): Markets, Firms, and the Process of Economic Development, in: Academy of Management Review, Vol. 24, pp. 390-412.

Mowery, D. C. (1983a): Economic Theory and Government Technology Policy. In: Policy Sciences, Vol. 16, pp. 27–43.

Mowery, D.C. (1983b): The Relationship between Intrafirm and Contractual Forms of Industrial Research in American Manufacturing, 1900-1940. In: Explorations in Economic History, Vol. 20 (4), pp. 351-374.

Mowery, D.C., Rosenberg, N. (1989): Technology and the Pursuit of Economic Growth. Cambridge (UK), Cambridge University Press.

Mowery, D.C., Rosenberg, N: (1998): Paths of Innovation. Cambridge (MA), Cambridge University Press.

Münt, G. (1996): Dynamik von Innovation und Außenhandel – Entwicklungen technologischer und wirtschaftlicher Spezialisierungsmuster. Band 21 der Schriftenreihe Technik, Wirtschaft und Politik des Fraunhofer-Instituts für Systemtechnik und Innovationsforschung (ISI). Heidelberg.

Murovec, N., Prodan, I. (2009): Absorptive Capacity, Its Determinants, and Influence on Innovation Output: Cross-cultural validation of the structural model. In: Technovation, Vol. 29, pp. 859–872.

Myers, S., Marquis. D.G. (1969): Successful Industrial Innovations: A study of factors underlying innovation in selected firms. NSF 69-17. Washington, DC: National Science Foundation.

N

Nahapiet, J., Ghoshal, S. (1998): Social Capital, Intellectual Capital, and the Organisational Advantage. In: Academy of Management Review, Vol. 23, pp. 242–266.

Nakamura, M. (2003): Research Alliances and Collaborations: introduction to the special issue. In: Managerial and Decision Economics, Vol. 24, pp.47–49.

Nascia, L., Perani, G. (2002): Diversity of Innovation in Europe. In. International Review of Applied Economics, Vol. 16 (3), pp. 277-293.

Nelson, R.R. (1973): Recent Exercises in Growth Accounting: New Understanding or Dead End? In: American Economic Review, Vol. 63 (3), pp. 462-468.

Nelson, R.R. (1990): Capitalism as an Engine of Progress. In: Research Policy, Vol. 19 (3), pp. 193-214.

Nelson, R.R. (1991): Why Firms Differ, and How Does It Matter?. In: Strategic Management Journal, Vol. 12, pp. 61-74.

Nelson, R.R. (1993): National Systems of Innovation: A Comparative Study. Oxford, Oxford University Press.

Nelson, R.R. (1994): An Agenda for Formal Growth Theory. Working Paper 94-85 (September 1994), International Institute for Applied Systems Analysis IIASA, Laxenburg, Austria, available at: http://www.iiasa.ac.at/Admin/PUB/Documents/WP-94-085.pdf (last access on 10th of September 2009).

Nelson, R.R. (2000): National Innovation Systems. In: Acs, Z.J. (ed.): Regional Innovation, Knowledge and Global Change. London, New York, Pinter Publishers, pp. 11-26.

Nelson, R.R. (2006): Economic development from the perspective of economic theory, The other Canon Foundation and Tallin University of Technology, Working Papers in Technology Governance and Economic Dynamics 02, TTU Institute of Humanities and Social Science.

NELSON, R.R., WINTER, S.G. (1974): Neoclassical vs. Evolutionary Theories of Economic Growth: Criticism and Prospectus. In: Economic Journal, Vol. 84, pp. 886-905.

NELSON, R.R., WINTER S.G.(1982): An Evolutionary Theory of Economic Change. Cambridge (MA.), Belknap Press.

NEUWEG, G.H. (1999): Könnerschaft und implizites Wissen. Münster, Waxmann.

NICKERSON, J.A., ZENGER, T.R. (2004): A Knowledge-Based Theory of the Firm - The Problem-Solving Perspective. In: Organisation Science, Vol. 15 (6), pp. 617-632.

NIELSEN, R.N., CHRISTENSEN, J.L., DAHL, M.S., ELIASEN, S., ØSTERGAARD, C.R. (2008): Innovation in Agricultural, Forest and Fishery. Paper for the DRUID Summer Conference 2008, "Entrepreneurship and Innovation", Copenhagen, CBS, Denmark, June 17-20, 2008.

NIJSSEN, E.J., HILLEBRAND, B., VERMEULEN, P.A.M., KEMP, R.G.M. (2006): Exploring product and service innovation similarities and differences. In: International Journal of Research in Marketing, Vol. 23, pp. 241-251.

NOLTE, H. (1998): Aspekte ressourcenorientierter Unternehmensführung. München, Mering.

NONAKA, I. (1991): The Knowledge Creating Company. In: Harvard Business Review, Vol. 69 (6), pp. 96-104.

NONAKA, I. (1994): A Dynamic Theory of Organisational Knowledge Creation. In: Organisation Science, Vol. 5, pp. 14-37.

NONAKA, I., TAKEUCHI, H. (1995): The Knowledge Creating Company. New York, Oxford University Press.

NONAKA, I., KONNO, N. (1998): The Concept of "Ba": Building a Foundation for Knowledge Creation. In: California Management Review, Vol. 40 (3), pp. 40-54.

NONAKA, I., TOYAMA, R. (2003): The Knowledge-creating Theory Revisited: Knowledge Creation as a Synthesizing Process. In: Knowledge Management Research & Practice, Vol. 1, pp. 2-10.

NOOTEBOOM, B. (1991): Entry, spending and firm size in a stochastic R&D race. In: Small Business Economics, Vol. 3, pp. 103–120.

NOOTEBOOM, B. (1992): Towards a Dynamic Theory of Transactions. In: Journal of Evolutionary Economics, Vol. 2, pp. 281-299.

NOOTEBOOM, B. (1999): Innovation and Inter-Firm Linkages: new implications for policy. In: Research Policy, Vol. 28 (8), pp. 793–805.

NOOTEBOOM, B. (2000): Learning and Innovation in Organisations and Economies. Oxford, Oxford University Press.

NOOTEBOOM, B. (2009): A Cognitive Theory of the Firm. Learning, Governance and Dynamic Capabilities. Northampton (MA.), Edward Elgar.

NORTH, D.C. (1981): Structure and Change in Economic History. New York, Norton.

NORTH, D.C. (1990): Institutions, Institutional Change and Economic Performance. New York, Cambridge University Press.

O

OECD (1963): The Measurement of Scientific and Technical Activities: Proposed Standard Practice for Surveys of R&D. Paris, OECD Publications.

OECD (1976): Science and Technology Indicators. DSTI/SPR/76.43.

OECD (1992): OSLO Manual: Proposed Guidelines for Collecting and Interpreting Technological Innovation Data. 1st Edition, Paris, OECD Publications.

OECD (1993): The Measurement of Scientific and Technical Activities: Proposed Standard Practice for Surveys of Research and Development. Paris, OECD Publications.

OECD (1994): Science and Technology Policy: Review and Outlook. Paris, OECD Publications.

OECD (1997): OSLO Manual: Proposed Guidelines for Collecting and Interpreting Technological Innovation Data. 2nd Edition, Paris, OECD Publications.

OECD (1999): Science, Technology and Industry Scoreboard 1999: Benchmarking Knowledge-Based Economies, Paris, OECD Publications.

OECD (2002): The Measurement of Scientific and Technical Activities: Proposed Standard Practice for Surveys of Research and Development. Paris, OECD Publications.

OECD (2005): OSLO Manual: Guidelines for Collecting and Interpreting Innovation Data. 3rd Edition, Paris, OECD Publications.

OECD (2009): Innovation in Firms. A microeconomic perspective. Paris, OECD Publications.

OELSNITZ, VON DER, D., GRAF, A. (2006): Inhalt und Aufbau interorganisationaler Kooperationskompetenz - Eine Konstruktbestimmung. In: Schreyögg, G., Conrad, P. (eds.): Management von Kompetenz. Managementforschung, Vol. 16, Wiesbaden, Gabler Verlag, pp. 83-120.

OERLEMANS, L.A.G., MEEUS, M.T.H., BOEKEMA, F.W.M. (1998): Do Networks Matter for Innovation: The usefulness of the economic network approach in analysing innovation. In: Journal of Economic and Social Geography, Vol. 89 (3), pp. 298–309.

OHLIN, B. (1933): Interregional and International Trade. Cambridge, Harvard University Press (revised edition 1967).

OKAMURO, H. (2007): Determinants of Successful R&D Cooperation in Japanese Small Businesses: the impact of organisational and contractual characteristics. In: Research Policy, Vol. 36 (10), pp. 1529–1544.

OLTRA, M., FLOR, M. (2003): The Impact of Technological Opportunities and Innovative Capabilities on Firms' Output Innovation. In: Creativity and Innovation Management, Vol. 12, pp. 137–144.

P

PARISI, M.L. (2001): Essays on Innovation, R&D Investment, and Productivity. Ph.D. Dissertation, Boston College.

PARUCHURI, S., NERKAR, A., HAMBRICK, D.C. (2006): Acquisition Integration and Productivity Losses in the Technical Core: Disruption of Inventors in Acquired Companies. In: Organisation Science, Vol. 17 (5), pp. 545–562.

PASINETTI, L.L. (1994): The Structure of Long-Term Development: Concluding Comments. In: Pasinetti, L.L.; Solow, R.M. (eds.): Economic Growth and the Structure of Long-Term Development. Proceedings of the IEA Conference held in Varenna, Italy. International Economic Association, St. Martin's Press, pp. 353-362.

PATEL, P., SOETE, L. (1988): Measuring the Economic Effects of Technology. In: STI Review, Vol. 4, pp. 121-66.

PAUTZKE, G. (1989): Die Evolution der organisatorischen Wissensbasis. Herrsching, Barbara Kirsch.

PAVITT, K. (1984): Sectoral patterns of technical change: towards a taxonomy and a theory. In: Research Policy, Vol. 13, pp. 343–373.

PAVITT, K., ROBSON, M., TOWNSEND, J. (1987): The Size Distribution of Innovating Firms in the UK: 1945-83'. In: The Journal of Industrial Economics, Vol. 35 (3), pp. 297-316.

PAVITT, K. (1991): Key Characteristics of the Large Innovating Firm. In: British Journal of Management, Vol. 2, pp. 41–50.

PAVITT, K. (1993): What Do Firms Learn from Basic Research? In: Foray, D., Freeman, C. (eds.): Technology and the Wealth of Nations. London, Pinter, pp. 29-40.

PAVITT, K. (2001a): Public Policies to Support Basic Research: What the rest of the world can learn from US theory and practice? (And what they should not learn). In: Industrial and Corporate Change, Vol. 10 (3), pp. 761-779.

PAVITT, K. (2001b): Research and development in organisations. In: Smelser, N.J., Baltes, P.B. (eds.): International Encyclopedia of the Social and Behavioral Sciences. Vol. 19; Amsterdam, Oxford, Elsevier, pp. 13220-13224.

PEDERSEN, T.E. (2005): Two types of 'Low-Tech' Sophistication: Production Techniques, Product Design and Formal Competence in Norwegian Mechanical Engineering. In: Hirsch-Kreinsen, H., Jacobson, D., Laestadius, S. (eds.): Low-Tech Innovation in the Knowledge Economy. Frankfurt a. Main, Berlin, New York, Peter Lang, pp. 253-284.

PEITZ, U. (2002): Struktur und Entwicklung von Beziehungen in Unternehmensnetzwerken. Wiesbaden.

PENEDER, M. (2001): Entrepreneurial Competition and Industrial Location. Cheltenham (UK), Edward Elgar.

PENEDER, M. (2002): Intangible Investment and Human Resources. In: Journal of Evolutionary Economics, Vol. 12, pp. 107–134.

PENEDER, M. (2003): Industry classifications. Aim, scope and techniques. In: Journal of Industry, Competition and Trade, Vol. 3 (1–2), pp. 109–129.

PENEDER, M. (2010): Technological regimes and the variety of innovation behaviour: Creating integrated taxonomies of firms and sectors. In: Research Policy, Vol. 39, pp. 323-334.

PENROSE, E.T. (1959): The Theory of the Growth of the Firm. New York, John Wiley & Sons.

PENROSE, E.T. (1985): The Theory of the Growth of the Firm Twenty-Five Years After. Acta Universitatis Upsaliensis, Studia Oeconomiae Negotiorum, Vol. 20, Uppsala.

PENTLAND, B.T., HAEREM, T., HILLISON, D. (2010): Comparing Organisational Routines as Recurrent Patterns of Action. In: Organisation Studies, Vol. 31 (7), pp. 917-940.

PESSOA, A. (2010): R&D and Economic Growth: How strong is the link? In: Economic Letters, Vol. 107, pp. 152-154.

PERETTO, P. (1998): Technological Change and Population Growth. In: Journal of Economic Growth, Vol. 3 (4), pp. 283-311.

PETERAF, M. A. (1993): The Cornerstones of Competitive Advantage. A Resource-Based View. In: Strategic Management Journal, Vol. 14, pp. 179-191.

PETERS, B. (2004): Employment Effects of Different Innovation Activities: Microeconomic evidence. ZEW Discussion Paper 04-73, Mannhein, Zentrum für Europäische Wirtschaftsforschung.

PHELPS, E. (1961): The Golden Rule of Accumulation: A Fable for Growthmen. In: American Economic Review, Vol. 51, pp. 638-643.

PIANTA. M. (2005): Innovation and Employment, In: Fagerberg, J., Mowery, D.C., Nelson, R.R. (eds.): The Oxford Handbook of Innovation. New York, Pxford University Press, pp. 568-598.

PICOT, A., REICHWALD, R., WIGAND, R. (2001): Die grenzenlose Unternehmung. Information, Organisation und Management, 4. Auflage, Wiesbaden, Gabler.

PIERSON, E.C. (1959): The Education of American Businessmen: A Study of University-College Programs in Business Administration. New York, McGraw-Hill.

PISANO, G.P., SHAN, W., TEECE, D.J. (1988): Joint Ventures and Collaboration in the Biotechnology Industry. In: Mowery, D. (ed.): International Collaborative Ventures in US Manufacturing. Cambridge (MA), Ballinger Publishing Co., pp. 183-222.

PISANO, G. (1990): The R&D boundaries of the Firm: an empirical analysis. In: Administrative Science Quarterly, Vol. 35, pp.153–176.

PITELIS, C. (2002): The Growth of the Firm; The Legacy of Edith Penrose. Oxford, Oxford University Press.

PITTAWAY, L., ROBERTSON, M., MUNIR, K., DENYER, D., NEELY, A. (2004): Networking and innovation: a systematic review of the evidence. In: International Journal of Management Reviews, Vol. 5/6 (3&4), pp. 137–168.

PIVA, M., VIVARELLI, M. (2002): The skill bias: comparative evidence and an econometric test. In: International Review of Applied Economics, Vol. 16 (3), pp. 347-358.

PIVA, M., VIVARELLI, M. (2004): Technological Change and Employment: Some micro evidence from Italy. In: Applied Economics Letters, Vol. 11 (6), pp. 373-376.

PIVA, M., VIVARELLI, M. (2005): Innovation and Employment: Evidence from Italian microdata. In: Journal of Economics, Vol. 86 (1), pp. 65-83.

PLEHN-DUJOWICH, J. (2007): Innovation, Firm Size, and RDSearch. In: Economics Bulletin, Vol. 12 (17), pp. 1-8.

PLESCHAK, F., SABISCH, H. (1996): Innovationsmanagement, Stuttgart, Schäffer-Poeschel.

POLANYI, M. (1958): Personal Knowledge: Towards a Post-Critical Philosophy. Chicago, University of Chicago Press.

POLANYI, M. (1966): The Tacit Dimension. New York, Anchor Day.

PORTER, M.E. (1979): How Competitive Forces Shape Strategy. In: Harvard Business Review, Vol. March/April, pp. 137-156.

PORTER, M.E. (1980): Competitive Strategy: Techniques for Analyzing Industries and Competitors. New York, Free Press.

PORTER, M.E. (1981): The Contributions of Industrial Organisation to Strategic Management. In: Academy of Management Review, Vol. 6 (4), pp. 609-620.

PORTER, M.E. (1985): Competitive Advantage: Creating and Sustaining Superior Performance. New York, Free Press.

PORTER,M. (1987): From Competitive Strategy to Cooperative Strategy. In: Harvard Business Review, Vol. 65 (3), pp. 43–59.

PORTER, M.E. (1990): The Competitive Advantage of Nations. New York, The Free Press.

PORTER, M.E. (1991): Towards a Dynamic Theory of Strategy. In: Strategic Management Journal, Vol. 12, pp. 95-117.

PORTER, M. E. (1996): What is strategy? In: Harvard Business Review, November-December, pp. 61–78.

PORTER, M.E. (1997): Wettbewerbsstrategie (Competitive Strategy). Methoden zur Analyse von Branchen und Konkurrenten. Frankfurt, Campus Verlag.

PORTER, M.E. (1999): Wettbewerbsstrategien. 10th edition, Frankfurt a. Main, Campus Verlag.

POSNER, M.V. (1961): International Trade and Technical Change. In: Oxford Economic Papers, Vol. 13, pp. 323-341.

PRAHALAD, C.K., BETTIS, R.A. (1986): The Dominant Logic: A New Linkage Between Diversity and Performance. In: Strategic Management Journal, Vol. 7, pp. 485-501.

PRAHALAD, C. K., HAMEL, G. (1990): The Core Competence of the Corporation. In: Harvard Business Review, Vol. 68 (3), pp. 79-91.

PREISSL, B. (2000): Service Innovation: What Makes it Different? In: Metcalfe, J. S., Miles, I. (eds.): Innovation Systems in the Service Economy. Kluwer Academic Publishers, Boston, Dordrecht, London, pp. 125-148.

PRICE,W. J., BASS, L.W. (1969): Scientific research and the innovative process. In: Science, Vol. 164 (3881): pp. 802–806.

PRIEM, R. L., BUTLER, J. E. (2001a): Tautology in the Resource-Based View and the Implications of Externally Determined Resource Value. Further comments. In: Academy of Management Review, Vol. 26 (1), pp. 57-66.

PRIEM, R. L., BUTLER, J. E. (2001b): Is the Resource-Based "View" a Useful Perspective for Strategic Management Research. In: Academy of Management Review, Vol. 26 (1), pp. 22-40.

PUGH, D.S., HICKSON, D.J., HININGS, C.R., MACDONALD, K.M., TURNER, C., LUPTON, T. (1963): A Conceptual Scheme for Organisational Analysis. In: Administrative Science Quarterly, Vol. 8 (3), pp. 289-315.

PUNJ, G., STEWART, D.W. (1983): Cluster Analysis in Marketing Research: Review and Suggestions For Application. In: Journal of Marketing Research, Vol. 20, pp. 134–148.

Q

QUINN, J.B. (2000): Outsourcing Innovation: The New Engine of Growth. In: Sloan Management Review, Vol. 41 (4), pp.13–28.

R

RAA (TEN), T.; MOHNEN, P. (2008): Competition and Performance: The Different Roles of Capital and Labor. In: Journal of Economic Behavior & Organisation, Vol. 65, pp. 573-584.

RAHMEYER, F. (1993): Technischer Wandel und sektorales Produktivitätswachstum. Jahrbuch für Nationalökonomie und Statistik, Vol. 211 (3-4), pp. 259-285.

RAHMEYER, F. (2005): Wirtschaftliche Entwicklung oder evolutorischer Wandel. Ein integrativer Versuch zur Fundierung der evolutorischen Ökonomik. Volkswirtschaftliche Diskussionsreihe, Beitrag Nr. 282, Dezember 2005, Universität Augsburg.

RAMMER, C., CZARNITZKI, D., SPIELKAMP, A. (2009): Innovation Success of Non-R&D-Performers: Substituting Technology by Management in SMEs. In: Small Business Economics, Vol. 33, pp. 35-58.

RAMMER, C., KÖHLER, C., MURMANN, M., PESAU, A., SCHWIEBACHER, F., KINKEL, S., KIRNER, E., SCHUBERT, T., SOM, O. (2011): Innovation ohne Forschung und Entwicklung. Eine Untersuchung zu Unternehmen, die ohne eigene FuE-Taätigkeit neue Produkte und Prozesse einführen. Studien zum deutschen Innovationssystem, Nr. 15-2011, Mannheim und Karlsruhe.

RASCHE, C. (1994): Wettbewerbsvorteile durch Kernkompetenzen. Ein ressourcenorientierter Ansatz. Wiesbaden.

RAYMOND, L., ST-PIERRE, J. (2010): R&D as a Determinant of Innovation in Manufacturing SMEs: An attempt at empirical clarification. In: Technovation, Vol. 30, pp. 48-56.

READ, A. (2000): Determinants of Successful Organisational Innovation: A Review of Current Research. In: Journal of Management Practice, Vol. 3 (1), pp. 95-119.

REBER, G. (1992): Organisationales Lernen. In: Freese, E. (ed.): Handwörterbuch der Organisation. 3rd Edition, Stuttgart, Poeschel, pp. 1240-1255.

REBELO, S. (1991): Long-Run Policy Analysis and Long-Run Growth. In: Journal of Political Economy, Vol. 99, pp. 500-521.

REED, R., DE FILIPPI, R. (1990): Casual Ambiguity, Barriers to Imitation and Sustainable Competitive Advantage. In: Academy of Management Journal, Vol. 15 (1), pp. 88-102.

REHME, G. (2007): Wissen und Neue Wachstumstheorie: Die Rolle von fachspezifischem Humankapital. Arbeitspapiere des Instituts für Volkswirtschaftslehre, Nr. 189, Technische Universität Darmstadt.

REINGANUM, J.F. (1981): Dynamic Games of Innovation. In: Journal of Economic Theory, Vol. 25, pp. 21-41.

REINGANUM, J.F. (1982): A Dynamic Game of R and D: Patent Protection and Competitive Behaviour. In: Econometrica, Vol. 50 (3), pp. 671-688.

REUER, J., ZOLLO, M. (2005): Termination Outcomes of Research Alliances. In: Research Policy, Vol. 34 (1), pp. 101-115.

RICARDO, D. (1951): On the principles of Political Economy and Taxation. Original 1817, reprint: Cambridge University Press, Cambridge.

RICHARD, P.J., DEVINNEY, T.M., YIP, G.S., JOHNSON, G. (2009): Measuring Organisational Performance: Towards Methodological Best Practice. In: Journal of Management, Vol. 35 (3), pp. 718-804.

RICHERSON, P.J., BETTINGER, R.L., BOYD, R. (2005): Evolution on a Restless Planet: Were Environmental Variability and Environmental Change Major Drivers of Human Evolution? In: Wuketits, F.M., Ayala, F.J: (eds): Handbook of Evolution, Vol. 2, The Evolution of Living Systems (including Hominids). Wiley-VCH, pp. 223-242.

RIGBY, D., ZOOK, D. (2002): Open-Market Innovation. In: Harvard Business Review, Vol. 80 (10), pp. 80–89.

ROBERTS, E.B., (1988): Managing Invention and Innovation. In: Research Management, Vol. 31, pp. 11–29.

ROBERTSON, P.L., PATEL, P.R. (2007): New wine in old bottles: technological diffusion in developed economies. In: Research Policy, Vol. 36, pp. 708–721.

ROBERTSON, P.L., SMITH, K. (2008): Distributed knowledge bases in low- and medium-technology industries. In: Hirsch-Kreinsen, H., Jacobson, D. (eds.): Innovation in Low-Tech Firms and Industries. Cheltenham (UK), Northampton (MA), Edward Elgar, pp. 93-117.

ROBSON, M.J. (2002): Partner Selection in Successful International Strategic Alliances: The Role of Co-Operation. In: Journal of General Management, Vol. 28 (1), pp. 1-15.

RÖPKE, J., STILLER, O. (2006): Einführung zum Nachdruck der 1. Auflage Joseph A. Schumpeters „Theorie der wirtschaftlichen Entwicklung". In: Röpke, J., Stiller, O. (eds.): Joseph Schumpeter. Theorie der wirtschaftlichen Entwicklung. Nachdruck der 1. Auflage von 1912. Berlin, Duncker & Humblot, Berlin, pp. V-XLI.

RÖTHELI, T.F. (1993): Exogenes und endogenes Wachstum: ein Streifzug. In: Duwendag, D., Rose, K., Wille, E. (eds.): Jahrbuch für Sozialwissenschaft. Zeitschrift für Wirtschaftswissenschaften, Vol. 44, Göttingen, Vanderhoeck & Ruprecht, pp. 1-10.

ROGERS, M. (1998): The Definition and Measurement of Innovation. Working Paper No. 19/98, Melbourne, Melbourne Institute of Applied Economic and Social Research, available at: http://melbourneinstitute.com/wp/wp1998n10.pdf (last access on 30th of June 2010).

ROGERS, M. (2004): Networks, Firm Size and Innovation. In: Small Business Economics, Vol. 22, pp. 141–153.

ROMER, P.M. (1986): Increasing Returns and Long-Run Growth. In: Journal of Political Economy, Vol. 94 (5), pp. 1002-1037.

ROMER, P.M. (1990): Endogenous Technological Change. In: Journal of Political Economy, Vol. 98 (5), pp. 71-102.

ROMER, P.M. (1994): The Origins of Endogenous Growth. In: Journal of Economic Perspectives, Vol. 8 (1), pp. 3-22.

ROPER, S. (1999): Underreporting of R&D in Small Firms: The Impact on International R&D Comparisons. In: Small Business Economics, Vol. 12, pp. 131-135.

ROPER, S., LOVE, J.H. (2002): Innovation and export performance: evidence from the UK and German manufacturing plants. In: Research Policy, Vol. 31 (7), pp. 1087-1102.

ROSA, J.M. (2004): Importance of Research and Development in the Business Services Sector. In: Innovation Analysis Bulletin (Statistics Canada), Vol. 6 (2), pp. 11-13.

ROSEN, S. (1976): A Theory of Life Earnings. In: Journal of Political Economy, Vol. 84 (4), pp. 45-67.

ROSENBERG, N. (1972): Technology and American Economic Growth. White Plains, New York, Sharpe.

ROSENBERG, N., BIRDZELL, L.E. (1986): How The West Grew Rich: The Economic Transformation Of The Industrial World. New York, Basic Books, Inc. Publishers.

ROSENBERG, N. (1990): Why do firms basic research (with their own money). In: Research Policy, Vol. 19, pp. 165-174.

ROSENBERG, N. (1994): Exploring the Black Box: Technology, Economics, and History. New York, Cambridge University Press.

ROSENTHAL, S.R. (1992): Effective Product Design and Development. Homewood (IL), Irwin.

ROTHWELL, R. (1992): Successful Industrial Innovation: Critical Factors for the 1990's. In: R&D Management, Vol. 22 (3), pp. 221-239.

ROTHWELL, R. (2003): Towards the Fifth-Generation Innovation Process. In: Jones, O., Tilley, F. (eds.): Competitive Advantage in SMEs. Organising for Innovation and Change. Chichester, John Wiley & Sons, pp. 115-135.

RUMELT, R.P (1984): Towards a strategic theory of the firm. In: Lamb, B. (ed.): Competitive strategic management. Englewood Cliffs, NJ: Prentice Hall, pp. 556-570.

RUMELT, R. P. (1987): Theory, Strategy, and Entrepreneurship. In: Teece, D. (ed.): The Competitive Challenge. Cambridge (MA), Ballinger, pp. 137-158.

RUMELT, R.P. (1991): How Much Does Industry Matter? In: Strategic Management Journal, Vol. 12, pp. 167-185.

RUMELT, R.P. (1994): Foreword. In: Hamel, G., Heene, A. (eds.): Competence-Based Competition. Chichester, John Wiley & Sons, pp. XV-XIX.

RYTTER, N.G., BOER, H., KOCH, C. (2007): Conceptualizing operations strategy processes. In: International Journal of Operations & Production Management, Vol. 27 (20), pp. 1093-1114.

S

SALA-I-MARTIN, X. (1990a): Lecture Notes on Economic Growth (I): Introduction to the Literature and Neoclassical Models. NBER Working Paper No. 3563.

SALA-I-MARTIN, X.X. (1990b): Lecture Notes on Economic Growth (II): Five Prototype Models of Endogenous Growth. NBER Working Paper No. 3564.

SALA-I-MARTIN, X.X. (1997): I Just Ran Two Million Regressions. In: American Economic Review, Vol. 87, pp. 178-183.

SAMSON, D., TERZIOVSKI, M. (1999): The Relationship between Total Quality Management Practices and Operational Performance. In: Journal of Operations Management, Vol. 17, pp. 393-409.

SAMUELSON, P.A., NORDHAUS, W.D. (1987): Volkswirtschaftslehre. Grundlagen der Makro- und Mikroökonomie. Vol. 2, 8th basically revised Edition (German), Köln, Bund-Verlag.

SANCHEZ, R. (1993): Strategic Flexibility, Firm Organisation, and Managerial Work in Dynamic Markets: A Strategic Options Perspectives. In: Shrivastava, P., Huff, A.S., Dutton, J.E. (eds.): Advances in Strategic Management. Vol. 9, Greenwich (CT), London, JAI Press, pp. 251-291.

SANCHEZ, R. (2001): Knowledge Management and Organisational Competence. Oxford, New York, Oxford University Press.

SANCHEZ, R. (1997): Strategic Management at the Point of Inflection: Systems, Complexity and Competence Theory. In: Long Range Planning, Vol. 30, pp. 939-946.

SANCHEZ, R. (2001): Building Blocks for Strategy Theory: Resources, Dynamic Capabilities and Competences. In: Volberda, H. W., Elfring, T. (eds.): Rethinking Strategy. London, Sage Publications, pp. 143-157.

SANCHEZ, R., HEENE, A. (1996): A Systems View of the Firm in Competence-Based Competition. In: Sanchez, R., Heene, A., Thomas, H. (eds.): Dynamics of Competence-based Competition: Theory and Practice in the New Strategic Management. Oxford, Elsevier, pp. 39-62.

SANCHEZ, R., HEENE, A. (2004): The New Strategic Management. New York, John Wiley & Sons.

SANCHEZ, R., HEENE, A., THOMAS, H. (1996): Introduction: Towards the Theory and Practice of Competence Based Competition. In: Sanchez, R., Heene, A., Thomas, H. (eds.): Dynamics of Competence-based Competition: Theory and Practice in the New Strategic Management. Oxford, Elsevier, pp. 1-35.

SANDVEN, T., SMITH, K., KALOUDIS, A. (2005): Structural Change, Growth and Innovation: The Roles of Medium and Low Tech Industries, 1980-2000. In: Hirsch-Kreinsen, H., Jacobson, D., Laestadius, S. (eds.): Low-Tech Innovation in the Knowledge Economy. Frankfurt a. M., Peter Lang, pp. 31-59.

SANTAMARÍA, L., NIETO, M.J., BARGE-GIL, A. (2009): Beyond formal R&D: Taking advantage of other sources of innovation in low- and medium technology industries. In: Research Policy, Vol. 38, pp. 507-517.

Santarelli, E., Sterlacchini, A. (1990): Innovation, Formal vs. Informal R&D, and Firm Size: some evidence from Italian manufacturing firms. In: Small Business Economics, Vol. 2, pp. 223-228.

Sargent, T.J., Wallace, N. (1976): Rational Expectations and the Theory of Economic Policy. In: Journal of Monetary Economies, Vol. 2, pp. 169-183.

Saviotti, P.P., Metcalfe, J. (1991): Present Development and Trends in Evolutionary Economics. In: ibid. (eds.): Evolutionary Theories of Economics and Technological Change, pp. 1-30.

Saviotti, P.P., Noteboom, B. (2000): Technology and Knowledge: From the Firm to Innovation Systems. Edward Elgar, Cheltenham.

Scheirer, W. (1991): Firm Size and Innovation: a comment. In: Small Business Economics, Vol. 3 (2), pp. 155–156.

Schendel, D. (1996): Introduction to the Winter 1996 Special Issue on 'Knowledge and the Firm'. In: Strategic Management Journal, Vol. 17 (Winter Special Issue), pp. 1-4.

Scherer, F.M. (1967): Research and Development Resource Allocation under Rivalry. In: Quarterly Journal of Economics, Vol. 81, pp. 359-394.

Schmidt, A. (2009): Relational View. In: Zeitschrift für Planung & Unternehmenssteuerung, Vol. 20, pp. 129-137.

Schmidt, T. (2005): What Determines Absorptive Capacity. In: DRUID Summer Conference 2005 on Dynamics of Industry and Innovation: Organisations, Networks and Systems. Copenhagen, Danish Research Unit For Industrial Dynamics

Schmiedeberg. C. (2008): Complementarities of Innovation Activities: An empirical analysis of the German manufacturing sector. In: Research Policy, Vol. 37, pp. 1492-1503.

Schmierl, K., Köhler, H.D. (2007): Organisationslernen in Lowtech- und Medium-Lowtech-Unternehmen. Wissens- und Personalmanagement. In: Abel, J., Hirsch-Kreinsen, H. (eds.): Lowtech-Unternehmen am Hightech-Standort. Berlin, edition sigma, pp. 21-56.

Schmookler, (1965): Technological Change and Economic Theory. In: American Economic Review, Vol. 55, pp. 333-341.

Schmookler, J. (1966): Invention and Economic Growth. Cambridge (MA), Harvard University Press.

Schneider, J. (1995): What's new and What's Old in New Growth Theory: Endogenous Technology, Microfoundation and Growth R. In: Zeitschrift für Wirtschafts- und Sozialwissenschaften, Vol. 115 (3), pp. 429-472.

Schneider, D. (1998): Vorläufer der Lehre von den Kernkompetenzen: von Xenophons Kyrupaedie über Babbage's Prinzip zu List's "Gesetz der Kräftevereinigung". In: Glaser, H., Schröder, E., Werder, A. v. (eds.): Organisation im Wandel der Märkte. Wiesbaden, pp. 343-358.

Schnell, R., Hill, P.B., Esser, E. (1995): Methoden der empirischen Sozialforschung. 5th Completed and Extended Edition, München, Wien, R.Oldenbourg Verlag.

Schreyögg, G. (1999): Strategisches Management - Entwicklungstendenzen und Zukunftsperspektiven. In: Die Unternehmung, Vol. 53 (6), pp. 387-407.

Schreyögg, G., Kliesch, M. (2005a): Dynamic Capabilities and the Development of Organisational Competencies. Diskussionsbeitrag No. 25/05 des Instituts für Management. Berlin: Freie Universität.

Schreyögg, G., Kliesch, M. (2005b): Individuelle und organisationale Kompetenzen im Rahmen des strategischen Managements. QUEM-Report, Schriften zur beruflichen Weiterbildung, No. 94, pp. 3-49.

Schuh, G., Friedli, T., Kurr, M.A. (2005): Kooperationsmanagement. München, Hanser Verlag.

SCHUMPETER, J.A. (1942): Geschichte der ökonomischen Analyse. Göttingen, Vandenhoeck & Ruprecht.

SCHUMPETER, J.A. (2006): Theorie der wirtschaftlichen Entwicklung. Reprint of the original first edition from 1912, edited and introduced by Jochen Röpke and Olaf Stiller, Berlin, Duncker & Humblot.

SCHUMPETER, J.A. (2008a): Konjunkturzyklen. German reprint of „the original edition of "Business Cycles. A Theoretical, Historical, and Statistical Analysis of the Capitalist Process" from 1939, Göttingen, Vandenhoeck & Ruprecht.

SCHUMPETER, J.A. (2008b): Capitalism, Socialism and Democracy. Reprint of the original first edition from 1942, extended by amendments of the 2nd (1947) and 3rd (1950) edition, introduced by Thomas K. McGraw, New York, Harper Perennial.

SCHWEBER, S. (1985): The Wider British Context in Darwin's Theorizing. In: Kohn, D. (ed.): The Darwinian Heritage. Princeton (NJ), pp. 35-69.

SCHWITALLA, B. (1993): Messung und Erklärung industrieller Innovationsaktivitäten mit einer empirischen Analyse für die westdeutsche Industrie. Heidelberg, Physica-Verlag.

SCOTT, M.F.G. (1989): A New View Of Economic Growth. Oxford, Clarendon Press.

SEGERSTROM, P.S. (1991): Innovation, Imitation and Economic growth. In: Journal of Political Economy 99 (4), pp. 807–827.

SELZNICK, P. (1957): Leadership in Administration. New York, Harper and Row.

SERTI, F., TOMASI, C. (2008): Self-Selection and Post-Entry Effects of Exports: Evidence from Italian Manufacturing Firms. In: Review of World Economics, Vol. 144 (4), pp. 660-694.

SHAPIRO, C., VARIAN, H. (1999): Information Rules. A Strategic Guide to the Network Economy. Boston, Harvard Business School Press.

SHEFER, D., FRENKEL, A. (2005): R&D, Firm Size and Innovation: an empirical analysis. In: Technovation, Vol. 25, pp. 25-32.

SIEG, G. (2007): Volkswirtschaftslehre. München, Wien, Oldenbourg.

SILVERBERG, G., VERSPAGEN, B. (1994): Learning, Innovation and Economic Growth: A Long Run Model of Industrial Dynamics. In: Industrial and Corporate Change, Vol. 3, pp. 199-223.

SIMON, H.A. (1991): Bounded Rationality and Organisational Learning. In: Organisation Science, Vol. 2, pp. 125–134.

SIMONIN, B.L. (1997): The Importance of Collaborative Know-how: An Empirical Test of the Learning Organisation. In: Academy of Management Journal, Vol. 40 (5), pp. 1150-1174.

SIMONIN, B.L. (1999): Ambiguity and the Process of Knowledge Transfer in Strategic Alliances. In: Strategic Management Journal, Vol. 20, pp. 595-623.

SIRILLI G. (1987): Patents and Inventors: An empirical study. In: Research Policy, Vol. 16, (2-4), pp. 157-74.

SITKIN, S. B. (1992): Learning Through Failure: The Strategy of Small Losses. In: Staw, B., Cummings, L. L. (eds.): Research in Organisational Behaviour, Vol. 14, Greenwich, JAI Press, pp. 231-266.

SKINNER, W. (1984): Operations Technology: Blind Spot in Strategic Management. In: Interfaces, Vol. 14 (1), pp. 116-125.

SMITH, A. (1937): An Inquiry into the Nature and the Causes of the Wealth of Nations. Original 1776, reprint: Random House. New York.

Smith, K. (2005): Measuring Innovation. In: Fagerberg, J., Mowery, D., Nelson, R. R. (eds.): The Oxford Handbook of Innovation. Oxford University Press, pp. 148-177.

Smolny, W. (1998): Innovation, Prices and Employment: A Theoretical Model and an Application for West German Manufacturing Firms. In: Journal of Industrial Economics, Vol. 46, pp. 359-381.

Smolny, W. (2000): Endogenous Innovations and Knowledge Spillovers. A Theoretical and Empirical Analysis. Heidelberg, Physica-Verlag.

Snow, C. C., Hrebiniak, L.G. (1980): Strategy, Distinctive Competence, and Organisational Performance. In: Administrative Science Quarterly, Vol. 25, pp. 317-336.

Snowdon, B., Vane, H.R. (1999): Paul M. Romer. In: Conversations with Leading Economists. Interpreting Modern Macroeconomics. Edward Elgar, pp. 292-313.

Soete, L. (1978): Inventive Activity, Industrial Organisation and International Trade, PhD thesis, University of Sussex.

Solow, R.M. (1956): A Contribution to the Theory of Economic Growth. In: Quarterly Journal of Economics, Vol. 70, pp. 65-94.

Solow, R.M. (1957): Technical Change and the Aggregate Production Function. In: Review of Economics and Statistics, Vol. 39, pp. 312-320.

Solow, R.M. (1959): Investment and Technical Progress. In: Arrow, K.J., Karlin, S., Suppes, P. (eds.): Mathematical Methods in the Social Sciences, Proceedings of the First Stanford Symposium 1959, Stanford, pp. 89-105.

Solow, R.M. (1994a): Perspectives on Growth Theory. In: Journal of Economic Perspectives, Vol. 8, pp. 45-54.

Solow, R.M. (1994b): Concluding Comments. In: Pasinetti, L.L.; Solow, R.M. (eds.): Economic Growth and the Structure of Long-Term Development. Proceedings of the IEA Conference held in Varenna, Italy. International Economic Association, St. Martin's Press, pp. 376-379.

Som, O., Kinkel, S., Kirner, E., Buschak, D., Frietsch, R., Jäger, A., Neuhäusler, P., Nusser, M., Wydra, S. (2010): Zukunftspotenziale und Strategien von nichtforschungsintensiven Industrien in Deutschland – Auswirkungen auf Wettbewerbsfähigkeit und Beschäftigung. Innovationsreport Nr. 140, Büro für Technikfolgen-Abschätzung beim Deutschen Bundestag, Berlin.

Sommerlatte, T. (1997): Die Evolution des Innovationsmanagements. In: Management von Innovation und Wachstum, Arthur D. Little (ed.), Wiesbaden: Gabler, pp. 155-162.

Sorenson, O., Rivkin, J.W., Fleming, L. (2006): Complexity, Networks and Knowledge Flow. In: Research Policy, Vol. 35 (7), pp. 994–1017.

Souitaris, V. (2002): Firm-specific Competencies Determining Technological Innovation: A survey in Greece. In: R&D Management, Vol. 32, pp. 61–77.

Specht, G., Beckmann, C., Amelingmeyer, J. (2002): F&E-Management. Kompetenz im Innovationsmanagement. 2nd edition. Stuttgart, Schäffer-Poeschel Verlag.

Spence, M.E. (1976): Product Selection, Fixed Costs, and Monopolistic Competition. In. Review of Economic Studies, Vol. 43, pp. 217-236.

Spender, J.C. (1989): Industry Recipes: The Nature and Sources of Managerial Judgment. Oxford, Blackwell.

Spender, J.C. (1996a): Making Knowledge the Basis of a Dynamic Theory of the Firm. In: Strategic Management Journal, Vol. 17 (special issue), pp.45-62.

SPENDER, J.C. (1996b): Competitive Advantage from Tacit Knowledge? Unpacking the Concept and his Strategic Implications. In: Moingeon, B., Edmondson, A. (eds.): Organisational Learning and Competitive Advantage. London, Sage Publications, pp. 56-73.

SPENDER, J.C., GRANT, R.M. (1996): Knowledge and the Firm: Overview. In: Strategic Management Journal, Vol. 17 (Special Issue), pp. 5-9.

SPERBER, D. (2000): An Objection to the Memetic Approach to Culture. In: Aunger, R. (ed.): Darwinizing Culture: The Status of Memetics as a Science. Oxford University Press, Oxford and New York, pp. 162-173.

SPITHOVEN, A., CLARYSSE, B., KNOCKAERT, M. (2010): Building Absorptive Capacity to Organise Inbound Open Innovation in Traditional Industries. In: Technovation, Vol. 30, pp. 130-141.

SRHOLEC, M., VERSPAGEN, B. (2008): The Voyage of the Beagle in Innovation Systems Land. Explorations on Sectors, Innovation, Heterogeneity and Selection. UNU-Merit Working Paper Series #2008-008, United Nations University, Maastricht.

STADLER, M. (1993): Die Modellierung des Innovationsprozesses – Ein integrative Mikro-Makro-Ansatz. In: ifo-Studien, Vol. 13, pp. 159-189.

STAHL, H. (1996): Beziehungskompetenz. In: Hinterhuber, H., Friedrich, S., Al-Ani, A., Handlbauer, G. (eds.): Das Neue Strategische Management. Wiesbaden, pp. 217-244.

STALK, G., EVANS, P., SHULMAN, L.E. (1992): Competing on Capabilities: the New Rules of Corporate Strategy. In: Harvard Business Review, Vol. 70 (2), pp. 57-69.

STERLACCHINI, A. (1994): Technological Opportunities, Intraindustry Spillovers and Firm R&D intensity. In: Economics of Innovation and New Technologies, Vol. 3, pp. 123-137.

STERLACCHINI, A. (2001): The Determinants of Export Performance: A firm level study of Italian manufacturing. In: Weltwirtschaftliches Archiv, Vol. 137 (3), pp. 450-472.

STERNBERG, R., ARNDT, O. (2001): The Firm or the Region: What determines the innovation behaviour of European firms? In: Economic Geography, Vol. 77 (4), pp. 364–382.

STOCK, G. N., GREIS, N. P., FISCHER, W. A. (2002): Firm Size and Dynamic Technological Innovation. In: Technovation, Vol. 22, pp. 537-549.

STONEMAN, P. (1995): Handbook of the Economics of Innovation and Technological Change. Oxford, Cambridge, Blackwell.

STRAUBE, M. (1972): Zwischenbetriebliche Kooperation. Wiesbaden.

STUART, T.E. (2000): Interorganisational Alliances and the Performance of Firms: A Study of Growth and Innovation. In: Strategic Management Journal, Vol. 21, pp. 791-812.

SUNDBO, J. (1996): The Balancing of Empowerment. A Strategic Resource Based Model of Organizing Innovation Activities in Service and Low-Tech Firms. In: Technovation, Vol. 16 (8), pp. 397-409.

SVEIBY, K.E. (1997): The New Organisational Wealth - Managing and Measuring Knowledge-Based Assets. San Francisco, Berrett-Koehler.

SWAN, T.W. (1956): Economic Growth and Capital Accumulation. In: Economic Record, Vol. 32, pp. 334-361.

SZULANSKI, G. (1996): Exploring Internal Stickiness: Impediments to the Transfer of Best Practice within the Firm. In: Strategic Management Journal, 17 (Special Issue), pp. 27-4.

T

Taylor, F.W. (2005): The Principles of Scientific Management. Unabridged facsimile of the original edition published in 1911 by Harper & Brothers (New York, London), Elibron Classics.

Teece, D.J. (1980): Economy of Scope and the Scope of the Enterprise. In: Journal of Economic Behaviour and Organisation, Vol. 1, pp. 223-245.

Teece, D. (1986): Profiting from technological innovation: Implications for integration, collaboration, licensing and public policy. In: Research Policy, Vol. 15 (6), pp. 285-305.

Teece, D. (1990): Contributions and impediments of economic analysis to the study of strategic management. In: Fredrickson, J. W. (ed.): Perspectives on Strategic Management. New York, Harper Business, pp. 39-80.

Teece, D., Pisano, G. (1994): The Dynamic Capabilities of Firms: an Introduction. In: Industrial and Corporate Change, Vol. 3, pp. 537-556.

Teece, D.J., Pisano, G., Shuen, A. (1997): Dynamic Capabilities and Strategic Management. In: Strategic Management Journal, Vol. 18 (7), pp. 509-533.

Teece, D.J. (2000): Managing Intellectual Capital. Organisational, Strategic, and Policy Dimensions. Oxford, New York, Oxford University Press.

Teece, D.J., Pisano, G., Shuen, A. (2000): Dynamic Capabilities and Strategic Management. In: Dosi, G., Nelson, R.R., Winter, S.G. (eds.): The Nature and Dynamics of Organisational Capabilities. Oxford, Oxford University Press, pp. 334-362.

Teece, D.J. (2007): Explicating Dynamic Capabilities: The Nature and Microfoundations of (Sustainable) Enterprise Performance. In: Strategic Management Journal, Vol.28, pp. 1319-1350.

Telser, L. G. (1980): A theory of self-enforcing agreements. In: Journal of Business, Vol. 53, pp. 27-44.

Tether, B. (2002): Who Co-Operates for Innovation, and Why: An Empirical Analysis. In: Research Policy, Vol. 31 (6), pp. 947–967.

Tether, B.S. (2003): What Is Innovation? – Approaches to Distinguishing New Products and Processes from Existing Products and Processes. Working Paper No. 12, Manchester, ESRC Centre for Research on Innovation and Competition (CRIC), available at: http://www.cric.ac.uk/cric/pdfs/wp12.pdf (last access on 30th June 2010).

Tether, B. (2004): Do Services innovate (differently)? CRIC Discussion Paper No. 66, University of Manchester.

Tether, B., Howells, J. (2007): Changing understanding of innovation in services. In: Department of Trade and Industry (DTI): Innovation in Services, DTI Occasional Paper No. 9, June 2007, pp. 21-60.

Thiele, M. (1997): Kernkompetenzorientierte Unternehmensstrukturen. Ansätze zur Neugestaltung von Geschäftsbereichorganisationen. Wiesbaden, Gabler.

Thornhill, S. (2006): Knowledge, Innovation and Firm Performance in High- and Low-Technology Regimes. In: Journal of Business Venturing, Vol. 21, pp. 687-703.

Tidd, J.,Trewhella,M. (1997): Organisational and Technological Antecedents for Knowledge Creation and Learning. In: R&D Management, Vol. 27, pp. 359–375.

Tidd, J., Bessant, J. (2009): Managing innovation. Integrating Technological, Market and Organisational Change. 4th edition, Chichester, Wiley.

Tiwari, A.K., Mohnen, P., Palm, F.C., van der Loeff, S. (2007): Financial Constraint and R&D Investment: Evidence from CIS. UNU-MERIT Working Paper No. 2007–011, Maastricht.

Tomiura, E. (2007): Effects of R&D and Networking on the Export Decision of Japanese Firms. In: Research Policy, Vol. 36 (5), pp. 758-767.

Totterdell, P., Leach, D., Birdi, K., Clegg, C., Wall, T. (2002): An investigation of the contents and consequences of major organisational innovations. In: International Journal of Innovation Management, Vol. 6 (4), pp. 343-368.

Tsai, W. (2000): Social Capital, Strategic Relatedness and the Formation of Intraorganisational Linkages. In: Strategic Management Journal, Vol. 21, pp. 925-939.

Tsai, W. (2001): Knowledge Transfer in Intraorganisational Networks: Effects of Network Position and Absorptive Capacity on Business Unit Innovation and Performance. In: Academy of Management Journal, Vol. 44, pp. 996-1004.

Tsai, W. (2002): Social Structure of "Coopetition" Within a Multiunit Organisation: Coordination, and Intraorganisational Knowledge Sharing. In: Organisation Science, Vol. 13, pp.179-190.

Tsai, K.H., Wang, J.C. (2009): External technology sourcing and innovation performance in LMT sectors: An analysis based on the Taiwanese Technological Innovation Survey. In: Research Policy, Vol. 38, pp. 518-526.

Tsoukas, H. (1996): The Firm as a Distributed Knowledge System: A Constructional Approach. In: Strategic Management Journal, Vol. 17 (Special Issue), pp. 11-25.

Tyler, B.B., Steensma, H.K. (1995): Evaluating Technological Collaborative Opportunities: a cognitive modelling perspective. In: Strategic Management Journal, Vol. 16, pp.43–70.

Tyler, B.B., Steensma, H.K. (1998): The Effects of Executives' Experiences and Perceptions on their Assessment of Potential Technological Alliances. In: Strategic Management Journal, Vol. 19 (10), pp. 939–965.

U

Ulrich, H. (1981): Die Betriebswirtschaftslehre als anwendungsorientierte Sozialwissenschaft. In: Geist, M., Köhler, R. (eds.): Die Führung des Betriebs. Stuttgart, pp. 1-25.

Ulrich, H., Hill, W. (1979): Wissenschaftstheoretische Aspekte ausgewählter betriebswirtschaftlicher Konzeptionen. In: Raffee, H., Abel, B. (eds.): Wissenschaftstheoretische Grundlagen der Wirtschaftswissenschaften. München, Vahlen, pp. 161-190.

Uzawa, H. (1965): Optimum Technical Change in an Aggregative Model of Economic Growth. In: International Economic Review, Vol. 6, pp. 18-31.

V

Vahs, D., Burmester, R. (2005): Innovationsmanagement, 3. Auflage, Stuttgart, Schäffel-Poeschel Verlag.

Vanberg, V. (1994): Kulturelle Evolution und die Gestaltung von Regeln. Walter Eucken Institut, Vorträge und Aufsätze, Vol. 144, Tübingen.

Vandenbussche, J., Aghio, P., Meghir, C. (2006): Growth, Distance to Frontier and the Composition of Human Capital. In: Journal for Economic Growth, Vol. 11, pp. 97-217.

VAN DEN HOOFF, B., DE LEEUW, VAN WEENEN, F. (2004): Committed to Share: Commitment and CMC Use as Antecedents of Knowledge Sharing. In: Knowledge and Process Management, Vol. 11 (1), pp. 13-24.

VAN LEEUWEN, G. (2002): Linking Innovation to Productivity Growth Using Two Waves of the Community Innovation Survey, OECD Science, Technology and Industry Working Papers, 2002/8, OECD Publishing. doi:10.1787/620221544571.

VAN REENEN, J. (1996): The Creation and Capture of Economic Rents: Wages and Innovation in a Panel of UK Companies. In: Quarterly Journal of Economics, Vol. 111 (1), pp. 195-226.

VAN REENEN, J. (1997): Technological Innovation and Employment in a Panel of British Manufacturing Firms. In: Journal of Labor Economics, Vol. 15 82), pp. 255-284.

VAN WIJK, R., JANSEN, J.J.P., LYLES, M.A. (2008): Inter- and Intra-Organisational Knowledge Transfer: a Meta-Analytic Review and Assessment of its Antecedents and Consequences. In: The Journal of Management Studies, Vol. 45 (4), pp. 830–853.

VDI NACHRICHTEN (2010): VDI nachrichten. Technik, Wirtschaft, Gesellschaft. Issue No. 28/29, 16th of July 2010, p. 10.

VEBLEN, T. (1898): Why is Economics not an Evolutionary Science? In: Quarterly Journal of Economics, Vol. 12, pp. 373-397.

VERNON, R. (1966): International Investment and International Trade in the Product Cycle. In: Quarterly Journal of Economics, Vol. 80, pp. 190-207.

VERSPAGEN, B. (1991): A new empirical approach to catching up or falling behind. In: Structural Change and Economic Dynamics, Vol. 2 (2), pp. 359–380.

VERSPAGEN, B. (2005): Innovation and Economic Growth. In: Fagerberg, J., Mowery, D.C., Nelson, R.R. (eds.): The Oxford Handbook of Innovation. New York, Oxford University Press, pp. 487-513.

VEUGELERS, R., CASSIMAN, B. (1999): Make and Buy in Innovation Strategies: Evidence from Belgian Manufacturing Firms. In: Research Policy, Vol. 28, pp. 63-80.

VINCETTI, W. (1990): What Engineers Know and How They Know It. Baltimore, Johns Hopkins University Press.

VOLBERDA, H.W. (1998): Building the Flexible Firm: How To Remain Competitive. Oxford, Oxford University Press.

VON HAYEK, F.A. (1945): The Use of Knowledge in Society. In American Economic Review, Vol. 35, pp. 519-532.

VON HAYEK, F.A. (1969): Freiburger Studien. Gesammelte Aufsätze, Tübingen, pp. 97-107.

VON HIPPEL, E. (1976): The Dominant Role of Users in the Scientific Instrument Innovation Process. In: Research Policy, Vol. 5, pp. 212-239.

VON HIPPEL, E. (1986): Lead Users: A Source of Novel Product Concepts. In: Management Science, Vol. 32 (7), pp. 791-805.

VON HIPPEL, E. (1988): The Source of Innovation. Cambridge, MIT Press.

VON HIPPE!, E. (1994): Sticky Information' and the Locus of Problem Solving: Implications for Innovation. In: Management Science, Vol. 40, pp. 429-439.

VON HIPPEL, E. (2001): Innovation by User Communities: Learning from Open-Source Software. In: MIT Sloan Management Review, Vol. 42 (4), pp. 82-86.

VON HIPPEL, E. (2004): Democratizing Innovation. Cambridge (MA), MIT Press.

VON HIPPEL, E., KATZ, R. (2002): Shifting Innovation to Users via Toolkits. In: Management Science, Vol. 48 (7), pp. 821-834.

VON KROGH, G., ROOS, J., KLEINE, D. (1998): Knowing in Firms: Understanding, Managing and Measuring Knowledge. London, Sage.

VON KROGH, G., GRAND, S. (2002): From Economic Theory Towards a Knowledge-based Theory of the Firm: Conceptual Building Blocks. In: Chun Wei, C., Bontis, N. (eds.): The Strategic Management of Intellectual Capital and Organisational Knowledge. Oxford, Oxford University Press, pp. 163-184.

VON KROGH, G., GRAND, S. (2004): Wissensmanagement. In: Schreyägg, G., von Werder, A. (eds.): Handwörterbuch Unternehmensführung und Organisation. 4th edition, Stuttgart, Poeschel, pp. 1647-1656.

VON TUNZELMANN, N., ACHA, V. (2005): Innovation in "low-tech" Industries. In: Fagerberg, J., Mowery, D.C., Nelson, R.R. (eds.), The Oxford Handbook of Innovation, Oxford University Press, Oxford, pp. 407-432.

VOSSEN, R.W., NOOTEBOOM, B. (1996): Firm size and participation in R&D. In: Kleinknecht, A. (Ed.): Determinants of Innovation: The Message from New Indicators. London, MacMillan, pp. 155–168.

VROMEN, J.J. (1995): Economic Evolution. An Inquiry into the Foundations of New Institutional Economics. London, New York, Routledge.

W

WAGNER, J. (2007a): Exports and Productivity. Comparable Evidence for 14 Countries. University of Nottingham, Research Paper #41. available at: http://papers.ssrn.com/sol3/papers.cfm?abstract_id=1048341 (last access on 13th of April 2010).

WAGNER, J. (2007b): Exports and Productivity: A Survey of the Evidence from Firm level Data. In: World Economy, Vol. 30 (1), pp. 60-82.

WAKELIN, K. (1998): Innovation and export behaviour at the firm level. In: Research Policy, Vol. 26 (7-8), pp. 829-841.

WAKELIN, K. (2001): Productivity Growth and R&D Expenditure in UK Manufacturing Firms. In: Research Policy, Vol. 30 (7), pp. 1079–1090.

WALSH, V. (1996): Design, Innovation and the Boundaries of the Firm. In: Research Policy, Vol. 25, pp. 509-529.

WEDER, R., GRUBEL, H.G. (1993): The New Growth Theory and Coasean Economics: Institutions to Capture Externalities. In: Weltwirtschaftliches Archiv, Vol. 129 (3), pp. 488-513.

WELGE, M.K., AL-LAHAM, A. (2008): Strategisches Management: Grundlagen - Prozess - Implementierung. 5. Auflage, Wiesbaden, Gabler.

WERNERFELT, B. (1984): A Resource-Based View of the Firm. In: Strategic Management Journal, Vol. 5, pp. 71-180.

WERNERFELT, B. (1995): The Resource-Based View of the Firm: Ten Years After. In: Strategic Management Journal, Vol. 16 (2), pp.171-174.

WERNERFELT, B., MONTGOMERY, C.A. (1986): What Is an Attractive Industry? In: Management Science, Vol. 32 (10), pp. 1223-1230.

Wheelwright, S. C. (1984): Manufacturing Strategy: Defining the missing Link, in: Strategic Management Journal, Vol. 5 (1), pp. 77-91.

Wheelwright, S.C., Clark, K.B. (1992): Revolutionizing Product Development: Quantum Leaps in Speed, Efficiency and Quality. The Free Press, New York.

Wiegand, M. (1996): Prozesse Organisationalen Lernens. Wiesbaden, Gabler.

Wildemann, H. (1997): Fertigungsstrategien. Reorganisationskonzepte für eine schlanke Produktion und Zulieferung, München, TCW-Verlag.

Williamson, O.E. (1975): Markets and Hierarchies: Analysis and Antitrust Implication. New York, Free Press.

Williamson, O.E. (1979): Transaction-Cost Economics: The Governance of Contractual Relations. In: Journal of Law & Economics, Vol. 22, pp. 233-261.

Williamson, O.E. (1985): The economic institutions of capitalism. New York, Free Press.

Willmore, L. (1992): Tansnationals and foreign trade: Evidence from Brazil. In: Journal of Development Studies, Vol. 28 (2), pp. 314–335.

Winter, S.G. (1964): Economic 'natural' selection and the theory of the firm. Yale Economic Essays, Vol. 4, pp. 225-272.

Winter. S.G. (1975): Optimization and evolution in the theory of the firm. In: Day, R., Groves, T. (eds.): Adaptive Economic Models. New York, pp. 73-118.

Winter, S.G. (1987): Knowledge and Competence as Strategic Assets. In: Teece, D.J. (ed.): The Competitive Challenge. Cambridge, Ballinger, pp. 159-184.

Winter, S.G. (1988): On Coase, Competence and the Corporation. In: Journal of Law, Economics and Organisation, Vol. 4, pp. 163-180.

Winter, S.G. (1990): Survival, Selection, and Inheritance in Evolutionary Theories of Organisation. In: Singh, J.V. (ed.): Organisational Evolution – New Directions. Newbury Park (CA), Sage, pp. 269-297.

Winter, S.G. (1995): Four Rs of profitability: rents, resources, routines, and replication. In: Montgomery, C. (ed.): Resource-Based and Evolutionary Theories of the Firm – Towards a Synthesis. Dordrecht, Kluwer, pp. 147-178.

Winter, S.G. (2003): Understanding Dynamic Capabilities. In: Strategic Management Journal, Vol. 24, pp. 991-995.

Witt, U. (1987): Individualistische Grundlagen der evolutorischen Ökonomik. Tübingen, JCB Mohr.

Witt, U. (1992a): Evolutionary Economics – an Interpretative Survey. Papers on Economics and Evolution, No. 9104 der Universität Freiburg, überarbeitete Version.

Witt, U. (1992b): Überlegungen zum gegenwärtigen Stand der evolutorischen Ökonomik. In: Biervert, B., Held, M. (eds.): Evolutorische Ökonomie. Frankfurt a. M., New York, Campus Verlag.

Witt, U. (1992c): Explaining Process and Change: Approaches to Evolutionary Economics. Ann Arbor, The University of Michigan Press.

Witt, U. (1996): A "Darwinian Revolution" in Economics? In: Journal of Institutional and Theoretical Economics, Vol. 152, pp. 707-716.

Witt, U. (1997): Economics and Darwinism. Papers on Economics and Evolution, No. 9705, Max-Planck-Institute for Research into Economic Systems, Jena.

WITT, U. (1999): Evolutionary Economics and Evolutionary Biology. In: Koslowski, P.: Sociobiology and Bioeconomics. Berlin, Heidelberg, pp. 279-298.

WITT, U. (2001): Evolutionary Economics: An Interpretative Survey. In: Dopfer, K. (ed.): Evolutionary Economics, Program and Scope. Boston, Dordrecht, London, pp. 45-88.

WITT, U. (2003): Evolutionary Economics and the Extension of Evolution to the Economy. In: Witt, U. (ed.): The Evolving Economy: Essays on the Evolutionary Approach to Economics, Edward Elgar, pp. 3-34.

WITT, U. (2005): The evolutionary perspective on organisational change and the theory of the firm. In: Dopfer, K. (ed.): The Evolutionary Foundations of Economics. Cambridge, Cambridge University Press, pp. 339-364.

WITT, H. (2008): Dynamic Capabilities im Strategischen Electronic Business-Management. Bestimmungsgrößen, Ausprägungen und Erfolgsfaktoren. Wiesbaden, Gabler Edition Wissenschaft.

WOLF, J. (2008): Organisation, Management, Unternehmensführung. Theorien, Praxisbeispiele und Kritik. 3. Auflage, Wiesbaden, Gabler.

WOLFF, E. (1996): Technology and Demand for Skills. In: Science Technology Industry, Vol. 18, pp. 95-124.

WOLLNIK, M. (1977): Die explorative Verwendung systematischen Erfahrungswissens – Plädoyer für einen aufgeklärten Empirismus in der Betriebswirtschaftslehre. In: Köhler, R. (ed.): Empirische und handlungstheoretische Forschungskonzeptionen in der Betriebswirtschaftslehre. Stuttgart, Poeschel, pp. 37-64.

WOMACK, J.P., JONES, D.T., ROOS, D. (1990): The Machine That Changed The World. Massachusetts Institute of Technology / International Motor Vehicle Program, New York, Maxwell Macmillan International.

WRIGHT, T.P. (1936): Factors affecting the costs of airplanes. In: Journal of the Aeronautical Sciences, Vol. 3, pp. 122– 128.

WU, L.Y. (2006): Resources, Dynamic Capabilities and Performance in a Dynamic Environment. In: Information and Management, Vol. 43, pp. 447-454.

Y

YOUNG, A. (1998): Growth without Scale Effects. In: Journal of Political Economy, Vol. 106, pp. 41–63.

Z

ZAHEER, A., MCEVILY, B., PERRONE, V. (1998): Does Trust Matter? Exploring the Effects of Interorganisational and Interpersonal Trust on Performance. In: Organisation Science, Vol. 9 (2), pp. 141-159.

ZAHRA, S., GEORGE, G. (2002): Absorptive Capacity: a Review, Reconceptualisation, and Extension. In: Academy of Management Review, Vol. 27, pp. 185–203.

ZANKER. C. (2011): Planung und Steuerung von Produktionssystemen im Kontext der strategischen Unternehmensplanung. Dortmund, Frankfurt am Main, Peter Lang Verlag (forthcoming).

ZIMMERMAN, K. (1991): The Employment Consequences of Technological Advance: Demand and Labour Costs in 16 German Industries. In: Empirical Economics, Vol. 16 (2), pp. 253-266.

ZOLLO, M, WINTER S.G. (2002): Deliberate Learning and the Evolution of Dynamic Capabilities. In: Organisation Science, Vol. 13, pp. 339-351.

Annex

A.1 Theoretical antecedents of the resource-based theory

Like any theory, the resource-based approach relies on prior theoretical work to develop its assumptions and predictions. In the case of resource-based theory, this prior basis mainly comes from the following strands of research (BARNEY/CLARK 2007: 4; WOLF 2008: 564f.).[244]

- Traditional studies of distinctive competencies (Gordon/Howell 1959; Pierson 1959) and business strategy (Learned et al. 1969; Andrews 1980; Chandler 1962, 1984)
- RICARDO's analysis of land rents (RICARDO 1951)
- The work of PENROSE (1959) and SELZNICK (1957)
- Studies of the antitrust implications of economics (DEMSETZ 1973)

Each of these prior theories is briefly discussed in the following.[245]

Before more economic approaches began to investigate the question of why some firms persistently outperform others (PORTER 1979), the research strand of firms' distinctive competencies had already focused on those firm level attributes (strengths and weaknesses) that enable some firms to pursue a strategy more efficiently and effectively than others (LEARNED ET AL. 1969; HREBINIAK/SNOW 1982; HITT/IRELAND 1985: 1986). Among the first identified distinctive competencies which determine persistent performance differences of firms, general management capability became the most important because of its significant impact on the strategies a firm decides to pursue and the ability of a firm to implement the strategies it develops (BARNEY/CLARK 2007: 5). Managerial writers like ANDREWS were interested in the role and functions of senior management and how they shape the future destiny of the firms by converting a distinctive competence into competitive advantage (ANDREWS

244 Other authors such as SCHNEIDER (1998: 353) or FREILING ET AL. (2006: 39) trace the roots of the resource-based view back even to FRIEDRICH LIST (1841) who supplemented the work of ADAM SMITH by adding aspects of human resources and competences.

245 For a more detailed discussion of this research work prior to resource-based theory, see BARNEY AND CLARK (2007: 4ff).

1980: 18f.). Unsurprisingly, it turns out that firms that have "high quality" general managers will usually perform better than firms with "low quality" managers. Thus, hiring such high quality general managers is one of the most important strategic choices that can be made by a firm.[246] This insight turned out to be fruitful, not only in the field of strategic management. For instance, in business history, explanations of the growth and success of firms were no more than the biographies of those who created and managed those firms (CHANDLER 1984). Today, there is little doubt that general managers with a set of certain individual key attributes can have a significant impact on firm performance (MACKEY 2006: 4ff.). But an exclusive emphasis on general managers as the key explanation of superior firm performance ignores a wide variety of other important firm attributes that may also be important for firm performance (e.g. financial or technological resources).[247] The sociologist PHILIP SELZNICK (1957) was among the first to recognise that general management skills are only one of several competencies that a firm may control. In his book "Leadership and Administration", he analysed the relationship between institutional leadership and distinctive competence of organisations and thereby widened the analytical focus of organisational competence analysis.

In contrast to the exclusive focus on general managers as possible explanations of differences in firm performance, the next major influence in the resource-based theory was interested in the economic consequences of the production input factor of land (RICARDO 1951). Unlike many other factors of production, the total supply of land is relatively fixed and cannot be simply increased in response to higher demand and prices (inelastic factor of production). In these settings, those that own higher-quality factors of production (high fertility land with low-cost of production) with inelastic supply can earn higher economic rents and thus superior performance than those who own lower-quality factors (low fertility land with high cost of production). Traditionally, most economists have implicitly assumed that relatively few factors of production are of an inelastic character. Most of their models assume that if prices for a factor rise, more of that factor will be produced. This results in an increasing supply of the desired production factor and the suppliers earning economic rents (see also chapter 2). The linkage between the economic

246 See, for example, GORDON/HOWELL (1959) and PIERSON (1959).

247 "In the end, general managers in organisations are probably similar to baseball managers: they receive too much credit when things go well and too much blame when things go poorly" (BARNEY/CLARK 2007: 7).

model of RICARDO and the resource-based theory lies in the inelastic nature of some production factors. As the resource-based theory suggests, numerous resources used by firms are inelastic in supply and are possible sources of economic rents. Highly firm-specific resources (like highly skilled, creative employees, teams or managers) and behavioural routines cannot be reproduced arbitrarily, transferred or copied by other firms. Thus, firms that own (or control) these kinds of resources may be able to earn economic rents by exploiting them (BARNEY/CLARK 2007: 10).

As briefly described in the second chapter, traditional economic models analysed firm growth by using the assumptions and tools of neoclassical microeconomics. They assumed that firms could adequately be modelled relatively simply by production functions. As NELSON AND WINTER put it, traditional economic models assumed that firms simply observe supply and demand conditions in the market and translate these into levels of production that maximise firm profits (NELSON/WINTER 1982: 201f.).

In her famous book titled "The Theory of the Growth of the Firm" (PENROSE 1959) which is often regarded as the cornerstone of resource-based theory (FOSS ET AL. 1995: 6; Foss 1997b: 14; WOLF 2008: 564f.), EDITH PENROSE criticised that this abstraction is not helpful. Instead, she argued that firms can, on the one hand, be understood as an administrative framework which links and coordinates the activities of its individual members and groups, and, on the other hand, represent a bundle of productive resources.[248] According to PENROSE, the growth of firms and their performance is limited by the productive opportunities that exist as a function of the bundle of productive resources controlled by a firm as well as by the administrative framework that coordinates the use of these resources (PENROSE 1959: 20). For PENROSE, it is not primarily the resources themselves that yield growth, but the responsible managers who discover new combinations and ways of deploying them in re-

248 "A firm is more than an administrative unit; it is also a collection of productive resources the disposal of which between different uses and over time is determined by administrative decision. When we regard the function of the private business firm from this point of view, the size of the firm is best gauged by some measure of the productive resources it employs" (PENROSE 1959: 25).

sponse to entrepreneurial views of opportunities. This provides the basis for ongoing growth of the firm (NOOTEBOOM 2009: 8).[249]

PENROSE'S analytical look inside the firm to identify the firms' ability to grow is of course one of the most important contributions which led to the resource-based theory. Her view of firm performance builds on managerial learning in the discovery and utilisation of as yet unutilised potential of existing organisational resources, in novel combinations between them, and suggests that this moves on to the development or adoption of new resources (NOOTEBOOM 2009: 2). Moreover, PENROSE paved the way for the resource-based view by her observation that these bundles of productive resources controlled by firms could vary significantly from firm to firm, even if they are located in the same industry. Thus, instead of controlling similar, homogeneous bundles of resources, firms differ fundamentally in their available productive resources. Therefore, PENROSE introduced a much broader definition of what may be considered as productive resources than traditional economists used to take as a basis for their models (BARNEY/CLark 2007: 12) by studying the competitive implications of inelastic productive resources like managerial teams, top management groups, and entrepreneurial skills. With this view of organisation-specific resources, the impact of PENROSE for the strand of research which has become the resource-based theory of the firm cannot be underappreciated (NOOTEBOOM 2009: 10).[250]

Last, but not least, another theoretical origin of the resource-based view comes from the research of social policy implications, especially from the area of antitrust regulation. Based on the conclusion that social welfare is maximised when markets are characterised by perfect competition, economists have developed various approaches to describe when an industry is less than perfectly competitive, what the social welfare implications are, and what remedies are available to re-increase competitiveness and restore social welfare (BARNEY/CLARK 2007: 12). For instance, in the conditional setting of a monopolistic or oligopolistic competition, prices will be higher than in the case of a

249 In his recent book "A Cognitive Theory of the Firm", BART NOOTEBOOM (2009: 8ff.) discusses the approach of PENROSE in a more detailed way and points also to some points of criticism. For instance, that she neglected the potential of smaller firms or the fact that she probably overestimated the role of firm managers by attaching to them the similar relevance as Schumpeter did when he created his heroic entrepreneur.

250 For a more detailed review of the legacy of EDITH PENROSE for the resource-based theory of firm growth, see the book by PITELIS (2002).

competitive market. In consequence, social welfare will be less than it would be in a more competitive market. This approach to analysing social welfare and antitrust flowed into the "Structure Conduct Performance" (SCP) paradigm which was already mentioned above in the context of market-based approaches of strategic management research (BAIN 1956). The SCP paradigm postulates that the structure of a firm's industry defines the range of activities that a firm can engage in, and, vice versa, the performance of firms in that industry. Following this line of argumentation, firms which operate in industries characterised by an imperfectly competitive market may have behavioural options that will enable them to obtain levels of performance that reduce social welfare. In the extreme, any persistently superior performance achieved by a firm must, by definition, reflect non-competitive firm behaviour (BARNEY/CLARK 2007: 13).

In the early 1970s, HAROLD DEMSETZ, an antitrust scholar, began to question this view by publishing an article that argued that industry structure was not the only determinant of a firm's performance (DEMSETZ 1973). He suggested that a firm which achieves persistently superior performance could not, a priori, be suspected of engaging in non-competitive activities. Instead, DEMSETZ claimed that some firms may achieve persistently superior performance either because they are lucky or - more likely - because of the fact these firms succeeded in establishing a reputation or goodwill that is difficult to separate from the firm itself, or it may be that the members of the employee team derive their higher productivity from the knowledge they possess about each other in the environment of the particular firm in which they work. Thus, "(...) superior competitive performance is unique to the firm, viewed as a team, and unobtainable to others except by purchasing the firm" (DEMSETZ 1973. 29).[251] Thereby, DEMSETZ emphasised the aspects of heterogeneity, different resource bundles as the sources of this heterogeneity and differential efficiencies of employee teams as the basis for differential competitive advantages. Moreover, with adducing the issues of team effects as well as talking about knowledge and learning effects, DEMSETZ touched on topics which have become core elements in today's resource-based theory and its extensions. Thus, it can be

---

251 "One such enterprise happens to "click" for some time while others do not. It may be very difficult for these firms to understand the reasons for this difference in performance or to know which inputs to attribute the performance of the successful firm. It is not easy to ascertain just why G.M. and I.B.M. perform better than their competitors. The complexity of these organisations defies easy analysis, so that the inputs responsible for success may be undervalued by the market for some time" (DEMSETZ 1973: 2).

summarised that DEMSETZ has anticipated some of the central aspects of the resource-based theory. Interestingly, he based his argumentation on his criticism of the SCP paradigm on which PORTER built his later theoretical work (PORTER 1979, 1980, 1985). Hence, DEMSETZ probably also sounded the starting signal for the following theoretical controversy that emerged between the supporters of the resource-based theory and the PORTER framework (BARNEY/CLARK 2007: 13).

But, in my view, the work of SCHUMPETER (2006, 2008a, 2008b, 1942) also represents an important influence on the ideas of the resource-based view of the firm which were not explicitly mentioned by BARNEY and CLARK. As described earlier, both neoclassic and evolutionary scholars like to refer their work to the universal ideas of SCHUMPETER. Hence, SCHUMPETER can indeed be designated as some kind of "border crosser" of economics. But referring to the resource-based approach of the firm, Schumpeter was the first to introduce a dimension of personal and behavioural skills as a predictor of firms' economic success by focussing on the creative entrepreneur and the continuous process of "creative destruction" and (re-) combination of production factors (SCHUMPETER 2006, 2008b). Economic growth and competitiveness are not considered as the result of exogenous factors, but as the result of entrepreneurial skills and, in that sense, as a result of internal factors of the firm.

Moreover, as it took up many of Schumpeter's core ideas, the emergence of the evolutionary economic theory (NELSON/WINTER 1982) was also offered strongly complementary to the developing resource-based approach, because it stresses the existence of fundamental heterogeneity and diversity among firms.[252] In the end, it is a firm's evolutionary path-dependent knowledge endowment, which is closely linked to processes of learning and imitation, that

252 In the evolutionary framework developed by NELSON and WINTER (1982) there was originally just little interest per se in the individual management strategies of firms. Instead, firms were primarily conceptualised as carriers of path-dependent knowledge bases whose bundle of routines is subject to the mechanism of selection working at the industry level (FOSS ET AL. 1995: 5). Therefore, attention has to be paid when using the term of "routines" introduced by NELSON and WINTER similarly to the "routines" of the resource-based approach (GERSCH ET AL 2005: 73 (footnote #264)). In the evolutionary theory, routines originally were conceptualised as a kind of "organisational gene" which is inherited from a successful firm population to another, while resource-based theory deployed them as a notion for stable, institutionalised patterns of behaviour (ZOLLO/WINTER 2002: 340). But with the later publications of NELSON (1991) and WINTER (1987, 1988), the perspective of the evolutionary theory became far more firm-level oriented.

distinguishes it from other enterprises (FOSS ET AL. 1995: 3; GERSCH ET AL. 2005: 72f.). Thus, the concepts of "routines" (NELSON/WINTER 1982), "capabilities" (TEECE ET AL. 1992), and "(core) competencies" (PRAHALAD/HAMEL 1990) refer to the reasons for the evolution and change of the competitive performance of firms. The evolutionary theory by NELSON and WINTER (1982) constructs a framework that incorporates innovation (in the sense of variation or mutation), market selection mechanisms, and - as a link to the resource-based perspective - the firm as a knowledge-bearing entity (heredity) which bequeaths its individual stock of knowledge and experiences through successful adaptation of its routines to environmental changes (FOSS ET AL. 1995: 5f.; WITT 2008: 45f.).

Besides these, let's say, more theoretical influences, the emergence of the resource-based view and the later resource-based theory was also substantially influenced by change process in the practical strategic management of firms (FOSS 1997a: 9f.) Through the economic turmoil of the 1970s, firms had to learn that the existing strategic planning tools with their strong rationalistic planning ideal[253] were not able to deliver the expected effects on the adaptiveness or even firm survival. Instead of ambitious long-range planning scenarios in strategic development at the corporate level, firms have put their emphasis more and more on the functional level by focussing on internal functional or dysfunctional management processes. As FOSS (1997a:10) states, the emergence and diffusion of management practices like total quality management[254] or the concept of learning organisation[255] can be seen as manifestations of the more introverted perspective of firms' strategic planning after the 1970s.

Thus, the failure of the established, rationalistic tools of strategic planning came to play a part in contributing to the acceptance and fostering of the resource-based perspective and its idea that firms are essentially different. The firms realised that the development of strategies for competitive advantage

253 For a detailed overview of the criticism of these traditional strategic management tools, see MINTZBERG (1994).

254 Total quality management (TQM) was developed and first introduced in the US around 1980, primarily in response to competitive challenges from Japanese companies and their concepts and management tools (SAMSON/TERZIOVSKI 1999: 394). The basic founders of TQM were DEMING (1982, 1986), JURAN (1988), ISHIKAWA (1985), JURAN/GYRNA (1993), CROSBY (1979, 1984), FEIGENBAUM (1983) and IMAI (1986).

255 See, for example, MARCH/OLSON (1976) and ARGYRIS/SCHÖN (1977).

has to start from this insight, rather than starting from the analysis of aggregate competitive market forces. Hence, this change of moods was the best opportunity for the development and acceptance of a resource-based perspective in strategic management.[256]

It became obvious that the resource-based approach did not emerge out of a single school of thought and is instead rooted in various theoretical origins in economics and sociology. With the development of the resource-based theory through the seminal contributions of WERNERFELT (1984), RUMELT (1984) and BARNEY (1986), these theoretical streams were connected and modified to develop what became known today as the resource-based theory of the firm.

256 To avoid the impression that firm-internal or resource-oriented perspectives played absolutely no role within the organisation and management research before 1984, it should be mentioned that some of the early tools of strategic management analysis were indeed developed earlier (WOLF 2008: 565). But I am sure that most colleagues would agree that the resources-based perspective did not diffuse into the broad scientific community and managerial practice until its revival that was triggered by the works of WERNERFELT, RUMELT, and BARNEY.

A.2 Classifications of industries based on R&D intensity

Tab. A-31: OECD - Classification of industries based on R&D intensity
(Source: OECD 1999: 106)

OECD classification	NACE	Sector
High-tech industries	24_4	Pharmaceuticals
≥5% BERD/production	30	Office, accounting, and computing machinery
	32	Radio, television, and communications equipment
	33	Medical, precision, and optical instruments
	35_3	Aircraft and spacecraft
Medium-high-tech industries	24 (excl. 24_4)	Chemicals
≥3-4.9% BERD/production	29	Machinery and equipment
	31	Electrical machinery and apparatus
	34	Motor vehicles and trailers
	35 (excl. 35_3)	Railroad and transport equipment
Medium-low-tech industries	23	Coke, refined petroleum products and nuclear fuel
≥1-2.9% BERD/production	25	Rubber and plastics products
	26	Other non-metallic mineral products
	27	Basic metals
	28	Fabricated metal products
Low-tech industries	15-16	Food products, beverages, and tobacco
>0-0.9% BERD/production	17-19	Textiles, textile products, leather and footwear
	20-22	Wood, pulp, paper, paper products, printing, and publishing
	36 (excl. 36_1)	Jewellery, music instruments, toys, sports equipment
	36_1	Furniture
	37	Recycling

Tab. A-32: Classification of industries based on R&D intensity
(Source: LEGLER/FRIETSCH 2007: 19ff.)

LEGLER/FRIETSCH CLASSIFICATION	NACE	SECTOR
High-tech industries	24_4	Pharmaceuticals
>7% BERD/production	30	Office, accounting, and computing machinery
	32	Radio, television, and communications equipment
	33	Medical, precision, and optical instruments
	35_3	Aircraft and spacecraft
Medium-tech industries	24 (excl. 24_4)	Chemicals
>2.5-7% BERD/production	29	Machinery and equipment
	31	Electrical machinery and apparatus
	34	Motor vehicles and trailers
	35 (excl. 35_3)	Railroad and transport equipment
	23	Coke, refined petroleum products and nuclear fuel
	25	Rubber and plastics products
	26	Other non-metallic mineral products
	27	Basic metals
	28	Fabricated metal products
Low-tech industries	15-16	Food products, beverages, and tobacco
>0-2.5% BERD/production	17-19	Textiles, textile products, leather and footwear
	20-22	Wood, pulp, paper, paper products, printing, and publishing
	36 (excl. 36_1)	Jewellery, music instruments, toys, sports equipment
	36_1	Furniture
	37	Recycling

A.3 Analysis of the macro-economic status of non-R&D-intensive sectors in German manufacturing industry

As to the best of my knowledge unfortunately no data exist which provide information about the macro-economic relevance of non-R&D-performing firms, I have to approach this question at the industry level, assuming that a large number of non-R&D-performing firms of the German manufacturing industry is affiliated to non-R&D-intensive sectors. This corresponds to the procedures of previous studies by SANDVEN ET AL. (2005), BENDER (2006), HIRSCH-KREINSEN ET AL. (2003), HAHN (2009), or HIRSCH-KREINSEN (2004).

As mentioned in chapter 2.6, in the early 1960s the OECD started to collect and publish international comparative data on science, technology and industry. The foundation for this activity has for a long time been the “Frascati Manual” the first edition of which was published in 1963 (OECD 1963) and today reached its present sixth edition (OECD 2002). The terms of low, medium, and high technology sectors were introduced by the OECD on the basis of this Frascati Manual during the 1980s and were widely adopted to classify the R&D intensity of manufacturing sectors and their goods (HIRSCH-KREINSEN ET AL. 2005: 17). This classification is based on the average share of sales spent on business R&D (BERD) in different industry sectors. Industrial sectors that spend less than 0.9 percent of their total sales are classified as "low-tech", sectors between 0.9% and 3% as "medium-low-tech", sectors between 3% and 5% as "medium-high-tech" and sectors that spend more than 5% of their total sales on R&D are labelled as "high-tech".[257] Medium-low-tech and low-tech sectors together form the group of non-R&D-intensive industries while medium-high- and high-tech sectors constitute the group of R&D-intensive sectors.

The German economy’s R&D activities take – qua definition – mainly place in R&D-intensive sectors. In 2006, almost 83 percent of all direct business R&D expenditures of German manufacturing industry were transacted within the R&D-intensive sectors. According to the sectoral OECD classification, non-

257 Due to structural and technological changes in economies, recent modifications of this typology (LEGLER/FRIETSCH 2007) differ in only using three categories of sectoral R&D intensity: less than 2.5% average sectoral expenditure for R&D in the case of low-tech sectors, between 2.5% and 7% for medium-tech sectors, and more than 7% for high-tech sectors. Because of its actuality, it would have been preferable to use this classification for the subsequent analysis. Unfortunately, this classification is not available on OECD level data.

R&D-intensive sectors within the German manufacturing industry are, for example, the manufacture of food products, beverages, tobacco, textiles, textile products and leather, wood and wooden products, basic metals and fabricated metal products, or the manufacture of pulp, paper and paper products. But when analysing the economic relevance of these non-R&D-intensive sectors of German manufacturing industry, the question arises if these sectors were always non-R&D-intensive, or if they have transformed from former R&D-intensive sectors to non-R&D-intensive sectors, which would mean that the composition of the non-R&D-intensive sector can change over time. In this case, today's observable economic performance could be biased either by their former high R&D intensity or by their recent transformation towards an R&D-intensive sector. Therefore, the subsequent subchapter addresses this issue, before the following paragraphs proceed with macro-economic performance data of non-R&D-intensive sectors.

A.3.1 Database

The subsequent analyses will be conducted on the basis of macro-economic data from the OECD database for structural analysis STAN.[258] The current edition of STAN indicators provide more than thirty indicators, dealing with production and employment structure, labour productivity and labour costs, investment, business research and development expenditures and international trade performance. The data are available for all OECD countries and cover the time period 1970-2008. However, the time coverage may vary considerably across countries, zones and indicators. For Germany, the OECD STAN data are very reliable because they are congruent with official statistical data from the German Federal Bureau of Statistics (GEHRKE/LEGLER 2010: 10). In order to answer the question about the today's economic relevance of non-R&D-intensive sectors in German manufacturing industry, this chapter draws on three main performance dimensions, namely a) value added, b) employment rate, and c) export intensity.

258 More information about the OECD STAN database is available at http://www.oecd.org/document/54/0,3343,en_2649_34445_21573686_1_1_1_1,00.html (last accessed on 15th of July 2010).

A.3.2 Non-R&D-intensive sectors in the German manufacturing industry in the course of time

In reflecting on the technological evolution of industries, WILLIAM J. ABERNATHY and JAMES M. UTTERBACK (1978) most prominently draw some analogies between the evolution of products/technical processes and industries. Their model describes the evolution of products and processes as a transition from an early "fluid" state to a state that is highly "specific" and rigid. In the early, fluid period of development, performance criteria for new products and processes are not well defined and market needs or process difficulties are approached through a variety of different product or process designs. In this stage, innovation is assumed to be relatively rapid and fundamental. The production processes must thus be highly flexible, relatively labour-intensive, and somewhat erratic in work flow.

> "New products which require reorientation of corporate goals or production facilities tend to originate outside organisations devoted to a "specific" production system; or, if originated within, to be rejected by them. (...) Many studies suggest that such new product innovations share common traits. They occur in disproportionate numbers in companies and units located in or near affluent markets with strong science-based universities or other research institutions and entrepreneurially oriented financial institutions. Their competitive advantage over predecessor products is based on superior functional performance rather than lower initial cost, and so these radical innovations tend to offer higher unit profit margins" (ABERNATHY/UTTERBACK 1978: 41f.)

As industrial development proceeds, technological diversity gives way to standardisation. Particular design approaches achieve dominance, production volumes increase, and performance criteria and processes are more clearly specified. This transition to the "specific" stage entails a change in the nature of innovation. In contrast to the rather radical changes made within the "fluid" stage, innovation in this phase is likely to alter only small aspects of the basic product, and any modifications are primarily intended to refine the established product design and performance. On the process side, work flow becomes rationalised, integrated and linear, unlike the fluid and flexible job characteristics of the early, "fluid" stage.

> "One distinctive pattern of technological innovation is evident in the case of established, high-volume products such as incandescent light-bulbs, paper, steel, standard chemicals, and internal-combustion engines, for examples. The markets for such goods are well defined; the product characteristics are well understood and often standardized; unit profit margins are typically low; production technology is efficient, equipment-intensive, and specialized to a particular product; and competition is primarily on the basis of price. Change is costly in such highly integrated systems because an alteration in any one attribute or process has ramifications for many others. In this environment innovation is typically incremental in nature, and it has a gradual, cumulative effect on productivity" (ABERNATHY/UTTERBACK 1978: 41).

Hence, the basic idea of industrial life cycle theory is that early phases of radical product innovation with high R&D intensity are followed by later, mature phases of incremental product and mainly process improvements, predominantly aiming at cost reduction. As KLEPPER (1997: 148f.) summarises, while distinguishing stages is somewhat arbitrary, the essence of industrial life cycles is that initially the market grows rapidly, many firms enter, and product innovation is fundamental. As the industry evolves, output growth slows, firm entry declines, the number of producers undergo a shakeout, product innovation becomes less significant, and process innovation rises.

It is obvious that the "specific", late or mature stage of industry life cycle closely corresponds to an innovation pattern of non-R&D-intensive which has been frequently mentioned in empirical literature. Thus, deploying the argumentation of industry life cycle theory on non-R&D-intensive sectors, it has to be assumed that non-R&D-intensive sectors represent industries which are in the "specific" stage of their life cycle. In the mature stage, R&D of industries may decrease either because of a lower level of innovation or the increase of value added and turnovers due to higher efficiency rates of production. In other words, a low R&D intensity of an industry could be used as an indicator to measure the technological maturity of an industry. However, this would also imply that these industries have not always been non-R&D-intensive as they have undergone the transition from the "fluid", R&D-intensive stage to the "specific" stage of low R&D intensity over time.

For this reason Figure A-22 provides some empirical data from the OECD about the development of R&D intensity of today's non-R&D-intensive sectors in the German manufacturing industry from the mid 1970s to 2006.[259] In this figure unlike the common definition of R&D intensity, R&D expenditures are not interrelated to the industry's total sales, but to its share of the gross domestic product, because the level of total sales strongly depends on the level of an industry's input which was significantly increased in some industries in the considered period.

First of all, it becomes apparent that the R&D intensity of German manufacturing industry in total significantly increased from 3.5% in 1975 to 7.6% in 2006.

259 Basically, it has to be mentioned that the inter-temporal comparability of these data is constrained because of changes in the underlying R&D definition or the classification of industrial sectors over time. Thus, the focus of interpretation should rather lie on the major trends which occur over time and not so much on the actual height or individual relationships between singular data points.

Corresponding to this, the examined non-R&D-intensive industries also show partial dynamic in their R&D intensity. This becomes particularly visible in the cases of manufacture of coke and refined petroleum products as well as the manufacture of textiles and textile products. In contrast, other industries show almost no changes in their R&D intensity over time like manufacture of pulp, paper and paper products or manufacture of wood and wooden products.

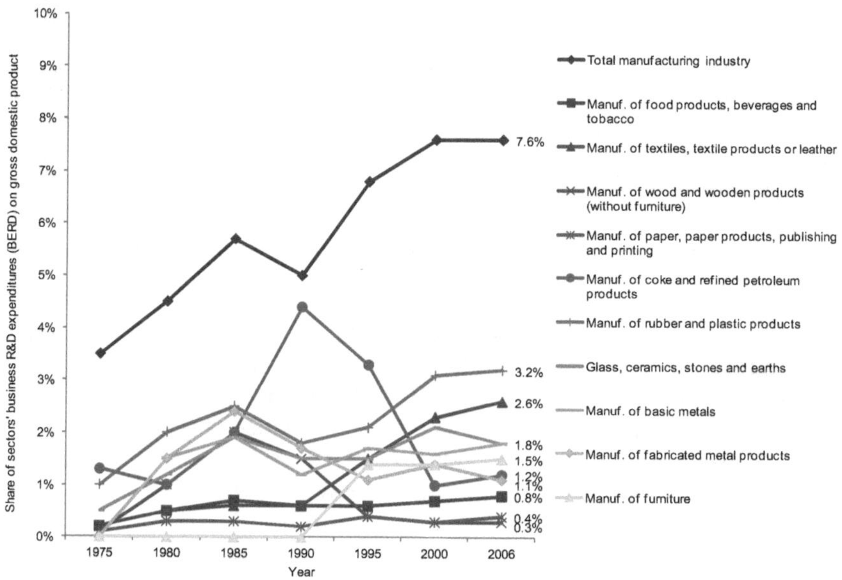

Fig. A-22: Trend in the R&D intensity of non-R&D-intensive sectors in German manufacturing industry from 1975 to 2006
(Source: own illustration on the basis of OECD STAN data)

But most interestingly, it appears from the figure that the non-R&D-intensive sectors of German manufacturing industry which were considered do not show the expected trends of the industry life cycle. That means that those sectors of the German manufacturing industry which were non-R&D-intensive in 1975 are – according to OECD classification - still non-R&D-intensive today.[260] Based on these data, there is no empirical evidence of any transition from former R&D-intensive sectors to non-R&D-intensive sectors. On the contrary, the overall R&D intensity of most of these non-R&D-intensive sectors has even slightly increased until today.

Thus, it can be summarised at this point that the set of non-R&D-intensive sectors in the German manufacturing industry remained stable for at least the last 30 years. Moreover, data show that these industries obviously managed to survive over time even with low R&D intensity, and non-R&D-intensive industries could not be assumed to disappear over time, as they do obviously not represent the "final stage" of the industry life cycle. With regard to the analysis of possible innovation patterns of non-R&D-performing firms this finding may also give a first hint that the suggested incremental and efficiency modes of innovation according to the theory must not necessarily be expected.

A.3.3 Value added of non-R&D-intensive sectors in German manufacturing industry

The second assumption that was made on the basis of endogenous growth theory suggested that non-R&D-intensive firms and industries would show a decreasing economic relevance due to their low competitiveness (see chapter 2.7). To address the question of the economic relevance of non-R&D-intensive sectors in German manufacturing industry, the following figure (Fig. A-23) projects the absolute value added of R&D-intensive and non-R&D-intensive sectors in proportion to a fixed starting point to underline the dynamics for each sector.

The figure shows that the value added increased in all aggregates of manufacturing industry. The value added of R&D-intensive sectors thereby lies considerably above the average of total manufacturing industry, while the value added of non-R&D-intensive sectors is below this average. R&D-intensive sectors achieve a value added that is about 1.7 times higher than that of non-R&D-intensive ones.

---

260 This result also holds for other developed OECD countries (SOM ET AL. 2010). However, it has to be mentioned that these data are only valid for the 2-digit NACE level, which means that within these aggregates, single product groups may nevertheless have changed from R&D intensity to non-R&D intensity. The same constraint accounts for the firms within the sectors. Thus, these data only reflect the overall R&D-level over time which obviously has not changed dramatically independently of single products or firms.

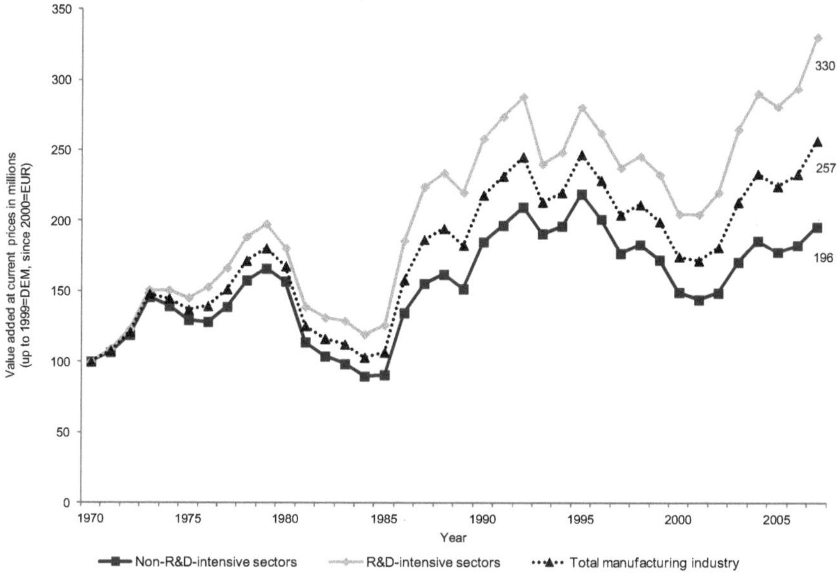

Fig. A-23: Value added of R&D- and non-R&D-intensive sectors of German manufacturing industry at current prices (in real terms) from 1970 to 2007
(Index: 1970=100)
(Source: Fraunhofer ISI on the basis of OECD STAN data)

But despite their lower level, the picture also shows that non-R&D-intensive sectors are still able to contribute increasingly to the overall value added of German manufacturing industry, as no sector-specific dynamics can be recognised over time. Thereby, the inter-temporal variations in the value added are congruent between R&D-intensive and non-R&D-intensive sectors, a finding which underlines the issue of economic interdependencies between both sectors. Moreover, it is also interesting to see that both sectors are equally affected by economic cycles, which means that value added of R&D-intensive sectors in-/decreases similarly to those of non-R&D-intensive sectors in economic up- or downturns.

The higher contribution of R&D-intensive sectors to the value added of manufacturing industry is also reflected in the next figure (Fig. A-24) which projects the share of direct value added that is contributed by both sectors to total manufacturing industry's value added. Here, the share of value added by non-R&D-intensive sectors constantly decreased from 55% in 1970 to 42% in 2006. But as the curve underlines this development is not necessarily a one-way street. Starting from 1981, the gap between R&D- and non-R&D-intensive sectors increased until 1992 before they almost met again in 1993.

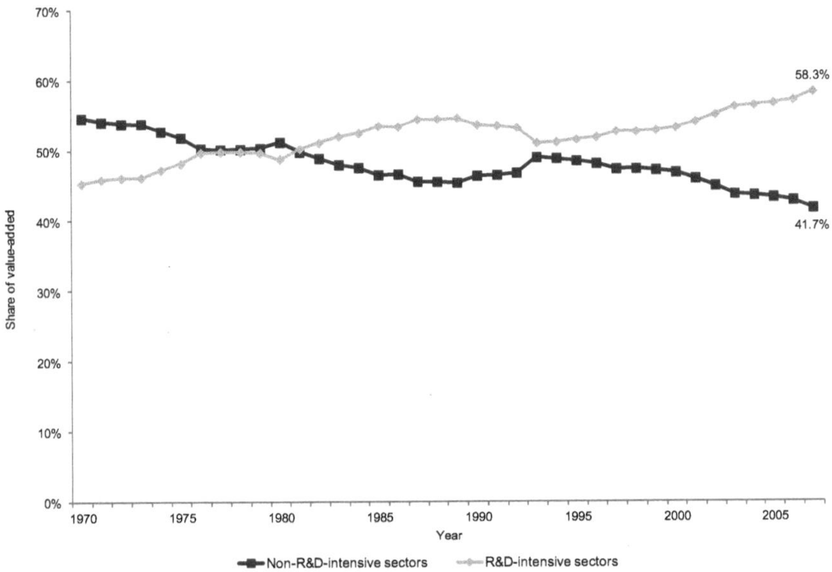

Fig. A-24: Share of R&D- and non-R&D-intensive sectors on the direct aggregate value added of German manufacturing industry from 1970 to 2007 (Source: Fraunhofer ISI on the basis of OECD STAN data)

But since then, the trend clearly shows a decreasing share of value added of non-R&D-intensive sectors which indicates a shrinking economic relevance of these sectors at least until 2006. However, it should be remarked that this decrease in value added of about 24% over the considered time period of 35 years is relatively small compared to what could have been expected on the basis of the theoretical argumentation. Against the background of this 42% share of direct value added, one could not seriously speak about the economic irrelevance of non-R&D-intensive sectors for German manufacturing industry. This is all the more important as the share of value added of only high-tech sectors (more than 5% R&D expenditure) is significantly lower at about 13% in 2007.

Besides this direct share of value added, one should also not forget the indirect share of value added by non-R&D-intensive sectors in their role of suppliers or customers of R&D-intensive sectors, which is not directly available or captured by statistical data but which was underlined in some empirical studies (e.g. ROBERTSON/PATEL 2007; HAUKNES/KNELL 2009; MENDONÇA (2009).

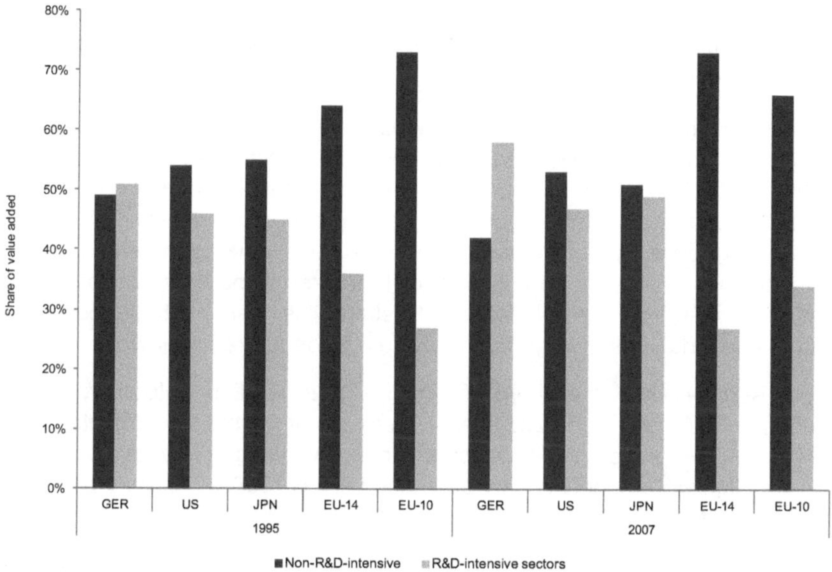

Fig. A-25: R&D- and non-R&D-intensive sectors' share of value added in international comparison
(Source: BELITZ ET AL. 2010: 37ff. own illustration)

Finally, an international comparison for the two years of 1995 and 2007 (Fig. A-25) shows that Germany's gross domestic product is strongly based on R&D-intensive sectors (and even increased from 1995 to 2007), which is probably due to the high share of characteristic medium-tech sectors like machinery and equipment and automotive industry.

Likewise, the decrease of non-R&D-intensive sectors' share of value added turns out to be weaker in the USA, Japan or the EU-10[261] than in German

261 EU-10 = "new" EU member states since May 2004 (Cyprus, Czech Republic, Estonia, Hungary, Lithuania, Latvia, Malta, Poland, Slovenia, Slovakia).

manufacturing industry. In the case of the EU-14[262], the share of non-R&D-intensive sectors even increased between 1995 and 2007. Thus, because of the special role of medium-tech industries in Germany, the numbers presented about the value added of non-R&D-intensive sectors may represent the lower end of the scale in an international comparison.

A.3.4 Employment rate of non-R&D-intensive sectors in the German manufacturing industry

A second indicator for the overall economic relevance of economic sectors is the share of employees working in the respective sector. As it was described in chapter 2.6.5.2, a higher R&D intensity leads to a higher output rate of innovative products, which in turn positively affects firm growth in terms of employees working there. Thus, the number of employees working in certain sectors and industries serves as an indicator for firm growth and thus economic success. Moreover, from the point of social welfare, it is desirable to support those sectors and industries that are supposed to provide future job growth.

Figure A-26 provides the share of employees (in relation to total workforce) working in R&D- and non-R&D-intensive sectors of German manufacturing industry. At a first glance, the share of employees working in total manufacturing industry slightly, but continuously decreased since 1996 which may be due to the simultaneous growth of the service sector in Germany.[263]

In 2006, about 22% of Germany's total working population were located in the manufacturing industry; whereby this share was almost equally distributed among R&D- and non-R&D-intensive sectors. Both shares were relatively stable during the time period considered. Concerning the non-R&D-intensive sectors, their share has slightly decreased from 13% in 1996 to 11% in 2006, while the share of R&D-intensive sectors has largely remained stable at 11%. This means that about the half of all employees in the manufacturing industry still work in non-R&D-intensive sectors. Thus, with regard to this indicator, there can be no talk of any drastically decreasing economic relevance of non-R&D-intensive sectors within the German manufacturing industry.

262 EU-14 = "old" EU member states without Germany (Austria, Belgium, Denmark, Spain, Finland, France, United Kingdom, Greece, Ireland, Italy, Luxemburg, Netherlands, Portugal, Sweden).

263 In 2006, almost 70% of the total working population worked in the service sector in Germany, which is an increase of almost 10% since 1996 (OECD STAN data).

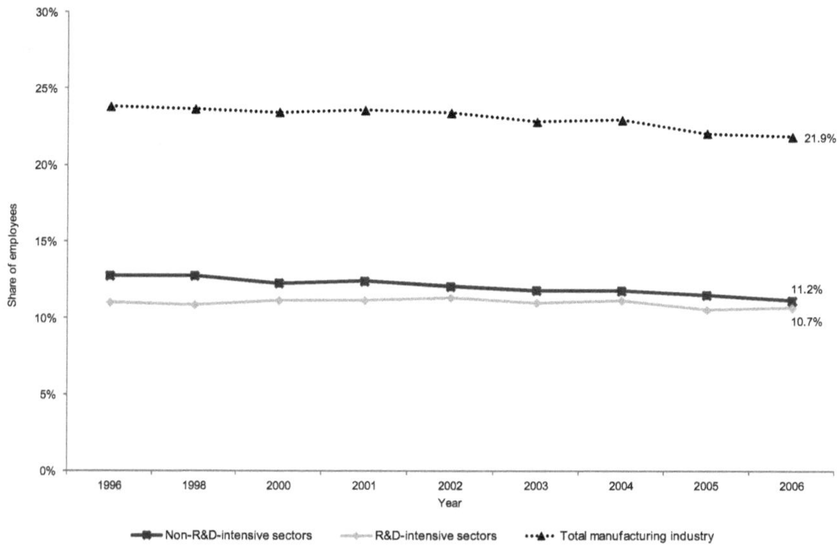

Fig. A-26: Share of employees in R&D- and non-R&D-intensive sectors compared to total German working population from 1996 to 2006
(Source: Fraunhofer ISI on the basis of German micro-census data)

When looking at academics and highly skilled employees, this picture of course changes. As industrial R&D departments are usually composed of academics and highly skilled employees (e.g. engineers, physicians, chemists), it is not very surprising to find that the share of these groups of employees is significantly larger in R&D-intensive sectors than in non-R&D-intensive sectors (Fig. A-27). In the data, the group of academics is defined as employees who successfully obtained a university degree or a PhD degree, while the group of highly skilled employees additionally comprises employees who have a master's degree or graduated as technicians. Due to this wider definition, the share of highly skilled employees is overall about 10% higher than that of academics.

As appears from the figure, the share of academics as well as highly skilled employees in R&D-intensive sectors is consistently twice as high as in non-R&D-intensive industries. But it is somewhat surprising that this difference obviously did not greatly increase over the years. This means that the attractiveness of non-R&D-intensive sectors for academics and highly skilled employees seems not to have dramatically declined over time. In the case of academics, their share has even minimally increased by about 1% in the years from 1996 to 2006. Empirically, some kind of "expert knowledge drain" in terms of an escape of academic or highly skilled employees from non-R&D-intensive industries cannot thus be observed on the basis of these data for the considered period of time.

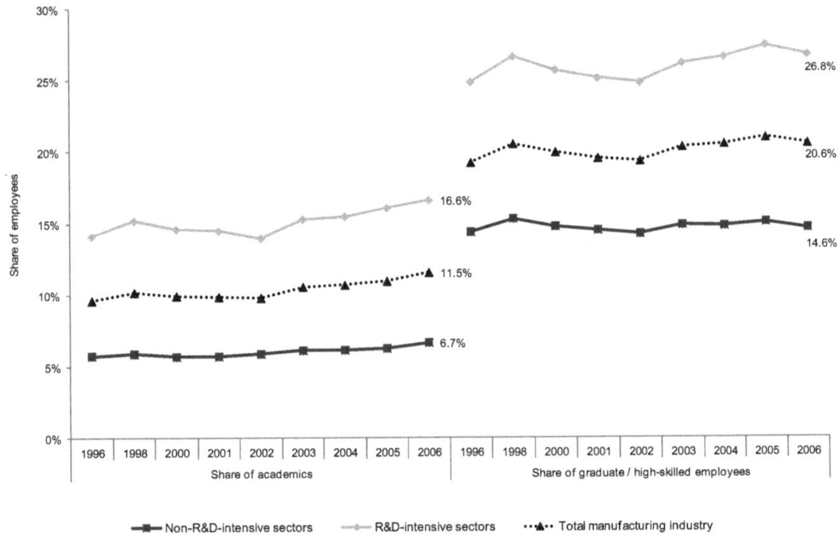

Fig. A-27: Share of academics and graduate/highly skilled employees in R&D- and non-R&D-intensive sectors in German manufacturing industry from 1996 to 2006 (Source: Fraunhofer ISI on basis of German micro-census data)

A.3.5 Export ratio of non-R&D-intensive sectors in the German manufacturing industry

The third performance dimension which was raised by endogenous growth theory and empirical research on the economic outcome of R&D, deals with the export intensity of firms and industries (see chapter 2.6.5.2). As international trade theories argue, the performance of industries becomes best visible when its actors directly compete with industries of other countries in international markets. Thereby, advances in the level of technology and quality of goods offered in international competition can compensate for competitive cost advantages of other economies. Especially in the case of high-wage countries such as Germany, an important reason for the economy's stable and successful performance is seen in the high share of exports of high-tech goods abroad. Therefore, it can be assumed that R&D-intensive sectors show a higher export intensity than their non-R&D-intensive counterparts.

This assumption is supported by the consideration of the absolute volume of exports for R&D- and non-R&D-intensive sectors of German manufacturing industry (Fig. A-28), again standardised by an index on the basis of 1980 to underline dynamics. As expected, the export ratio of R&D-intensive sectors is significantly higher than that of non-R&D-intensive sectors. Today, it is almost four times higher than in 1980 for the R&D-intensive sectors, while the export volume of non-R&D-intensive sectors has approximately increased by the factor 3 within the same time.[264] Moreover, it can be seen in the graph that the increase of export volume did not evolve continually over time. There are several ups and downs, for instance, in the years of 1993, 1996, and 1999. This stresses the fact that the export volume indicator as such is subject to economic trends and the dynamics of exchange rates (GEHRKE/LEGLER 2010: 9).

Nevertheless, with regard to the question of the economic relevance of non-R&D-intensive sectors in Germany, it seems remarkable that they were also able to significantly increase their direct export volume. This becomes especially apparent in the very recent development between the years 2001 and 2006. Within this period, the export volume index of non-R&D-intensive sectors climbed from the factor of about 1.4 to 2.8 compared to the basis of 1980. Thus, with respect to these absolute numbers, the export volume of non-R&D-intensive sectors is indeed below that of R&D-intensive sectors, but it cannot be stated that products and goods produced by non-R&D-intensive sectors are not competitive in international markets. And, in direct contrast to the reasoning of international trade theory, this is not necessarily due to their superior cost performance, but may also be rooted in the superior quality level of their products (GEHRKE/LEGLER 2010: 24).

Despite this positive trend in absolute numbers, the relative export share of non-R&D-intensive sectors declined between 1980 and 2007 from about 38% to 30% (Fig. A-29). Conversely, the export share of R&D-intensive sectors increased in the same period from about 63% to 70%. But it is interesting to see

264 Since the export ratio has been increasing more strongly than the import ratio, the foreign trade balance (exports-imports) of non-R&D-intensive sectors has been positive since 2002.

that since 2004 this trend has stopped since 2004. Since then, the export share of non-R&D-intensive sectors slightly increased by about 2%.

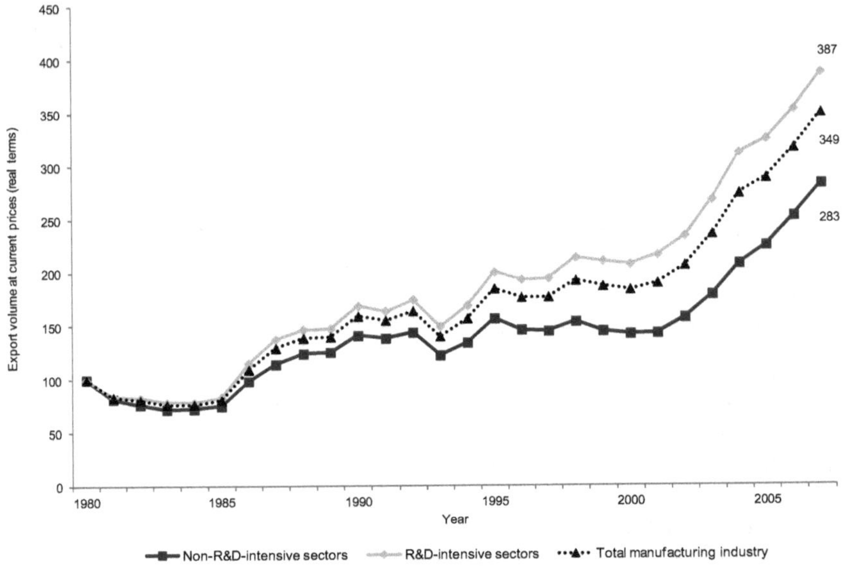

Fig. A-28: Exports of German goods at current prices (in real terms) from 1980 to 2007 (Index: 1980=100)
(Source: Fraunhofer ISI on the basis of OECD STAN data)

While for a long time R&D-intensive sectors were believed to be the main drivers of the economy and growth, the past years rather show a trend of consolidation and increasing importance of non-R&D-intensive sectors regarding the export performance of German manufacturing industry. However, it remains unanswered if this recent development is just an intermediate phenomenon due to the economic cycle or if this could be interpreted as a first sign of changing preferences in international markets (e.g. due to changing shortages, demand, protectionism) which may favour exports of non-R&D-intensive sectors in the future (GEHRKE/LEGLER 2010: 23).

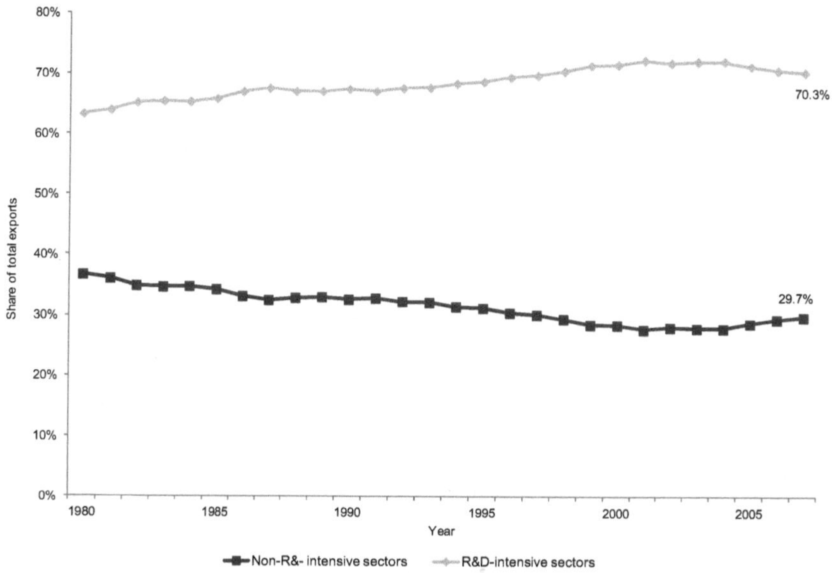

Fig. A-29: Share of R&D- and non-R&D-intensive sectors on total export of German manufacturing industry from 1980 to 2007
(Source: Fraunhofer ISI on the basis of OECD STAN data)

Finally, the international comparison reveals (Fig. A-30) that Germany underperforms in terms of the export share of R&D-intensive sectors compared to the USA and Japan, but outperforms the average share of the EU-14 or EU-10. Nevertheless, the time period from 1995 to 2007, it becomes apparent that almost all considered geographical entities are characterised by a declining export share of non-R&D-intensive sectors. This means that competition in international markets is being more and more shaped by R&D-intensive sectors.

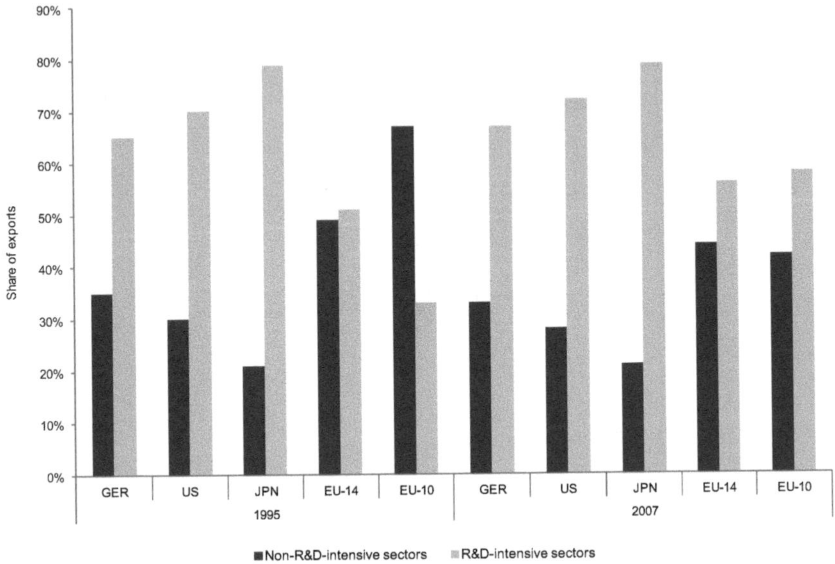

Fig. A-30: R&D- and non-R&D-intensive sectors' share of export in international comparison (Source: Belitz et al. 2010: 33, own illustration)

But this picture changes slightly if one looks at the export ratio of both sectors in terms of export as a share of gross domestic product (Fig. A-31). The international comparison at the two points in time depicts that this export ratio significantly increased for both R&D- and non-R&D-intensive sectors. This means that the importance of non-R&D-intensive sectors' products and goods on international markets became more important. In 2007 the export ratio of German non-R&D-intensive sectors was at 45%, which is lower than the EU-14 average, but much higher by far than those of the United States and Japan.

To conclude, the export indicators discussed above clearly provide empirical evidence of the high importance of R&D-intensive sectors for the international competitiveness of German manufacturing industry. But this is not to say that non-R&D-intensive sectors are not able to successfully compete in international markets. As was shown, they were able to increase their export volume as well as R&D-intensive sectors, and, within the last few years, there are indications that non-R&D-intensive sectors enhanced their export performance while R&D-intensive sectors tended to stagnate.

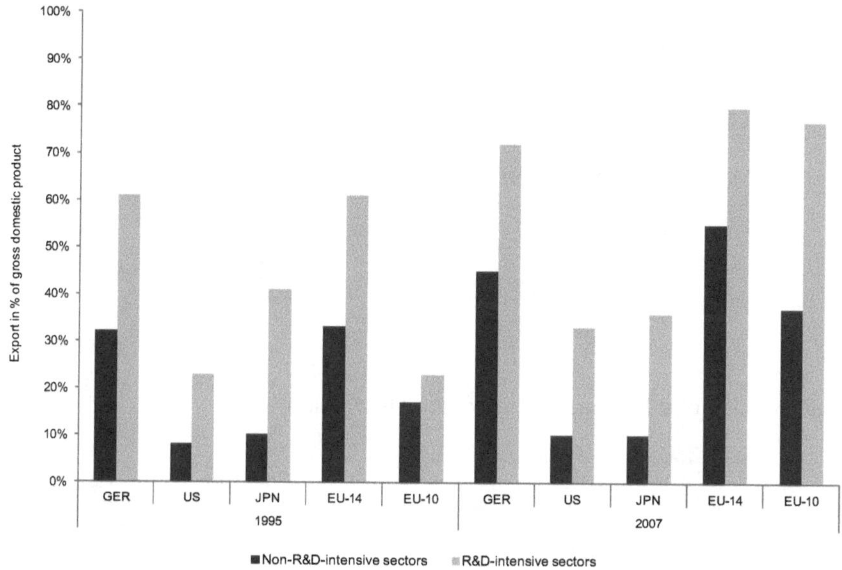

Fig. A-31: Export share of gross domestic product in international comparison (Source: own illustration, data from BELITZ ET AL. 2010: 33)

Additionally, it is important to mention that these indicators only represent the directly measurable export performance. Again, due to multiple interdependencies and interrelations between R&D-intensive and non-R&D-intensive sectors in terms of supplier and customer relations, it can be assumed that the indirect contribution of non-R&D-intensive sectors as suppliers of high quality intermediate products to the overall export performance is much higher than these findings suggest. Thus, one should not forget the fact that R&D-intensive export goods strongly rely on high quality intermediate products which are often produced by non-R&D-intensive sectors. If these intermediate products did not meet the high requirements of international competition, R&D-intensive sectors would be forced to import these precursors and investment goods from foreign manufacturers.

A.4 Extract of indicators and variables derived from *German Manufacturing Survey* 2009

A.4.1 Most important competitive factor

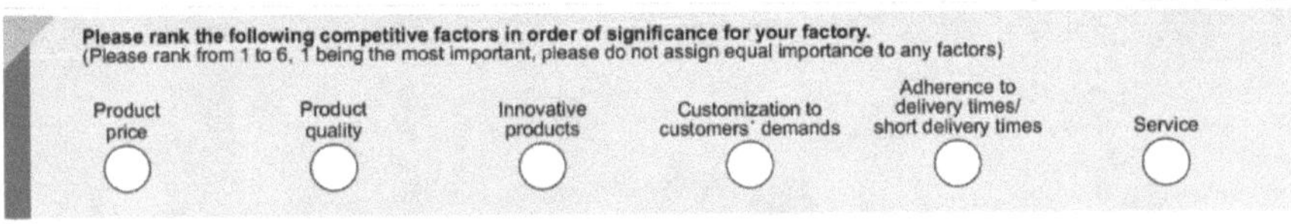

Please rank the following competitive factors in order of significance for your factory.
(Please rank from 1 to 6, 1 being the most important, please do not assign equal importance to any factors)

Product price ◯ | Product quality ◯ | Innovative products ◯ | Customization to customers' demands ◯ | Adherence to delivery times/ short delivery times ◯ | Service ◯

A.4.2 Use of advanced manufacturing technique

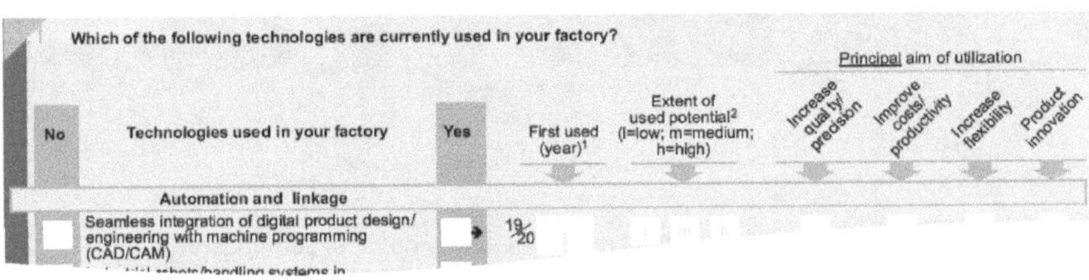

Which of the following technologies are currently used in your factory?

No	Technologies used in your factory	Yes	First used (year)[1]	Extent of used potential[2] (l=low; m=medium; h=high)	Principal aim of utilization: Increase quality/ precision	Improve costs/ productivity	Increase flexibility	Product innovation
	Automation and linkage							
☐	Seamless integration of digital product design/ engineering with machine programming (CAD/CAM)	☐	19/20					

Following this scheme, the item battery includes the following techniques:

- Seamless integration of digital product design / engineering with machine programming (CAD/CAM)
- Industrial robots/handling systems in manufacturing and assembly
- (Process-)integrated quality control
- Laser as a tool
- Dry processing/minimum quantity lubrication system
- Rapid prototyping or tooling
- Virtual reality and/or simulation in product development and/or manufacturing
- Control system for shutdown of machines in off-peak periods
- Electric motors with rotation speed regulation
- Application of highly efficient pumps
- Low temperature joining processes
- Retrieval of kinetic and process energy
- Combined cold, heat and power

Additional open categories:

Are any other innovative technologies used in your factory?

No | Yes → Which one? 1) ______ First used (year) 19/20 ____

2) ______ First used (year) 19/20 ____

A.4.3 Use of organisational concepts and management tools

Which of the following organisational concepts are currently used in your factory?

No	Organizational concept	Yes	First used (year)[1]	Extent of used potential[2] (l=low; m=medium; h=high)	Principal aim of utilization: Increase in quality/ precision	Improve costs/ productivity	Increase in flexibility	Product innovation
	Organization of work							
☐	Team work in production	☐ →	19/20	l m h				
☐	Task integration (planning, controlling or	☐						

Following this scheme, the item battery includes the following organisational and management concepts:

- Team work in production
- Task integration
- Temporary cross-functional project teams
- Customer- or product-focussed lines/cells in the factory
- Internal zero-buffer principle
- Quality circles
- Knowledge-base systems
- Collective arrangement for flexible working hours
- Financial participation by employees eligible for all employee groups
- Regular individual appraisal interviews
- Personnel training programmes as a special function on human resources

Additional open categories:

Are any other innovative organizational concepts used in your factory?

No | Yes → Which one? 1) ______ First used (year) 19/20 ____

2) ______ First used (year) 19/20 ____

A.4.4 Offer of new product-related services and share of sales with these new product-related services

Since 2006, has your factory offered new product-related services that were completely new to your factory or contained significant improvements?

☐ No ☐ Yes → Please indicate the share of turnover with these since 2006 newly offered product-related services, directly and indirectly generated in 2008. approx. ☐ %

A.4.5 Average share of product-related services that have been directly and indirectly invoiced

If you offer product-related services, please estimate their share of the total turnover in 2008.

Share of total turnover of services in 2008 directly invoiced approx. ☐ %

Share of total turnover of services in 2008 indirectly invoiced (via product price) approx. ☐ %

A.4.6 Participation in innovation cooperation

Do you cooperate in any of the following four innovation areas with other companies (resp. with other establishments)?

No	Cooperation in innovation	Yes	Cooperation with ...: Customers	Suppliers	Competitors	Service providers	Research institutions, universities	Frequency of collaboration: Seldom	From time to time	Often
☐	New products	☐ →	☐	☐	☐	☐	☐	☐	☐	☐
☐	New technical production processes	☐ →	☐	☐	☐	☐	☐	☐	☐	☐
☐	New service offers	☐ →	☐	☐	☐	☐	☐	☐	☐	☐
☐	New organizational concepts	☐ →	☐	☐	☐	☐	☐	☐	☐	☐

A.4.7 Knowledge-sources for innovation

Which of the following information sources are most relevant for your company in terms of triggering innovation activities in the following areas? (Please tick maximally 3 boxes per line)

Innovation area	Own R&D department	Other employees	Customers	Suppliers	Competitors	Research institutions, universities	Conferences, trade fairs, other professional events
New products	☐	☐	☐	☐	☐	☐	☐
New techical production processes	☐	☐	☐	☐	☐	☐	☐
New services	☐	☐	☐	☐	☐	☐	☐
New organizational concepts	☐	☐	☐	☐	☐	☐	☐

A.4.8 Qualification structure of employees

Please indicate the formal qualification level of your employees.

national items for LEVEL 3 LEVEL 2 LEVEL 1

Graduate degree, PhD	approx.	%	
Technicians	approx.	%	
Employees with commercial or technical/industrial training	approx.	%	=100%
Semiskilled and unskilled workers	approx.	%	
Technical/industrial or commercial apprentices	approx.	%	

A.4.9 Distribution of employees over functional areas within the firm

Please indicate the distribution of your personnel over the following areas.

Research and development	approx.	%	
Configuration, design	approx.	%	
Manufacturing and assembly	approx.	%	=100%
Customer service	approx.	%	
Other (administration, purchase, sales, maintenance, production planning etc.)	approx.	%	

A.4.10 Positioning in the value chain

With regard to your main line of products, is your factory <u>predominantly</u> a producer of finished products or a supplier of systems, components, or jobshops? (Please tick one box only!)

Producer of finished goods:		Supplier		Contract manufacturer
for consumers	for industrial business	system supplier	parts/components supplier	contract manufacturer (such as turning, coating, welding, grinding)

A.4.11 Type of product development

Product development (Please tick one box only)

- According to customers' specification
- As a standardised basic programme into which customer specific options are implemented
- For a standard programme from which the customer can select
- Does not exist in this factory

A.4.12 Product complexity

Product complexity (Please tick one box only)

- Simple products
- Products with medium complexity
- Complex products

A.4.13 Manufactured batch or lot size

Batch or lot size (Please tick one box only

- Single unit production
- Small or medium batch/lot
- Large batch/lot

A.4.14 Processual performance measures

Please answer the following questions on the production of your main product (your main line or group of products).

Which is the average manufacturing lead time of your main product? (Manufacturing lead time: from receipt of order (at shop floor) until it's ready to ship) approx. ___ work days or ___ hrs.

What is the average time span between receipt of customer's order and delivery (delivery time)? approx. ___ calendar days

What is the percentage of orders delivered on time (according to confirmed time of delivery)? approx. ___ %

What is the percentage of products or semifinished products that have to be scrapped or reworked because of quality problems? approx. ___ . ___ %

A.4.15 Turnover, number of employees, temporary workers

Please characterize your factory:

Annual turnover 2008 . Mio. Euro 2006 . Mio. Euro

Number of employees (excl. temporary agency workers) 2008 2006

Did your factory employ temporary agency workers in 2008? No Yes → How many temporary agency workers have been employed in your factory on average in 2008? approx. Number

A.4.16 Return on sales, year of establishment

Return on sales (before tax, 2008) negative 0 up to 2% >2 up to 5% >5 up to 10% >10%

Year of establishment of the factory

A.4.17 Performance of formal R&D

Did your factory in 2008 perform Research and Development (R&D) or order R&D activities from external partners?

No Yes → Expenditures on R&D as share of turnover of the year 2008: approx. %

A.4.18 Export intensity

Please estimate from where your factory has received its input in 2008 and where it has sold its products.

Input (parts, materials, etc.) generated in: Home country % ; Abroad % (=100%)

Products sold to: Home country % ; Abroad % (=100%)

A.4.19 Sector affiliation

Please indicate your industry and the main product or line of products produced at your factory.

Industry (e.g. textile industry, chemical industry, machinery etc.):

Main line of products

A.4.20 Ownership structure

Who is the majority or exclusive owner of the firm your factory belongs to?

Private person / Family; Financial investor (e.g. venture capital); Other firm (not financial investor); Foundation; Other owners; No major owner

→ Is the family represented in the management board? No Yes

A.5 Additional descriptive analysis

A.5.1 Prior goals of AMT implementation

Tab. A-33: Prior goals of AMT implementation of non-R&D-performing firms (Source: GMS 2009, own calculations)

	R&D-performing firms (>0 % R&D expenditure)	Non-R&D-performing firms (0 % R&D expenditure)
Average share (%) of AMT that were primarily introduced to...		
...increase quality	27.0	25.9
...reduce costs	42.2	37.1 +
...increase flexibility	19.0	15.8 +
...promote product innovativeness	14.7	7.9 ***
	(n=791)	(n=675)

Significance levels: *** = p<0.000 / ** = p<0.005 / * = p<0.01 / + = p<0.1

AMT = advanced manufacturing techniques

(-) =empty cells due to insufficient number of cases

A.5.2 Prior goals of OMC implementation

Tab. A-34: Prior goals of OMC implementation of non-R&D-performing firms (Source: GMS 2009, own calculations)

	R&D-performing firms (>0 % R&D expenditure)	Non-R&D-performing firms (0 % R&D expenditure)
Average share (%) of OMC that were primarily introduced to...		
...increase quality	35.5	31.9 +
...reduce costs	40.2	38.6
...increase flexibility	30.7	28.6
...promote product innovativeness	7.1	4.1 ***
	(n=791)	(n=675)

Significance levels: *** = p<0.000 / ** = p<0.005 / * = p<0.01 / + = p<0.1

AMT = advanced manufacturing techniques

(-) =empty cells due to insufficient number of cases

A.5.3 Contingency tables: Distribution of employees across functional area within the firm by qualification structure

Tab. A-35: Distribution of low and unskilled employees across different functional areas within the firm
(Source: GMS 2009, own calculations)

	Share of low and unskilled employees - empirical thirds		
R&D performing firms (>0 % R&D expenditure) (n=791)	Lowest third	Middle third	Highest third
share (%) of employees in R&D	8.8	7.3	5.9
share (%) of employees in engineering, construction, design	10.4	7.0	4.9
Share (%) of employees in manufacturing and assembling	51.2	54.2	64.5
Share (%) of employees in customer service	7.8	7.2	4.8
Non-R&D performing firms (0 % R&D expenditure) (n=675)			
share (%) of employees in R&D	1.9	2.2	1.8
share (%) of employees in engineering, construction, design	7.4	5.7	3.6
Share (%) of employees in manufacturing and assembling	63.9	63.1	71.0
Share (%) of employees in customer service	9.1	9.7	6.1

Tab. A-36: Distribution of graduate / highly skilled employees across different functional areas with the firm
(Source: GMS 2009, own calculations)

	Share of graduate / highly skilled employees - empirical thirds		
R&D performing firms (>0 % R&D expenditure) (n=791)	Lowest third	Middle third	Highest third
share (%) of employees in R&D	3.9	4.9	10.0
share (%) of employees in engineering, construction, design	6.1	3.8	9.5
Share (%) of employees in manufacturing and assembling	63.1	65.2	50.5
Share (%) of employees in customer service	6.2	4.7	7.4
Non-R&D performing firms (0 % R&D expenditure) (n=675)			
share (%) of employees in R&D	1.0	1.6	3.8
share (%) of employees in engineering, construction, design	4.1	3.2	8.4
Share (%) of employees in manufacturing and assembling	70.4	67.2	58.7
Share (%) of employees in customer service	8.3	6.7	8.0

A.5.4 Innovation output measures controlled for product complexity

Tab. A-37: R&D- and non-R&D-performing firms' innovation output in the field of technical and non-technical process innovation
(Source: GMS 2009, own calculations)

	Non-R&D-performing firms (0 % R&D expenditure)				
	Simple products	Products with medium complexity	Complex products	Total	
Turnover with new product-related services					
Average share (%) of sales with new product-related services	6.7	5.6	6.6	**6.4**	(n=76)
Time to market					
Average number of months	10.3 **	13.2 ***	11.7 ***	**12.1** ***	(n=219)
Manufacturing lead time					
Average number of working days à 8h	10.7	15.1 +	60.2	**23.1** **	(n=609)
Delivery time					
Average number of calendar days	16.3	25.5	59.8 **	**29.3** ***	(n=641)
Adherence to delivery dates					
Percentage of orders delivered on time	91.7	92.0	90.0	**91.6** +	(n=641)
Rework and scrap rate					
Percentage of products or semi-products that have to be scrapped or reworked due to quality problems	2.3	2.8 +	3.6	**2.8**	(n=633)

	R&D-performing firms (>0 % R&D expenditure)				
	Simple products	Products with medium complexity	Complex products	Total	
Turnover with new product-related services					
Average share (%) of sales with new product-related services	4.2	4.7	5.2	**5.0**	(n=175)
Time to market					
Average number of months	16.4	19.5	20.6	**19.6**	(n=595)
Manufacturing lead time					
Average number of working days à 8h	11.3	19.4	65.3	**36.4**	(n=761)
Delivery time					
Average number of calendar days	16.3	28.8	91.0	**51.7**	(n=761)
Adherence to delivery dates					
Percentage of orders delivered on time	93.4	90.9	88.5	**90.3**	(n=761)
Rework and scrap rate					
Percentage of products or semi-products that have to be scrapped or reworked due to quality problems	2.1	2.2	4.0	**3.1**	(n=742)

Significance levels: *** = p<0.000 / ** = p<0.005 / * = p<0.01 / + = p<0.1

AMT = advanced manufacturing techniques

(-) =empty cells due to insufficient number of cases

A.6 Cluster stop analysis

A.6.1 Calinski/Harabasz pseudo-F

Tab.A-38: Cluster stop analysis – CALINSKI/HARABASZ pseudo-F
(Source: GMS 2009, own calculation)

Number of clusters	Calinski/ Harabasz pseudo-F
2	34.96
3	29.41
4	26.67
5	23.93
6	22.47
7	21.19
8	20.16
9	19.47
10	18.79
11	18.29
12	17.85
13	17.34
14	16.84
15	16.39

→ Indicates 2 cluster solution

A.6.2 Duda-Hart

Tab. A-39: Cluster stop analysis – DUDA/HART
(Source: GMS 2009, own calculation)

Number of clusters	Je(2)/Je(1)	pseudo T-squared
2	0.916	21.36
3	0.9053	20.72
4	0.8701	17.01
5	0.9131	12.56
6	0.7651	19.34
7	0.9028	10.66
8	0.8794	10.01
9	0.8623	10.22
10	0.892	9.93
11	0.8533	9.8
12	0.865	8.27
13	0.8531	6.2
14	0.8166	7.64
15	0.8101	7.73

→ Indicates 5 cluster solution

A.7 Additional descriptive analysis of non-R&D-performing firms innovation patterns identified by cluster analysis

A.7.1 Sectoral distribution of non-R&D-performing firms' innovation patterns

Tab. A-40: Sectoral distribution of non-R&D-performing firms' innovation patterns (Source: GMS 2009, own calculations)

Sectoral affiliation	Cluster I 1	Cluster II 2	Cluster III 3	Cluster IV 4	5	Total	
Manufacturing industries (NACE 2-digit)*							
15/16 - Manuf. of food products, beverages and tobacco	12.5	9.7	**24.3**	7.8	3.5	12.9	(n=56)
17-19 - Manuf. of textiles, textile products or leather	3.1	1.1	2.7	2.6	**5.2**	2.8	(n=12)
21/22 - Manuf. of paper products, publishing or printing	6.3	**15.1**	7.2	13.0	6.9	9.7	(n=42)
24 - Chemical industries	3.1	2.2	**3.6**	1.3	1.7	2.5	(n=11)
25 - Manuf. of rubber and plastic products	10.4	**17.2**	6.3	5.2	13.8	10.3	(n=45)
27/28 - Manuf. of basic metals and fabricated metal products	14.6	24.7	24.3	**41.6**	27.6	25.8	(n=112)
29 - Manuf. of machinery und equipment	**24.0**	7.5	3.6	6.5	19.0	11.5	(n=50)
30-32 - Manuf. of electrical and optical equipment (excl. NACE 33)	5.2	3.2	1.8	3.9	**6.9**	3.9	(n=17)
33 - Manuf. of medical, precision, and optical instruments etc.	**8.3**	1.1	2.7	2.6	0.0	3.2	(n=14)
34/35 - Manuf. of transport equipment	0.0	**5.4**	2.7	3.9	3.5	3.0	(n=13)
Other sectors (20/23/26/36/37)	12.5	12.9	**20.7**	11.7	12.1	14.5	(n=63)
	100.0	100.0	100.0	100.0	100.0	100.0	(n=435)
Sectoral classification (LEGLER/FRIETSCH 2007)*							
Low-Tech sectors (average share of BERD < 2.5%)	65.6	82.8	**87.4**	83.1	81.0	80.0	(n=348)
Medium-Tech sectors (average share of BERD 2.5 - 7%)	**25.0**	12.9	9.0	14.3	19.0	15.6	(n=68)
High-Tech sectors (average share of BERD > 7%)	**9.4**	4.3	3.6	2.6	0.0	4.4	(n=19)
	100.0	100.0	100.0	100.0	100.0	100.0	(n=435)

All values are percentages of non-R&D-performing firms / Bold = highest mean/percentage for each variable
Significance levels: *** = p < 0.001 / ** = p < 0.005 / * = p < 0.01
Key: Cluster 1: Knowledge-intensive product developers / Cluster 2: Customer-driven, technical process specialists / Cluster 3: Occasional B2C product developers / Cluster 4: Low-innovative, labour-intensive manufacturers / Cluster 5: Volume-flexible specialised suppliers

A.7.2 Orders received in German manufacturing industry (2005=100)

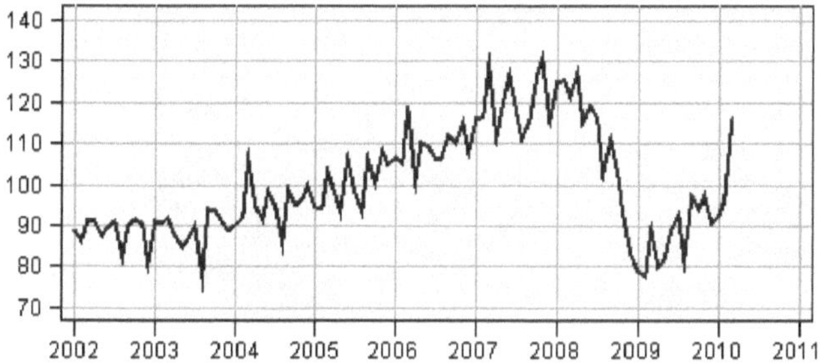

Fig. A-32: Orders received in German manufacturing industry (volume index; 2005=100) (Source: German Federal Statistical Office; available at www.destatis.de)

A.7.3 Total turnover of German manufacturing industry (2005=100)

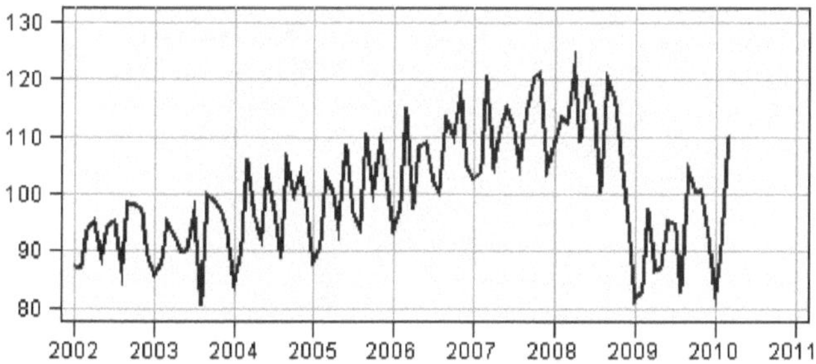

Fig. A-33: Turnover of German manufacturing industry (volume index; 2005=100) (Source: German Federal Statistical Office; available at www.destatis.de)

Kolibri 360 GmbH
Ferdinand-Jühlke-Straße 7
99095 Erfurt, Deutschland
produktsicherheit@kolibri360.de